Lecture Notes in Computer Science　　　　10047

Commenced Publication in 1973
Founding and Former Series Editors:
Gerhard Goos, Juris Hartmanis, and Jan van Leeuwen

More information about this series at http://www.springer.com/series/7409

Emma Spiro · Yong-Yeol Ahn (Eds.)

Social Informatics

8th International Conference, SocInfo 2016
Bellevue, WA, USA, November 11–14, 2016
Proceedings, Part II

 Springer

Editors
Emma Spiro
University of Washington
Seattle, WA
USA

Yong-Yeol Ahn
Indiana University
Bloomington, IN
USA

ISSN 0302-9743   ISSN 1611-3349   (electronic)
Lecture Notes in Computer Science
ISBN 978-3-319-47873-9   ISBN 978-3-319-47874-6   (eBook)
DOI 10.1007/978-3-319-47874-6

Library of Congress Control Number: 2016954470

LNCS Sublibrary: SL3 – Information Systems and Applications, incl. Internet/Web, and HCI

Printed on acid-free paper

This Springer imprint is published by Springer Nature
The registered company is Springer International Publishing AG
The registered company address is: Gewerbestrasse 11, 6330 Cham, Switzerland

Preface

This volume contains the papers presented at SocInfo 2016, the 8th International Conference on Social Informatics, held during November, 2016, in Bellevue, WA, USA. After the conferences in Warsaw, Poland, in 2009, Laxenburg, Austria, in 2010, Singapore in 2011, Lausanne, Switzerland, in 2012, Kyoto, Japan, in 2013, Barcelona, Spain, in 2014, and Beijing, China in 2015, the International Conference on Social Informatics came to United States for the first time.

SocInfo is an interdisciplinary venue for researchers from diverse fields including computer science, informatics, and the social sciences to share ideas and opinions, and present results from their research at the intersection of social sciences and information sciences. The ultimate goal of social informatics is to facilitate and promote multidisciplinary research that transcends the boundaries between social sciences, computer science, and information sciences, so that researchers can better leverage the power of informatics, computing, and social theories to advance our understanding of society and social phenomena. We envision SocInfo as a venue that attracts open-minded researchers who can cross the disciplinary boundaries and talk to other researchers regardless of their background and training. In doing so, we have invited and selected highly interdisciplinary keynote speakers and papers, which integrate social concepts and theories with large-scale datasets, algorithms, or other concepts and methods of computing.

We were delighted to present a strong technical program, which was a result of the hard work of the authors, reviewers, and conference organizers. We received 120 submissions, an increase from the last SocInfo. From these, 36 papers were accepted as full papers (30.0 %), and 39 were accepted as poster papers (32.5 %). This year, we decoupled the presentation format and the paper format; papers that are accepted as posters are published as is (with the same page limit as the full papers), without enforcing the shorter page limit. We also allowed the authors of accepted papers to opt for a "presentation only" mode with no inclusion in the proceedings: The authors of eight papers chose that option. Finally, a lightning talk option was offered to all authors of papers accepted as posters, to give interested authors the opportunity to present results with a brief oral control initial.

We were also pleased to have Joshua Blumenstock (University of California, Berkeley), Meeyoung Cha (KAIST and Facebook), Tina Eliassi-Rad (Northeastern University), Adam Russell (DARPA), Matthew Salganik (Princeton University), and Hanna Wallach (Microsoft Research and University of Massachusetts Amherst) give exciting keynote talks.

This year we hosted eight satellite workshops, namely, on Data Visualization (SocInfo VIZ: Actionable I From Visualization to Research Narratives); Virality and Memetics; Cultural Analytics; Activity-Based Networks; Social Media for Older Adults (SMOA); Urban Homelessness and Wise Cities; Web, Social Media, and Cellphone Data for Demographic Research; Computational Approaches to Social Modeling (ChASM), and Online Experimentation with Large and Diverse Samples.

We would like to thank all authors and participants for making the conference and the workshops a success. We express our gratitude to the Program Committee members and reviewers for their hard and dedicated work that ensured the highest-quality papers were accepted for presentation. We are extremely grateful to the program co-chairs, Y.Y. Ahn and Emma Spiro, for their tireless efforts in putting together a high-quality program and for directing the activity of the Program Committee. We owe special thanks to Nathan Hodas, our local co-chair, who had a vital role in all the stages of the organization. We thank our publicity chairs, Munmun De Choudhury and Brian Keegan, our Web chair Farshad Kooti, and workshop chairs, Tim Weninger and Emilio Zagheni. Also, last but not least we are grateful to Adam Wierzbicki for his continuous support.

Lastly, this conference would not be possible without the generous help of our sponsors and supporters: Leidos, University of Washington eScience Institute, Facebook, Microsoft Research, and MDPI.

September 2016

Emilio Ferrara
Kristina Lerman
Katherine Stovel

Organization

Organizing Committee

General Co-chairs

Emilio Ferrara	University of Southern California, Los Angeles, USA
Kristina Lerman	University of Southern California, Los Angeles, USA
Katherine Stovel	University of Washington, Seattle, USA

Steering Committee Chair

Adam Wierzbicki	Polish-Japanese Institute of Information Technology, Poland

Program Co-chairs

Yong-Yeol Ahn	Indiana University, Bloomington, USA
Emma Spiro	University of Washington, Seattle, USA

Workshop Co-chairs

Tim Weninger	University of Notre Dame, South Bend, USA
Emilio Zagheni	University of Washington, USA

Publicity Co-chairs

Munmun De Choudhury	Georgia Institute of Technology, Atlanta, USA
Brian Keegan	University of Colorado, Boulder, USA

Local Chair

Nathan Hodas	PNNL, Richland, WA, USA

Web Chair

Farshad Kooti	University of Southern California, Los Angeles, USA

Program Committee

Palakorn Achananuparp	Singapore Management University, Singapore
Luca Maria Aiello	Yahoo Labs, UK
Leman Akoglu	Stony Brook University, USA
Harith Alani	KMi, The Open University, UK
Fred Amblard	IRIT, University of Toulouse 1 Capitole, France

Jisun An	Qatar Computing Research Institute, Qatar
Pablo Barberá	New York University, USA
Fabricio Benevenuto	Federal University of Minas Gerais (UFMG), Brazil
Andras A. Benczur	Hungarian Academy of Sciences, Hungary
Alex Beutel	Carnegie Mellon University, USA
Arnim Bleier	GESIS-Leibniz Institute for the Social Sciences, Germany
Javier Borge-Holthoefer	Internet Interdisciplinary Institute (IN3-UOC), Spain
Piotr Bródka	Wroclaw University of Technology, Poland
Matthias Brust	Singapore University of Technology and Design, Singapore
Carlos Castillo	Eurecat, Spain
Ciro Cattuto	ISI foundation, Italy
James Caverlee	Texas A&M University, USA
Fabio Celli	University of Trento, Italy
Nina Cesare	University of Washington, USA
Munmun De Choudhury	Georgia Institute of Technology, USA
Freddy Chong Tat Chua	Hewlett Packard Labs, USA
Giovanni Luca Ciampaglia	Indiana University, USA
David Corney	Signal Media, UK
Rense Corten	Utrecht University, The Netherlands
Michele Coscia	Harvard University, USA
David Crandall	Indiana University, USA
Negin Dahya	University of Washington, USA
Simon Dedeo	Carnegie Mellon University, USA
Guillaume Deffuant	Laboratoire d'Ingénierie des Systèmes Complexes, France
Bruce Desmarais	University of Massachusetts Amherst, USA
Jana Diesner	University of Illinois at Urbana-Champaign, USA
Victor M. Eguiluz	IFISC (CSIC-UIB), Spain
Young-Ho Eom	IMT Institute for Advanced Studies Lucca, Italy
Andreas Flache	ICS University of Groningen, Germany
Alessandro Flammini	Indiana University, USA
Fabian Flöck	GESIS Cologne, Germany
Vanessa Frias-Martinez	University of Maryland, USA
Wai-Tat Fu	University of Illinois at Urbana-Champaign, USA
Matteo Gagliolo	Université libre de Bruxelles (ULB), Belgium
Manuel Garcia-Herranz	UNICEF, USA
Maria Giatsoglou	Aristotle University of Thessaloniki, Greece
Scott Golder	Context Relevant, USA
Bruno Gonçalves	New York University, USA
Przemyslaw Grabowicz	Max Planck Institute for Software Systems (MPI-SWS), Germany
Jeff Hemsley	Syracuse University, USA
Marco Janssen	Arizona State University, USA
Adam Jatowt	Kyoto University, Japan
Marco Alberto Javarone	Università degli studi di Cagliari, Italy

Kyomin Jung	Seoul National University, South Korea
Andreas Kaltenbrunner	Barcelona Media, Spain
Kazuhiro Kazama	Wakayama University, Japan
Brian Keegan	University of Colorado, USA
Andreas Koch	University of Salzburg, Germany
Farshad Kooti	USC Information Sciences Institute, USA
Tobias Kuhn	VU University Amsterdam, The Netherlands
Haewoon Kwak	Qatar Computing Research Institute, Qatar
Sang Hoon Lee	Korea Institute for Advanced Study, South Korea
David Liben-Nowell	Carleton College, USA
Yu-Ru Lin	University of Pittsburgh, USA
Huan Liu	Arizona State University, USA
Yabing Liu	Northeastern University, USA
Vera Liao	University of Illinois at Urbana-Champaign, USA
Jared Lorince	Indiana University, USA
Tyler McCormick	University of Washington, USA
Matteo Magnani	Uppsala University, Sweden
Winter Mason	Facebook, USA
Pasquale De Meo	VU University, Amsterdam, The Netherlands
Rosa Meo	University of Turin, Italy
Eric Meyer	University of Oxford, UK
Stasa Milojevic	Indiana University, USA
Mikolaj Morzy	Poznan University of Technology, Poland
Tsuyoshi Murata	Tokyo Institute of Technology, Japan
Mirco Musolesi	University College London, UK
Shinsuke Nakajima	Kyoto Sangyo University, Japan
Keiichi Nakata	University of Reading, UK
Alice Oh	KAIST, South Korea
Diego Fregolente Mendes de Oliveira	Northwestern University, USA
Nuria Oliver	Telefonica Research, Spain
Anne-Marie Oostveen	University of Oxford, UK
Symeon Papadopoulos	Information Technologies Institute, Greece
Luca Pappalardo	University of Pisa, Italy
Jaimie Park	KAIST, South Korea
Orion Penner	École polytechnique fédérale de Lausanne, Switzerland
Ruggero G. Pensa	University of Turin, Italy
Nicola Perra	Northeastern University, USA
Alexander Petersen	IMT Lucca Institute for Advanced Studies, Italy
Georgios Petkos	Information Technologies Institute/CERTH, Greece
Giovanni Petri	ISI Foundation, Italy
Gregor Petrič	University of Ljubljana, Slovenia
Hemant Purohit	George Mason University, USA
Alessandro Provetti	University of Messina, Italy
Jose J. Ramasco	IFISC, Spain
Georgios Rizos	CERTH-ITI, Greece

John Robinson	University of Washington, USA
Fabio Rojas	Indiana University, USA
Giancarlo Ruffo	University of Turin, Italy
Mostafa Salehi	University of Tehran, Iran
Claudio Schifanella	RAI research centre, Italy
Rossano Schifanella	University of Turin, Italy
Xiaoling Shu	UC Davis, USA
Philipp Singer	GESIS - Leibniz Institute for the Social Sciences, Germany
Frank Schweitzer	ETH Zurich, Switzerland
Tom Snijders	University of Oxford, UK
Rok Sosic	Stanford University, USA
Steffen Staab	University of Koblenz-Landau, Germany
Bogdan State	Stanford University, USA
Markus Strohmaier	University of Koblenz-Landau, Germany
Cassidy Sugimoto	Indiana University Bloomington, USA
Bart Thomee	Yahoo Labs, USA
George Valkanas	University of Athens, Greece
Daniel Villatoro	IIIA-CSIC, Spain
Claudia Wagner	GESIS-Leibniz Institute for the Social Sciences, Germany
Dashun Wang	Northwestern University, USA
Ingmar Weber	Qatar Computing Research Institute, Qatar
Katrin Weller	GESIS Leibniz Institute for the Social Sciences, Germany
Brooke Foucault Welles	Northeastern University, USA
Robert West	Stanford University, USA
Christo Wilson	Northeastern University, USA
Emilio Zagheni	University of Washington, USA
Li Zeng	University of Washington, USA
Arkaitz Zubiaga	University of Warwick, UK

Sponsors

Leidos (www.leidos.com/)
University of Washington eScience Institute (escience.washington.edu/)
Facebook (www.facebook.com)
MDPI (www.mdpi.com)
Microsoft Research (www.microsoft.com)

Contents – Part II

Poster Papers: Networks, Communities and Groups

Towards Understanding User Participation in Stack Overflow
Using Profile Data 3
    Ifeoma Adaji and Julita Vassileva

Identifying Correlated Bots in Twitter 14
    Nikan Chavoshi, Hossein Hamooni, and Abdullah Mueen

Predicting Online Extremism, Content Adopters,
and Interaction Reciprocity. 22
    Emilio Ferrara, Wen-Qiang Wang, Onur Varol, Alessandro Flammini,
    and Aram Galstyan

Content Centrality Measure for Networks: Introducing Distance-Based
Decay Weights. 40
    Takayasu Fushimi, Tetsuji Satoh, Kazumi Saito, Kazuhiro Kazama,
    and Noriko Kando

A Holistic Approach for Link Prediction in Multiplex Networks 55
    Alireza Hajibagheri, Gita Sukthankar, and Kiran Lakkaraju

Twitter Session Analytics: Profiling Users' Short-Term Behavioral Changes . . . 71
    Farshad Kooti, Esteban Moro, and Kristina Lerman

Senior Programmers: Characteristics of Elderly Users from Stack Overflow . . . 87
    Grzegorz Kowalik and Radoslaw Nielek

Predicting Retweet Behavior in Online Social Networks Based on Locally
Available Information 97
    Guanchen Li and Wing Cheong Lau

Social Influence: From Contagion to a Richer Causal Understanding 116
    Dimitra Liotsiou, Luc Moreau, and Susan Halford

Influence Maximization on Complex Networks with Intrinsic
Nodal Activation. 133
    Arun V. Sathanur and Mahantesh Halappanavar

Applicability of Sequence Analysis Methods in Analyzing Peer-Production
Systems: A Case Study in Wikidata 142
    To Tu Cuong and Claudia Müller-Birn

Network-Oriented Modeling and Its Conceptual Foundations   157
  Jan Treur

Poster Papers: Politics, News, and Events

Social Contribution Settings and Newcomer Retention in Humanitarian
Crowd Mapping   179
  Martin Dittus, Giovanni Quattrone, and Licia Capra

A Relevant Content Filtering Based Framework for Data
Stream Summarization   194
  Cailing Dong and Arvind Agarwal

Relevancer: Finding and Labeling Relevant Information in Tweet
Collections   210
  *Ali Hürriyetoğlu, Christian Gudehus, Nelleke Oostdijk,
  and Antal van den Bosch*

Analyzing Large-Scale Public Campaigns on Twitter   225
  *Julia Proskurnia, Ruslan Mavlyutov, Roman Prokofyev, Karl Aberer,
  and Philippe Cudré-Mauroux*

Colombian Regulations for the Implementation of Cognitive Radio
in Smart Grids   244
  *Julián Giraldo Torres, Brayan S. Reyes Daza,
  and Octavio J. Salcedo Parra*

Using Demographics in Predicting Election Results with Twitter   259
  Eric Sanders, Michelle de Gier, and Antal van den Bosch

On the Influence of Social Bots in Online Protests: Preliminary
Findings of a Mexican Case Study   269
  *Pablo Suárez-Serrato, Margaret E. Roberts, Clayton Davis,
  and Filippo Menczer*

What am I not Seeing? An Interactive Approach to Social Content
Discovery in Microblogs   279
  Byungkyu Kang, Nava Tintarev, Tobias Höllerer, and John O'Donovan

Poster Papers: Markets, Crowds, and Consumers

Targeted Ads Experiment on Instagram   297
  Heechul Kim, Meeyoung Cha, and Wonjoon Kim

Exploratory Analysis of Marketing and Non-marketing E-cigarette
Themes on Twitter   307
  Sifei Han and Ramakanth Kavuluru

Obtaining Rephrased Microtask Questions from Crowds   323
    Ryota Hayashi, Nobuyuki Shimizu, and Atsuyuki Morishima

To Buy or Not to Buy? Understanding the Role of Personality
Traits in Predicting Consumer Behaviors   337
    Zhe Liu, Yi Wang, Jalal Mahmud, Rama Akkiraju, Jerald Schoudt,
    Anbang Xu, and Bryan Donovan

What Motivates People to Use Bitcoin?   347
    Masooda Bashir, Beth Strickland, and Jeremiah Bohr

Spiteful, One-Off, and Kind: Predicting Customer Feedback
Behavior on Twitter   368
    Agus Sulistya, Abhishek Sharma, and David Lo

Poster Papers: Privacy, Health and Well-being

Validation of a Computational Model for Mood and Social Integration   385
    Altaf Hussain Abro and Michel C.A. Klein

PPM: A Privacy Prediction Model for Online Social Networks   400
    Cailing Dong, Hongxia Jin, and Bart P. Knijnenburg

Privacy Inference Analysis on Event-Based Social Networks   421
    Cailing Dong and Bin Zhou

Empirical Analysis of Social Support Provided via Social Media   439
    Lenin Medeiros and Tibor Bosse

User Generated vs. Supported Contents: Which One Can Better Predict
Basic Human Values?   454
    Md. Saddam Hossain Mukta, Mohammed Eunus Ali, and Jalal Mahmud

An Application of Rule-Induction Based Method in Psychological
Measurement for Application in HCI Research   471
    Maria Rafalak, Piotr Bilski, and Adam Wierzbicki

A Language-Centric Study of Twitter Connectivity   485
    Priya Saha and Ronaldo Menezes

Investigating Regional Prejudice in China Through the Lens of Weibo   500
    Xi Wang, Zhiya Zuo, Yang Zhang, Kang Zhao, Yung-Chun Chang,
    and Chin-Shun Chou

Author Index   515

Contents – Part I

Networks, Communities, and Groups

How Well Do Doodle Polls Do?. 3
Danya Alrawi, Barbara M. Anthony, and Christine Chung

Bring on Board New Enthusiasts! A Case Study of Impact of Wikipedia
Art + Feminism Edit-A-Thon Events on Newcomers. 24
Rosta Farzan, Saiph Savage, and Claudia Flores Saviaga

The Social Dynamics of Language Change in Online Networks 41
Rahul Goel, Sandeep Soni, Naman Goyal, John Paparrizos,
Hanna Wallach, Fernando Diaz, and Jacob Eisenstein

On URL Changes and Handovers in Social Media 58
Hossein Hamooni, Nikan Chavoshi, and Abdullah Mueen

Comment-Profiler: Detecting Trends and Parasitic Behaviors
in Online Comments. 75
Tai-Ching Li, Abdullah Mueen, Michalis Faloutsos,
and Huy Hang

On Profiling Bots in Social Media 92
Richard J. Oentaryo, Arinto Murdopo, Philips K. Prasetyo,
and Ee-Peng Lim

A Diffusion Model for Maximizing Influence Spread in Large Networks 110
Tu-Thach Quach and Jeremy D. Wendt

Lightweight Interactions for Reciprocal Cooperation
in a Social Network Game. 125
Masanori Takano, Kazuya Wada, and Ichiro Fukuda

Continuous Recipe Selection Model Based on Cooking History 138
Shuhei Yamamoto, Noriko Kando, and Tetsuji Satoh

Politics, News, and Events

Examining Community Policing on Twitter: Precinct Use
and Community Response 155
Nina Cesare, Emma S. Spiro, Hedwig Lee, and Tyler McCormick

The Dynamics of Group Risk Perception in the US After Paris Attacks. 168
Wen-Ting Chung, Kai Wei, Yu-Ru Lin, and Xidao Wen

Determining the Veracity of Rumours on Twitter 185
  Georgios Giasemidis, Colin Singleton, Ioannis Agrafiotis,
  Jason R.C. Nurse, Alan Pilgrim, Chris Willis, and D.V. Greetham

PicHunt: Social Media Image Retrieval for Improved Law Enforcement. 206
  Sonal Goel, Niharika Sachdeva, Ponnurangam Kumaraguru,
  A.V. Subramanyam, and Divam Gupta

TwitterNews+: A Framework for Real Time Event Detection
from the Twitter Data Stream 224
  Mahmud Hasan, Mehmet A. Orgun, and Rolf Schwitter

Uncovering Topic Dynamics of Social Media and News:
The Case of Ferguson 240
  Lingzi Hong, Weiwei Yang, Philip Resnik, and Vanessa Frias-Martinez

Identifying Partisan Slant in News Articles and Twitter During
Political Crises 257
  Dmytro Karamshuk, Tetyana Lokot, Oleksandr Pryymak,
  and Nishanth Sastry

Predicting Poll Trends Using Twitter and Multivariate
Time-Series Classification 273
  Tom Mirowski, Shoumik Roychoudhury, Fang Zhou,
  and Zoran Obradovic

Inferring Population Preferences via Mixtures of Spatial Voting Models 290
  Alison Nahm, Alex Pentland, and Peter Krafft

Contrasting Public Opinion Dynamics and Emotional Response
During Crisis 312
  Svitlana Volkova, Ilia Chetviorkin, Dustin Arendt,
  and Benjamin Van Durme

Social Politics: Agenda Setting and Political Communication
on Social Media 330
  Xinxin Yang, Bo-Chiuan Chen, Mrinmoy Maity, and Emilio Ferrara

Markets, Crowds, and Consumers

Preference-Aware Successive POI Recommendation with Spatial
and Temporal Influence 347
  Madhuri Debnath, Praveen Kumar Tripathi, and Ramez Elmasri

Event Participation Recommendation in Event-Based Social Networks. 361
  Hao Ding, Chenguang Yu, Guangyu Li, and Yong Liu

An Effective Approach to Finding a Context Path in Review Texts
Using Pathfinder Scaling 376
   Erin Hea-Jin Kim and SuYeon Kim

How to Find Accessible Free Wi-Fi at Tourist Spots in Japan. 389
   Keisuke Mitomi, Masaki Endo, Masaharu Hirota, Shohei Yokoyama,
   Yoshiyuki Shoji, and Hiroshi Ishikawa

Privacy, Health and Wellbeing

Mobile Communication Signatures of Unemployment 407
   Abdullah Almaatouq, Francisco Prieto-Castrillo, and Alex Pentland

Identifying Stereotypes in the Online Perception of Physical Attractiveness. ... 419
   Camila Souza Araújo, Wagner Meira Jr., and Virgilio Almeida

Analysing RateMyProfessors Evaluations Across Institutions, Disciplines,
and Cultures: The Tell-Tale Signs of a Good Professor 438
   Mahmoud Azab, Rada Mihalcea, and Jacob Abernethy

Detecting Coping Style from Twitter 454
   Jennifer Golbeck

User Privacy Concerns with Common Data Used in Recommender Systems. ... 468
   Jennifer Golbeck

How a User's Personality Influences Content Engagement in Social Media ... 481
   Nathan O. Hodas, Ryan Butner, and Court Corley

Semi-supervised Knowledge Extraction for Detection of Drugs
and Their Effects 494
   Fabio Del Vigna, Marinella Petrocchi, Alessandro Tommasi,
   Cesare Zavattari, and Maurizio Tesconi

Using Social Media to Measure Student Wellbeing: A Large-Scale Study
of Emotional Response in Academic Discourse...................... 510
   Svitlana Volkova, Kyungsik Han, and Courtney Corley

EmojiNet: Building a Machine Readable Sense Inventory for Emoji 527
   Sanjaya Wijeratne, Lakshika Balasuriya, Amit Sheth, and Derek Doran

Author Index 543

Poster Papers: Networks, Communities and Groups

Towards Understanding User Participation in Stack Overflow Using Profile Data

Ifeoma Adaji[(✉)] and Julita Vassileva

Department of Computer Science, University of Saskatchewan, Saskatoon, Canada
{Ifeoma.adaji,Julita.vassileva}@usask.ca

Abstract. When designing a Q&A social network, it is essential to know what profile elements are necessary to build a complete profile for a user. Using data from Stack Overflow, we examined the profile data of users in order to determine the relationship between a complete profile (one that has values for each profile element: *website URL, location, about me, profile image* and *age*) and their contribution to the network in terms of reputation scores and quality of question and answer posts. Our analysis shows that most users do not have a complete profile, however the average reputation earned by users with complete profiles is significantly higher than that earned by users with incomplete profiles. In addition, users with complete profiles post higher quality question and answers, hence are more useful to the network. We also determined that, of the five profile elements studied, *location* and *about me* have a higher correlation than the others. This research is a step in determining what profile elements are important in a typical Q&A social network and which of these elements should regularly be used together.

Keywords: User profiles · Social network analysis · Stack overflow

1 Introduction

User profiles in social networks are an integral part of the network, as they describe who the user is and what connections or activities the user wants to take part in. Since social networks differ in structure and purpose, the profile elements will likely differ from one network to the other. For example, Facebook's objective is to connect the world[1] hence contains profile information to enhance that. Facebook's user profile includes information such as *work and education, places you've lived, contact and basic info, family and relationships, details about you* and *life events*. These details are used by Facebook in making recommendations about friends to add and what content to show a particular user. LinkedIn on the other hand is a professional network whose objective is to connect professionals together making them more productive and successful[2]. For this reason, a user's profile information contains information such as *profile picture, work experience, skills and endorsements, education* and *certifications*. The objective of a question and

---

[1] http://newsroom.fb.com/news/2016/04/marknote/.
[2] https://www.linkedin.com/about-us.

© Springer International Publishing AG 2016
E. Spiro and Y.-Y. Ahn (Eds.): SocInfo 2016, Part II, LNCS 10047, pp. 3–13, 2016.
DOI: 10.1007/978-3-319-47874-6_1

answer (Q&A) social network like Stack Overflow[3] is to provide a platform where users can ask specific IT related questions while other users provide answers, while earning reputation and rewards in the process. In Stack Overflow, a user's profile contains information such as *profile image, age, website URL, location, reputation score, badges earned* and *about me.*

In Q&A social networks, profiles are important as they give an overview of a user's participation on the network. A user's profile tells a story about the user's level of expertise through his/reputation score, rewards earned, the number of questions and answers posted and what they look like. Knowing how the existence of these profile elements relate to the user's participation is important as this could lead to a better network with quality questions and answers. For example, if users with complete profiles are the high achievers, they could be recommended to answer difficult questions on the network, as research has shown that users are currently not matched with questions that meet their level of expertise [1]. In addition, in order to remain active in the network, there could be specific rewards or persuasive strategies targeted at users with complete profiles. Furthermore, knowing what profile elements to include in a Q&A social network will help developers build networks that are successful.

In this paper we aim to answer the following research questions regarding Q&A social networks, in particular, Stack Overflow:

RQ1. Do profiles matter? Are users with a complete profile more helpful to the community?
RQ2. What profile elements should be used together?
RQ3. Do users that post helpful questions and answers have a complete profile?

To answer these questions, we studied user profile data in Stack Overflow in two ways; using reputation score of the users and using scores earned by question and answer posts. Using reputation score, we computed the average daily reputation score of the top earners (users who have earned at least 10,000 reputation points) and selected all users who have earned that score. On the other hand, we identified the question and answer posts that have earned at least 100 points and studied the profile of the users that made these posts.

Using the available profile elements: *website URL, location, about me, profile image* and *age,* we identified the number of users who are missing these profile elements and their average reputation scores. We compared these to the average reputation scores of users who were not missing these profile elements. Our analysis shows that the users with complete profile elements on the average have earned more reputation scores that those with incomplete profile elements. We also analyzed the profiles of users who have posted questions and answers that scored over 100 in order to identify the common profile elements that exist among these users. Our analysis shows that most question and answer posts with high reputation scores were posted by users with a complete profile. In addition, the average score earned by posts made by users with complete profile was higher than those made by users with incomplete profiles. Finally, in order

---

[3] http://stackoverflow.com/.

to determine what profile elements are better connected, we computed the correlation coefficient of the five profile elements *website URL, location, about me, profile image* and *age*. Or results show that *location* and *about me* have a higher correlation than the other profile elements.

Our contribution to ongoing research is to determine how profile elements of users in a Q&A social network impact the contributions of those users; we do not aim to identify experts or influential users. To the best of our knowledge, there is currently no research that studies the relationship between user profiles and performance of users in the network. This research, though work in progress, can act as a guide to social network developers in modelling profiles and profile elements in Q&A social networks to make the network successful.

2   Literature Review

2.1   Stack Overflow

Stack Overflow[4] is a Q&A platform where users can ask and answer specific IT related questions. Question and answer posts can be upvoted or downvoted by members of the community based on the clarity and quality of such posts. The upvotes and downvotes are used to compute the final score of a post and are also a means through which users earn reputation. By posting high quality questions and useful answers, users can gain reputation[5]. For example, users can gain 5 reputation points when their question is voted up or 10 points when their answer is voted up. Users can also gain 15 reputation points when their answer is marked "accepted" and 2 reputation points when a change they propose to an existing question or answer is approved. The higher the reputation score, the more privileges the user can earn. Privileges control what users can do in Stack overflow[6]. For example, only users with the "create tags" privilege can add new tags to the site. In order to earn this privilege, a user must have earned a reputation score of at least 1,500. Another example of privileges is the "edit questions and answers" privilege which users can earn when they have at least 2,000 reputation score. Users with this privilege can edit any question or answer and have the changes visible to other users without the need for further authorization from the site's administrator. Stack Overflow currently has over 5 million users with over 11 million questions.

2.2   User Profiles in Social Networks

Strano [2] studied the choices people make when selecting their profile image in Facebook. Their study, which differs from ours, was based on self-presentation and impression management. Though we studied profile pictures, in our study, the overall aim was to understand the level of participation and usefulness of users based on their profile elements.

---

[4] http://stackoverflow.com.

[5] http://stackoverflow.com/help/whats-reputation.

[6] http://stackoverflow.com/help/privileges.

In their study of social networks, Zhou et al. [3] exploited the use of profile information to rank answers in Yahoo! Answers. Instead of applying the conventional ranking method of using machine learning techniques, they used user profile details. Their study differs from ours in various ways. First, they studied a different Q&A social network; their study was on Yahoo! Answers while ours is on Stack Overflow. Secondly, the profile elements in Yahoo! Answer differs from those in Stack Overflow. Thirdly, their categorization of users differs from ours. Finally, their contribution to research differs from our; while they ranked answers, our research focused on the contribution of users based on the completeness of their profiles.

Though [4] studied user profiles in Stack Overflow, their research differs from ours in several ways. Their study was on discovering experts by user profiling and ranking answers by machine learning algorithm. They used profile elements that were not used in our study like *user name* which is a mandatory field in Stack Overflow, hence everyone has a *user name*. In addition, the focus of their study was on developing an answer ranking model using the Ranking Support Vector Machine classifier, which differs from our paper's objective.

Using Quickstep and Foxtrot as their experimental systems, Middleton et al. [5] developed a new ontological approach to recommending research papers online using user profiling. Though their research studied user profiles, it differs from ours due to the context of the research; recommender systems versus Q&A social networks.

Yerva et al. [6] explored techniques for creating user profiles based on their activities in various social networks. They also indicated the advantages of maintaining social network user profiles. Though we worked with user profiles, our study used existing profiles of Stack Overflow, unlike Yerval et al.'s study where profiles were created based on users' activities.

Although Dijk et al. [7] used Stack Overflow as their case study, their research was on early detection of experts in the network based on textual, behavioral and time-aware features of the users. Our study differs from theirs as our contribution to current research is different. While they focused on predicting which new users will eventually become experts, we focused on the difference in the level of participation between users with complete and incomplete profiles.

Though there have been studies on understanding user influence and expertise in Stack Overflow using rewards earned, the focus of this paper is specifically on user profiles and how the completeness or otherwise of a profile relates to the level of participation of that user. Adaji and Vassileva [8] identified expert respondents in Stack Overflow using in-degree and the rewards earned by respondents. They went further to predict the churn of these experts using various data mining algorithms. Unlike that study, this paper investigates the profile elements of users and the relationship between complete/incomplete profiles and user participation in the network. In their study on modelling problem difficulty and expertise in Stack Overflow, Hanrahan et al. [9] came up with indicators that identify the difficulty level of questions and who experts are. To identify experts, the authors used reputation score, z-score of users and the average delta between upvotes and downvotes of answer posts of users. This study also differs from ours because the authors did not study the profile elements of users.

To the best of our knowledge, there is currently no research that studies the relationship between user profiles and performance of users in the network using the profile elements *website URL, location, about me, profile image* and *age.*

3 Data Collection

In order to determine if profile elements are important to the success of a Q&A social network, we used Stack Overflow as our case study. Stack Overflow data is currently available through its public data explorer[7]. Stack Overflow currently has various profile elements including *display name, profile image, role and company, age, website URL, location, reputation score, badges earned* and *about me.* For this study we excluded *display name* because it is mandatory. We also excluded *reputation score* and *badges earned* because they are computed automatically by the system and are not provided by the user. Of the profile elements in Stack Overflow, we *used profile image, age, website URL, location,* and *about me.* These are not mandatory and are not dependent on the system, but rather on the user.

For this paper, we determined our test data set in two ways; (1) using reputation earned by users, and (2) using scores earned by question and answer posts. We selected these categories because posting quality questions and answers while earning reputation is what keeps a Q&A social network like Stack Overflow active [10].

For the first category of data, we identified the users with reputation score of at least 10,000. We selected 10,000 because it is the minimum score one has to earn in Stack Overflow in order to be rewarded one of the top privileges in the network, a *site moderator* and it is not easy to attain. We then computed the average daily reputation score of these users to be 18, therefore we used 18 as our benchmark. Hence for this study, we only considered users who have earned an average daily reputation of 18 points irrespective of their total reputation score. We did this to ensure that we did not include only users with really high or too low reputation scores. We identified 16,547 users that met this criteria.

For the second category, in order to study the profile of users that post helpful questions and answers, we selected only question and answer posts that have earned at least 100 points. Any score higher than 100 would mean fewer question and answer posts. There were 45,827 posts that were made by 25,942 users that met this criteria.

For each user in both categories, we reviewed their profile data to check for completeness or otherwise. We had two categories of users based on the completeness of their profiles. We defined a complete profile as one having all the profile elements: *profile image, age, website URL, location,* and *about me,* and an incomplete profile as a profile with no profile element or a profile missing at least one profile element.

---

[7] https://data.stackexchange.com/stackoverflow/queries.

4 Analysis and Results

Of the 16,547 user profiles we studied in the first category of data selection, 274 users had complete profiles; they had information for each profile element *profile image, age, website URL, location,* and *about me.* On the other hand, 6,041 users were missing at least one profile element, while 10,142 users did not have any of all profile elements *profile image, age, website URL, location,* and *about me.* Figure 1 shows the breakdown of users based on the completeness of their profile elements.

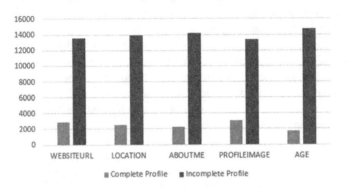

Fig. 1. Breakdown of users bassed on completeness of profile elements

In this section, we present our findings based on the two categories of selecting our test data.

4.1 Using Reputation Earned by Users

We computed the average reputation score of users with complete profile elements and those without complete profile elements. The group of users with complete profile elements, though a lot smaller in number, had a significantly higher average reputation score of 57,513, while the group of users with incomplete profile elements had an average reputation score of 8,092. Hence, one can conclude that the users with complete profiles were more helpful to the community than the users with incomplete profile elements since they earned more reputation. Based on this result, we were able to answer RQ1 as follows.

RQ1. Do profiles matter? Are users with a complete profile more helpful to the community?

Answer to RQ1. From the result of our analysis, profiles do matter as users with a complete profile earned significantly higher reputation than users without a complete profile. Since reputation scores are earned by being helpful in the Stack Overflow community[8], and the users with complete profiles earned significantly higher reputation

---

[8] http://stackoverflow.com/help/whats-reputation.

that those without complete profiles, one can conclude that users with complete profiles are more helpful to the Stack overflow community than those with incomplete profiles.

Of the users with incomplete profile elements, we sought to understand the spread of reputation scores among these profile elements. For example, do users with missing *profile image* earn higher reputation score than users with missing *about me* details? Or do users with *website URL* information (with other missing profile elements) earn higher reputation scores than users with say *location* information. This is important in determining what profile elements could be made mandatory when users are creating their profiles.

To this end, we computed the average reputation score of the users with missing profile elements for *profile image, age, website URL, location,* and *about me* and compared it to the average reputation score of users without missing profile elements. For example, we computed the average reputation score of users who had a *profile image* and compared it to the average reputation score of users who didn't have a *profile image* but had other profile elements. For each profile element, the average reputation score of users that had the element present was significantly higher than the average score for users without the corresponding profile element. Of the five profile elements, users with *age* scored higher than users with other profile elements, while users with *profile image* scored least. Hence we concluded that of the users with incomplete profile elements, users that had profile elements *age* and *about me* were more useful to the community than users with other profile elements. Figure 2 shows the result of our analysis.

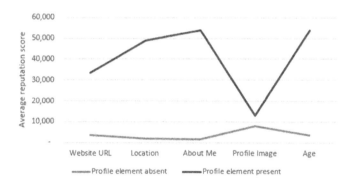

Fig. 2. Average reputation score of users based on profile elements

In order to determine what profile elements should be used together, we calculated the Pearson correlation coefficient of the profile elements using the average reputation score earned by users. Of the five profile elements, *location* and *about me* had the highest correlation of 0.999798 while *about me* and *profile image* had the lowest correlation of 0.662645. Table 1 shows the correlation between the five profile elements. Based on this result, we answered RQ2.

RQ2. What profile elements could be used together?

Answer to RQ2. From the computation of the correlation coefficient of the 5 profile elements, *location* and *about me* have the highest correlation, hence should be used together. In order words, if *location* is a mandatory field during profile creation, *about me* should be made mandatory also.

Table 1. Correlation of profile elements

	Website URL	Location	About Me	Profile Image	Age
Website URL	1				
Location	0.995126	1			
About Me	0.99323	0.999798	1		
Profile Image	0.741	0.673367	0.662645	1	
Age	0.993753	0.998831	0.999167	0.679334	1

4.2 Using Scores Earned by Question and Answer Posts

Since quality questions and answers are key to a successful Q&A network [11], it is imperative to study the profile elements of the users that post helpful questions and answers. We determined the quality of a post by the score the post earned. We used 100 as our threshold[9], hence, we only included question and answer posts with a score of at least 100 points. There were 45,827 posts that met this criteria and these were posted by 25,942 users.

We separated the posts into 2 groups; one of users with complete profiles and the other was of users with incomplete profiles. There were 1403 users with complete profiles and 24,539 users with incomplete profiles. The group of users with complete profiles had an average reputation score of 17,236, while their posts had an average score of 262. On the other hand, the users with incomplete profiles had an average reputation score of 11,145 and the average score earned by their posts was 249. Table 2 summarizes the findings from the study on posts. From this result, we concluded that the users with complete profiles posted more useful questions and answers in the community.

Table 2. Breakdown of posts dataset

	Complete Profile	Incomplete Profile
Average reputation score of users	17,236	11,145
Average score of posts by users	262	249

Based on this result, we answered RQ3 as follows.

---

[9] This threshold was set in order to have a sizeable dataset to work with.

RQ3. Do users that post helpful questions and answers have a complete profile?

Answer to RQ3. Since helpful posts are determined by how much score they earn, and the questions and answers posted by users with complete profiles on the average scored higher than those posted by users with incomplete profiles, we concluded that the users with complete profiles post more helpful questions and answers. We also came to this conclusion because the average reputation score of the users with complete profile that posted these questions and answers was higher than that of users with incomplete profiles.

5 Discussion

Based on the result of our study, having a complete profile is typically a sign of commitment to a community or network. Hence, it is not surprising that users with more developed profiles have shown more commitment in participation, have been careful in their posts and care about their reputation. With regards to the low average reputation score of users with *profile image* compared to other profile elements (as described in Fig. 2), we concluded that it could be as a result of the type of social network Stack Overflow is. Unlike a network like Facebook where the physical appearance of a user can contribute to his/her network of friends, or be a cause of narcissism [12] (hence a profile image might be an important profile element to a user), in Stack Overflow, a user's physical appearance is largely immaterial. Rather, the user's ability to answer questions and earn reputation scores is more important [8, 13]. Hence users might not see the need to have profile pictures.

6 Conclusion and Future Work

This paper studied profiles in Stack Overflow with the aim of understanding how helpful users with complete profiles are compared to users with incomplete profiles. We defined a complete profile as users having all of the five profile elements *profile image, age, website URL, location,* and *about me*, while incomplete profiles were user profiles missing all or at least one profile element. We studied the profiles of two categories of people; users with a daily average reputation score of over 18 points and users who posted questions and answers that have earned at least 100 points. Based on the result of our analysis, users with complete profile elements had a higher average reputation score than users with incomplete profile elements. In addition, the posts made by users with complete profile elements scored higher than those made by users with incomplete profiles. Furthermore, the users that posted quality questions and answers in the network were users with complete profile elements. Based on these results, we concluded that user profiles are important and that users with complete profiles are more useful to the community than users with incomplete profile elements. We also computed the Pearson's correlation coefficient to determine what profile elements go together in the network. *Location* and *about me* had the highest correlation while *about me* and *profile image* had the least correlation.

In the future, we plan to analyze the evolution of user profiles over time, in particular, to compare the performance of users in the network before and after they had complete profiles, assuming the users did not start out with complete profiles. We also want to compare the influence and expertise level of users with complete profiles to users identified as experts using other methods like [8, 9]. In addition, we plan to carry out further statistical analysis on the dataset to determine the statistical significance of our result. Furthermore, we plan to conduct a user study to understand the importance of the various profile elements from a user's perspective and if users regard other users with complete profile elements as authority figures. We will also investigate the level of trust users give fellow users based on the completeness of their profiles.

References

1. Riahi, F., Zolaktaf, Z., Shafiei, M., Milios, E.: Finding expert users in community question answering. In: Proceedings of the 21st International Conference on World Wide Web - WWW 2012, p. 791 (2012)
2. Strano, M.M.: User descriptions and interpretations of self-presentation through Facebook profile images. Cyberpsychology J. Psychosoc. Res. Cybersp. **2**(2), 5 (2008)
3. Zhou, Z.-M., Lan, M., Niu, Z.-Y., Lu, Y.: Exploiting user profile information for answer ranking in cQA. In: Proceedings 21st International Conference World Wide Web - WWW, pp. 767–774 (2012)
4. Ginsca, A., Popescu, A.: User profiling for answer quality assessment in Q&A communities. In: Proceedings of the 2013 Workshop on Data-Driven User Behavioral Modelling and Mining from Social Media, pp. 25–28 (2013)
5. Middleton, S., Shadbolt, N., De Roure, D.: Ontological user profiling in recommender systems. ACM Trans. Inf. Syst. **22**(1), 54–88 (2004)
6. Yerva, S.R., Catasta, M., Demartini, G., Aberer, K.: Entity disambiguation in tweets leveraging user social profiles. In: 2013 IEEE 14th International Conference on Information Reuse and Integration (IRI), pp. 120–128 (2013)
7. van Dijk, D., Tsagkias, M., de Rijke, M.: Early detection of topical expertise in community question answering. In: Proceedings of the 38th International ACM SIGIR Conference on Research and Development in Information Retrieval - SIGIR 2015, pp. 995–998 (2015)
8. Adaji, I., Vassileva, J.: Predicting churn of expert respondents in social networks using data mining techniques: a case study of stack overflow. In: Proceedings of 14th IEEE International Conference on Machine Learning and Applications (ICMLA) (2015)
9. Hanrahan, B.V., Convertino, G., Nelson, L.: Modeling problem difficulty and expertise in stackoverflow. In: Proceedings of the ACM 2012 conference on Computer Supported Cooperative Work Companion, pp. 91–94 (2012)
10. Movshovitz-Attias, D., Movshovitz-Attias, Y., Steenkiste, P., Faloutsos, C.: Analysis of the reputation system and user contributions on a question answering website. In: Proceedings of the 2013 IEEE/ACM International Conference on Advances in Social Networks Analysis and Mining - ASONAM 2013, pp. 886–893 (2013)
11. Dror, G., Pelleg, D., Rokhlenko, O., Szpektor, I.: Churn prediction in new users of Yahoo! answers. In: Proceedings of the 21st International Conference on World Wide Web, pp. 829–834 (2012)

12. Kapidzic, S.: Narcissism as a predictor of motivations behind Facebook profile picture selection. Cyberpsychol. Behav. Soc. Netw. **16**(1), 14–19 (2013)
13. Pal, A., Chang, S., Konstan, J.A.: Evolution of experts in question answering communities. In: Proceedings of the 6th International Conference on Web and Social Media (ICWSM) (2012)

Identifying Correlated Bots in Twitter

Nikan Chavoshi$^{(\boxtimes)}$, Hossein Hamooni, and Abdullah Mueen

University of New Mexico, Albuquerque, USA
chavoshi@unm.edu

Abstract. We develop a technique to identify abnormally correlated user accounts in Twitter, which are very unlikely to be human operated. This new approach of bot detection considers cross-correlating user activities and requires no labeled data, as opposed to existing bot detection techniques that consider users independently, and require large amount of recently labeled data. Our system uses a lag-sensitive hashing technique and a warping-invariant correlation measure to quickly organize the user accounts in clusters of abnormally correlated accounts. Our method is 94 % precise and detects unique bots that other methods cannot detect. Our system produces daily reports on bots at a rate of several hundred bots per day. The reports are available online for further analysis.

1 Introduction

Automated accounts, called bots, are common in social media. Although all bots are not bad, bots are easy means to engage in unethical and illegal activities in social media. Examples of such activities include selling accounts [18], spamming inappropriate content [1], and participating in sponsored activities [7]. Many social metrics are calculated based on social media data [3,15]. The significant presence of bots in social media will make many of these metrics useless. The exact number of bots is dynamic and unknown. The range of the estimates is between 3 % [18] to 7 % [14]. Social media sites, such as Twitter, regularly suspend abusive bots [19]. Yet, the number of bots is growing because of almost zero-cost in creating new bots.

Existing bot detection methods are not capable of fighting such evolving set of bots. There are several reasons. Current methods are mostly non-adaptive, require supervised training, and consider accounts independently [6,20]. Typical features used in some of the methods need a long duration of activities (e.g. weeks) [21] which makes the detection process useless, as the bots can initiate a fair amount of harm before being detected. Moreover, bots are becoming smarter. They mimic humans to avoid being detected and suspended, and increase throughput by creating many accounts. We take a novel *unsupervised* approach of *cross-correlating* account activities, that can detect such dynamic bots as soon as two hours after starting their activities.

Our *novelty* is in using activity correlation as an absolute indicator of bot behavior. Millions of users interact in social media at any time. Even at this large scale, human users are not expected to have highly correlated activities in social

© Springer International Publishing AG 2016
E. Spiro and Y.-Y. Ahn (Eds.): SocInfo 2016, Part II, LNCS 10047, pp. 14–21, 2016.
DOI: 10.1007/978-3-319-47874-6_2

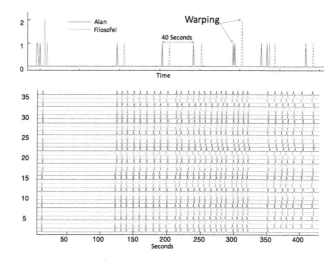

Fig. 1. (top) Two highly correlated activity sequence (six minutes) of two Twitter users: Alan and Filosofei. Warping-invariant correlation between them is 0.99, while cross-correlation is 0.72 and Pearson's correlation is 0.07. (bottom) A group of 35 correlated bots' activity sequence. Note that slight misalignment in the timestamps.

media for hours. However, in Twitter, large groups of such correlated user accounts are actively operating. A video capture of two completely unrelated (no one follows the other) and yet perfectly correlated Twitter accounts is shown in [2]. You can find several examples of activity sequence of correlated Twitter accounts in Fig. 1 (better in high resolution). Such correlation in tweeting activities is only possible if the accounts are controlled automatically, indicating that the accounts are bots. We provide mathematical significance of our approach and empirically achieve 94 % precision of our approach.

In the rest of the paper, we discuss the level of significance of correlated bots and show empirical evaluation. We omit technical details of our method, named DeBot, due to limited presentation scope. The daily reports of bots and an expanded paper are available in [2].

2   Significance of Correlation in Bot Detection

In this section, we analyze the significance of correlation in detecting bots. We first assume each user tweets independently and then relax the restriction.

We estimate the probability of two users having n posts at identical timestamps among m seconds when there are N such active users. We assume the users are independently tweeting. There are $M = m^n$ possible ways a user can post n actions in m seconds. Let us estimate the probability $\hat{p}$ that no two users have n identical timestamps under user independence.

$$\hat{p} = 1 \times \frac{M-1}{M} \times \frac{M-2}{M} \times \ldots \times \frac{M-N+1}{M}$$

The probability p of at least two users posting at the same n seconds in m seconds is simply $1 - \hat{p}$.

$$p = 1 - \frac{M!}{M^N(M-N)!}$$

Note that, if $N > M$ then $p = 1$, as there are more trials (i.e. users) than possible options (i.e. combination of seconds). If we realistically set $N = 10^9$ and $m = 3600$, p sharply goes down from one to zero, when we move from $n = 6$ to $n = 8$. Therefore, observing two users with seven or more identical posting timestamps is an extremely unlikely event when users are independent.

Let us now consider the warped instance of the above estimation. If the warping constraint is w, then we can pessimistically assume that any pair of the n tweets are more than $2w$ apart. This ensures that, for each of the n tweets, there can be a maximum of $W = 2w + 1$ locations available for an equivalent tweet. The new expression for $\hat{p}$ is the following.

$$\hat{p} = 1 \times \frac{M-W^n}{M} \times \frac{M-2W^n}{M} \times \ldots \times \frac{M-NW^n}{M}$$

Similar to the exact matching, in case of warped matching, $p = 1 - \hat{p}$ tends to zero for $n = 13$ when $w = 20$ seconds, $N = 10^9$ and $m = 3600$.

Let us now consider the dependent case where the Twitter users react to similar news/events in similar ways. Let us assume q is the probability of a user reacting to any tweet within $\pm w$ seconds of the relevant tweet. The probability of none of the n tweets of a user fall within $\pm w$ of n tweets from another user is $1 - q^n$. The expression for $\hat{p}$ becomes the following.

$$\hat{p} = 1 \times (1 - q^n) \times (1 - 2q^n) \times \ldots \times (1 - Nq^n)$$

Note that, in the equal probability case, $q = \frac{2w}{m}$, which is identical to the $\hat{p}$ for warped correlation. In an extreme scenario, if users are perfectly in sync, $q = 1$ ensures $\hat{p} = 0$ and $p = 1$. If $q = 0.25$, p tends to zero for $n = 40$ and if $q = 0.5$, p tends to zero for $n = 80$. However, $q = 0.25$ is an extremely high probability. To elaborate, consider how many tweets/posts, that a user sees, is retweeted or shared. For an average user, it may be one in every few. Now consider how many a user shares within w seconds of *seeing*, which should be much less. Then consider how many a user shares within w seconds of another user *authoring* the tweets or retweets, which should be even smaller.

Thus, even for this unlikely high probability of a user tweeting or retweeting within ± 20 seconds ($q = 0.25$) of another tweet, the probability of two users with forty or more matching tweets in an hour is close to zero. Our system, therefore, considers users with at least forty tweets in an hour and identifies highly correlated (≥ 0.995) users as bots because of their extreme unlikelihood of being humans. This approach of identifying bots is highly precise with almost *no false-positive*.

One may think that evading detection by this simple approach is a very easy task. It is indeed very simple to evade such detection by inserting unbounded random time delays among the same tweet from many accounts. However, such randomization will severely damage the throughput of a bot-master, making it worthless to maintain large pool of uncontrolled bots. Moreover, although evasion is fairly easy, we have detected hundreds of thousands of unique correlated bots that are freely operating in absence of such a simple detection system.

We do not claim that correlated bot detection is the solution to bot related problems in social media. Detecting benign or malicious bot is out of the scope of this work. We simply suggest that detecting correlated bots has a potential to improve the performance of suspension systems that safeguard large social networks, eventually increasing the cost of bot operation and maintenance.

A pathological argument against correlated bot detection is that a human user may be identified as bot if some bots *mimic* the human user. If a human user is mimicked by bots, it is an urgent matter to take some action, such as blocking all of the accounts and asking all the users to prove their humanity once again. Naturally, only the human user can prove it while the bot mimickers will just remain blocked.

3   Empirical Evaluation

As per the discussion in the previous section, synchronized behavior in a sequence of forty activities is a near absolute indicator of automated accounts. In this section, we show empirical evaluation in comparison to other bot detection approaches.

We calculate relative support from other methods to the bots detected by our system. We compare against five methods. We have run bot discovery in every 4 hours for sixteen days (May 18 - June 3, 2015) and merged all the clusters into one consolidated set of clusters using *friend-of-friend* approach. We picked the top ten clusters in size that contained a total of 9,134 bot accounts to form our **base set** to compare against other methods.

- We compare the support to our method by Twitter's suspension process. We first ask the question, *how many bots that we detect are later suspended by Twitter?* If Twitter suspends them, we are certain that the bots were bad ones. On June 12, 2015, we began tracking these accounts via Twitter API to check whether or not they were suspended. We checked every few days until August 28, 2015. Twitter increasingly suspended more bots that we had detected months ahead. Twitter suspended 2,491 accounts in the very first probe and reached to 4,126 in the last probe. This means that roughly **45 %** of the bots were suspended by Twitter in 12 weeks.
- A successful existing technique developed in the Truthy project [6] is *Bot or Not?*, which is a supervised technique to estimate the probability of an account being bot. It uses account features, network features and content features to train a model [6] and estimates a probability of *"being bot"* for a

given account. We set a threshold of 50 % or more to classify an account as bot and found that **59 %** of the bots in our *base set* were also flagged by *Bot or Not?* on June 12, 2015. We probed *Bot or Not?* for the *base set* two more times and noticed no significant change in detection performance.

- We compare our method to an existing per-user method [21] which uses the dependence between minute-of-an-hour and second-of-a-minute as an indicator for bot accounts. The method in [21] tests the independence of these two quantities using the χ^2 test and declares an account bot if there is any dependence. The method fails for user *alan26oficial* (the same Alan as in Fig. 1) because of independence among the quantities, while our method can detect *alan26official* because of its correlation with *FrasesFilosofos* (the same Filosofei in Fig. 1). We calculate the relative support from the χ^2 test method and identify **76 %** of the bots are supported by the χ^2 test.

- We evaluate the bots using contextual information such as tweet content and cross-user features. We investigate whether the synchronously aligned tweets have identical texts and authors. We define the *"botness"* of a group of accounts as the average of the botness of all the pairs of accounts in the cluster. For a given pair, botness is the percentage of aligned tweets that also match in their content (e.g. author, text). The higher the botness score the more successful DeBot is. We achieve an average of **78 %** botness when we match text and/or authors of the tweets. Simply put, the aligned tweets have identical text and authors **78 %** of the time. Note that there is a very little difference between **and** and **or** configuration. This suggests that most of the time tweets and authors match.
  Less botness score does not necessarily mean that our method is detecting false positives. We see many bot accounts that correlate in time perfectly, but do not have identical tweets.

- We investigate if approximate text matching would increase botness by employing human judges in Amazon Mechanical Turk. We ask the judges to determine whether fifty random pairs of accounts are showing similar text (may not be exact), URLs, authors and languages. We then calculate the botness. DeBot achieves up to 94 % botness score from the contextual information. Simply put, **94 %** of the tweets are not only synchronized in time, but also share the same information (Table 1).

Table 1. Relative support of different tests of DeBot

	Twitter	BotOrNot?	χ^2 Test	Text &Author	Text || Author	Human Judgment
Relative Support	45 %	59 %	76 %	78 %	79 %	94 %

3.1   Recall Estimates

It is impossible to calculate the exact recall of a bot detection technique because a complete list of known bots does not exist. we estimate the recall of three bot detection methods by a simple approach. First, we listen to the Twitter streaming API for 30 min and pick those user that have more than 1 activity to be able to calculate DTW distances. In 30 min we filter out 8600 user accounts, on average. We test these accounts using *Bot or Not?* and χ^2 test methods. We apply DeBot to identify the bots based on temporal correlation.

The final results, which are the average of three rounds of our experiments, show the highest recall rate of 6.3 % for DeBot, which is very close to the true bot ratio (8.5 %) estimated and disclosed by Twitter recently [17]. *Bot or Not?* achieves 3.4 % bot detection rate.

4 Related Work

Real-time correlation monitoring has been a well-researched topic for over a decade now. One of the first works is StatStream [22], which can monitor thousands of signals. In [16], authors show a method to monitor lagged correlation in streaming fashion for thousands of signals. In [5], authors develop a sketch (i.e. random projection) based correlation monitoring algorithm that does not consider time warping. Twitter stream can provide tweets of millions of users which are at least an order of magnitude more in number, and an order of magnitude less in density than the method in [5], and time warping exists in Twitter. Such warped sparseness has not been addressed previously for correlation monitoring.

A good characterization of spammers in Twitter is presented in [10]. Authors concluded that 92 % of the accounts that Twitter suspends for spamming activities are suspended within three days of the first post. Therefore, if a spamming bot survives one week, it is very likely to survive a long time. Our work identifies bots that are tweeting for months, if not years. In [18], authors characterize the spam detection strategies very well. Spam detection methods that analyze social graph properties, characterize contents and rates of postings, and identify common spam redirect paths, are typically *at-abuse* methods. Such methods find the spam after the spam has done the harm. In contrast, our method can detect accounts registered by account merchants which will eventually be sold to miscreants, and thus, our method detects these bots *soon-after-registration* to prevent future abuse. *Detecting bots by correlating users is our novelty.*

Other relevant works include detecting campaign promoters in Twitter [12]. Correlating user activity across sites (e.g. Yelp and Twitter) can provide useful information about linked-accounts, and thus, form a basis of privacy attack [9]. In [8], authors perform offline analysis to discover *link-farming* by which spammers acquire a large number of followers. In [13], authors develop a fast algorithm to mine millions of co-evolving signals and find anomalies. In [4], authors find temporally coherent collaborative Liking of Facebook pages. The authors in [11], present a method to characterize groups of malicious users. They consider three features such as individual information, and social relationships to provide deep understanding of these groups. As opposed to most of these works, our focus is to correlate within the same site to identify bot accounts that already are or will potentially become spammers.

5   Conclusion

We introduce a real-time method that detects bots by correlating their activities. Our method can detect hundreds of bot accounts everyday, which now have aggregated to hundreds of thousands of bots in eight months. Human judges in Amazon Mechanical Turk have found the detected bots are highly similar to each other. Our method, DeBot, is identifying bots at a higher rate than the rate Twitter is suspending them. In comparison to per-user methods, our cross-user temporal method detects more bots with strong significance.

References

1. How twitter bots fool you into thinking they are real people. http://www.fastcompany.com/3031500/how-twitter-bots-fool-you-into-thinking//-they-are-real-people
2. Supporting web page containing video, data, code and daily report. http://www.cs.unm.edu/~chavoshi/debot/
3. Asur, S., Huberman, B.A.: Predicting the future with social media. In: 2010 IEEE/WIC/ACM International Conference on Web Intelligence and Intelligent Agent Technology, vol. 1, pp. 492–499. IEEE, Aug. 2010
4. Beutel, A., Xu, W., Guruswami, V., Palow, C., Faloutsos, C.: Copycatch: stopping group attacks by spotting lockstep behavior in social networks. In: Proceedings of the 22nd International Conference on World Wide Web, pp. 119–130. International World Wide Web Conferences Steering Committee (2013)
5. Cole, R., Shasha, D., Zhao, X.: Fast window correlations over uncooperative time series. In: Proceeding of the Eleventh ACM SIGKDD International Conference on Knowledge Discovery in Data Mining - KDD 2005, p. 743 (2005)
6. Ferrara, E., Varol, O., Davis, C., Menczer, F., Flammini, A.: The rise of social bots. arXiv preprint 2014. arXiv:1407.5225
7. Galán-García, P., de la Puerta, J.G., Gómez, C.L., Santos, I., Bringas, P.G.: Supervised machine learning for the detection of troll profiles in twitter social network: application to a real case of cyberbullying. In: Herrero, Á., et al. (eds.) International Joint Conference SOCO'13-CISIS'13-ICEUTE'13. AISC, vol. 239, pp. 419–428. Springer, Heidelberg (2014)
8. Ghosh, S., Viswanath, B., Kooti, F., Sharma, N.K., Korlam, G., Benevenuto, F., Ganguly, N., Gummadi, K.P.: Understanding and combating link farming in the twitter social network. In: Proceedings of the 21st International Conference on World Wide Web - WWW 2012, p. 61. ACM Press, New York, April 2012
9. Goga, O., Lei, H., Parthasarathi, S.H.K., Friedland, G., Sommer, R., Teixeira, R.: Exploiting innocuous activity for correlating users across sites, pp. 447–458, May 2013
10. Grier, C., Thomas, K., Paxson, V., Zhang, M.: @spam: the underground on 140 characters or less. In: Proceedings of the 17th ACM Conference on Computer and Communications Security - CCS 2010, p. 27. ACM Press, New York, October 2010
11. Jiang, J., Shan, Z.-F., Wang, X., Zhang, L., Dai, Y.-F.: Understanding sybil groups in the wild. J. Comput. Sci. Technol. **30**(6), 1344–1357 (2015)
12. Li, H., Mukherjee, A., Liu, B., Kornfield, R., Emery, S.: Detecting campaign promoters on twitter using markov random fields. In: 2014 IEEE International Conference on Data Mining (ICDM), pp. 290–299 (2014)

13. Matsubara, Y., Sakurai, Y., Ueda, N., Yoshikawa, M.: Fast and exact monitoring of co-evolving data streams. In: 2014 IEEE International Conference on Data Mining, pp. 390–399. IEEE, December 2014

14. Morstatter, F., Carley, K.M., Liu, H.: Bot detection in social media: networks, behavior, and evaluation. In: ASONAM - Tutorial, August 2015

15. Ruiz, E.J., Hristidis, V., Castillo, C., Gionis, A., Jaimes, A.: Correlating financial time series with micro-blogging activity. In: Proceedings of the Fifth ACM International Conference on Web Search and Data Mining - WSDM 2012, p. 513. ACM Press, New York, February 2012

16. Sakurai, Y., Papadimitriou, S., Faloutsos, C.: Braid: Stream mining through group lag correlations. In: Proceedings of the 2005 ACM SIGMOD International Conference on Management of Data, p. 610 (2005)

17. Subrahmanian, V., Azaria, A., Durst, S., Kagan, V., Galstyan, A., Lerman, K., Zhu, L., Ferrara, E., Flammini, A., Menczer, F., Waltzman, R., Stevens, A., Dekhtyar, A., Gao, S., Hogg, T., Kooti, F., Liu, Y., Varol, O., Shiralkar, P., Vydiswaran, V., Mei, Q., Huang, T.: The darpa twitter bot challenge. IEEE Comput. 1, 38–46 (2016). (In press)

18. Thomas, K., Paxson, V., Mccoy, D., Grier, C.: Trafficking fraudulent accounts: the role of the underground market in twitter spam and abuse trafficking fraudulent accounts. In: USENIX Security Symposium, pp. 195–210 (2013)

19. Twitter. The Twitter Rules. https://support.twitter.com/articles/18311

20. Wang, A.H.: Detecting spam bots in online social networking sites: a machine learning approach. In: Foresti, S., Jajodia, S. (eds.) DBSec 2010. LNCS, vol. 6166, pp. 335–342. Springer, Heidelberg (2010). doi:10.1007/978-3-642-13739-6_25

21. Zhang, C.M., Paxson, V.: Detecting and analyzing automated activity on twitter. In: Spring, N., Riley, G.F. (eds.) PAM 2011. LNCS, vol. 6579, pp. 102–111. Springer, Heidelberg (2011). doi:10.1007/978-3-642-19260-9_11

22. Zhu, Y., Shasha, D.: StatStream: statistical monitoring of thousands of data streams in real time. In: Proceedings of the 28th International Conference on Very Large Data Bases, volume 54 of VLDB 2002, pp. 358–369 (2002)

Predicting Online Extremism, Content Adopters, and Interaction Reciprocity

Emilio Ferrara[1]([✉]), Wen-Qiang Wang[1], Onur Varol[2],
Alessandro Flammini[2], and Aram Galstyan[1]

[1] University of Southern California, Los Angeles, CA, USA
emilio.ferrara@gmail.com
[2] Indiana University, Bloomington, IN, USA

Abstract. We present a machine learning framework that leverages a mixture of metadata, network, and temporal features to detect extremist users, and predict content adopters and interaction reciprocity in social media. We exploit a unique dataset containing millions of tweets generated by more than 25 thousand users who have been manually identified, reported, and suspended by Twitter due to their involvement with extremist campaigns. We also leverage millions of tweets generated by a random sample of 25 thousand regular users who were exposed to, or consumed, extremist content. We carry out three forecasting tasks, (i) to detect extremist users, (ii) to estimate whether regular users will adopt extremist content, and finally (iii) to predict whether users will reciprocate contacts initiated by extremists. All forecasting tasks are set up in two scenarios: a *post hoc* (time independent) prediction task on aggregated data, and a simulated real-time prediction task. The performance of our framework is extremely promising, yielding in the different forecasting scenarios up to 93 % AUC for extremist user detection, up to 80 % AUC for content adoption prediction, and finally up to 72 % AUC for interaction reciprocity forecasting. We conclude by providing a thorough feature analysis that helps determine which are the emerging signals that provide predictive power in different scenarios.

Keywords: Social media · Online extremism · Radicalization prediction

1 Introduction

Researchers are devoting increasing attention to the issues related to online extremism, terrorist propaganda and radicalization campaigns [31,34]. Social media play a central role in these endeavors, as increasing evidence from social science research suggests [7,18]. For example, a widespread consensus on the relationship between social media usage and the rise of extremist groups like the Islamic State of Iraq and al-Sham (viz. ISIS) has emerged among policymakers and security experts [11,35,41]. ISIS' success in increasing its roster to thousands of members has been related in part to a savvy use of social media for propaganda and recruitment purposes. One reason is that, until recently, social

© Springer International Publishing AG 2016
E. Spiro and Y.-Y. Ahn (Eds.): SocInfo 2016, Part II, LNCS 10047, pp. 22–39, 2016.
DOI: 10.1007/978-3-319-47874-6_3

media platform like Twitter provided a public venue where single individuals, interest groups, or organizations, were given the ability to carry out extremist discussions and terrorist recruitment, without any form of restrictions, and with the possibility of gathering audiences of potentially millions. Only recently, some mechanisms have been put into place, based on manual reporting, to limit these abusive forms of communications. Based on this evidence, we argue in favor of developing computational tools capable of effectively analyzing massive social data streams, to detect extremist users, to predict who will become involved in interactions with radicalized users, and finally to determine who is likely to consume extremist content. The goal of this article is to address the three questions above by proposing a computational framework for detection and prediction of extremism in social media. We tapped into Twitter to obtain a relevant dataset, leveraged expert crowd-sourcing for annotation purposes, and then designed, trained and tested the performance of our prediction system in static and simulated real-time forecasts, as detailed below.

Contributions of this work

The main contributions of our work can be summarized as:

- We formalize three different forecasting tasks related to online extremism, namely the detection of extremist users, the prediction of adoption of extremist content, and the forecasting of interaction reciprocity between regular users and extremists.
- We propose a machine prediction framework that analyzes social media data and generates features across multiple dimensions, including user metadata, network statistics, and temporal patterns of activity, to perform the three forecasting tasks above.
- We leverage an unprecedented dataset that contains over 3 millions tweets generated by over 25 thousand extremist accounts, who have been manually identified, reported, and suspended by Twitter. We also use around 30 million tweets generated by a random sample of 25 thousand regular users who were exposed to, or consumed, extremist content.
- For each forecasting task, we design two variants: a post-hoc (time independent) prediction task performed on aggregated data, and a simulated real-time forecast where the learning models are trained as if data were available up to a certain point in time, and the system must generate predictions on the future.
- We conclude our analysis by studying the predictive power of the different features used for prediction, to determine their role in the three forecasts.

2   Data and Preliminary Analysis

In this section we describe our dataset, the curation strategy yielding the annotations, and some preliminary analysis.

2.1 Sample Selection and Curation

In this work we rely on data and labels constructed by using a procedure of manual curation and expert verification. We retrieved on a public Website a list of over 25 thousands Twitter accounts whose activity was labeled as supportive of the Islamic State by the crowd-sourcing initiative called *Lucky Troll Club*. The goal of this project was to leverage annotators with expertise in Arabic languages to identify ISIS accounts and report them to Twitter. Twitter's anti-abuse team manually verifies all suspension requests, and grants some based on the active violation of Twitter's Terms of Service policy against terrorist- or extremist-related activity. Here we focus on the 25,538 accounts that have been all suspended between January and June 2015 by Twitter as a consequence of evidence of activity supporting the Islamic State group. For each account, we also have at our disposal information about the suspension date, and the number of followers of that user as of the suspension date.

2.2 Twitter Data Collection

The next step of our study consisted in collecting data related to the activity of the 25,538 ISIS supporters on Twitter. To this purpose, we leveraged the Twitter *gardenhose* data source (roughly 10 % of the Twitter stream) collected by Indiana University [15]. We decided to collect not only the tweets generated by these accounts prior to their suspension, but also to build a dataset of their targets. In particular, we are concerned with accounts unrelated to ISIS with whom the ISIS supporters tried to establish some forms of interaction. We therefore constructed the following two datasets:

ISIS accounts: this dataset contains 3,395,901 tweets generated in the time interval January-June 2015 by the 25,538 accounts identified by Twitter as supporters of ISIS. This is a significant portion of all the accounts suspended by Twitter in relation to ISIS.[1]

Users exposed to ISIS: this dataset contains 29,193,267 tweets generated during January-June 2015 by a set of 25 thousand users randomly sampled among the larger set of users that constitute ISIS accounts' followers. This set is by choice of equal size to the former one, to avoid introducing class imbalance.

For prediction purposes, we will use as positive and negative labels the *ISIS accounts* group and the accounts in the *users exposed to ISIS*, respectively.

3 Methodology

In this section we discuss the learning models and the features adopted by our framework. The complete prediction pipeline (learning models, cross validation,

---

[1] The Guardian recently reported that between April 2015 and February 2016, Twitter's anti-terror task force suspended about 125,000 accounts linked to ISIS extremists: http://www.theguardian.com/technology/2016/feb/05/twitter-deletes-isis-accounts-terrorism-online.

feature selection, and performance evaluation) is developed using Python and the scikit-learn library [28].

3.1 Learning Models

We adopt two off-the-shelf learning models as a proof of concept for the three classification tasks that we will discuss later (see Sect. 4):

Logistic Regression: The first implemented algorithm is a simple Logistic Regression (LR) with LASSO regularization. The advantage of this approach is its scalability, which makes it very effective to (possibly real-time) classification tasks on large datasets. The only parameter to tune is the loss function C. We expect that LR will provide the baseline prediction performance.

Random Forests: We also use a state-of-the-art implementation of *Random Forests* (RF) [9]. The vectors fed into the learning models represent each user's features. *Random Forests* are trained using 100 estimators and adopting the Gini coefficient to measure the quality of splits. Optimal parameters setting is obtained via cross validation (see Sect. 3.1).

Note that the goal of this work is not to provide new machine learning techniques, but to illustrate that existing methods can provide promising results. We also explored additional learning models (e.g., SVM, Stochastic Gradient Descent, etc.), which provide comparable prediction performance but are less computationally efficient and scalable.

Cross Validation. The results of our performance evaluation (see Sect. 4) are all obtained via k-fold cross validation. We adopt $k = 5$ folds, and therefore use 80 % of data for training, and the remainder 20 % for testing purpose, averaging performance scores across the 5 folds. We also use 5-fold cross validation to optimize the parameters of the two learning algorithms (LR and RF), by means of an exhaustive cross-validated grid search on the hyperparameter space.

Evaluation Scores. We benchmark the performance of our system by using four standard prediction quality measures, namely Precision, Recall, F1 (harmonic mean of Precision and Recall), and AUC—short for Area Under the Receiver Operating Characteristic (ROC) curve [23].

3.2 Feature Engineering and Feature Selection

We manually crafted a set of 52 features belonging to three different classes: user metadata, timing features, and network statistics, as detailed below.

User metadata and activity features: User metadata have been proved pivotal to model classes of users in social media [16,27]. We build user-based features leveraging the metadata provided by the Twitter API related to the author of each tweet, as well as the source of each retweet. User features include the number of tweets, followers and friends associated to each users, the frequency of

Table 1. List of 52 features extracted by our framework

User metadata & activity	Number of followers
	Number of friends (i.e., followees)
	Number of posted tweets
	Number of favorite tweets
	Ratio of retweets / tweets
	Ratio of mentions / tweets
	Avg number of hashtags
	(avg, var) number of retweets
	Avg. number of mentions
	Avg. number of mentions (excluding retweets)
	Number of URLs in profile description
	(avg, std, min, max, proportion) URLs in tweets
	Length of username
Timing	(avg, var) number of tweets per day
	(avg, std, min, max) interval between two consecutive tweets
	(avg, std, min, max) interval between two consecutive retweets
	(avg, std, min, max) interval between two consecutive mentions
Netw. stats	(avg, std, min, max) distribution of retweeters' number of followers
	(avg, std, min, max) distribution of retweeters' number of friends
	(avg, std, min, max) distribution of mentioners' number of followers
	(avg, std, min, max) distribution of mentioners' number of friends
	(avg, std, min, max) number of retweets of the tweets by others

adoption of hashtags, mentions, and URLs, and finally some profile descriptors. In total, 18 user metadata and activity features are computed (cf. Table 1).

Timing features: Important insights may be concealed by the temporal dimension of content production and consumption, as illustrated by recent work [19,36]. A basic timing feature is the average number of tweets posted per day. Other timing features include statistics (average, standard deviation, minimum, maximum) of the intervals between two consecutive events, e.g., two tweets, retweets, or mentions. Our framework generates 14 timing features (cf. Table 1).

Network statistics: Twitter content spreads from person to person via retweets and mentions. We expect that the emerging network structure carries important information to characterize different types of communication. Prior work shows that using network features significantly helps prediction tasks like social bot detection [15,16,36], and campaign detection [17,30]. Our framework focuses on two types of networks: (i) retweet, and (ii) mention networks. Users are nodes of such networks, and retweets or mentions are directed links between pairs of users. For each user, our framework computes the distribution of followers and friends of all users who retweet and mention that user, and extracts some descriptive statistics (average, standard deviation, minimum, maximum) of these distributions. Our system builds 20 network statistics features (cf. Table 1).

Greedy Feature Selection. In our predictions, some features exhibit more predictive power than others: temporal dependencies introduce strong correlations, thus some possible redundancy. Among the different existing ways to select the most relevant features for the prediction task at hand [22], in the interest of computational efficiency, we adopted a simple greedy forward feature selection method, as follows: (i) initialize the set of selected features $S = \emptyset$; (ii) for each feature $f \in F - S$, consider the union set $U = S \cup f$; (iii) train the classifier using the features in U; (iv) test the average performance of the classifier trained on this set; (v) add to S the feature providing the best performance; (vi) repeat (ii)–(v) as long as a significant performance increase is yield.

4   Experiments

Task I (T1): Detection of extremist supporters. The first task that our system will face is a binary classification aimed to detect ISIS accounts and separate them from those of regular users. The problem is to test whether any predictive signal is present in the set of features we designed to characterize social media activity related to extremism, and serves as a yardstick for the next two more complex problems.

Task II (T2): Predicting extremist content adoption. The set of 25 thousand users we randomly sampled among followers of ISIS accounts can be leveraged to perform the prediction of extremist content adoption. We define as a positive instance of adoption in this context when a regular user retweets some content s/he is exposed to that is generated by an ISIS account.

Task III (T3): Predicting interactions with extremists. The third task presents likely the most difficult challenge: predicting whether a regular user will engage into interactions with extremists. A positive instance of interaction is represented by a regular user replying to a contact initiated by an ISIS account.

Static versus real-time predictions. For each of the three prediction tasks described above, we identified two modalities, namely a static (time independent) and a simulated real-time prediction. In the former scenario, a static prediction ignores temporal dependencies in that the system aggregates all data available across the entire time range (January-June 2015), and then performs training and testing using the 5-fold cross validation strategy by randomly splitting datapoints into the 5 folds and averaging the prediction performance across folds. In the latter scenario, a real-time prediction is simulated in which data are processed for training and testing purposes by respecting the timeline of content availability: for example, the system can exploit the first month of available data (January 2015) for training, and then producing predictions for the remainder 5 months (Feb-Jun 2015), for which performance is tested.

The performance of our framework in the three tasks, each with the two prediction modalities, is discussed in the following. The section concludes with the analysis of feature predictive power (see Sect. 4.4).

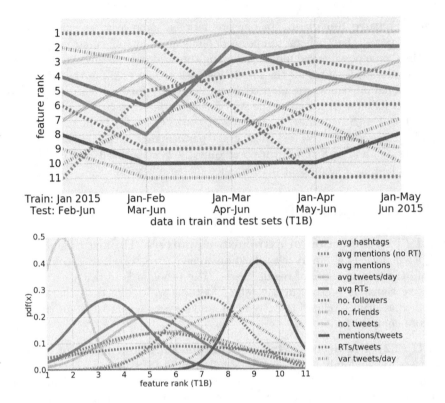

Fig. 1. T1B: Feature selection analysis and feature rank distribution (top 11 features)

4.1 T1: Detection of Extremist Supporters

In the following we discuss the static (T1A) and real-time (T1B) scenarios for the first prediction task, namely detecting extremist accounts on Twitter.

T1A: Time-Independent Detection. The detection of extremist user accounts is the most natural task to start the performance evaluation of our framework. Our analysis aims at verifying that the 52 features we carefully hand-crafted indeed carry some predictive signal useful to separate extremist users from regular ones. The dataset at hand contains two roughly equal-sized classes (about 25 thousand instances each), where ISIS accounts are labeled as positive instances, and regular users as negative ones. Each instance is a characterized by a 52-dimensional vector, and positive and negative examples are fed to the two learning models (LR and RF). The first task, in short T1A, is agnostic of time dependencies: data are aggregated throughout the entire 6 months period (January–June 2015) and training/testing is performed in a traditional 5-fold cross-validated fashion (cf. Sect. 3.1). Table 2 summarizes the performance of LR and RF according to the four quality measures described above (cf. 3.1):

Table 2. Extremists detection (T1A)

	Precision	Recall	F1	AUC
Logistic Regression	0.778	0.506	0.599	0.756
Random Forests	0.855	0.893	0.874	**0.871**

Both models perform well, with *Random Forests* achieving an accuracy above 87 % as measured by AUC. These results are encouraging and demonstrate that simple off-the-shelf models can yield good performance in T1A.

T1B: Simulated Real-Time Detection. A more complex variant of this prediction task is by taking into account the temporal dimension. Our prior work has demonstrated that accounting for temporal dependencies is very valuable in social media prediction tasks and significantly improves prediction performance [17]: therefore we expect that the performance of our framework in a simulated real-time prediction task will exceed that of the static scenario.

In this simulated real-time prediction task, T1B, we divide the available data into temporal slices used separately for training and prediction purposes. Table 3 reports five columns, each of which defines a scenario where one or more months of data are aggregated for training, and the rest is used for prediction and performance evaluation. For example, in the first column, the learning models are trained on data from January 2015, and the prediction are performed and evaluated on future data in the interval February-June 2015.

Random Forests greatly benefits from accounting for temporal dependencies in the data, and the prediction performance as measured by AUC ranges between 83.8 % (with just one month of training data) to an excellent 93.2 % (with five months of training data). Figure 1(left) illustrates the ranking of the top 11 features identified by feature selection, as a function of the number of months of data in the training set. Figure 1(right) displays the distributions of the rankings of each feature across the 5 different temporal slices. For the extremist users detection task, the most predictive features are (1) number of tweets, (2) average number of hashtags, and (3) average number of retweets. One hypothesis is that extremist users are more active than average users, and therefore exhibit distinctive patterns related to volume and frequency of activity.

4.2 T2: Predicting Extremist Content Adoption

The second task, namely predicting the adoption of extremist content by regular users, is discussed in the static (T2A) and real-time (T2B) scenarios follows.

T2A: Time-Independent Prediction. The first instance of T2 is again on the time-aggregated datasets spanning January-June 2015. Predicting content adoption is a known challenging task, and a wealth of literature has explored the

Table 3. Real-time extremists detection (T1B)

Training:	Jan	Jan-Feb	Jan-Mar	Jan-Apr	Jan-May
Testing:	Feb-Jun	Mar-Jun	Apr-Jun	May-Jun	Jun
AUC (LR)	0.743	0.753	0.655	0.612	0.602
Precision (LR)	0.476	0.532	0.792	0.816	0.796
Recall (LR)	0.629	0.675	0.377	0.289	0.275
F1 (LR)	0.542	0.595	0.511	0.427	0.409
AUC (RF)	**0.838**	**0.858**	**0.791**	**0.942**	**0.932**
Precision (RF)	0.984	0.922	0.868	0.931	0.910
Recall (RF)	0.679	0.733	0.649	0.957	0.959
F1 (RF)	0.804	0.817	0.743	0.944	0.934

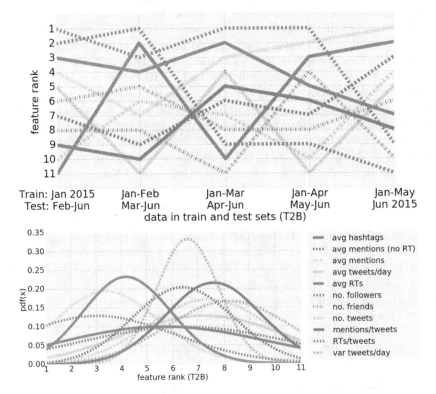

Fig. 2. T2B: Feature selection analysis and feature rank distribution (top 11 features)

factors behind online information contagion [25]. In this scenario, we aim to predict whether a regular user will retweet a content produced by an ISIS account. Positive instances are represented by users who retweeted at least one ISIS tweet

Table 4. Content adoption prediction (T2A)

	Precision	Recall	F1	AUC
Logistic Regression	0.433	0.813	0.565	0.755
Random Forests	0.745	0.615	0.674	**0.771**

in the aggregated time period, while negative ones are all users exposed to such tweets who did not retweet any of them. Table 4 summarizes the performance of our models: *Random Forests* emerges again as the best performer, although the gap with *Logistic Regression* is narrow, and the latter provides significantly better Recall. Overall, T2A appears clearly more challenging than T1A, as the top performance yields a 77.1 % AUC score. The results on the static prediction are promising and set a baseline for the real-time prediction scenario.

T2B: Simulated Real-Time Prediction. We again consider temporal data dependencies to simulate a real-time prediction for T2. Similarly to T1B, in T2B we preserve the temporal ordering of data, and divide the dataset in training and testing according to month-long temporal slices, as summarized by Table 5. *Random Forests* again seems to benefit from the temporal correlations in the data, and the prediction performance at peak improves up to 80.2 % AUC. *Logistic Regression* fails again at exploiting temporal information, showing some performance deterioration if compared to T2A. Figure 2 shows that, for the content adoption prediction, the ranking of the top 11 features in T2B is less stable than that of T1B. The top three most predictive features for this task are (1) ratio of retweets over tweets, (2) number of tweets, and (3) average number of retweets. Note that the latter two top features also appear in the top 3 of the previous task, suggesting an emerging pattern of feature predictive dynamics.

Table 5. Real-time content adoption prediction (T2B)

Training:	Jan	Jan-Feb	Jan-Mar	Jan-Apr	Jan-May
Testing:	Feb-Jun	Mar-Jun	Apr-Jun	May-Jun	Jun
AUC (LR)	0.682	0.674	0.673	0.703	0.718
Precision (LR)	0.188	0.240	0.148	0.116	0.043
Recall (LR)	0.814	0.367	0.345	0.725	0.362
F1 (LR)	0.305	0.290	0.207	0.199	0.077
AUC (RF)	**0.565**	**0.598**	**0.676**	**0.779**	**0.802**
Precision (RF)	0.433	0.384	0.266	0.205	0.070
Recall (RF)	0.087	0.070	0.336	0.648	0.813
F1 (RF)	0.145	0.119	0.297	0.311	0.130

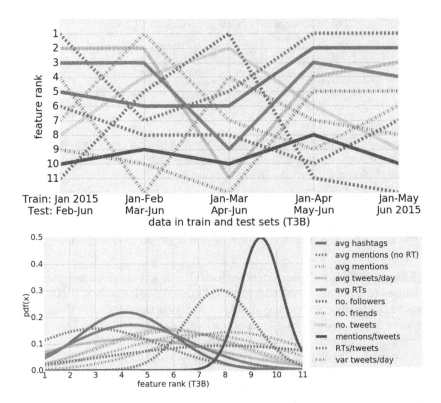

Fig. 3. T3B: Feature selection analysis and feature rank distribution (top 11 features)

4.3 T3: Predicting Interactions with Extremists

Our third and last task, namely the prediction of interactions between regular users and extremists, is discussed in the following, again separately for the static (T3A) and real-time (T3B) scenarios.

T3A: Time-Independent Prediction. We expect the interaction prediction task to be the most challenging among the three tasks we proposed. Similarly to content adoption prediction, recent literature has explored the daunting challenge of predicting interaction reciprocity and intensity in social media [20]. Consistently with the prior two tasks, our first approach to interaction prediction is time agnostic: we plan to test whether our system is capable to predict whether a regular user who is mentioned by an extremist account will reply back or not. In this case, positive instances are represented by users who reply to at least one contact initiated by ISIS in the aggregated time period (January–June 2015), whereas negative instances are those regular users who did not reply to any ISIS contact. Table 6 reports the prediction performance of our two models: overall, the task proves challenging as expected, being *Random Forests* the best

Table 6. Interaction reciprocity prediction (T3A)

	Precision	Recall	F1	AUC
Logistic Regression	0.697	0.690	0.693	0.658
Random Forests	0.686	0.830	0.751	**0.692**

performer, yielding excellent Recall and 69.2 % AUC. *Logistic Regression* performance fairly with 65.8 % AUC, and both Precision and Recall around 69 %.

T3B: Simulated Real-Time Prediction. The final task discussed in this
paper is the interaction prediction with temporal data. Given the complexity
of this problem, as demonstrated by T3A, we plan to test whether incorporating the temporal dimension will help our models achieve better performance.
Table 7 shows that this appears to be the case: *Random Forests* exhibits an
improved temporal prediction performance, boasting up to 72.6 % AUC, using
the first 5 months of data for traning, and the last month for prediction and evaluation. *Logistic Regression* improves as well, jumping to a 68.3 % AUC score.
Both models provide very good Precision/Recall performance, if one considers
the challenging nature of predicting interaction reciprocity within our context.
Figure 3 summarizes the top 11 features ranking, this time showing a more clear
division among top features. The top three features in the interaction reciprocity
prediction are (1) ratio of retweets over tweets, (2) average number of hashtags,
and (3) average number of retweets. Note that all three features already occurred
in the top features of the two previous tasks, reinforcing the notion of a clear
pattern of feature predictive power, discussed next.

Table 7. Real-time interaction reciprocity prediction (T3B)

Training:	Jan	Jan-Feb	Jan-Mar	Jan-Apr	Jan-May
Testing:	Feb-Jun	Mar-Jun	Apr-Jun	May-Jun	Jun
AUC (LR)	0.610	0.589	0.618	0.638	0.683
Precision (LR)	0.562	0.560	0.574	0.553	0.367
Recall (LR)	0.720	0.775	0.813	0.783	0.647
F1 (LR)	0.631	0.650	0.672	0.649	0.468
AUC (RF)	**0.628**	**0.633**	**0.649**	**0.671**	**0.726**
Precision (RF)	0.614	0.627	0.603	0.637	0.542
Recall (RF)	0.779	0.676	0.641	0.717	0.765
F1 (RF)	0.687	0.650	0.621	0.675	0.634

4.4   Feature Predictive-Power Analysis

We conclude our analysis by discussing the predictive power of the features adopted by our framework. First, the choice to focus on the top 11 features, rather than the more traditional top 10, is justified by the occurrence of two *ex aequo* in the final ranking of top features, displayed in Table 8. Here, we report the ranking of the top 11 features in the three tasks above. Feature selection is performed on the real-time prediction tasks (not on the time-aggregated ones). This analysis captures the essence of the predictive value of our hand-crafted features in the context of real-time predictions. A clear pattern emerges: (1) ratio of retweets over tweets, (2) average number of hashtags, (2 *ex-aequo*) number of tweets, and (4) average number of retweets, consistently ranked in the top features for the three different prediction tasks. This insight is encouraging: all these features can be easily computed from the metadata reported by the Twitter API, and therefore could be potentially implemented in a real-time detection and prediction system operating on the Twitter stream with unparalleled efficiency.

Table 8. Feature ranking across the 3 prediction tasks

Feature	Rank: T1B	T2B	T3B	Final
Ratio of retweets/tweets	4	1	1	1
Avg number of hashtags	2	4	2	2
Number of tweets	1	2	5	=
Avg number of retweets	3	3	3	4
Avg tweets per day	5	8	4	5
Avg no. mentions (w/out retweets)	8	5	8	6
Number of followers	7	6	9	7
Number of friends	6	11	7	8
Avg number of mentions	11	9	6	9
Var tweets/day	9	7	10	=
Ratio of mentions/tweets	10	10	11	11

5   Related Literature

Two relevant research trends recently emerged in the *computational social sciences* and in the *computer science* communities, discussed separately as follows.

Computational social sciences research. This research line is concerned more with understanding the social phenomena revolving around extremist propaganda using online data as a proxy to study individual and group behaviors. Various recent studies focus on English- and Arabic-speaking audiences online to study the effect of ISIS' propaganda and radicalization. One example of the former is the work by Berger and collaborators that provided quantitative evidence of ISIS' use of social media for propaganda. In a 2015 study [5], the authors characterized the Twitter population of ISIS supporters, quantifying its size, provenance, and organization. They argued that most of ISIS' success on Twitter is due to a restricted number of highly-active accounts (500–1000 users). Our analysis illustrates that indeed a limited number of ISIS accounts achieved a very high visibility and followership. Berger's subsequent work [6] however suggested that ISIS' reach (at least among English speakers) has stalled for months as of the beginning of 2016, due to more aggressive account suspension policies enacted by Twitter. Again, a limited amount of English accounts sympathetic to ISIS was found (less than one thousand), and these users were mostly interacting with each other, while being only marginally successful at acquiring other users' attention. This analysis suggests a mechanism of diminishing returns for extremist social media propaganda. Using Twitter data as a historical archive, some researchers [26] recently tried to unveil the roots of support for ISIS among the Arabic-speaking population. Their analysis seems to suggest that supporters of the extremist group have been discussing about Arab Spring uprisings in the past significantly more than those who oppose ISIS on Twitter. Although their method to separate ISIS supporters from opposers is simplistic, the findings relating narrative framing and recruitment mechanisms are compatible with the literature on social protest phenomena [12,13,21,39]. A few studies explored alternative data sources: one interesting example is the work by Vergani and Bliuc [40] that uses sentiment analysis (Linguistic Inquiry and Word Count [38]) to investigate how language evolved across the first 11 Issues of Dabiq, the flagship ISIS propaganda magazine. Their analysis offers some insights about ISIS radicalization motives, emotions and concerns. For example, the authors found that ISIS has become increasingly concerned with females, reflecting their need to attract women to create their utopia society, not revolving around warriors but around families. ISIS also seems to have increased the use of internet jargon, possibly to connect with the identities of the youth.

Computer science research. This research stream concerns with the machine learning and data aspects, to model, detect, and/or predict social phenomena such as extremism or radicalization often with newly-developed methods. One of the first computational frameworks, proposed by Bermingham *et al.* [8] in 2009, combined social network analysis with sentiment detection tools to study the agenda of a radical YouTube group: the authors examined the topics discussed within the group and their polarity, to model individuals' behavior and spot signs of extremism and intolerance, seemingly more prominent among female users. The detection of extremist content (on the Web) was also the focus of a 2010 work by Qi *et al.* [29]. The authors applied clustering to extremist Web pages to divide them into different categories (religious, politics, etc.). Scanlon and Gerber proposed the first method to detect cyber-recruitment efforts in 2014 [33]. They exploited data retrieved from the Dark Web Portal Project [10], a repository of posts compiled from 28 different online fora on extremist religious discussions (e.g., Jihadist) translated from Arabic to English. After annotating a sample of posts as recruitment efforts or not, the authors use Bayesian criteria and a set of textual features to classify the rest of the corpus, obtaining good accuracy, and highlighted the most predictive terms. Along the same trend, Agarwal and Sureka proposed different machine learning strategies [1,3,4,37] aimed at detecting radicalization efforts, cyber recruitment, hate promotion, and extremist support in a variety of online platforms, including YouTube, Twitter and Tumblr. Their frameworks leverage features of contents and metadata, and combinations of crawling and unsupervised clustering methods, to study the online activity of Jihadist groups on those platforms. Concluding, two very recent articles [24,32] explore the activity of ISIS on social media. The former [32] focuses on Twitter and aims at detecting users who exhibit signals of behavioral change in line with radicalization: the authors suggest that out of 154 K users only about 700 show significant signs of possible radicalization, and that may be due to social homophily rather than the mere exposure to propaganda content. The latter study [24] explores a set of 196 pro-ISIS aggregates operating on VKontakte (the most popular Russian online social network) and involving about 100 K users, to study the dynamics of survival of such groups online: the authors suggest that the development of large and potentially influential pro-ISIS groups can be hindered by targeting and shutting down smaller ones. We refer the interested reader to two recent literature reviews on this topic [2,14].

6   Conclusions

In this article we presented the problem of predicting online extremism in social media. We defined three machine learning tasks, namely the detection of extremist users, the prediction of extremist content adoption, and the forecasting of interactions between extremist users and regular users. We tapped into the power of a crowd-sourcing project that aimed at manually identifying and reporting suspicious or abusive activity related to ISIS radicalization and propaganda agenda, and collected annotations to build a ground-truth of over 25 thousand suspended Twitter accounts. We extracted over three million tweets related to the activity of these accounts in the period of time between January and June 2015. We also randomly identified an equal-sized set of regular users exposed to the extremist content generated by the ISIS accounts, and collected almost 30 million tweets generated by the regular users in the same period.

By means of state-of-the-art learning models we managed to accomplish predictions in two types of scenarios, a static one that ignores temporal dependencies, and a simulated real-time case in which data are processed for training and testing by respecting the timeline of content availability. The two learning models, and the set of 52 features that we carefully crafted, proved very effective in all of the six combinations of forecasts (three prediction tasks each with two prediction modalities, static and real-time). The best performance in terms of AUC ranges between 72 % and 93 %, depending on the complexity of the considered task and the amount of training data available to the system. We concluded our analysis by investigating the predictive power of different features. We focused on the top 11 most significant features, and we discovered that some of them, such as the ratio of retweets to tweets, the average number of hashtags adopted, the sheer number of tweets, and the average number of retweets generated by each user, systematically rank very high in terms of predictive power. Our insights shed light on the dynamics of extremist content production as well as some of the network and timing patterns that emerge in this type of online conversation.

Our work is far from concluded: for the future, we plan to identify more realistic and complex prediction tasks, to analyze the network and temporal dynamics of extremist discussion, and to deploy a prototype system that allows for real-time detection of signatures of abuse on social media.

References

1. Agarwal, S., Sureka, A.: A focused crawler for mining hate and extremism promoting videos on youtube. In: Proceedings of the 25th ACM Conference on Hypertext and Social Media, pp. 294–296 (2014)
2. Agarwal, S., Sureka, A.: Applying social media intelligence for predicting and identifying on-line radicalization and civil unrest oriented threats. arXiv preprint (2015). arXiv:1511.06858
3. Agarwal, S., Sureka, A.: Using KNN and SVM based one-class classifier for detecting online radicalization on twitter. In: Natarajan, R., Barua, G., Patra, M.R. (eds.) ICDCIT 2015. LNCS, vol. 8956, pp. 431–442. Springer, Heidelberg (2015). doi:10.1007/978-3-319-14977-6_47
4. Agarwal, S., Sureka, A.: Spider and the flies: Focused crawling on tumblr to detect hate promoting communities. arXiv preprint (2016). arXiv:1603.09164
5. Berger, J., Morgan, J.: The ISIS twitter census: Defining and describing the population of isis supporters on twitter. The Brookings Project on US Relations with the Islamic World 3(20) (2015)
6. Berger, J., Perez, H.: The Islamic States diminishing returns on Twitter. GW Program on extremism 2–16 (2016)
7. Berger, J., Strathearn, B.: Who matters online: measuring influence, evaluating content and countering violent extremism in online social networks. Int. Centre Study Radicalisation (2013)
8. Bermingham, A., Conway, M., McInerney, L., O'Hare, N., Smeaton, A.F.: Combining social network analysis and sentiment analysis to explore the potential for online radicalisation. In: 2009 International Conference on Advances in Social Network Analysis and Mining (ASONAM), pp. 231–236. IEEE (2009)
9. Breiman, L.: Random forests. Mach. Learn. 45(1), 5–32 (2001)
10. Chen, H., Chung, W., Qin, J., Reid, E., Sageman, M., Weimann, G.: Uncovering the dark web: A case study of jihad on the web. J. Am. Soc. Inf. Sci. Technol. 59(8), 1347–1359 (2008)
11. Cockburn, P.: The rise of Islamic State: ISIS and the new Sunni revolution. Verso Books, London (2015)
12. Conover, M.D., Davis, C., Ferrara, E., McKelvey, K., Menczer, F., Flammini, A.: The geospatial characteristics of a social movement communication network. PloS One 8(3), e55957 (2013)

13. Conover, M.D., Ferrara, E., Menczer, F., Flammini, A.: The digital evolution of occupy wall street. PloS One **8**(5), e64679 (2013)
14. Correa, D., Sureka, A.: Solutions to detect and analyze online radicalization: a survey. arXiv preprint (2013). arXiv:1301.4916
15. Davis, C.A., Varol, O., Ferrara, E., Flammini, A., Menczer, F.: Botornot: A system to evaluate social bots. In: Proceedings of the 25th International Conference Companion on World Wide Web, pp. 273–274. International World Wide Web Conferences Steering Committee (2016)
16. Ferrara, E., Varol, O., Davis, C., Menczer, F., Flammini, A.: The rise of social bots. Commun. ACM **59**(7), 96–104 (2016)
17. Ferrara, E., Varol, O., Menczer, F., Flammini, A.: Detection of promoted social media campaigns. In: Proceedings of the 10th International Conference on Web and Social Media (2016)
18. Fisher, A.: How jihadist networks maintain a persistent online presence. Perspect. Terrorism **9**(3), 3–20 (2015)
19. Ghosh, R., Surachawala, T., Lerman, K.: Entropy-based classification of retweeting activity on twitter. In: Proceedings of KDD workshop on Social Network Analysis (SNA-KDD), August 2011
20. Gilbert, E., Karahalios, K.: Predicting tie strength with social media. In: Proceedings of the SIGCHI Conference on Human Factors in Computing Systems, pp. 211–220. ACM (2009)
21. González-Bailón, S., Borge-Holthoefer, J., Rivero, A., Moreno, Y.: The dynamics of protest recruitment through an online network. Sci. Rep. **1**, 197 (2011)
22. Guyon, I., Elisseeff, A.: An introduction to variable and feature selection. J. Mach. Learn. Res. **3**, 1157–1182 (2003)
23. Hastie, T., Tibshirani, R., Friedman, J., Franklin, J.: The elements of statistical learning: data mining, inference and prediction. Math. Intell. **27**(2), 83–85 (2005)
24. Johnson, N.F., Zheng, M., Vorobyeva, Y., Gabriel, A., Qi, H., Velasquez, N., Manrique, P., Johnson, D., Restrepo, E., Song, C., Wuchty, S.: New online ecology of adversarial aggregates: Isis and beyond. Science **352**(6292), 1459–1463 (2016)
25. Lerman, K., Ghosh, R.: Information contagion: an empirical study of the spread of news on digg and twitter social networks. In: Proceedings of the 4th International AAAI Conference on Weblogs and Social Media, pp. 90–97 (2010)
26. Magdy, W., Darwish, K., Weber, I.: #failedrevolutions: Using Twitter to study the antecedents of ISIS support. First Monday **21**(2), 1481–1492 (2016)
27. Mislove, A., Lehmann, S., Ahn, Y.Y., Onnela, J.P., Rosenquist, J.N.: Understanding the demographics of twitter users. In: Proceedings of the 5th International AAAI Conference on Weblogs and Social Media (2011)
28. Pedregosa, F., Varoquaux, G., Gramfort, A., Michel, V., Thirion, B., Grisel, O., Blondel, M., Prettenhofer, P., Weiss, R., Dubourg, V., et al.: Scikit-learn: Machine learning in python. J. Mach. Learn. Res. **12**, 2825–2830 (2011)
29. Qi, X., Christensen, K., Duval, R., Fuller, E., Spahiu, A., Wu, Q., Zhang, C.Q.: A hierarchical algorithm for clustering extremist web pages. In: 2010 International Conference on Advances in Social Networks Analysis and Mining (ASONAM), pp. 458–463 (2010)
30. Ratkiewicz, J., Conover, M., Meiss, M., Goncalves, B., Flammini, A., Menczer, F.: Detecting and tracking political abuse in social media. In: Proceedings of the 5th International AAAI Conference on Weblogs and Social Media, pp. 297–304 (2011)
31. Reardon, S.: Terrorism: science seeks roots of terror. Nature **517**(7535), 420–421 (2015)

32. Rowe, M., Saif, H.: Mining pro-ISIS radicalisation signals from social media users. In: Proceedings of the 10th International Conference on Web and Social Media (2016)
33. Scanlon, J.R., Gerber, M.S.: Automatic detection of cyber-recruitment by violent extremists. Secur. Inf. **3**(1), 1–10 (2014)
34. Schiermeier, Q.: Terrorism: Terror prediction hits limits. Nature **517**(7535), 419 (2015)
35. Stern, J., Berger, J.M.: ISIS: The state of terror. Harper, New York (2015)
36. Subrahmanian, V., Azaria, A., Durst, S., Kagan, V., Galstyan, A., Lerman, K., Zhu, L., Ferrara, E., Flammini, A., Menczer, F.: The DARPA Twitter bot challenge. Computer **49**(6), 38–46 (2016)
37. Sureka, A., Agarwal, S.: Learning to classify hate and extremism promoting tweets. In: 2014 IEEE Joint Intelligence and Security Informatics Conference (JISIC), pp. 320–320. IEEE (2014)
38. Tausczik, Y.R., Pennebaker, J.W.: The psychological meaning of words: LIWC and computerized text analysis methods. J. Lang. Soc. Psychol. **29**(1), 24–54 (2010)
39. Varol, O., Ferrara, E., Ogan, C.L., Menczer, F., Flammini, A.: Evolution of online user behavior during a social upheaval. In: Proceedings of the 2014 ACM Conference on Web Science, pp. 81–90. ACM (2014)
40. Vergani, M., Bliuc, A.M.: The evolution of the ISIS' language: a quantitative analysis of the language of the first year of dabiq magazine. Secur. Terrorism Soc. **1**(2), 217–224 (2015)
41. Weiss, M., Hassan, H.: ISIS: Inside the army of terror. Simon and Schuster, New York (2015)

Content Centrality Measure for Networks: Introducing Distance-Based Decay Weights

Takayasu Fushimi[1]([✉]), Tetsuji Satoh[1], Kazumi Saito[2], Kazuhiro Kazama[3], and Noriko Kando[4]

[1] Faculty of Library, Information and Media Science,
University of Tsukuba, 1-2 Kasuga, Tsukuba-city, Ibaraki 305-8550, Japan
{fushimi,satoh}@ce.slis.tsukuba.ac.jp
[2] School of Management and Information, University of Shizuoka, 52-1 Yada,
Suruga-ku, Shizuoka 422-8526, Japan
k-saito@u-shizuoka-ken.ac.jp
[3] Faculty of Systems Engineering, Wakayama University, Sakaedani 930,
Wakayama-city, Wakayama 640-8510, Japan
kazama@ingrid.org
[4] Information and Society Research Division, National Institute of Informatics,
2-1-2 Hitotsubashi, Chiyoda-ku, Tokyo 101-8430, Japan
kando@nii.ac.jp

Abstract. We propose a novel centrality measure that is called Content Centrality for a given network that considers the feature vector of each node generated from its posting activities in social media, its own properties and so forth, in order to extract nodes who have neighbors with similar features. We assume that nodes with similar features are located near each other and unevenly distributed over a network, and the density gradually or rapidly decreases according to the distance from the center of the feature distribution (node). We quantify the degree of the feature concentration around each node by calculating the cosine similarity between the feature vector of each node and the resultant vector of its neighbors with distance-based decay weights, then rank all the nodes according to the value of cosine similarities. In experimental evaluations with three real networks, we confirm the validity of the centrality rankings and discuss the relation between the estimated parameters and the nature of nodes.

1 Introduction

In social media and such SNSs as review sites, weblog sites, Twitter, and Facebook, many interactions exist among users. By analyzing these interations as networks, various useful results can be obtained (Newman et al. 2002, Newman 2003, Gruhl et al. 2004, Domingos 2005). In such networks, users share common features with others who are connected (Newman 2002, Ting et al. 2013). For example, a user of a review site about cosmetics is expressed by a feature vector whose element stands for the review score of the corresponding cosmetic item. Even though the vectors of the connected nodes tend to be relatively similar, not all of the nodes in a network

© Springer International Publishing AG 2016
E. Spiro and Y.-Y. Ahn (Eds.): SocInfo 2016, Part II, LNCS 10047, pp. 40–54, 2016.
DOI: 10.1007/978-3-319-47874-6_4

have similar feature vectors. Some change points of feature distributions exist somewhere in a network. In addition, although connected nodes tend to possess similar features, such tendencies vary based on each node.

In this paper, we quantify the degree of concentration of contents to find the modes of the feature distributions[1] over a network structure (Fig. 1). Since features like words or expressions tend to be reused by neighborhood nodes in a conversational context, nodes who located in the modes of the feature distribution are supposed to have strong influence for neighborhood nodes. In order to extract such nodes, we propose a novel measure. One representative feature space analysis method for finding the maxima of distributions is mean shift clustering (Cheng 1995), which uses the kernel density estimation technique (Parzen 1962); however, no method focuses on distributions over a network. Assuming content distribution over a network, we propose a novel centrality measure, Content Centrality. For each node, which is called a target node, we define the similarity as cosine similarity between the feature vector (Hereafter, we call content vector) of the target node and the resultant vector of its neighbors. However, we must consider not only the directly connected nodes but also the neighbor nodes that can be reached in a few steps from the target node. Then it is not obvious which distant nodes we should calculate resultant vector with. Moreover, it is naturally conceivable that the contents of the very distant nodes have almost no influence; distant nodes are generally less influential. To reflect such effects, we calculate them with multiplying weights that decay based on the distance from the target node. Since it is natural that the degree of decay varies with each node, we multiply by a strong (weak) decay weight if similar nodes are narrowly (widely) distributed around the node. For each node, we have to estimate the parameter value that realizes the optimal decay weight. In this paper, we estimate the distance-based decay parameter of each node to maximize the cosine similarity, which defines the cosine similarity under the estimated parameter as the node's content centrality score. The estimated parameter value implicitly stands for the size of the region where similar neighbors are distributed. Although we can utilize the CNM method (Clauset et al. 2004) to extract communities as neighbor nodes, there are no guarantees that the boundary of communities is coincident with that of contents. Furthermore, not all node regions with similar content vectors can be hard-partitioned, so we express continuous boundaries using a gradually decay weight.

The rest of this paper is organized as follows. After citing related works in Sect. 2, we describe the details of our proposed centrality measure and estimation method of decay parameters in Sect. 3. Then with three networks, in Sects. 4, we discuss the relation between highly ranked nodes by our measure and the existing centrality measures. Finally, we provide a conclusion in Sect. 5.

---

[1] We use the term "content distribution" in the same sense with "feature distribution".

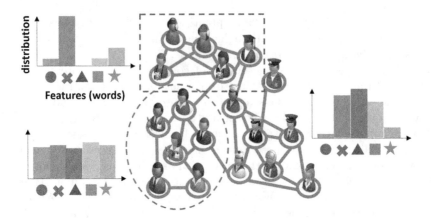

Fig. 1. Feature distribution of each node groups. For example, each node belonging to the blue square group frequently uses the feature (word) "crosses," but no "triangles"; each node in the red circle group tends to almost equally utilize five kinds of words. (Color figure online)

2   Related Work

There are many previous studies on centrality measures that are only based on network structure, and community detection methods based on both structural and semantic features. To the best of our knowledge, the centrality measure proposed in this paper is the first that focuses on content similarity of connected nodes for node ranking. Therefore, we describe some related works on community detection methods using node features.

Yang et al. (Yang et al. 2009) proposed a discriminative model, called the PCL-DC model, to overcome two intrinsic shortcomings of generative models: failing to consider the additional factors that might affect communities and failing to isolate contents that are irrelevant to communities. Their scheme incorporates two models, the Popularity-based Conditional Link model that estimates the link-existence probability for node pairs using the popularity and community belongingness of each node, and the Discriminative Content model that calculates community-belonging probability using weighted feature (content) vectors. As a result, the PCL-DC model can detect communities with significantly high accuracy.

Zhou et al. (Zhou et al. 2009) proposed a graph clustering algorithm, called the SA-Cluster, based on both structural and attribute similarities through a unified distance measure. They, first define the attribute augmented graph where augmented nodes correspond to attribute values of each original node. Then they define the distance measure as the number of paths of random walk over the augmented graph. The random walk works according to the weighted transition probability, where each weight means the contribution of attribute similarity in the distance measure. In their paper, they estimate adequate contribution weights of each attribute node and take a K-medoids clustering to partition the graph into K clusters which have both cohesive intra-cluster structures and homogeneous attribute values.

Sun et al. (Sun et al. 2009) focused on content-based network tensors which are typically obtained through communication flows like emails in the triplet form of sender, recipient and message. In their method, they first conduct the tensor decomposition and second cluster the dimensions along each mode of tensor. After dividing the dimensions into some clusters, they find the correlated cluster across different modes. Finally leveraging the clustering results, they produced the hierarchical visualization to explorer the data at certain granularity of detail.

Kuramochi et al. (Kuramochi et al. 2012) proposed a new community detection method that extracts overlapping communities using intersection graph notions and calculates the weights among communities. These weights consist of two similarities: the overlap of the community members and the similarity of the content information possessed by each node. Before calculating the content similarities, the method aggregates the feature (content) vectors of the nodes in the community. Similar to this method, we also calculate the resultant vector of neighbors; however, we add the decay weights based on the distance between nodes and tune the decay parameter with respect to each node.

Natarajan et al. (Natarajan et al. 2013) proposed a generative model, called the Link-Content model, for detecting topic based communities modeled as multinomial distributions over node set like our measure, and efficient Gibbs sampling algorithm to infer the model. Our measure differs from this model in assuming the center of content distribution is located in a node and the density of distribution gradually or rapidly decrease according to the distance from center.

For a given network referred to as a physical network, Wu et al. (Wu et al. 2015) constructed a conceptual network with content-based edge weights and extracted the Densest Connected Subgraphs with the largest density in the conceptual network that are also connected in the physical network. Their method reduced the search space of the densest subgraphs by pruning the low degree nodes.

Unlike the above studies, our method calculates the centrality score for each node using the cosine similarity between the content vector of the node and the resultant vector of its neighbors. Especially, our measure differs substantially from these methods excluding the Natarajan's model in assuming content distributions over the network structure and finding modes of the distributions like the mean shift clustering.

3 Methodology

In this section, we propose a novel centrality measure, *Content Centrality*, that quantifies the density of similar contents as the centrality score of each node, and ranks all the nodes by their scores. First, we describe the concept and the assumption of our measure, next the calculation method of the content centrality score for each node, and finally the estimation method of the decay parameters.

3.1   Concept and Assumption

In real networks, there are some common features among the users who are connected to each other (Newman 2002). Although connected nodes tend to possess similar features, all the nodes in a network have similar features. That is, features are unevenly distributed over the network. We assume that the features are obtained from the node's activity, the posted content, and so forth. We also assume that nodes with similar features are unevenly distributed over the network, and the density gradually or rapidly decreases according to the distance from the center of the distribution (node) (Fig. 2). From this assumption, it is naturally conceivable that the contents of the very distant nodes have almost no influence; distant nodes are generally less influential. To reflect such effects, we introduce some decay functions based on the distance among nodes. We define the degree of the concentration of the contents as the content centrality, and thus a highly scored node means that some sort of similar contents are distributed around it.

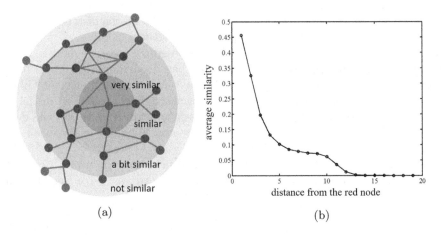

(a)   (b)

Fig. 2. A sketch of similarity decreasing in our hypothetic situation. (a) shows a network structure around the red node. (b) shows average similarities with respect to the distance from the red nodes. For example, orange nodes, who directly connected to the red node would be very similar, but blue nodes, who located in distant from the red node would not be similar in content vector. (Color figure online)

3.2   Score Calculation

In a given, simple undirected network structure $G = (V, E)$ where V and E are the node set and link set, respectively, each node $u \in V$ has J-dimensional content vector $\mathbf{x}_u$. Let $d(u, v)$ be the shortest path length between nodes u and v, and $d(u, v) = d(v, u), d(u, u) = 0$. We define u's neighbor node set with distance d as $\Gamma_d(u) = \{v : d(u, v) = d\} \subset V$.

Assuming content distribution, the very distant nodes naturally exert almost no influence. To express such effects, we introduce two types of decay functions. The first is an exponential decay function defined by

$$\rho(d; \lambda) = \exp(-\lambda d),$$

where λ is a parameter that controls the decay power. Another natural one is a power-law decay function defined by

$$\rho(d; \lambda) = \exp(-\lambda \log d).$$

Now, for each node, we calculate the resultant vector with the distance-based decay weight as follows:

$$\mathbf{y}_u = \sum_{d=1}^{D_u} \rho(d; \lambda) \sum_{v \in \Gamma_d(u)} \mathbf{x}_v = \sum_{v \in V \setminus \{u\}} \rho(d(u, v); \lambda) \mathbf{x}_v, \tag{1}$$

where $D_u = \max_{v \in V} d(u, v)$. We call this the Resultant Vector with a Decay weight of node u (RVwD). The RVwD of node u is appended to the vectors of the near nodes, including the directly connected nodes with strong weights and those of the distant nodes with weak weights. Therefore, the vector is somewhat smoothed.

In order to quantify the density expressing how many similar nodes exist near each node, we calculate the cosine similarity between each node and its neighbors, and define this value as content centrality score of each node:

$$\mathrm{CDC}(u) = \langle \mathbf{x}_u, \mathbf{y}_u \rangle = \left\langle \mathbf{x}_u, \frac{\sum_{d=1}^{D_u} \rho(d; \lambda) \sum_{v \in \Gamma_d(u)} \mathbf{x}_v}{\| \sum_{d=1}^{D_u} \rho(d; \lambda) \sum_{v \in \Gamma_d(u)} \mathbf{x}_v \|} \right\rangle, \tag{2}$$

where original content vector $\mathbf{x}_u$ is normalized as the L2 norm to 1. When this value $\mathrm{CDC}(u)$ exceeds the other nodes, node u is a highly ranked node of content centrality, which means that similar contents are concentratedly distributed around it.

3.3   Parameter Estimation

The aforementioned RVwD of node u, $\mathbf{y}_u$ is constructed by adding the content vectors of node u's neighbors with decay weight based on the distance from node u. However, since not all the contents are equally distributed around each node, we must tune the degree of decay.

In this paper, for each node u, we set parameter λ_u to maximize the cosine similarity between content vector $\mathbf{x}_u$ and RVwD $\mathbf{y}_u$. We set the L2 norm of content vector $\mathbf{x}_u$ to 1 and define the following objective function:

$$F_u(\lambda_u) = \mathbf{x}_u^T \frac{\sum_{v \in V \setminus \{u\}} \rho(d(u, v); \lambda_u) \mathbf{x}_v}{\| \sum_{v \in V \setminus \{u\}} \rho(d(u, v); \lambda_u) \mathbf{x}_v \|}. \tag{3}$$

Next we explain the procedure of maximizing the objective function (3) with exponential decay function, $\rho(d; \lambda) = \exp(-\lambda d)$, to solve adequate parameter λ_u. However, by replacing d with $\log(d)$, the derived results that used the exponential decay are consistent with those that used the power-law decay without a loss of generality. For each node u, we define the resultant vector of the nodes with distance d as

$$\mathbf{f}_{u,d} = \sum_{v \in \Gamma_d(u)} \mathbf{x}_v,$$

and the inner product with the content vector of u as

$$g_{u,d} = \langle \mathbf{x}_u, \mathbf{f}_{u,d} \rangle.$$

Now, we sum up the inner products of the pairs of resultant vectors $\mathbf{f}_{u,d_1}$ and $\mathbf{f}_{u,d_2}$ whose sum of distance becomes d, i.e., $d = d_1 + d_2$, as follows:

$$h_{u,d} = \sum_{d_1+d_2=d} \langle \mathbf{f}_{u,d_1}, \mathbf{f}_{u,d_2} \rangle.$$

Then we can rewrite our objective function (3):

$$F_u(\lambda_u) = \frac{\sum_{d=1}^{D_u} \exp(-\lambda_u d) g_{u,d}}{\sqrt{\sum_{d=2}^{2D_u} \exp(-\lambda_u d) h_{u,d}}}.$$

For calculation simplicity, we regard the following logarithmic function:

$$\log F_u(\lambda_u) = \log \sum_{d=1}^{D_u} \exp(-\lambda_u d) g_{u,d} - \frac{1}{2} \log \sum_{d=2}^{2D_u} \exp(-\lambda_u d) h_{u,d}. \tag{4}$$

We define the posterior probability function as

$$r_{u,d} = \frac{\exp(-\lambda_u d) g_{u,d}}{\sum_{d'=1}^{D_u} \exp(-\lambda_u d') g_{u,d'}}$$

and transform Eq. (4) as follows:

$$\log F_u(\lambda_u) = \sum_{d=1} \bar{r}_{u,d}\{(-\lambda_u d) + \log g_{u,d}\} - \sum_{d=1} \bar{r}_{u,d} \log r_{u,d} - \frac{1}{2} \log \sum_{d=2}^{2D_u} \exp(-\lambda_u d) h_{u,d}.$$

We also remove the irrelevant terms from parameter λ_u and obtain

$$Q_u(\lambda_u) = -\lambda_u \sum_{d=1} \bar{r}_{u,d} \cdot d - \frac{1}{2} \log \sum_{d=2}^{2D_u} \exp(-\lambda_u d) h_{u,d},$$

and its first-order derivative becomes

$$\frac{dQ_u(\lambda_u)}{d\lambda_u} = -\sum_{d=1} \bar{r}_{u,d} \cdot d + \frac{\sum_{d=2}^{2D_u} \exp(-\lambda_u d) \cdot d \cdot h_{u,d}}{2 \sum_{d=2}^{2D_u} \exp(-\lambda_u d) h_{u,d}}.$$

Here after defining the following value,

$$s_{u,d} = \frac{\exp\left(-\lambda_u d\right) h_{u,d}}{\sum_{d'=2}^{2D_u} \exp\left(-\lambda_u d'\right) h_{u,d'}},$$

we can express the second-order derivative as

$$\frac{d^2 Q_u(\lambda_u)}{d\lambda_u^2} = -\frac{1}{2}\left\{ \sum_{d=2}^{2D_u} s_{u,d} \cdot d^2 - \left(\sum_{d=2}^{2D_u} s_{u,d} \cdot d\right)^2 \right\},$$

and this value is guaranteed to be less than or equal to zero because the value in the brace is always non-negative, just as in the second-order moment. Since the first-order derivative cannot be written in a closed form with respect to the parameter, we solve the optimal parameter with the Newton iteration.

The higher the value of the estimated parameter, the stronger the decay power behaves, in effect, only considering the content vectors of extremely near neighbor nodes and almost ignoring those of distant nodes. Conversely, when the value is close to zero, the content vectors of almost every node are equally treated.

Let J and $\bar{D}$ be the number of dimensions of a content vector and the average shortest path length (distance), respectively. The dominant computational time of estimating the parameters of all nodes needs $O(|V| \times 2\bar{D} \times J)$ for calculating $h_{u,d}$. We can reduce the time by utilizing dimensionality reduction techniques.

4 Experiment

In this paper, we conducted several experimental studies. First, we discuss our comparison of the estimated parameters with the semantic features of each node. The parameters stand for the size of the tail of the content distributions. Second, we confirm the adequacy of the centrality rankings and compare them with other centrality measures and the semantic features of each node. Before going into the details, we describe our datasets.

4.1 Datasets

We utilized three network datasets. The first is the web hyperlink site of a computer science department at a Japanese university, where we obtained a network by crawling it on Aug. 2010[2]. We defined the web pages and hyperlinks as nodes and links and constructed a network without direction or multiplicity. The content vector of each node consists of the numbers of times specific nouns occurred in the correspondent web page. As shown below in Fig. 3(a), many content distributions exist, such as nouns relevant to mathematics, computer science, entrance examinations, syllabus organizations, classes, and so forth. The

---

[2] http://cis.k.hosei.ac.jp/.

number of nodes, links, and dimensions of the content vector (equal to the number of different nouns) were 600, 1,299, and 4,412, respectively. Hereafter, we refer to this network as the web network.

The second is a network of people that was derived from a "list of people" on Japanese Wikipedia[3]. We extracted the maximal connected components of the undirected graph obtained by linking two people from the "list of people" if they co-occur in six or more Wikipedia pages. The content vector of each node consists of the numbers of times nouns occurred in the correspondent wiki page. As shown below in Fig. 3(b), many content distributions exist, such as nouns relevant to history, politics, entertainments, sports, and so forth. The number of nodes, links, and dimensions of the content vector (equal to the number of different nouns) were 9,481, 122,522, and 20,411, respectively. This is the wiki network.

The last is a user relation network of a Japanese recipe site[4]. We defined the users and the follow relations as nodes and links and extracted the maximal connected component of the undirected simple graph obtained by linking two users if one posts more than ten comments to another. The content vector of each node consists of the numbers of times the representative foodstuffs were used for the recipes of the correspondent user. As shown below in Fig. 3(c), many content distributions exist, such as foodstuffs prepared for baby food, Japanese-style foods, western foods, and so forth. The number of nodes, links, and dimensions of the content vector (equal to the number of different foodstuffs) were 7,815, 40,569, and 4,171, respectively. This is the recipe network.

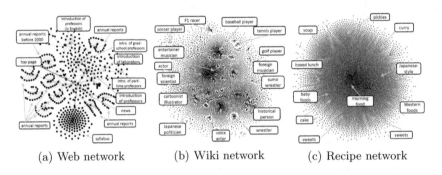

| (a) Web network | (b) Wiki network | (c) Recipe network |

Fig. 3. Visualization results with manually labels added

4.2 Results of Estimated Parameters

In this subsection, we discuss the value of estimated parameter $\hat{\lambda}$ and compare them with the semantic features of each node. We rank all the nodes by their estimated values and qualitatively evaluate the high and low ranked nodes. First, we evaluate the rank correlation between the estimated parameters of the

---

[3] https://ja.wikipedia.org/.
[4] http://cookpad.com/.

exponential and power-law decays. For all three networks, Spearman's rank correlation coefficients indicate values over 0.8. Next we discuss the exponential decay results.

In the web network, many nodes with large parameter value $\hat{\lambda} > 1$ are "introduction of teaching staff" pages that contain the biographical and career information of professors and research topics for students. The neighbor nodes also contain such contents with similar nouns. These pages can be reached in a few steps from each other, even though these contents are not contained in such distant pages as those about syllabi and classes. Some contents are densely and narrowly distributed around each of these nodes. Many nodes with small parameter value $\hat{\lambda} \simeq 0$ are "university news and topics" and "announcement" pages, which contain information about prize-winning students and professors. Both neighbor and distant nodes lack these contents with similar nouns.

In the wiki network, many nodes with large parameter value $\hat{\lambda} > 1$ are "entertainer" pages for individual members of bands, comedian duos, teen idols, and so forth. They are connected to each other, and these wikipedia pages contain very similar contents like television or radio programs. On the other hand, the parameters of the "actors/actresses" pages are estimated as lower values than those of the above entertainers because there is much background information about their careers and their various roles in the actor or actress pages. These contents are generally different, even though they starred in the same drama. That is to say, even if two actor nodes were connected (starred in the same drama several times), these wikipedia pages contain relatively different contents. Many nodes with small parameter value $\hat{\lambda} \simeq 0$ are pages containing peculiar nouns or with few nouns. Since there are almost no pages similar to them not only near them but also in the whole network, our method had to evenly add various content vectors, regardless of the distance, to raise the cosine similarities with their own content vectors.

In the recipe network, many nodes with large parameter value $\hat{\lambda} > 1$ are users who belong to relatively small recipe communities that are relevant to "soup," "baby food," "beverages," and so forth. In each of these communities, many users post recipes with similar ingredients. On another front, since there are few similar users outside of the community, the parameters of these nodes are estimated at higher values. Many nodes with small parameter value $\hat{\lambda} \simeq 0$ are users who post recipes in various categories. These users exist among some categories and use various ingredients, and so our method had to evenly add various content vectors, not only of neighbor nodes but also of distant nodes, to raise the cosine similarities with their own content vectors. These users cannot be categorized into specific categories.

Figure 4(a) shows the similarities between nodes with large or small parameter values and their neighbors. From Fig. 4(a), we can confirm that similarities with respect to the distance from nodes with large parameter values (solid line) drastically decrease. In contrast, similarities of nodes with small parameter values (dotted line) decay very little regardless of high and low similarities. These facts indicate that the estimated parameter value $\hat{\lambda}$ depends on the breadth of the contents distribution around these nodes.

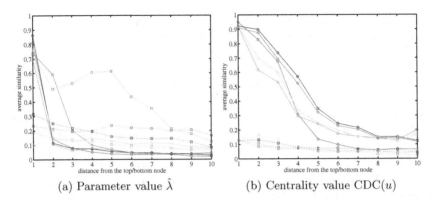

(a) Parameter value $\hat{\lambda}$ (b) Centrality value CDC(u)

Fig. 4. Average similarities between a specific node and its neighbors with respect to the distance. In each figure, solid lines and dotted lines depict similarities of nodes with large parameter/centrality values and small parameter/centrality values, respectively. Due to limitations of space, we only show the results of the wiki network, however we obtained consistent results from the web and the recipe networks.

4.3 Results of Centrality Rankings

In this subsection, we discuss the node rankings based on our proposed content centrality measure and compare it with structure-based centrality measures. First, we show the top ten nodes in Tables 1 and plot highly ranked nodes with red color in Fig. 5.

From the second column of Table 1, we can see that all the top nodes are syllabus pages, including such content distributions shown in Fig. 5(a) in the web network, and the content centrality ranks nodes in the upper level over which largest contents distributions, like "syllabus", are located. In fact, the web pages of the top nodes contain many similar nouns.

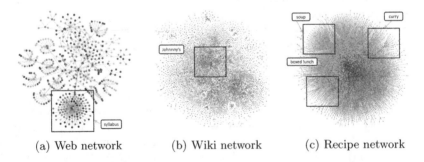

(a) Web network   (b) Wiki network   (c) Recipe network

Fig. 5. Visualization results with centrality rankings. In these figures, highly ranked nodes are colored with red end of the spectrum, conversely, low ranked nodes with blue end of the spectrum. (Color figure online)

Table 1. Highly ranked nodes of content centrality

	Web network	Wiki network	Recipe network
1	Scientific Computing (syllabus)	Kazama Shunsuke (idol)	soup
2	Functional Language (syllabus)	Tohshin Yoshikazu (idol)	soup
3	Introduction to DB (syllabus)	Kohda Kumi (musician)	boxed lunch
4	Applied AI (syllabus)	Ikuta Tohma (actor)	boxed lunch
5	Introduction to AI (syllabus)	Matsumoto Jun (idol)	soup
6	Basic Natural Science (syllabus)	Ohno Satoshi (idol)	curry
7	Practical Programming (syllabus)	Tegoshi Yuya (idol)	soup
8	Technical Writing (syllabus)	Yamashita Tomohisa (idol)	soup
9	Scientific English (syllabus)	Kamenashi Kazuya (idol)	soup
10	Discrete Structure (syllabus)	Inoue Kohsei (athlete)	curry

From the third column of Table 1 and Fig. 5(b), almost all the top nodes belong to a collection of Japanese boy-bands called "Johnny's". Those who belong to the same group are connected with each other and their wikipedia pages contain similar nouns. Content centrality detects large content distributions where many nodes with similar content vectors are connected.

In the fourth column of Table 1 and Fig. 5(c), we show the ranking results without node names for privacy concerns. Almost all of the top nodes belong to communities relevant to such food as "soup," "boxed lunches," and "curry". The ingredients used in each dish tend to be similar, e.g., in the curry community, potatoes, carrots, curry powder, and onions are used in almost every curry.

Figure 4(b) shows the similarities between nodes with large or small centrality values and their neighbors. From Fig. 4(b), we can confirm that similarities with respect to the distance from nodes with large centrality values (solid line) have lots of similar nodes in their neighborhood about 1 to 4 distance. In contrast, similarities of nodes with small centrality values (dotted line) have relatively dissimilar nodes not only in their neighborhood but also in the whole network. These facts indicate that the proposed centrality value $CDC(u)$ of node u can reflect concentration level of contents around the node u, which is our aim.

With community detection techniques, we can extract densely connected node groups. However, there is no guarantee that the nodes belonging to these groups have similar content vectors. On the other hand, by quantifying the degree of the concentration of contents around each node, the content centrality can detect nodes that are located in the mode of a content distribution (Table 1).

Next, to evaluate the relations between content centrality and other structure-based centrality measures, Degree centrality, Closeness centrality, which targets the distance between other nodes, Betweenness centrality, which targets the frequency of the existence between other node pairs, Eigenvector centrality (Bonacich 1987), which considers the centrality scores of adjacent nodes and recursively calculates them, Community centrality (Newman 2006) which

Table 2. Rank correlation coefficients

	Web network	Wiki network	Recipe network
Community centrality	0.76	0.78	0.53
Closeness centrality	0.66	0.73	0.43
Degree centrality	0.47	0.52	0.19
PageRank	0.49	0.53	0.22
Betweenness centrality	0.28	0.19	0.30
Eigenvector centrality	0.27	0.15	0.20
Dimension of contents	0.14	0.02	0.18

focuses on the degree of belonging to the communities, and PageRank (Langville and Meyer 2004) as one of the well known ranking algorithm. Since these centrality measures aim at only the structural features, they cannot reflect the content-based features obtained from node activities.

For all three networks, we observe the following relations.

- Each ranking of the closeness and community centrality is correlated with the content centrality. This result is consistent with our intuition that the content distribution mode is located in almost the center of a community, and the community and closeness centrality of such nodes became relatively high.
- Each ranking of the degree centrality and PageRank is somewhat correlated with the content centrality. Although the degree centrality and PageRank intrinsically have correlation, the cosine similarity of the high degree node becomes larger because its RVwD is added relatively large number of content vectors with large weights.
- Almost no correlation with the content centrality can be observed in each ranking of the betweenness centrality, eigenvector centrality and dimension of content vector.

5   Conclusion

In this paper, we proposed Content Centrality, which considers content vectors generated from node activities. We assumed that nodes with similar contents are unevenly distributed, and the density of contents distribution decreases according to the distance from the center of the distribution. For each node, we quantified the degree of the concentration of the contents near the node by employing the cosine similarity between the content vector and the resultant vector of neighbors with distance-based decay weights and ranked all the nodes. By experimental evaluation using three real networks, we confirmed following results: (1) optimal decay parameter value depends on the size of the content distributions and the variance of the contents in node groups; (2) the content centrality ranks nodes in the upper level over which largest contents distributions are located.

Future work will consider other decay functions and investigate the performance of our proposed measure on different types of datasets.

Acknowledgements. This work was supported by JSPS KAKENHI Grant No. 15J00735 and by NII's strategic open-type collaborative research. In our experiments, we used recipe data provided by Cookpad and the National Institute of Informatics.

References

Bonacich, P.: Power and centrality: a family of measures. Am. J. Sociol. **92**(5), 1170–1182 (1987)

Cheng, Y.: Mean shift, mode seeking, and clustering. IEEE Trans. Pattern Anal. Mach. Intell. **17**(8), 790–799 (1995)

Clauset, A., Newman, M.E.J., Moore, C.: Finding community structure in very large networks. Phys. Rev. E **70**(6), 66111 (2004)

Domingos, P.: Mining social networks for viral marketing. IEEE Intell. Syst. **20**(1), 80–82 (2005)

Gruhl, D., Guha, R., Liben-Nowell, D., Tomkins, A.: Information diffusion through blog space. In: Proceedings of the 13th International Conference on World Wide Web, WWW 2004, pp. 491–501. ACM, New York (2004)

Kuramochi, T., Okada, N., Tanikawa, K., Hijikata, Y., Nishida, S.: Community extracting using intersection graph and content analysisin complex network. In: Proceedings of the 2012 IEEE/WIC/ACM International Joint Conferences on Web Intelligence and Intelligent Agent Technology, vol. 1 WI-IAT 2012, pp. 222–229. IEEE Computer Society, Washington, DC, USA (2012)

Langville, A.N., Meyer, C.D.: Deeper inside page rank. Int. Math. **1**(3), 335–380 (2004)

Natarajan, N., Sen, P., Chaoji, V.: Community detection in content-sharing social networks. In: Proceedings of the 2013 IEEE/ACM International Conference on Advances in Social Networks Analysis and Mining, ASONAM 2013, pp. 82–89. ACM, New York, NY, USA (2013)

Newman, M.E.J.: Assortative mixing in networks. Structure **2**(4), 5 (2002)

Newman, M.E.J.: The structure and function of complex networks. SIAM Rev. **45**, 167–256 (2003)

Newman, M.E.J.: Finding community structure in networks using the eigenvectors ofmatrices. Phys. Rev. E **74**(3), 36104 (2006)

Newman, M.E.J., Forrest, S., Balthrop, J.: Email networks and the spread of computer viruses. Phys. Rev. E **66**, 035101 (2002)

Parzen, E.: On estimation of a probability density function and mode. Ann. Math. Statist. **33**(3), 1065–1076 (1962)

Sun, J., Papadimitriou, S., Lin, C.-Y., Cao, N., Liu, S., Qian, W.: Multivis: Content-based social network exploration through multi-wayvisual analysis. In: SIAM International Conference on Data Mining, pp. 1064–1075. SIAM (2009)

Ting, I.-H., Wang, S.-L., Chi, H.-M., Wu, J.-S.: Content matters: A study of hate groups detection based on social networks analysis and web mining. In: Proceedings of the 2013 IEEE/ACM International Conference on Advances in Social Networks Analysis and Mining, ASONAM 2013, pp. 1196–1201. ACM, New York, NY, USA (2013)

Wu, Y., Jin, R., Zhu, X., Zhang, X.: Finding dense and connected subgraphs in dual networks. In: Proceedings of the IEEE 31st International Conference on Data Engineering (ICDE2015), pp. 915–926 (2015)

Yang, T., Jin, R., Chi, Y., Zhu, S.: Combining link and content for community detection: A discriminative approach. In: Proceedings of the 15th ACM SIGKDD International Conferenceon Knowledge Discovery and Data Mining, KDD 2009, pp. 927–936. ACM, New York, NY, USA (2009)

Zhou, Y., Cheng, H., Yu, J.X.: Graph clustering based on structural/attribute similarities. Proc. VLDB Endow. **2**(1), 718–729 (2009)

A Holistic Approach for Link Prediction in Multiplex Networks

Alireza Hajibagheri[1], Gita Sukthankar[1]([✉]), and Kiran Lakkaraju[2]

[1] University of Central Florida, Orlando, FL, USA
{alireza,gitars}@eecs.ucf.edu
[2] Sandia National Labs, Albuquerque, NM, USA
klakkara@sandia.gov

Abstract. Networks extracted from social media platforms frequently include multiple types of links that dynamically change over time; these links can be used to represent dyadic interactions such as economic transactions, communications, and shared activities. Organizing this data into a dynamic multiplex network, where each layer is composed of a single edge type linking the same underlying vertices, can reveal interesting cross-layer interaction patterns. In coevolving networks, links in one layer result in an increased probability of other types of links forming between the same node pair. Hence we believe that a holistic approach in which all the layers are simultaneously considered can outperform a factored approach in which link prediction is performed separately in each layer. This paper introduces a comprehensive framework, MLP (Multiplex Link Prediction), in which link existence likelihoods for the target layer are learned from the other network layers. These likelihoods are used to reweight the output of a single layer link prediction method that uses rank aggregation to combine a set of topological metrics. Our experiments show that our reweighting procedure outperforms other methods for fusing information across network layers.

1 Introduction

As social media platforms offer customers more interaction options, such as *friending*, *following*, and *recommending*, analyzing the rich tapestry of interdependent user interactions becomes increasingly complicated. In this paper, we study two types of online societies: (1) players in a massively multiplayer online game (Travian) [1] (2) dialogs between Twitter users before, during, and after an exceptional event [2]. Although standard social network analysis techniques [3] offer useful insights about these communities, there is relatively little theory from the social sciences on how to integrate information from multiple types of online interactions.

Rather than organizing this data into social networks separately chronicling the history of different forms of user interaction, dynamic multiplex networks [4] offer a richer formalism for modeling the social fabric of online societies. A multiplex network is a multilayer network that shares the same set of vertices across

© Springer International Publishing AG 2016
E. Spiro and Y.-Y. Ahn (Eds.): SocInfo 2016, Part II, LNCS 10047, pp. 55–70, 2016.
DOI: 10.1007/978-3-319-47874-6_5

all layers. This network can be modeled as a graph $G = < V, E >$ where V is the set of vertices and E is the set of edges present in the graph. The dynamic graph $G = \{G_0, G_1, ..., G_t\}$ represents the state of the network at different times. The network is then defined as: $G_t = < V, E_t^1, ..., E_t^M >$ with $E_t^\alpha \subseteq V \times V$, $\forall \alpha \in \{1, ..., M\}$, where each set E_t^α corresponds to the edge set of a distinct layer at time t. Thus a dynamic multiplex network is well suited for representing diverse user activities over a period of time.

In this paper, we address the problem of predicting future user interactions from the history of past connections. Assuming the data is represented as a graph, our goal is to predict the structure of graph G_t using information from previous snapshots as well as other layers of the network. Link prediction algorithms [5–9] have been implemented for many types of online social networks, including massively multiplayer online games and location-based social networks. These systems offer great value to social networking services due to their practical applicability for friend recommendations and social network bootstrapping. Although user profiles can be mined for additional data, topological approaches (1) perform well in many networks (2) preserve user privacy since they do not rely on actor information and (3) can be combined with node content approaches to enhance prediction performance.

Despite the fact that link prediction is a well studied problem, few link prediction techniques specifically address the problem of simultaneously predicting links across multiple networks [10–13]. Basu et al. [14] note that there are many real-world cases where interdependencies between processes cause the layers of a multiplex network to coevolve, resulting in a higher number of overlapping edges between the same node pair in different network layers. In this paper, we explore the role of overlapping edges towards improving the performance of link prediction; our aim is to leverage the cross-layer link co-occurrence history to model coevolution in a multiplex network. Our contributions can be summarized as follows:

- We introduce a framework for multiplex link prediction (MLP) that integrates complementary information sources, including topological metrics, network dynamics, and overlapping edges.
- MLP uses a likelihood based method for learning cross-layer dependencies and a temporal decay function to model the network dynamics. Rank aggregation is then employed to collect information from multiple topological metrics into one scoring matrix for ranking potential links.
- Extensive experiments are conducted on datasets collected from different types of social networks, and the proposed model is shown to outperform state-of-the-art link prediction methods.

In the next section, we present related work on link prediction. The proposed framework is described in Sect. 4. Section 5 presents a comparison of our method vs. two other approaches for fusing information across network layers. We conclude in Sect. 7 with a description of possible directions for future work.

2   Related Work

A variety of computational approaches have been employed for predicting links in single layer networks, including supervised classifiers, statistical relational learning, matrix factorization, metric learning, and probabilistic graphical models (see surveys by [15–17] for a more comprehensive description). Regardless of the computational framework, topological network measures are commonly used as features to describe node pairs and can be combined in a supervised or unsupervised fashion to do link prediction [6]. In this paper, we aggregate several of these metrics (listed in the next section), but our framework can be easily generalized to include other types of features.

The primary focus of this paper is leveraging cross-layer information to improve link prediction in multiplex networks, although we also introduce our own single layer link prediction technique. This process of using cross-layer information can be treated as a transfer learning problem where information is learned from a source network and applied to improve prediction performance the target network. Tang et al. [10] introduced a transfer-based factor graph (TranFG) model which incorporates social theories into a semi supervised learning framework. This model is then used to transfer supervised information from a source network to infer social ties in the target network.

Another strategy is to create more general versions of the topological measures that capture activity patterns in multilayer networks. Davis et al. [11] introduced a probabilistically weighted extension of the Adamic/Adar measure for these networks. Weights are calculated by doing a triad census to estimate the probability of different link type combinations. The extended Adamic/Adar metric is then used, along with other unsupervised link predictors, as input for a supervised classifier. Similarly, Hristova et al. [12] extend the definition of network neighborhood by considering the union of neighbors across all layers. These multilayer features are then combined in a supervised model to do link prediction. One weakness with the above mentioned models is their inability to use temporal information accrued over many snapshots, rather than relying on a single previous snapshot. In this paper, we evaluate two versions of our MLP framework, a version that only uses topological metrics calculated from one time slice vs. multiple snapshots. Rossetti et al. [13] combined multidimensional versions of Common Neighbors and Adamic/Adar with predictors that are able to utilize temporal information. However, like the standard version of these metrics, these extended versions do not necessarily generalize to networks generated from different processes.

Conversely, there are a number of approaches that ignore cross-layer network dependencies, while using the history of changes between snapshots to predict future network dynamics. We have experimented with two types of techniques: time series forecasting [5,18] and decay models [19]. Soares and Prudêncio [18] investigated the use of time series within both supervised and unsupervised link prediction frameworks. The core concept of their approach is that it is possible to predict the future values of topological metrics with time series; these values can either be used in an unsupervised fashion or combined in a supervised way with a classifier. In previous work, we introduced a rate prediction model [5] that uses time series to predict the rate of link formation. Our proposed framework, MLP, both models the rate of link formation in each layer and uses a decay model to account for changes in the topological metrics over time. In our results, we compare the improvements achieved by temporal vs. cross-layer modeling.

However, incorporating more features is not helpful, without an effective information fusion procedure. Pujari et al. [20] employed computational social choice algorithms for aggregating multiple topological features. They evaluated the performance of two well-known rank aggregation methods, Borda and Kemeny, for single layer link prediction. In their method, weights are learned for each voter participating in the rank aggregation, where each topological metric is treated as a voter. These weights are tuned to maximize the identification of positive examples or minimize negative examples. To extend their method to multiplex networks [21], the authors compute topological attributes for each network layer and combine them using (1) a simple aggregation of these scores across all layers or (2) an entropy-aggregation of values. These combinations are then used as a series of features in a decision tree model. In this paper, we use rank aggregation to fuse our features and compare our procedure to their aggregation methods. Another example of a supervised framework that uses rank aggregation is *RankMerging* [22]. During a learning phase, weights are assigned to each unsupervised method using a training set of node pairs. The contribution of each ranking to the merged ranking is then computed using sliding indices. At each step, the aim is to identify the ranking with the highest number of true predictions in the upcoming steps. Rank aggregation methods can be highly effective, but the more complex social choice algorithms can suffer from high computational complexity, making them less effective for large datasets. For this reason, we opted to use the Borda rank aggregation procedure in MLP.

3   Node Similarity Metrics

This section provides a brief description of the topological and path-based metrics for encoding node similarity that are used within our MLP framework to create ranked score lists for each node pair. These techniques are often used in isolation as unsupervised methods for link prediction. Note that $\Gamma(x)$ stands for the set of neighbors of vertex x while $w(x, y)$ represents the weight assigned to the interaction between node x and y. More details about these metrics could be found in [23].

- **Number of Common Neighbors (CN)**
  The CN measure is defined as the number of nodes with direct relationships with both evaluated nodes x and y [24].
- **Jaccard's Coefficient (JC)**
  The JC measure assumes higher values for pairs of nodes who share a higher proportion of common neighbors relative to their total neighbors.

– **Preferential Attachment (PA)**
The PA measure assumes that the probability that a new link originates from node x is proportional to its node degree. Consequently, nodes that already possess a high number of relationships tend to create more links [25].
– **Adamic-Adar Coefficient (AA)**
This metric [26] is closely related to Jaccard's coefficient in that it assigns a greater importance to common neighbors who have fewer neighbors. Hence, it measures the exclusivity of the relationship between a common neighbor and the evaluated pair of nodes.
– **Resource Allocation (RA)**
RA was first proposed in [27] and is based on physical processes of resource allocation.
– **Page Rank (PR)**
The PageRank algorithm [28] measures the significance of a node based on the significance of its neighbors. We use the weighted PageRank algorithm proposed in [29].
– **Inverse Path Distance (IPD)**
The Path Distance measure for unweighted networks simply counts the number of nodes along the shortest path between x and y in the graph. Note that $PD(x, y) = 1$ if two nodes x and y share at least one common neighbor. In this article, the Inverse Path Distance is used to measure the proximity between two nodes, where:

$$IPD(x, y) = \frac{1}{PD(x, y)} \tag{1}$$

– **Product of Clustering Coefficient (PCF)**

The clustering coefficient of a vertex v is defined as:

$$PCF(v) = \frac{3 \times \#\text{of triangles adjacent to v}}{\#\text{of possible triples adjacent to v}} \tag{2}$$

To compute a score for link prediction between the vertex x and y, one can multiply the clustering coefficient score of x and y.

Section 5 compares MLP vs. unsupervised versions of these approaches.

4   Proposed Method

MLP is a hybrid architecture that utilizes multiple components to address different aspects of the link prediction task. We seek to extract information from all layers of the network for the purpose of link prediction within a specific layer known as the target layer. To do so, we create a weighted version of the original target layer where interactions and connections that exist in other layers receive higher weights. After reweighting the layer, we employ the collection of node similarity metrics described in the previous section on the weighted network.

To express the temporal dynamics of the network, we use a decay model on the time series of similarity metrics to predict future values. Finally, the Borda rank aggregation method is employed to combine the ranked lists of node pairs into a single list that predicts links for the next snapshot of the target network layer. Each component of the model is explained in more detail in the following sections.

4.1 Multiplex Likelihood Assignment and Edge Weighting

This component leverages information about cross-layer link co-occurrences. During the coevolution process, links may be engendered due to activity in other network layers. Some layers may evolve largely independently of the rest of the network, whereas links in other layers may be highly predictive of links in the target layer. In our proposed method, a weight is assigned to each layer based on its influence on the target layer. Weights are calculated using a likelihood function: where L^i and w_i represent the ith layer and the weight calculated for it respectively. L^{Target} indicates the target layer for which we want to predict future links. The *Likelihood* function computes the similarity between the target layer and the ith layer; to do this, we use the current ratio of overlapping edges. Next, we calculate weights for every node pair by checking the link correspondence between two layers using the likelihood of a link being present in the target layer given the existence of the link in the other layer at any other previous snapshot. This orders other layers in terms of their relative importance for a specific target layer. The process assigns higher weights to node pairs which occur in more than one layer (multiplex edges). The rate of link formation is incorporated into the model as the first term of the edge weight. Algorithm 1 shows the process of assigning likelihoods to layers and reweighting the adjacency matrix.

Algorithm 1. Likelihood Assignment and Edge Weighting

1: Input: Edge sets $(E^1, ..., E^M)$ for M layers where E^α is the edge set of target layer
2: Output: E^α_w weighted adjacency matrix for layer α (target layer)
   //Calculate weights for the layers
3: **for** $i \in \{1, 2, ..., M\} - \{\alpha\}$ **do**
4:     $w_i = Likelihood(\text{Link in } L^\alpha | \text{Link in } L^i)$
5: **end for**
   //Weighting target layer
6: **for** edge $e \in E^\alpha$ **do**
7:     $w_e = rate + \sum_{i=1 \& i \neq \alpha}^{M} w_i \times linkExist(e)$
8: **end for**

The term *rate* is defined as the average value of the source node's out-degree over previous timesteps. Function *linkExist* is used to obtain information about a link's existence in other layers during previous snapshots. It checks each layer for the presence of an edge and returns 1 if an edge is present in that layer.

4.2   Temporal Link Structure

Given the network history for T time periods, we need to capture the temporal dependencies of the coevolution process. To do so, our framework uses a weighted exponentially decaying model [30]. Let $\{Sim_t(i,j), t = t_0 + 1, ..., t_0 + T\}$ be a time series of similarity score matrices generated by a node similarity metric on a sliding window of T successive temporal slices. An aggregated weighted similarity matrix is constructed as follows:

$$Sim_{(t_0+1)\sim(t_0+T)}(i,j) = \sum_{t=t_0+1}^{t_0+T} \theta^{t_0+T-t} Sim_t(i,j) \tag{3}$$

where the parameter $\theta \in [0,1]$ is the smoothing weight for previous time periods. Different values of θ modify the importance assigned to the most or least recent snapshots before current time $t+1$. This procedure generates a composite temporal score matrix for every node similarity metric. $Sim_{(t_0+1)\sim(t_0+T)}$ (shortened to Sim) is used by the algorithm as a summary of network activity, encapsulating the temporal evolution of the similarity matrix.

4.3   Rank Aggregation

Before describing the final step of our approach, let us briefly discuss existing methods for *ranked list aggregation/rank aggregation*. List merging or list aggregation refers to the process of combining a number of lists with the same or different numbers of elements in order to get one final list including all the elements. In rank aggregation, the order or rank of elements in input lists is also taken into consideration. The input lists can be categorized as *full, partial, or disjoint lists*. Full lists contain exactly the same elements but with a different ordering, partial lists may have some of the elements in common but not all, and disjoint lists have completely different elements. In this case, we are only dealing with full lists since each similarity metric produces a complete list for the same set of pairs, differing only in ordering.

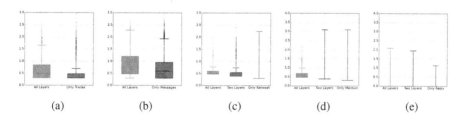

(a)   (b)   (c)   (d)   (e)

Fig. 1. Log scale box-whisker plots for user interactions in different layers of the network: (a) Travian (Trades) (b) Travian (Messages) (c) Cannes2013 (Retweets) (d) Cannes2013 (Mentions) (e) Cannes2013 (Replies)

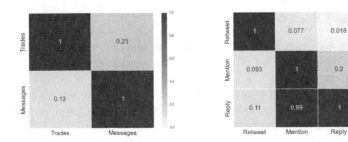

Fig. 2. Heatmap representing the edge overlap between pairs of layers for datasets (a) Travian (b) Cannes2013

Several rank aggregation methods are described in [31], including Borda's, Markov chain, and median rank methods. Borda's method is a *rank-then-combine* method originally proposed to obtain a consensus from a voting system. Since it is based on the absolute positioning of the rank elements and not their relative rankings, it can be considered a truly positional method. For every element in the lists, a Borda score is calculated and elements are ranked according to this score in the aggregated list. For a set of complete ranked lists $L = [L_1, L_2, L_3, ..., L_k]$, the Borda score for an element i and a list L_k is given by:

$$B_{L_k}(i) = \{count(j)|L_k(j) < L_k(i) \& j \in L_k\} \qquad (4)$$

The total Borda score for an element is given as:

$$B(i) = \sum_{t=1}^{k} B_{L_t(i)} \qquad (5)$$

Borda's method is computationally cheap, which is a highly desirable property for link prediction in large networks.

Algorithm 2 shows our proposed framework which incorporates edge weighting, the temporal decay model, and rank aggregation to produce an accurate prediction of future links in a dynamic multiplex network. The Borda function produces the final output of the MLP framework. Results of the proposed algorithm are compared with other state-of-the-art techniques in the next section.

5   Experimental Study

This paper evaluates the MLP framework on networks extracted from two real-world datasets, Travian and Cannes2013. To investigate the impact of each component of our proposed method, not only do we compare our results with two other approaches for fusing cross-layer information, but we also analyze the performance of ablated versions of our method. The complete method, MLP (Hybrid), is compared with MLP (Decay Model + Rank Aggregation) and MLP (Weighted + Rank Aggregation). All of the algorithms were implemented in

Algorithm 2. Multiplex Link Prediction Framework (MLP)

1: Input: Weighted edge sets of the target layer for T previous snapshots
2: Output: Temporal aggregated score matrix S for the target layer
3: **for** each node similarity metric u **do**
4:    **for** $t \in \{1, ..., T\}$ **do**
5:        Calculate score matrix Sim_{t0+t}^{u}
6:    **end for**
7:    Calculate temporal similarity matrix Sim^{u}
8: **end for**
9: Final score matrix $S = Borda(Sim^{1}, ..., Sim^{u})$

Python and executed on a machine with the Intel(R) Core i7 CPU and 24GB of RAM for the purpose of fair comparison. Our implementation uses Apache Spark to speed the link prediction process.

5.1 Datasets

We use two real-world dynamic multiplex networks to demonstrate the performance of our proposed algorithm. These networks are considerably disparate in structure and were selected from different domains (a massively multiplayer online game (MMOG) and an event-based Twitter dataset). Table 1 provides the network statistics for each of the datasets:

– **Travian MMOG** [1] Travian is a browser-based, real-time strategy game in which the players compete to create the first civilization capable of constructing a Wonder of the World. The experiments in this paper were conducted on a 30 day period in the middle of the Travian game cycle. In Travian, players can execute different game actions including: sending messages, trading resources, joining alliances, and attacking enemy villages. In this research, we focus on networks created from trades and messages.
– **Twitter Interactions** [2] This dataset consists of Twitter activity before, during, and after an "exceptional" event as characterized by the volume of communications. Unlike most Twitter datasets which are built from follower-followee relationships, links in this multiplex network correspond to retweeting, mentioning, and replying to other users. The Cannes2013 dataset was created from tweets about the Cannes film festival that occurred between May 6,2013 to June 3, 2013. Each day is treated as a separate network snapshot.

5.2 Evaluation Metrics

For the evaluation, we measure receiver operating characteristic (ROC) curves for the different approaches. The ROC curve is a plot of the *true positive rate (tpr)* against the *false positive rate (fpr)*. These curves show achievable true positive rates (TP) with respect to all false positive rates (FP) by varying the

Table 1. Dataset Summary: Number of edges, nodes, and snapshots for each network layer

Dataset		Travian		Cannes2013
No. of Nodes		2,809		438,537
No. of Snapshots		30		29
Layers/No. of Edges	Trades	87,418	Retweet	496,982
	Messages	44,956	Mention	411,338
			Reply	83,534

decision threshold on probability estimations or scores. For all of our experiments, we report area under the ROC curve (AUROC), the scalar measure of the performance over all thresholds. Since link prediction is highly imbalanced, straightforward accuracy measures are well known to be misleading; for example, in a sparse network, the trivial classifier that labels all samples as missing links can have a 99.99 % accuracy.

5.3 Analysis of Cross-Layer Interaction

Figure 1 shows log scale box-whisker plots that depict the frequency of interactions between users who are connected across multiple layers. We compare the frequency of interactions in cases where the node pair is connected on all layers vs. the frequency of being connected in a single layer (Travian) or less than all layers (for Cannes which has three layers). As expected, in cases where users are connected on all layers, the number of interactions (trades, messages, retweets, mentions and replies) is higher. The heatmap of the number of overlapping edges between different network layers (Fig. 2) suggests that a noticeable number of edges are shared between all layers. This clearly indicates the potential value of cross-layer information for the link prediction task on these datasets. Our proposed likelihood weighting method effectively captures the information revealed by our analysis.

5.4 Performance of Multiplex Link Prediction

For our experiments, we adopted a moving-window approach to evaluate the performance of our temporal multiplex link prediction algorithm. Given a specified window size T, for each time period $t(t > T)$, graphs of T previous periods $(G_{t-T}, ..., G_{t-1})$ (where each graph consists of M layers) are used to predict links that occur at the target layer α in the current period (G_t^α). To assess our proposed framework and study the impact of its components, we compare against the following baselines:

- **MLP (Hybrid)**: incorporates all elements discussed in the framework section. It utilizes the likelihood assignment and edge weighting procedure to extract cross-layer information. Node similarity scores are modified using the temporal decay model and combined with Borda rank aggregation.

- **MLP (Likelihood + Rank Aggregation)**: This method only uses the aggregated scores calculated from the graphs weighted with cross-layer information. It does not consider the temporal aspects of network coevolution.
- **MLP (Decay Model + Rank Aggregation)**: This method does not use the cross-layer weighting scheme and relies on temporal information alone to predict future links. The final aggregated score matrix is calculated based on forecast values at time t for each node similarity metric using the decay model.
- **Likelihood**: Weights generated by the cross-layer likelihood assignment procedure are treated as scores for every node pair. We then sort the pairs based on their score and calculate the AUROC.
- **Rank Aggregation**: This method is a simple aggregated version of all unsupervised scoring methods using the Borda's rank aggregation method applied to node similarity metrics from the target layer.
- **Unsupervised Methods**: The performance of our proposed framework is compared with eight well-known unsupervised link prediction methods described in Sect. 3. All unsupervised methods are applied to the binary static graph from time 0 to $t-1$ in order to predict links at time t. Only the structure of the target layer is used.
- **Average Aggregation**: In order to extend the rank aggregation model to include information from other layers of the network, we use the idea proposed in [21]. Node similarity metrics are aggregated across all layers. So for attribute X (Common Neighbors, Adamic/Adar, etc.) over M layers the following is defined:

$$X(u, v) = \frac{\sum_{\alpha=1}^{M} X(u, v)^{\alpha}}{M} \qquad (6)$$

where $X(u, v)$ is the average score for nodes u and v across all layers and $X(u, v)^{\alpha}$ is the score at layer α. Borda's rank aggregation is then applied to the extended attributes to calculate the final scoring matrix.

- **Entropy Aggregation**: Entropy aggregation is another extended rank aggregation model proposed in [21] where $X(u, v)$ is defined as follows:

$$X(u, v) = -\sum_{\alpha=1}^{M} \frac{X(u, v)^{\alpha}}{X_{total}} \log(\frac{X(u, v)^{\alpha}}{X_{total}}) \qquad (7)$$

where $X_{total} = \sum_{\alpha=1}^{M} X(u, v)^{\alpha}$. The entropy based attributes are more suitable for capturing the distribution of the attribute value over all dimensions. A higher value indicates a uniform distribution of attribute values across the multiplex layers.

- **Multiplex Unsupervised Methods**: Finally, using the definition of core neighborhood proposed in [12], we extend four unsupervised methods (Common Neighbors, Preferential Attachment, Jaccard Coefficient and Adamic/Adar) to their multiplex versions.

Table 2 shows the results of different algorithms on the Travian and Cannes2013 datasets. With 30 days of data from Travian and 27 days for Cannes2013, we were able to extensively compare the performance of the proposed methods and the impact of using different elements. Bold numbers indicate the best results on each target layer considered; MLP (Hybrid) is the best performing algorithm in all cases.

6   Discussion

In this section, we discuss the most interesting findings:

Does rank aggregation improve the performance of the unsupervised metrics? As shown in Table 2, although the aggregated scores matrix produced by Borda's method achieves better results than unsupervised methods in some cases (Travian message, Cannes2013 retweet and mention networks) and comparable results on others (Travian trade and Cannes2013 reply networks), it is not able to significantly outperform all unsupervised methods in any of the networks. As discussed before, we are using the simple Borda method for the rank aggregation which does not consider the effect of each ranker on the final performance. While adding weights to the rankers or using more complex rank aggregation models such as Kemeny might achieve better results, it has been shown that those approaches have high computational complexity which makes them less suitable for large real-world networks [20,22]. Despite the fact that the rank aggregation alone does not significantly improve the overall performance of the link prediction task, it enables us to effectively fuse different kinds of information (edge and node features, nodes similarity, etc.).

On the other hand, the Average and Entropy Aggregation methods, which are designed to consider attribute values from other layers, are able to outperform regular Rank Aggregation and MLP (Decay Model + Rank Aggregation). However, both methods use the static structure of all snapshots from time 0 to $t-1$, while MLP (Decay Model + Rank Aggregation) only incorporates the past T snapshots which makes it more suitable for large networks.

Does the likelihood assignment procedure outperform the unsupervised scores? To study the ability of our likelihood weighting method to model the link formation process, we generate results for two methods: using likelihood explicitly as a scoring method as well as using the values to generate a weighted version of the networks. First, the *Likelihood* method is used in isolation to demonstrate the prediction power of its weights as a new scoring approach. Table 2 shows significant improvements on unsupervised scores as well as the aggregated version of them. As expected, the more overlap between the target layer and predictor layers, the more performance improvement *Likelihood* achieves. As an example, Likelihood achieves $\sim 7\%$ of improvement on Travian (Trade) compared with $\sim 5\%$ of improvement on Travian (Message). Not only is there a lower rate of overlapping edges between those layers, but also the number of interactions is higher than the two other layers. The same holds true for Cannes2013 (Retweet) compared with the mention and reply layers.

Table 2. AUROC performances for a target layer averaged over all snapshots with a sliding time window of $T = 3$ for Travian layers and $T = 5$ for Cannes2013 layers used in the decay model. Variants of our proposed framework are shown at the top of the table, followed by standard unsupervised methods. The algorithms shown in the bottom half of the table are techniques for multiplex networks proposed by other research groups. The best performer is marked in bold.

Algorithms / Networks	Trade	Message	Retweet	Mention	Reply
MLP (Hybrid)	0.821±0.001	0.803±0.002	0.812±0.002	0.834±0.003	0.839±0.002
MLP (LH/RA)	0.802±0.001	0.790±0.0021	0.809±0.003	0.814±0.004	0.816±0.003
MLP (DM/RA)	0.722±0.002	0.731±0.002	0.727±0.002	0.728±0.003	0.733±0.002
Likelihood	0.770±0.033	0.760±0.041	0.752±0.022	0.781±0.052	0.757±0.042
Rank Aggregation	0.694±0.001	0.712±0.001	0.700±0.002	0.706±0.002	0.700±0.003
Common Neighbors	0.656±0.002	0.667±0.002	0.699±0.002	0.705±0.003	0.699±0.001
Jaccard Coefficient	0.628±0.002	0.680±0.003	0.594±0.002	0.733±0.002	0.711±0.003
Preferential Attachment	0.709±0.002	0.637±0.001	0.584±0.002	0.612±0.002	0.587±0.003
Adamic/Adar	0.635±0.003	0.700±0.003	0.700±0.002	0.642±0.002	0.516±0.003
Resource Allocation	0.625±0.005	0.690±0.003	0.597±0.002	0.622±0.002	0.672±0.003
Page Rank	0.595±0.0016	0.687±0.002	0.660±0.002	0.630±0.003	0.613±0.002
Inverse Path Distance	0.572±0.003	0.650±0.003	0.631±0.003	0.641±0.002	0.561±0.004
Clustering Coefficient	0.580±0.002	0.633±0.003	0.570±0.020	0.621±0.011	0.522±0.004
Average Aggregation	0.744±0.030	0.752±0.020	0.740±0.003	0.737±0.011	0.761±0.003
Entropy Aggregation	0.731±0.004	0.763±0.020	0.75±0.0030	0.758±0.031	0.744±0.002
Multiplex CN	0.729±0.0040	0.643±0.013	0.672±0.003	0.716±0.003	0.733±0.002
Multiplex JC	0.666±0.031	0.619±0.012	0.580±0.003	0.736±0.002	0.722±0.002
Multiplex PA	0.722±0.010	0.646±0.012	0.580±0.003	0.640±0.003	0.621±0.003
Multiplex AA	0.671±0.010	0.690±0.031	0.671±0.003	0.669±0.003	0.552±0.003

On the other hand, the method introduced in Algorithm 1 generates a weighted version of input graphs which is used to generate a weighted version of unsupervised methods to produce the final scoring matrix. This paired with the rank aggregation method generates significantly better average AUROC performance compared with other proposed methods. Also, when temporal information from previous snapshots of the network is included, MLP (Hybrid) outperforms other variants of MLP as well as well-known unsupervised methods. This indicates the power of overlapping links in improving the performance of link prediction in coevolving multiplex networks.

Does including temporal information improve AUROC performance? The importance of incorporating temporal information into link prediction has been discussed in our previous work [5]. However, here we are interested in analyzing the impact of this information on improving the performance of MLP. For that purpose, first, the decay model is employed in MLP (Decay Model + Rank Aggregation) to determine whether it improves the results generated by the aggregated score matrix. The final aggregated score matrix is calculated based on forecast values at time t for each unsupervised method using the decay model. As expected, this version of MLP is able to achieve up to $\sim 3\%$ of

AUROC improvement using only information from the last three and five snapshots of the Travian and Cannes2013 networks respectively. On the other hand, we observed the same pattern when the decay model was added to MLP (Hybrid) along with likelihood and rank aggregation. Using the scores generated by our hybrid approach outperformed all other proposed and existing methods. The results presented here have been obtained using $T = 3$ for the Travian dataset and $T = 5$ for Cannes2013. These values are based on experiments performed on both datasets. While for Travian layers, increasing the value of T tends to improve the prediction performance slightly until $T = 3$; higher values of T may decrease the performance. The same pattern occurs for Cannes2013 layers when $T = 5$. Similarly, the value of θ is set to 0.4 for both datasets.

In summary, MLP (Decay Model + Rank Aggregation) is able to achieve results comparable to other baseline methods except Average and Entropy Aggregation since they benefit from the entire graph structure. Although rank aggregation by itself is not able to significantly improve the performance of unsupervised methods, paired with decay models and taking temporal aspects of the network, it can achieve better performance. On the other hand, the multiplex versions of the neighborhood based unsupervised methods are able to improve average AUROC performance, however the results are inconsistent and they achieve lower performance in many cases. Finally, both MLP (Hybrid) and MLP (Likelihood + Rank Aggregation) achieve higher performance compared with all other methods, illustrating the importance of the cross-layer information created by the network coevolution process. A paired two-sample t-test is used to indicate the significance of the results produced by each method where the p-value is smaller than 0.0001. It is worth mentioning that, even though MLP (Hybrid) is able to outperform all other methods, its performance is not significantly better than MLP (Likelihood + Rank Aggregation) in the case of Travian (Message) and Cannes (Retweet).

7   Conclusion and Future Work

In this paper, we introduce a new link prediction framework, MLP (Multiplex Link Prediction), that employs a holistic approach to accurately predict links in dynamic multiplex networks using a collection of topological metrics, the temporal patterns of link formation, and overlapping edges created by network coevolution. Our analysis on real-world networks created by a variety of social processes suggests that MLP effectively models multiplex network coevolution in many domains.

The version of Borda's method used in this research assigns the same weight to all rankers. However, for different networks, each scoring method might add differing value to the final scoring matrix. In future work, it would be interesting to use weighted Borda to calculate final scores. Also, while using more network features often increases the performance of a link prediction algorithm, this might not be true for all networks. Thus it may be useful to employ a feature selection algorithm to identify the best subset of unsupervised methods to be used in MLP, based on performance improvements in early snapshots. Finally, in this research it is implicitly assumed that the existence of links in other layers increases the probability of link formation in the target layer. A promising direction for future research would be to modify our reweighting procedure to account for negative cross-layer influences, where connections in one layer lower the target link likelihood.

Acknowledgments. Sandia National Laboratories is a multi-program laboratory managed and operated by Sandia Corporation, a wholly owned subsidiary of Lockheed Martin Corporation, for the U.S. Department of Energy's National Nuclear Security Administration under contract DE-AC04-94AL85000. The Travian dataset was provided by Drs. Rolf T. Wigand and Nitin Agarwal (University of Arkansas at Little Rock, Department of Information Science); their research was supported by the National Science Foundation and Travian Games GmbH, Munich, Germany.

References

1. Hajibagheri, A., Lakkaraju, K., Sukthankar, G., Wigand, R.T., Agarwal, N.: Conflict and communication in massively-multiplayer online games. In: Agarwal, N., Xu, K., Osgood, N. (eds.) SBP 2015. LNCS, vol. 9021, pp. 65–74. Springer, Heidelberg (2015). doi:10.1007/978-3-319-16268-3_7
2. Omodei, E., De Domenico, M., Arenas, A.: Characterizing interactions in online social networks during exceptional events. arXiv preprint (2015). arXiv:1506.09115
3. Scott, J.: Social Network Analysis. Sage, London (2012). https://www.amazon.com/Social-Network-Analysis-John-Scott/dp/1446209040
4. Kivela, M., Arenas, A., Barthelemy, M., Gleeson, J., Moreno, Y., Porter, M.: Multilayer networks. J. Complex Netw. **2**, 203–271 (2014)
5. Hajibagheri, A., Sukthankar, G., Lakkaraju, K.: Leveraging network dynamics for improved link prediction. In: Proceedings of the International Conference on Social Computing, Behavioral-Cultural Modeling, and Prediction, Washington, D.C., June 2016
6. Liben-Nowell, D., Kleinberg, J.: The link-prediction problem for social networks. J. Am. Soc. Inf. Sci. Technol. **58**(7), 1019–1031 (2007)
7. Menon, A.K., Elkan, C.: Link prediction via matrix factorization. In: Gunopulos, D., Hofmann, T., Malerba, D., Vazirgiannis, M. (eds.) ECML PKDD 2011. LNCS (LNAI), vol. 6912, pp. 437–452. Springer, Heidelberg (2011). doi:10.1007/978-3-642-23783-6_28
8. Scellato, S., Noulas, A., Mascolo, C.: Exploiting place features in link prediction on location-based social networks. In: Proceedings of the ACM SIGKDD International Conference on Knowledge Discovery and Data Mining, pp. 1046–1054 (2011)
9. Beigi, G., Tang, J., Liu, H.: Signed link analysis in social media networks. arXiv preprint (2016). arXiv:1603.06878
10. Tang, J., Lou, T., Kleinberg, J.: Inferring social ties across heterogenous networks. In: Proceedings of the ACM International Conference on Web Search and Data Mining, pp. 743–752 (2012)
11. Davis, D., Lichtenwalter, R., Chawla, N.V.: Supervised methods for multi-relational link prediction. Soc. Netw. Anal. Min. **3**(2), 127–141 (2013)

12. Hristova, D., Noulas, A., Brown, C., Musolesi, M., Mascolo, C.: A multilayer approach to multiplexity and link prediction in online geo-social networks. arXiv preprint (2015). arXiv:1508.07876
13. Rossetti, G., Berlingerio, M., Giannotti, F.: Scalable link prediction on multidimensional networks. In: 2011 IEEE 11th International Conference on Data Mining Workshops, pp. 979–986. IEEE (2011)
14. Basu, P., Dippel, M., Sundaram, R.: Multiplex networks: A generative model and algorithmic complexity. In: IEEE/ACM International Conference on Advances in Social Networks Analysis and Mining (ASONAM), pp. 456–463 (2015)
15. Lü, L., Zhou, T.: Link prediction in complex networks: A survey. Phys. A: Stat. Mech. Appl. **390**(6), 1150–1170 (2011)
16. Al Hasan, M., Zaki, M.J.: A survey of link prediction in social networks. In: Aggarwal, C.C. (ed.) Social Network Data Analytics, pp. 243–275. Springer, Heidelberg (2011)
17. Zhang, J., Philip, S.Y.: Link prediction across heterogeneous social networks: A survey (2014)
18. Soares, P.R.d.S., Prudêncio, R.B.C.: Time series based link prediction. In: International Joint Conference on Neural Networks, pp. 1–7. IEEE (2012)
19. Gao, S., Denoyer, L., Gallinari, P.: Temporal link prediction by integrating content and structure information. In: Proceedings of the ACM International Conference on Information and Knowledge Management, pp. 1169–1174 (2011)
20. Pujari, M., Kanawati, R.: Supervised rank aggregation approach for link prediction in complex networks. In: Proceedings of the International World Wide Web Conference, pp. 1189–1196 (2012)
21. Pujari, M., Kanawati, R.: Link prediction in multiplex networks. Netw. Heterogen. Media **10**(1), 17–35 (2015)
22. Tabourier, L., Bernardes, D.F., Libert, A.S., Lambiotte, R.: Rankmerging: A supervised learning-to-rank framework to predict links in large social network. arXiv preprint (2014). arXiv:1407.2515
23. Wang, X., Sukthankar, G.: Link prediction in heterogeneous collaboration networks. In: Missaoui, R., Sarr, I. (eds.) Social Network Analysis - Community Detection and Evolution. LNCS, pp. 165–192. Springer, Heidelberg (2014)
24. Newman, M.E.J.: Clustering and preferential attachment in growing networks. Phys. Rev. E **64**, 25102 (2001)
25. Barabási, A.L., et al.: Scale-free networks: a decade and beyond. Science **325**(5939), 412 (2009)
26. Adamic, L.A., Adar, E.: Friends and neighbors on the web. Soc. Netw. **25**(3), 211–230 (2003)
27. Zhou, T., Lü, L., Zhang, Y.C.: Predicting missing links via local information. Eur. Phys. J. B **71**(4), 623–630 (2009)
28. Brin, S., Page, L.: Reprint of: The anatomy of a large-scale hypertextual web search engine. Comput. Netw. **56**(18), 3825–3833 (2012)
29. Ding, Y.: Applying weighted PageRank to author citation networks. J. Am. Soc. Inf. Sci. Technol. **62**(2), 236–245 (2011)
30. Acar, E., Dunlavy, D.M., Kolda, T.G.: Link prediction on evolving data using matrix and tensor factorizations. In: Workshops at IEEE International Conference on Data Mining, pp. 262–269 (2009)
31. Sculley, D.: Rank aggregation for similar items. In: SIAM International Conference on Data Mining, pp. 587–592 (2007)

Twitter Session Analytics: Profiling Users' Short-Term Behavioral Changes

Farshad Kooti[1](✉), Esteban Moro[2], and Kristina Lerman[1]

[1] USC Information Sciences Institute, Marina Del Rey, USA
kooti@usc.edu
[2] Universidad Carlos III de Madrid, Madrid, Spain

Abstract. Human behavior shows strong daily, weekly, and monthly patterns. In this work, we demonstrate online behavioral changes that occur on a much smaller time scale: minutes, rather than days or weeks. Specifically, we study how people distribute their effort over different tasks during periods of activity on the Twitter social platform. We demonstrate that later in a session on Twitter, people prefer to perform simpler tasks, such as replying and retweeting others' posts, rather than composing original messages, and they also tend to post shorter messages. We measure the strength of this effect empirically and statistically using mixed-effects models, and find that the first post of a session is up to 25 % more likely to be a composed message, and 10–20 % less likely to be a reply or retweet. Qualitatively, our results hold for different populations of Twitter users segmented by how active and well-connected they are. Although our work does not resolve the mechanisms responsible for these behavioral changes, our results offer insights for improving user experience and engagement on online social platforms.

1   Introduction

Understanding people's online behavior can motivate the design of human-computer interfaces that enhance user experience, increase engagement, and reduce cognitive load. Until recently, most of the research in this area focused on web search and browsing. Researchers found it useful to segment online activity into sessions, defined as periods of time that the user is actively engaged with the platform and usually has a single intent [15,17]. For example, Kumar and Tomkins found that about half of all web page views during a typical session are of inline content, one-third are communications, and the remaining one-sixth are search [20]. Search sessions have also been studied on Twitter to compare search on Twitter and web. Researchers found that search sessions on Twitter tend to be shorter and include fewer queries compared to web search [28]. Similarly, Benevenuto et al. analyzed activity sessions on an online social network aggregator to understand how frequently and for how long people use different social networking platforms, and what sequence of actions they take during a session [16].

© Springer International Publishing AG 2016
E. Spiro and Y.-Y. Ahn (Eds.): SocInfo 2016, Part II, LNCS 10047, pp. 71–86, 2016.
DOI: 10.1007/978-3-319-47874-6_6

In this paper, we carry out a study of user activity sessions on Twitter to document short-term behavioral changes occurring over the course of a single session. Similar to earlier studies of web search, we segment the time series of an individual's activity on Twitter into sessions, where each session is a series of consecutive interactions—tweeting, retweeting, or replying—without a break longer than a specified threshold. (We experimented with different ways of defining sessions and different thresholds, and our findings are qualitatively very similar with different definitions of session.) We find that most sessions are short, but there are considerable number of sessions that span hours. Despite their short duration, we find that significant behavioral changes occur over the course of a single session, with people preferring easier interactions later in the session. Specifically, people tend to compose longer tweets at the beginning of a session, and reply and retweet more later in the session, and also when there is a short time period between consecutive interactions. While Twitter population is highly heterogeneous, these patterns hold across different subsets of the population, e.g., for both highly connected and poorly connected users, as well as for highly-active and less-active users.

Earlier studies have shown strong daily, weekly, and monthly patterns in social activity. For example, Foursquare check-ins, mobile phone calls, or tweets show strong daily and weekly patterns corresponding to food consumption and nightlife [13], different social contexts [1], economical activity [21], or worldwide daily and seasonal mood variations in Twitter [9]. In this work, we find patterns that occur in far shorter time scales of only a few minutes, compared to daily and monthly patterns of earlier work. While long term patterns can be explained by the circadian cycles, work schedules, and other global macroscopic forces, the behavioral changes we study appear to be qualitatively different, arising from the individual decisions (perhaps unconscious) to allocate attention and effort. To our knowledge, this is the first demonstration of short-term behavioral changes on Twitter.

The main contributions of our work are as follows:

- We present a detailed analysis of user activity sessions on Twitter. We show that most of the sessions are very short; however, while large fraction of sessions include only one type of tweet, most of the sessions are mixture of different types of tweets (e.g., normal tweets, replies, and retweets) (Sect. 2).
- We show that later in a session people tend to perform easier or more socially rewarding interactions, such as replying or retweeting, instead of composing original tweets. Also, they tend to compose shorter tweets later in a session (Sect. 3).
- We divide people based on their characteristics, such as position in the follower graph or activity, and show that people with higher activity or more friends behave differently (Sect. 4).

Several mechanisms could explain our observations. First, deterioration of performance following a period of sustained mental effort has been documented in a variety of settings, including data entry [14] and exerting self-control [23], and led researchers to postulate cognitive fatigue [2] as the explanation. On

Twitter, as people become fatigued over the course of a session, they may switch to easier tasks that require less cognitive effort, such as retweeting instead of composing original tweets. Alternately, our observations could be explained by growing boredom or loss of motivation. It is plausible that social interactions are highly motivating, and the fact that users continue to reply to others, even when they are less likely to create original tweets, appears to indicate that they shift their effort to the more engaging tasks, such as social interactions. Still other explanations are possible, such as users' choice to strategically shift their attention to other tasks. While our work does not address the causes of these behavioral changes, our findings are significant in that they can be used to predict users' future actions, which could, in turn, be leveraged to improve user online experience on social platforms.

2   Methods

Our Twitter dataset includes more than 260 M tweets posted by 1.9 M randomly selected users and all their tweets, using Twitter's API. Twitter is known to include lots of spammers. To eliminate spammers from our dataset, we took the approach of [8] and classified users as spammers or bots based on entropy of content generated and entropy of time intervals between tweets (spammers and bots tend to have low entropy of content and tweeting time intervals).

User online activity can be segmented into sessions, usually characterized by a single intent [15,17]. We apply a similar idea to our Twitter data. To construct activity sessions from the time series of user's tweets, we examine the time interval between successive tweets and consider a break between sessions to be a time interval greater than some threshold. Following [17], we use a 10-min threshold. Thus, all tweets posted by a user within 10 min of his or her previous tweet are considered to be in the same session, and the first tweet posted following a time period longer than 10 min starts a new session (Fig. 1). We experimented with different time thresholds and the results remain robust. Due to the heavy-tailed distribution of inter-tweet time interval, increasing the threshold only merges a very small fraction of sessions. Figure 2 shows the probability (PDF) and cumulative (CDF) distribution of time between consecutive tweets. This distribution is very similar to the distribution of time between phone calls a person makes [25]. There is no clear cut-off and the plot drops gradually. This figure also shows that increasing the 10 min threshold to 30 min, only affects 6 % of the sessions.

To understand sessions, we look at the distribution of session length (time interval between the first and last tweet of the session) and number of tweets posted in the session. While these distributions would change if a different time threshold was used, as explained above, the change is not significant. Most of the sessions include few tweets: 64 % of sessions include only two tweets, and only 1 % include 12 or more tweets. Moreover, sessions tend to be very short: 99 % of sessions are only 1 min long, even if we only consider sessions that include 5 tweets or more, 98 % of them are still only 1 min long.

We also analyze the types of tweets that are posted in a session. We classify tweets into three main types:

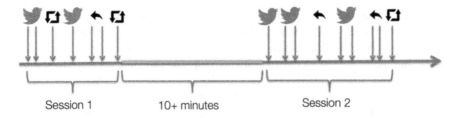

Fig. 1. Timeline of user activity on Twitter segmented into sessions. The timeline is a time series of tweets, including normal tweets, retweets, and replies. These activities fall into sessions. A period between consecutive tweets lasting longer than 10 min indicates a break between sessions.

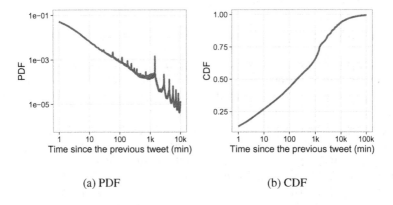

(a) PDF   (b) CDF

Fig. 2. Distribution of the time interval between consecutive tweets.

reply a message directed to another user, usually starting with an @mention.
retweet an existing message that is re-shared by the user, sometimes preceded by an 'RT'
normal all other tweets; typically composed tweets, which may include urls and hashtags

Considering all sessions, 59 % of sessions include only one type of tweet. This percentage is very high because a large fraction of sessions include only two tweets, so there is a very low probability of diversity. Considering only sessions that include more than five tweets, then only 35 % of the sessions include one type of tweet, 41 % include two types of tweets, and the remaining 24 % include all three types of tweets. To better understand the diversity of sessions, we consider sessions that include 10 tweets and cluster them based on the fraction of normal tweets, replies, and retweets. We use the X-means algorithm from Weka[1] that automatically detects the number of clusters. The algorithm creates three clusters, where in each cluster one type of tweet is dominant. 44 % of sessions belong to the cluster where majority of tweets are normal, 31 % are sessions with

---

[1] http://weka.sourceforge.net/doc.packages/XMeans/weka/clusterers/XMeans.html.

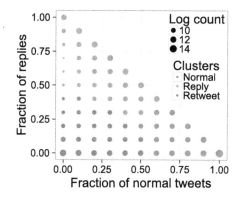

Fig. 3. Visualization of clustering of sessions using the fraction of normal tweets, replies, and retweets. (Color figure online)

many replies, and 25 % of the sessions include mostly retweets. Figure 3 shows a visualization of the sessions with each color representing a cluster and the size of dots representing the number of sessions with that fractions of tweet types. The x-axis shows the fraction of normal tweets in the session, and y-axis shows the fraction of replies in the session. Each cluster could be found in the plot by considering the fractions, e.g., the red circles belong to replies, because they have high fraction of replies, and the green circles belong to the retweet cluster, because they have low fraction of normal tweets and replies. As it is shown in the figure, these clusters are not clearly separated and there is a spectrum of sessions with different fraction of tweet types. This means there is no clear users or sessions that have a particular purpose, and most of the sessions include a mixture of different types of tweets.

3 Session-Level Behavioral Changes

In this section, we present evidence for changes in user behavior over the course of a single session on Twitter. We focus on three types of behaviors: (i) the type of the message (tweet) a user posts on Twitter, (ii) the length of the message the user composes, and (iii) the number of spelling errors the user makes. Since sessions are typically short, with the vast majority lasting only a few minutes, the demonstrated behavioral changes take place on far faster time scales than those previously reported in literature (e.g., diurnal and seasonal changes).

3.1 Time to Next Tweet

The type of a tweet a user posts depends on how much time has elapsed since the user's previous interaction on Twitter. As shown in Fig. 4, 30 % of the tweets posted 10 s after another tweet are normal tweets, whereas more than 50 % of tweets posted two minutes or more following a previous tweet are normal tweets.

In general, the longer the period of time since a user's last action on Twitter, the more likely the new tweet is to be a normal tweet. Note that we excluded tweets posted within 10 s of the previous tweet, because they are likely to have been automatically generated, e.g., by a Twitter bot. Despite the filtering, our data still contains some machine-generated activity, as evidenced by spikes at 60 s, 120 s, etc. The shorter the time delay from the previous tweet, the more likely the tweet is to be a retweet. Replies are initially similar to normal tweets: the more time elapsed since the previous tweet, the more likely the new tweet is to be a reply, but unlike normal tweets, their probability saturates and even decreases slightly with longer delays.

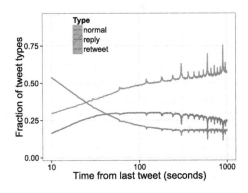

Fig. 4. Fraction of different tweet types given the time from the user's last tweet.

To understand these temporal patterns, we segment a user's activity into sessions, as described in the previous section. We can characterize sessions along two dimensions: (a) the number of tweets produced during the session and (b) the length of the session in terms of seconds or minutes, i.e., the time period between the first and last tweet of the session. Each of these dimensions plays an important role in the types of the tweets that are produced during the session. For example, short sessions with many tweets are very intense, and the user may not have enough time to compose original tweets; hence, the tweets are likely to be replies. On the other hand, a long session with few tweets is more likely to include more normal tweets, because the user has had enough time to compose them. The fraction of tweets that are replies is shown in Fig. 5, which shows these trends: users are more likely to reply as sessions become longer (in time), or there are fewer tweets posted during sessions of a given duration.

We can study the behavioral change with respect to either the position of the tweet in the session or the time elapsed since the beginning of the session. Our preliminary analysis showed that the number of tweets in a session plays a more significant role compared to the time since the first tweet of the session. Hence, in the following analyses, we study changes with respect to the position of the tweet within a session and not with respect to the time since the first tweet. In general, the trends are similar but weaker if we consider the time since the first tweet of the session.

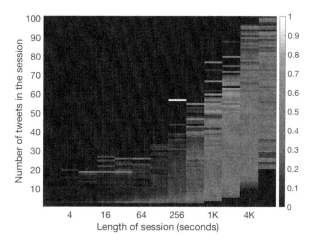

Fig. 5. Fraction of tweets that are replies posted during sessions of a given length in time and number of tweets in the session. The data was binned and only bins with more than 100 sessions are included.

3.2 Changes in Tweet Type

Next, we study the types of tweets that are posted at different times during a session. Since user behavior during longer sessions could be systematically different from their behavior during shorter sessions, we aggregate sessions by their length, which we define as the number of tweets posted. Then for each tweet position within a session, we calculate the fraction of tweets that belong to each of our three types. Figure 6 shows that tweets are more likely to be normal tweets early in a session, and later in a session, users prefer cognitively easier (i.e., retweet) or socially more rewarding (i.e., reply) interactions.

Since user population on Twitter is highly heterogenous, these observations could result from non-homogeneous mixing of different user populations. Kooti et al. show an example of this, where a specific population of users is over-represented on one side of the plot (e.g., early during a session), producing a trend that does not actually exist [18]. One way to test for this effect is through a *shuffle test*. In a shuffle test, we randomize the data and conduct analysis on the randomized (i.e., shuffled) data. If the analysis of the shuffled data yields a similar result as of the original data, then the trend is simply an artifact of the analysis and does not exist in the data. If trends disappear completely, it suggests that the original analysis is meaningful.

To shuffle the data, we reorder the tweets within each session, keeping the time interval between them the same. Figure 7 shows results of the analysis on the shuffled data. Flat lines indicate that the factions of all tweet types do not change over the course of the shuffled session. This suggests that the trends observed in the original data have a behavioral origin.

We use values in Fig. 7 as baseline to normalize the average fraction of tweets types in Fig. 6. Figure 8 shows the change in the fraction of tweet types relative

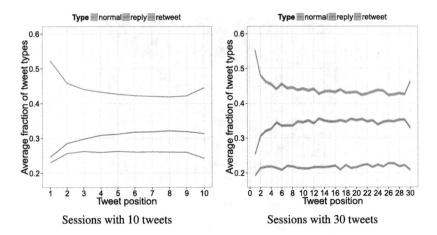

Fig. 6. Change in the fraction of tweets of each type over the course of sessions in which users posted 10 or 30 tweets.

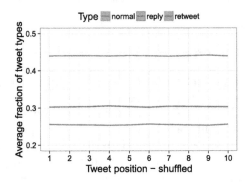

Fig. 7. Change in the fraction of tweets of each type over the course of sessions of length 10 in shuffled data.

to the baseline and clearly shows that the first tweets of a session are up to 30 % relatively more likely to be normal tweets, and 10–20% less likely to be replies or retweets. The time when a normal tweet becomes less likely than the baseline (red line crossing zero) is later during longer sessions, and it happens after ∼30 % of the tweets are posted, i.e., at the 3rd position for sessions with 10 tweets and at the 10th position in sessions with 30 tweets.

What explains the observed trends? To partially address this question, we focus on the fraction of replies. As explained above, users are more likely to reply later in a session rather than compose an original tweet. This may arise because some sessions are extended by the ongoing conversations the user has with others. To test this hypothesis, we calculate the fraction of replies at each position within the session that are in response to a tweet that was posted since the start of that session. In other words, we calculate the fraction of replies in

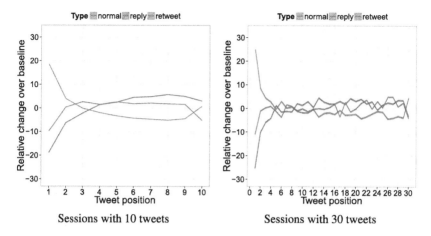

Fig. 8. Relative change in the fraction of tweets of each type over the course of sessions with 10 or 30 tweets.

conversations initiated during that session. Figure 9 shows this fraction: replies that are posted later in the session are much more likely to belong to an ongoing conversation. This means that some part of the trend found above could be explained by users extending their sessions to interact with others.

3.3   Change in Tweet Length

Next, we study the change in the length of tweets posted over the course of a session. We exclude retweets from this analysis, because length of the retweets does not represent the effort needed to compose them. First, we calculate the average length of the tweet at each position in the session, but there is too much variation in tweet length to produce any statistically significant trends. Instead,

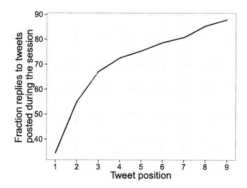

Fig. 9. Fraction of tweets that are replies to tweets posted since the beginning of the same session (for sessions with 10 tweets) .

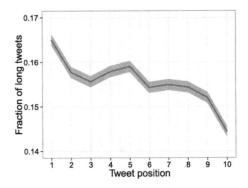

Fig. 10. Fraction of long tweets posted over the course of sessions of a given length (10 tweets). Long tweets are defined as non-reply tweets that are longer than 130 characters.

we divide tweets into long (longer than 130 characters) and short tweets (shorter than 130 characters), and measure the fraction of long tweets over the course of the session. We find a statistically significant trend, wherein tweets posted later in the session are more likely to be short, compared to tweets posted earlier in the session (Fig. 10). We choose a high threshold for the long tweets, because when a user is reaching the 140 character limit imposed by Twitter, they usually have to make an effort to shorten their tweet by rephrasing and abbreviating the message. We believe that this results in a stronger signal for analysis, compared to the situation where the user is just typing a few more characters e.g., 30 characters vs. 35 characters. To ensure that the drop in the fraction of long tweets is a real trend, we perform the shuffle test and obtain a flat line. This suggests that users are less likely to devote the effort to compose long tweets later in a session. We exclude tweets including URLs and repeat the analysis again, and we achieve very similar results. Similarly, considering only normal tweets and replies results in the similar trend.

3.4 Change in the Number of Spelling Mistakes

Finally, we consider the percentage of words that are spelled incorrectly in a tweet. Earlier studies have shown that when people are tired their judgment is impaired [3], and it is harder for them to solve problems correctly [14]. We hypothesize that we can observe this effect in terms of number of spelling errors that users make. To this end, for each tweet we calculate the percentage of words that are spelled incorrectly (i.e., typos) and calculate the average percentage of typos at each tweet position in a session. We exclude retweets, non-English tweets, and punctuations and use a dictionary that includes all forms of a word, e.g., including the past tense of the verbs and the plural of the nouns.

Figure 11 shows that there is a small but statistically significant increase in the percentage of typos made in tweets over the course of a session. This percentage rises quickly initially, but saturates later in the session. Overall, there

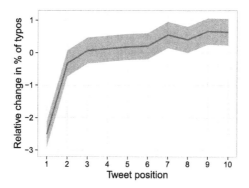

Fig. 11. Percentage of change of spelling errors made in tweets over the course of session relative to shuffled data.

is a 3 % relative increase in the probability of making a spelling mistake later in the session, compared to first tweets of the session. The same trend exists for replies and normal tweets when considered individually.

3.5 Modeling

The results presented above strongly suggest that tweeting behavior changes over the course of a session. To make these findings more quantitative, we model the trends statistically. One challenge for statistical analysis is that the data samples are not independent, as we have multiple sessions from the same user. In addition, there is significant heterogeneity among the users, with some users posting mostly normal tweets, while the others mostly retweeting. As a result, our conclusions, which are based on data aggregated over the entire population, could be affected by the heterogeneous mixture of different populations (Simpson's paradox). To resolve this issue, we model the tweeting activity using mixed-effects models, which consider the individual differences.

The mixed-effects models include two main components: (*i*) fixed effects, which are constant across different user populations, e.g., the index or position of the tweet in the session, and (*ii*) random effects, which vary across different users, e.g., reflecting user's preference to post tweets of a particular type. The random effect enables us to consider individual differences among users to identify the role of the fixed effects.

We model each tweet type independently as a binary response. The model determines if a tweet is a particular tweet type given the position of the tweet in the session, the session length, and considering the user who has posted the tweet. This model can be written as *tweet type* $\sim 1 + tweet\ index + session\ length + (1|user)$. We represent the intercept of the model by 1, and the next two terms are the fixed effects that we are interested in, and finally the particular user is also considered. In modeling the normal tweets, the coefficient of the tweet index is -0.0148, meaning that tweets posted later in the session are less likely to be a normal tweet. On the other hand, in the model for replies, the tweet index coefficient is $+0.0149$, confirming our earlier findings and showing that tweets that are posted later in the session are more likely to be a reply. For retweets, the index coefficient is -0.0001, which is very small and negative, meaning retweeting is slightly less likely later in the session. This is due to the strong over-representation of replies later in the sessions, and if we consider only normal tweets and retweets, then the index coefficient becomes positive. The median scaled residuals for the three models are only -0.07 for modeling normal tweets, and -0.19 for modeling replies and retweets, showing that the model has a very low rate of errors.

In short, we considered the individual differences by modeling the tweet types using mixed-effects models. The results of the modeling confirmed that the results of our empirical analyses are not due to aggregating over different user population.

4   User Characteristics

In this section, we investigate how differences between users may contribute to behavioral changes. We split users based on their characteristics and carry out analysis described in the previous section within subpopulations of users.

4.1   User Connectivity

One of the main characteristics of Twitter users is the number of friends they have, i.e., the number of other Twitter users they follow. This number is highly correlated with the amount of information users receive and the number of inter-actions they have with other users. We rank users based on the number of friends and compare the session-level behavioral differences of the bottom 20 % with the top 20 %. In both cases, we measure how the fraction of tweet types change relative to the baseline, over the course of a session. Figure 12 shows that users with many friends retweet significantly more compared to users who follow few others. This is perhaps not surprising, as the well-connected users tend to receive many more tweets and have more opportunities for retweeting. These users also tend to be very active, and as users become more active, they tend to retweet more (arguably because it takes less effort). However, even though the fraction of tweet types is different in the two groups, the change over the course of a session is very similar. Therefore, we conclude that users with different numbers of friends act differently in general, but their behavior changes the same way over the course of a session. We verify that the results are not an artifact of the analysis by performing the shuffle test.

4.2   User Activity

Next, we divide users into different classes based on their activity, i.e., the rate of tweeting. We order users based on the average number of tweets in a month,

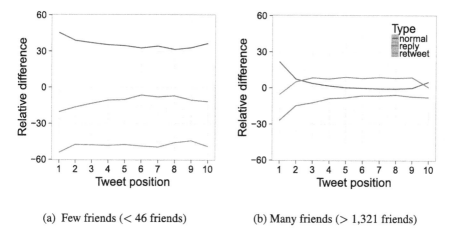

(a) Few friends (< 46 friends)   (b) Many friends (> 1,321 friends)

Fig. 12. Relative change in the tweet type throughout a session for users with few friends and many friends. The change is relative to shuffled sessions with 10 tweets.

and compare the top 20 % of the most active users to the bottom 20 % of the users. We find that the less active users tend to compose more original (normal) tweets, and are more likely to do it than users with most tweets. In contrast, the more active users produce many more retweets and replies, compared to users with lower levels of activity (Fig. 13). And, unlike previous analysis that divided users based on the number of friends, the change in the fraction of replies shows a higher increase for more active users. We again conduct a shuffle test to ensure that the observed effect is real.

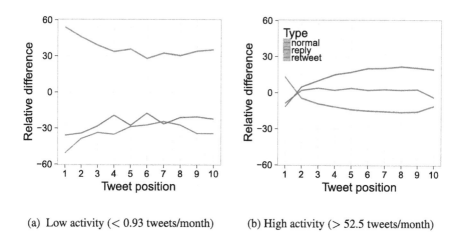

(a) Low activity (< 0.93 tweets/month)   (b) High activity (> 52.5 tweets/month)

Fig. 13. Relative change in tweet type throughout a session for users with low and high activity.

We conclude that part of what makes users active is their willingness to engage in social interactions on Twitter. Users extend their session to carry on conversations with others. People appear to prioritize their online activity on Twitter, and social interactions appear to be preferable, especially more active users, later in the session.

5   Related Work

Sessions of activity have shown to be an effective way to characterize people's online behavior, by segmenting a person's activity to meaningful smaller sections that are easier to study and analyze [5,24,26]. In the research community, sessions are usually constructed in two ways: a series of actions that serve a single intent [6,17], or more commonly, a period of time without a break longer than a given threshold [11,27], which is our definition of session.

Sessions have been studied extensively in context of browsing and search behavior [15,17,20]. In the recent years, sessions of activity have been also used for understanding users' behavior in online social networks. Benevenuto et al. created sessions of activity from a social network aggregator to understand users' behavior in high-level, e.g. how frequently and for how long the social networks are used [16]. On Twitter, Teevan et al. studied sessions to compare Twitter search with web search [28]. And more recently on Facebook, Grinberg et al. studied the effect of content production on length and number of sessions [12].

The changes in behavior of users over the course of a session could be attributed to fatigue or cognitive depletion. These concepts have been studied extensively in the offline world by psychologists. They have shown that there is a temporal component in cognitive performance. Mental effort makes it more difficult for people to perform cognitively demanding tasks at a later time, whether to solve problems correctly [14], make a decision [3], or exercise self-control [7,22]. The phenomenon of lower cognitive ability after sustained mental effort is generally referred to as "ego depletion" [4]. Although there have been multiple proposals for various mechanisms of ego depletion and they are still debated, there is consensus among researchers that cognitive performance declines over a period of continuous mental effort. Our study is another evidence for this phenomena.

Our study presents behavioral changes that occur on a very small time scale; only in order of minutes. Multiple studies have shown daily, weekly, monthly, and yearly patterns of activity in offline and online world: people make more donations in the mornings [19], strong daily and weekly patterns of food consumption exist in Foursquare checkins [13], there are significant seasonal patterns in communications among college students on Facebook [10], or diurnal and seasonal trends affect people's sentiment expressed on Twitter posts [9].

6   Conclusion

In this work, we analyzed user behavior during activity sessions on Twitter. We found that users engage with Twitter usually for short periods of time, what we refer to as activity sessions, that are on the order of minutes and include only a few tweets. The tweets posted during these times tend to be diverse tweets, including original (composed) messages, retweets of others' messages, and replies to other users. Despite its short duration, users' behavior changes over the course of a session, as they appear to prioritize different types of interactions. The longer they are on Twitter, the more they prefer to perform easier or more socially engaging tasks, such as retweeting and replying, rather than harder tasks, such as composing an original tweet. This effect is quite large: at the beginning of the session, the tweets are up to 25 % more likely to be original tweets than near the end of the session.

We also found that tweets tend to get shorter later in the session, and people tend to make more spelling mistakes. All these results could be explained by people becoming cognitively fatigued, or perhaps careless due to loss of motivation. If we divide users into classes based on the number of friends they follow, or their activity level (i.e., the number of tweets they posted), we find that while these user classes behave differently in general, in terms of the types of tweets they tend to post, all classes manifest similar behavioral changes over the course of the session. While our work does not resolve the mechanisms responsible for these behavioral changes, our findings are significant in that they can be used to forecast dynamics of user behavior, which could, in turn, be leveraged to improve user online experience on social platforms.

References

1. Aledavood, T., López, E., Roberts, S.G., Reed-Tsochas, F., Moro, E., Dunbar, R.I., Saramäki, J.: Daily rhythms in mobile telephone communication. PloS One **10**(9), e0138098 (2015)
2. Baumeister, R.F., Bratslavsky, E., Muraven, M., Tice, D.M.: Ego depletion: is the active self a limited resource? J. Pers. Soc. Psychol. **74**(5), 1252 (1998)
3. Baumeister, R.F., Sparks, E.A., Stillman, T.F., Vohs, K.D.: Free will in consumer behavior: self-control, ego depletion, and choice. J. Consum. Psychol. **18**(1), 4–13 (2008)
4. Baumeister, R.F., Vohs, K.D.: Self-regulation, ego depletion, and motivation. Soc.Pers. Psychol. Compass **1**(1), 115–128. http://dx.doi.org/10.1111/j.1751-9004.2007.00001.x
5. Daoud, M., Tamine-Lechani, L., Boughanem, M., Chebaro, B.: A session based personalized search using an ontological user profile. In: Proceedings of the 2009 ACM Symposium on Applied Computing, pp. 1732–1736. ACM (2009)
6. Eickhoff, C., Teevan, J., White, R., Dumais, S.: Lessons from the journey: a query log analysis of within-session learning. In: Proceedings of the 7th ACM International Conference on Web Search and Data Mining, pp. 223–232. ACM (2014)
7. Gailliot, M.T., Baumeister, R.F., DeWall, C.N., Maner, J.K., Plant, E.A., Tice, D.M., Brewer, L.E., Schmeichel, B.J.: Self-control relies on glucose as a limited energy source: willpower is more than a metaphor. J. Pers. Soc. Psychol. **92**(2), 325 (2007)
8. Ghosh, R., Surachawala, T., Lerman, K.: Entropy-based classification of retweeting activity on twitter. In: Proceedings of KDD Workshop on Social Network Analysis (SNA-KDD), August 2011

9. Golder, S.A., Macy, M.W.: Diurnal and seasonal mood vary with work, sleep, and daylength across diverse cultures. Science **333**(6051), 1878–1881 (2011)
10. Golder, S.A., Wilkinson, D.M., Huberman, B.A.: Rhythms of social interaction: Messaging within a massive online network. In: Communities and Technologies 2007, pp. 41–66. Springer (2007)
11. Goševa-Popstojanova, K., Singh, A.D., Mazimdar, S., Li, F.: Empirical characterization of session-based workload and reliability for web servers. Empirical Softw. Eng. **11**(1), 71–117 (2006)
12. Grinberg, N., Dow, P.A., Adamic, L.A., Naaman, M.: Extracting diurnal patterns of real world activity from social media. In: CHI (2016)
13. Grinberg, N., Naaman, M., Shaw, B., Lotan, G.: Extracting diurnal patterns of real world activity from social media. In: ICWSM (2013)
14. Healy, A.F., Kole, J.A., Buck-Gengle, C.J., Bourne, L.E.: Effects of prolonged work on data entry speed and accuracy. J. Exp. Psychol. Appl. **10**(3), 188–199. http://view.ncbi.nlm.nih.gov/pubmed/15462620
15. Huang, J., Efthimiadis, E.N.: Analyzing and evaluating query reformulation strategies in web search logs. In: Proceedings of the 18th ACM Conference on Information and Knowledge Management, pp. 77–86. ACM (2009)
16. Jin, L., Chen, Y., Wang, T., Hui, P., Vasilakos, A.V.: Understanding user behavior in online social networks: a survey. IEEE Commun. Mag. **51**(9), 144–150 (2013)
17. Jones, R., Klinkner, K.L.: Beyond the session timeout: automatic hierarchical segmentation of search topics in query logs. In: Proceedings of the 17th ACM Conference on Information and Knowledge Management, pp. 699–708. ACM (2008)
18. Kooti, F., Lerman, K., Aiello, L.M., Grbovic, M., Djuric, N., Radosavljevic, V.: Portrait of an online shopper: understanding and predicting consumer behavior. In: Proceedings of the 9th ACM International Conference on Web Search and Data Mining (WSDM 2016), San Francisco, USA, February 2016
19. Kouchaki, M., Smith, I.H.: The morning morality effect the influence of time of day on unethical behavior. Psychol. Sci. **25**(1), 95–102 (2013). 0956797613498099
20. Kumar, R., Tomkins, A.: A characterization of online browsing behavior. In: Proceedings of the 19th International Conference on World Wide Web, pp. 561–570. ACM (2010)
21. Llorente, A., Garcia-Herranz, M., Cebrian, M., Moro, E.: Social media fingerprints of unemployment. PloS One **10**(5), e0128692 (2015)
22. Muraven, M., Tice, D., Baumeister, R.: Self-control as a limited resource: regulatory depletion patterns. J. Pers. Soc. Psychol. **74**(3), 774 (1998)
23. Muraven, M., Baumeister, R.F.: Self-regulation and depletion of limited resources: does self-control resemble a muscle? Psychol. Bull. **126**(2), 247 (2000)
24. Rose, D.E., Levinson, D.: Understanding user goals in web search. In: Proceedings of the 13th International Conference on World Wide Web, pp. 13–19. ACM (2004)
25. Saramäki, J., Moro, E.: From seconds to months: an overview of multi-scale dynamics of mobile telephone calls. Eur. Phys. J. B **88**(6), 1–10 (2015)
26. Smith, B.R., Linden, G.D., Zada, N.K.: Content personalization based on actions performed during a current browsing session, uS Patent 6,853,982, 8 February 2005
27. Spiliopoulou, M., Mobasher, B., Berendt, B., Nakagawa, M.: A framework for the evaluation of session reconstruction heuristics in web-usage analysis. Inf. J. Comput. **15**(2), 171–190 (2003)
28. Teevan, J., Ramage, D., Morris, M.R.: # twittersearch: a comparison of microblog search and web search. In: Proceedings of the Fourth ACM International Conference on Web Search and Data Mining, pp. 35–44. ACM (2011)

Senior Programmers: Characteristics of Elderly Users from Stack Overflow

Grzegorz Kowalik[✉] and Radoslaw Nielek

Polish-Japanese Academy of Information Technology,
Koszykowa 86, 02-008 Warsaw, Poland
{grzegorz.kowalik,nielek}@pjwstk.edu.pl

Abstract. In this paper we present results of research about elderly users of Stack Overflow (Question and Answer portal for programmers). They have different roles, different main activities and different habits. They are an important part of the community, as they tend to have higher reputation and they like to share their knowledge. This is a great example of possible way of keeping elderly people active and helpful for society.

Keywords: Elderly · Stack Overflow · Q&A · Aging · Online communities · Programmers · Crowdsourcing

1 Introduction

When we think about programmers, we usually imagine relatively young people sitting in front of Hi-Tech computers. They learn fast and are able to follow rapid changes in technology and new trends. But do they get older? Some of them become senior programmers but this phrase is related to the experience and not age. On the other hand computers and programming languages have been with us already for over sixty years, thus, there should exist also somewhere people who operated computer systems in sixties and seventies. What are they doing now. Do they learn new skills and languages?

In this paper we take a closer look on elderly users of Stack Overflow[1] – the biggest online community for programmers according to Alexa rating[2] with over 5.7 million registered users and 31 million posts.

The very first question that appears is whether older adults use the Stack Overflow at all. Galit Nimrod [13] identified 40 online communities where older adults discuss health, tourism or retirements related issues, so there is no reason to assume that technical savvy seniors avoid Q&A web sites devoted to programming but what roles do they play in the community? Do they overwhelmingly look for information or maybe they are experts in long-forgotten technologies and programming languages? How different their habits are in comparison with younger users?

---

[1] http://stackoverflow.com.

[2] http://www.alexa.com/siteinfo/stackoverflow.com.

© Springer International Publishing AG 2016
E. Spiro and Y.-Y. Ahn (Eds.): SocInfo 2016, Part II, LNCS 10047, pp. 87–96, 2016.
DOI: 10.1007/978-3-319-47874-6_7

Three things make Stack Overflow a particularly interesting place for studying older adults behaviors. First, there are users of very different age range, starting from thirteen or fourteen years of age, thus a whole spectrum of intergenerational behaviors might emerge. Second, it is a community purely oriented on knowledge exchange in a fast pacing topics, so being active requires either an in-depth knowledge or readiness to learn. Finally, all data are openly available and coupled with meta-information such as up and downvotes, timestampes and tags for posts and reputation for users.

Understanding how older adults can benefit from online communities and how they can contribute to the communities' goals is crucial for long-term development of online Q&A sites in the light of dramatically increasing proportion of elderly in society. Active participation in Q&A sites is also a sign of long-life learning which is important to keep people in labor market.

The remaining part of the paper is organized as follow: next section presents related work, dataset used in this paper is described in Sect. 3. Data analyses and results are presented in Sect. 4. Possible further studies and conclusion are gathered in Sect. 5.

2   Related Work

The aging and programming skills was researched before using Stack Overflow data in [12]. Their research shows the differences between older and younger users, and prove that elderly users have better reputation. They also found that elderly users are still learning new technologies.

The concept of deeper differences between older and younger groups is a well known issue for sociologists. There are even studies dedicated to the generations related to technology [11,17]. Results show that we can describe IT workers from some technological inventions as "generations" of some technology and they also find themselves a part of it.

About activation and using knowledge and skills of elderly people in crowd sourcing, there were experiments and research showing that they might not be interested in doing small tasks on demand in exchange for small remuneration, although they are interested in "being useful" for society, like proofreading [8–10] or Amazon mTurk small tasks [4]. Some works were focused on similar topic as this paper - knowledge sharing [7].

Stack Overflow is also an example of an online community. Regarding elderly people in such communities, there were some research about them like: [15] where authors show different types of users (information swappers, aging-oriented, socializers), roles of online communities for elderly [3], showing that they can give them both entertainment and valuable information, or [13], where authors show Natural Language Processing method of measuring users engagement. Research in [16] shows that being a part of an online community can enhance elderly people well-being. Topic of positive influence of IT technologies on elderly people well being is also researched in [6]. Other topics of discussion than programming were researched in [14] (tourism), where we can also see knowledge and information sharing, similar to Stack Overflow.

3 Data Sources

All results presented in this paper are based on the analysis of one of two data sources (sometimes on both):

- Stack Exchange API[3] – tool shared by an administrator that makes possible to submit SQL-like queries and receives results directly from Stack Overflow database; all elements existing in the web site can be gathered that way, i.a.: reputation, users public profiles, posts, topics, tags; the only limitation is a cap of 50 thousands rows per query, thus in some cases we had to do calculation on a sample (always clearly stated in text if applicable);
- Stack Overflow Survey – a survey of Stack Overflow users is conducted every year; the most recent results available are from year 2015[4]; among 25831 users who answered questionnaire only 0.5 % were 60+, therefore in some cases it was impossible to calculate the significance tests or verify hypothesis requiring intersection of many variables.

4 Results

4.1 Do Older Adults Use Stack Overflow?

As "seniors" (elderly people) we call users that are 60 or more years old. This is mainly because of the limitations of survey data where age was given in range. Figure 1 shows the distribution of age for surveys conducted in 2015 and 2016[5]. Bins reflect possible answers in this question. There is 0.5 % – 0.8 % of users with age 60+.

Next to the survey age can also be extracted from users profiles but we have to be aware that providing birth year is not mandatory and users may also lie. We therefore excluded all users without birth year in their profile, assuming that providing age is not correlated with any important variables. Moreover, we have taken a closer look at age distribution from Stack Overflow users profile data. Age distribution from both sources look similar with one exception. Closer look at Fig. 2 reveals surprisingly many users over that are over 90. We should expect age to have a more Gaussian shape, thus we can assume that these profiles are fake. We decided to exclude profiles with age over 90.

After excluding users with "too high" age, we received 562795 users (0.8 % of them have a age 60 or more). This is very similar to the survey results (Fig. 1) in 2016, so we can assume that our filtering was justified. To further confirm that there are real older adults in on Stack Overflow we investigated users self-descriptions published on users profile web pages. In-depth insight can be found in Sect. 4.4.

---

[3] https://data.stackexchange.com/.
[4] http://stackoverflow.com/research/developer-survey-2015.
[5] In the rest of the paper we use data from 2015 instead of 2016 as not all data are still available for the newest survey.

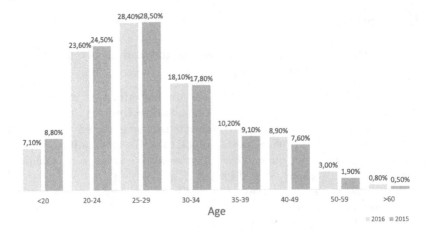

Fig. 1. Age (survey data). N=55338 (2016) and N=25831 (2015)

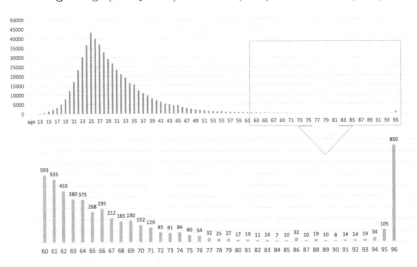

Fig. 2. Age distribution (number of users) among Stack Overflow users

4.2 Do They Teach or Do They Learn?

If we assume that older adults have more experience and are more knowledgeable, we should observe that they are posting overwhelmingly more answers instead of questions. How much effort do they put in developing the community? We try to answer these questions by comparing frequency of post types, length of profile texts and responses about motivation (taken from survey) between two age groups.

Post Types: Main Activity. In Stack Overflow users can be active in several ways: writing posts, comments, upvoting, downvoting etc. The most important user activity is posting. Posts can have different types: question or answer. Asking questions we can interpret as learning – looking for knowledge, and by answering – teaching, sharing knowledge. Our hypothesis is, that elderly people are less likely to learn, and are more interested in sharing their knowledge – comparing to younger users.

To verify this, we used post type table and frequency of different post types in each age group. As you can see in Table 1, answers are the most popular post type in both groups. This is probably due to the fact that for each question you can have more than one answer, so it is not surprising that both groups have more answers than questions in their posting activity. Moreover, instead of posting a new question users can always read answers for similar questions that are already posted and answered.

Table 1. Post types frequency in each age group. Remain four types of posts has been omitted because are extremely rare in Stack Overflow – less than 1 %.

PostTypeId	Juniors	Seniors
Question	25.04 %	13.54 %
Answer	74.61 %	86.26 %

It is the difference that is important here – "Seniors" have significantly higher percentage of answers in their posts than "juniors" (Significance t-test for difference of answers ratio in both groups have p-value below 0.0001). This confirms our hypothesis – "seniors" more often (more than 10 percentage points) gives answers, share their knowledge, than ask questions. They more teach than learn.

Profile Texts: Effort. We also expect elderly people to be more engaged in Stack Overflow community. This is already somehow confirmed in the previous point, as giving more answers is also creating a content for portal and being responsible for it. In addition, we checked their profile texts. We assume that if they put more effort in writing about themselves, they care more about the community.

As we can see in Table 2, "Seniors" have significantly (verified by T-Test), higher average of characters used in their profile texts – almost two times higher. This confirms our hypothesis that they put more effort into their activity.

Motivation. In survey, there was a question regarding motivation of posting answers to questions in Stack Overflow. Results, divided into our age groups, are presented in Fig. 3. We can learn that "helping programmers in need" and "future programmers" are most important for both groups. We also observe the differences in motivation with older adults biased more toward sense of responsibility but because of small sample differences are not statistically significant (sample of seniors that given answer is too small).

Table 2. Profile text length (number of characters)

	Seniors	Juniors
Mean	180	92
Standard Deviation	342.22	209.40
N	4273	528847

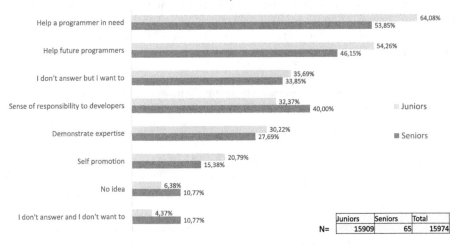

Fig. 3. Motivation to post answers (survey data)

Table 3. Reputation overview

Age	Mean	St. Dev
Juniors	645	5630.17
Seniors	938	11165.99

4.3   Earned Reputation or Gamed Reputation?

Older adults share their knowledge by posting more answers than younger users but are their answers valuable for the community? In order to check it, we decided to take a closer look on the reputation earned by two group of users. According to Stack Overflow *"Reputation is a rough measurement of how much the community trusts you; it is earned by convincing your peers that you know what youre talking about."*[6]. As we can see in Table 3, older adults have higher reputation (significance test confirms it). "Seniors" have higher mean of reputation, but also, as we could see also in paper [12], much higher standard deviation.

---

[6] http://stackoverflow.com/help/whats-reputation.

Reputation have very high standard deviation in general - as it is not manda-tory to post, a lot of users have low activity and low reputation. Only around a half have the reputation higher than 1. This way we need to look closer on the results than mean.

Table 4. Reputation percentiles (for "Juniors" we used sampling (around 20 % of total))

Percentile	Juniors*	Seniors
20	1	1
40	5	1
60	37	11
80	276	101
90	833.9	456
100	265086	446919

As we can see in Table 4, reputation is distributed in different ways among both groups. Among "Juniors", there are more active users, around 40 % of them have reputation higher than 1. For "seniors" its around 50 %. After this point, we can see the reputation distributed in a more equal way among "juniors" and less equal in "seniors" group. That also match the higher variation in "seniors" group.

In addition, we also checked how often answers of "seniors" and "juniors" were marked as "accepted" (best answers for question). "Seniors" have 32.98 % answers marked as accepted, "juniors" have 36.04 %. As we can see juniors have higher percentage of accepted answers, but this is not a high difference. This might have many reasons, as accepted answers are usually those provided fast or younger users have some better strategies [1].

4.4   Profile Texts: Recognizing Themselves as Elderly

As we were processing profile texts, we noticed that elderly people describe themselves using age-related words. Below there are few examples from profiles of older adult users with reputation bigger than 1:

- *"I'm an old guy who likes programming, photography, chess, science fiction, and music.",*
- *"Old-ish IT Geezer, young at heart, memoir fanboy",*
- *"retired, learning python to help save what grey cells are left.".*
- *"Started programming in 1980 at SAC headquarters in Omaha, NE. I worked on 3D applications for B-52 and Cruise Missile mission planning. (...)",*
- *"Software Architect, aspiring writer. Programmer for well over 30 years, about 70 % of my working life.",*
- *"Grandfather, programmer, vegan."*

We clearly see, that their age in profiles is not fake, as they identify with old age in their profile texts. They use words like "old", "ancient", "grandfather", etc.

Survey: Different Habits - IDE. We were also looking for different habits related to programming among elderly users. There is significant difference in preferred IDE (Integrated Development Environment), as we can see in Table 5, elderly users prefer light IDE (light/white background with dark/black fonts), and younger prefers opposite.

Table 5. Preferred IDE (survey data)

	Seniors	Juniors
Dark	9.23 %	52.65 %
Light	64.62 %	7.61 %
Don't use IDE	26.15 %	39.76 %

This might be an important information for tool developers, but also significant example of differences between younger and elder programmers. This can be explained by medical aspects, considering sight problems, but also with their habits. There is a consensus saying that for elderly people there should be a high contrast in the interface and the best is the combination of black and white. However, if the background should be black or white and font opposite, there is less consent.

Some papers show that black background is better, like in [18]. Other say opposite like [5]. White background is also recommended in medical articles, like in [2]. However, we must remember that technology is changing rapidly, and a lot depends on screen type etc. In some cases this might be also a lack of ability to change IDE, but it is not likely for such group like programmers. We are not sure if they prefer it because of old habits or medical issues.

5   Conclusions

To sum up, we can say that elderly users constitutes an important part of the community and they feel more engaged in it. We had no possibility of measuring factors like well-being, but we can expect, according to similar cases in [16] and [6], that it positively influences it. They are recognized and awarded by reputation, their posts are up voted, they feel rewarded and important - and in fact, they are, because with their answers they build a portal (crowd sourcing way) that is used worldwide. Everyday programmers looks for answers in Stack Overflow and use their knowledge.

As computers, mobile technology and internet are becoming more and more popular, the next generations of older adults will be more willing to be a part of such online communities. It might be a great way of keeping the elderly people active.

Acknowledgements. This project has received funding from the European Unions Horizon 2020 research and innovation programme under the Marie Sklodowska-Curie grant agreement No. 690962.

References

1. Adamska, P., Juwin, M.: Study of the temporal-statistics-based reputation models for Q&A systems. Comput. Sci. **16**(3), 253 (2015)
2. Bauer, D., Cavonius, C.R.: Improving the legibility of visual display units through contrast reversal. Ergon. Aspects Vis. Display Terminals **39**, 137–142 (1980)
3. Berdychevsky, L., Nimrod, G.: "let's talk about sex": Discussions in seniors' online communities. J. Leisure Res. **47**(4), 467 (2015)
4. Brewer, R., Morris, M.R., Piper, A.M.: Why would anybody do this?: Understanding older adults' motivations and challenges in crowd work. In: Proceedings of the 2016 CHI Conference on Human Factors in Computing Systems, pp. 2246–2257. ACM (2016)
5. Carmien, S., Manzanares, A.G.: Elders using smartphones – a set of research based heuristic guidelines for designers. In: Stephanidis, C., Antona, M. (eds.) UAHCI 2014. LNCS, vol. 8514, pp. 26–37. Springer, Heidelberg (2014). doi:10.1007/978-3-319-07440-5_3
6. Dickinson, A., Gregor, P.: Gregor.: Computer use has no demonstrated impact on the well-being of older adults. Int. J. Hum. Comput. Stud. **64**(8), 744–753 (2006)
7. Hiyama, A., Nagai, Y., Hirose, M., Kobayashi, M., Takagi, H.: Question first: Passive interaction model for gathering experience and knowledge from the elderly. In: 2013 IEEE International Conference on Pervasive Computing and Communications Workshops (PERCOM Workshops), pp. 151–156. IEEE (2013)
8. Itoko, T., Arita, S., Kobayashi, M., Takagi, H.: Involving senior workers in crowdsourced proofreading. In: Stephanidis, C., Antona, M. (eds.) UAHCI 2014. LNCS, vol. 8515, pp. 106–117. Springer, Heidelberg (2014). doi:10.1007/978-3-319-07446-7_11
9. Kobayashi, M., Arita, S., Itoko, T., Saito, S., Takagi, H.: Motivating multi-generational crowd workers in social-purpose work. In: Proceedings of the 18th ACM Conference on Computer Supported Cooperative Work and Social Computing, pp. 1813–1824. ACM (2015)
10. Kobayashi, M., Ishihara, T., Itoko, T., Takagi, H., Asakawa, C.: Age-based task specialization for crowdsourced proofreading. In: Stephanidis, C., Antona, M. (eds.) UAHCI 2013. LNCS, vol. 8010, pp. 104–112. Springer, Heidelberg (2013). doi:10.1007/978-3-642-39191-0_12
11. McMullin, J.A., Comeau, T.D., Jovic, E.: Generational affinities and discourses of difference: a case study of highly skilled informationtechnology workers. Brit. J. Sociol. **58**(2), 297–316 (2007)
12. Morrison, P., Murphy-Hill, E.: Is programming knowledge related to age? An exploration of stack overflow. In: 2013 10th IEEE Working Conference on Mining Software Repositories (MSR), pp. 69–72, May 2013
13. Nimrod, G.: Seniors online communities: A quantitative content analysis. Gerontologist **50**(3), 382–392 (2010)

14. Nimrod, G.: Online communities as a resource in older adults tourism. J. Community Inf. **8**(1), 33–43 (2012)
15. Nimrod, G.: Probing the audience of seniors online communities. J. Gerontol. Ser. B: Psychol. Sci. Soc. Sci. **68**(5), 773–782 (2013)
16. Nimrod, G.: The benefits of and constraints to participation in seniors online communities. Leisure Stud. **33**(3), 247–266 (2014)
17. Robat, C.: The History of Computing Project TimeLine History of Computing. Technical report, Accessed 02/03/2006) (2006). http://www.thocp.net/timeline/timeline.htm
18. Slavicek, T., Balata, J., Mikovec, Z.: Designing mobile phone interface for active seniors: User study in Czech Republic. In: 2014 5th IEEE Conference on Cognitive Infocommunications (CogInfoCom), pp. 109–114, November 2014

Predicting Retweet Behavior
in Online Social Networks
Based on Locally Available Information

Guanchen Li[(✉)] and Wing Cheong Lau

The Chinese University of Hong Kong, Shatin, Hong Kong
leeguanchen@gmail.com, wclau@ie.cuhk.edu.hk

Abstract. Behavior prediction in online social networks (OSNs) has attracted lots of attention due to its vast applications. However, most previous work needs global network information to train classifiers. Due to the large data volume and privacy concern, it is infeasible to obtain global network information for every OSN. We propose a decentralized framework, named REPULSE, to predict whether a target user will retweet a message relayed by his friends. We also identify a new set of community-related features that improve retweet prediction accuracy considerably.

To demonstrate the value of community-related features, we propose another framework named HOTPIE to predict tweets popularity. Utilizing community-related features can boost the F1 score of popularity prediction from 0.43 to 0.55. To the best of our knowledge, this is the first work which systematically studies the impact of global vs. locally observable information on the prediction of retweet behavior in OSNs.

1 Introduction

Behavior prediction [17,29] in OSNs has attracted lots of attention. Most of the existing work requires the access to global network information using some APIs provided by the platform service providers. However, this is practically infeasible for most large-scale OSNs. Besides, with more and more people concerned about the privacy protection for OSN users, platform providers tend to restrict the scope of information accessible by third parties, including individual users and researchers. Such trends motivate us to investigate whether it is feasible to accurately predict message diffusion behavior using only locally observable information. In particular, we are interested in quantifying the trade-offs in prediction accuracy and the constraint of using only locally available information for prediction.

Towards this end, we propose a decentralized prediction framework named REtweet Prediction Using Localized information SolEly (REPULSE), and focus on the retweet prediction problem when only locally observable information is used. The problem considers the case when one of a target user's (the so-called ego node) friends retweets a new tweet. The objective is to predict whether the target user will forward that tweet too. The prediction is solely based on

© Springer International Publishing AG 2016
E. Spiro and Y.-Y. Ahn (Eds.): SocInfo 2016, Part II, LNCS 10047, pp. 97–115, 2016.
DOI: 10.1007/978-3-319-47874-6_8

locally observable information inside the ego network of the target user. Besides, to demonstrate the value of the newly identified information source, namely retweet-paths, we formulate another localized prediction problem and propose a framework named HOt Tweets Prediction by Individual Ego (HOTPIE). Here, the goal is to predict tweet popularity within a subgraph of the online social network. More specifically, when a message is forwarded by a node within the subgraph for the first time, the node will help to predict whether this message will become popular in the subgraph using only locally available information. In both prediction cases, our focus is to study the impact on prediction performance if only locally observable information can be used.

We take the features-based approach to tackle the two problems described above. Both problems are formulated as supervised learning problems, and a series of features is selected. Then for each problem, we train a Support Vector Machine (SVM) based classifier to make the prediction. In addition to using features extracted from the ego profile and content of tweets, we leverage a new source of information, namely the information diffusion paths [10] (also known as the retweet-paths), to boost prediction accuracy. One example of retweet is shown in Fig. 1a, and users in rectangles form the retweet-path. A retweet-path provides information on how a message is spread throughout the network. Figure 1b provides another view of a retweet-path. Each retweet-path observed by the target node is just a branch of the message's diffusion tree. Notice that the retweet-path information is embedded as part of the message header of every tweet in OSNs, such as Renren and Sina Weibo. As such, it is observable to the downstream users along the path. The retweet-paths are expected to contain some valuable features to improve the prediction performance. It is just like when a user follows a new friend, it is because he is interested in some content the friend is publishing. For users in the same retweet-path, they may share some common interests. Users interested in the same topic can be treated as if they belong to the same community, and thus, it becomes an overlapping community detection problem using retweet-paths information. We recast the overlapping community detection problem to the short text topic modeling problem. More details are presented in Sect. 4. It turns out that features extracted in this way can boost the performance in both prediction tasks considerably.

In summary, the technical contributions of this paper are as follows:

- We propose and implement two frameworks named REPULSE and HOTPIE, to predict retweet behavior using the locally observable information solely. Our findings indicate that it is feasible to predict retweet behavior without relying on global OSN information.
- We identify and leverage a new and valuable source of information, namely the retweet-paths. Features are extracted from the locally observable retweet-paths for our prediction tasks. Such information empowers us to improve the accuracy of retweet prediction even without analyzing the content of tweets.
- We provide a large-scale dataset obtained by crawling from the Renren OSN, one of the largest Facebook-like OSNs in China. The dataset contains 4,840,843 user profiles and the detailed activity records of 21,499 users including their tweet

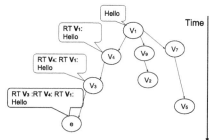

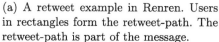

(a) A retweet example in Renren. Users in rectangles form the retweet-path. The retweet-path is part of the message.

(b) Retweet-path in the diffusion tree. The retweet-path observed by node e is: v_1, v_4, v_3, e.

Fig. 1. Illustration of the retweet-path.

history, content of each tweet as well as their friend lists. In terms of tweets, the dataset has 7,512,356 tweets and 1,807,152 retweets.

The rest of the paper is organized as follows. In Sect. 2, a literature review of related work is provided. Section 3 formulates the two prediction problems: Sect. 3.2 describes the problem of localized retweet prediction and proposes the REPULSE framework as a solution; Sect. 3.3 formulates another problem of localized popular tweets prediction and proposes the HOTPIE framework to tackle the problem. We tackle both problems using the features-based method. Section 4 discusses all features used in this paper and the corresponding methods to extract them. To compare the performance with HOTPIE and REPULSE, we define a baseline predictor using global information in Sect. 5. In Sect. 6, a series of extensive experiments is conducted to compare the prediction accuracy when global vs. locally observable information is used. We conclude our work and propose some future work in Sect. 7.

2   Related Work

Behavior prediction using user activity data in OSNs can be categorized to internal and external predictions. External prediction covers topics like earthquake prediction [9], stock market prediction [4,28] and election prediction [22] using online social network data. Internal prediction can be classified as macro-level and micro-level predictions.

Common objectives of macro-level prediction include tweet popularity and cascade size predictions. However, such predictions often do not consider how information got propagated in a hop-by-hop manner within the network. To perform the aforementioned predictive analytics, some works [2,7,11,12,15,21,25] use features-based methods, e.g. deploying Support Vector Machine or Decision Trees for classification after extracting a list of features. Others use sophisticated probabilistic models to conduct the prediction, like the Bayesian model in [26] and the self-exciting model in [29]. Most of algorithms developed in these works require global network information of the OSN. In contrast, our prediction of tweet popularity using HOTPIE as described in Sect. 3.3, is solely based on features extracted from the ego network. While [8] also tried to discover potential popular tweets using only locally observable information, their prediction performance is critically contingent on the availability of accurate estimates of critical parameters such as network centrality and community membership. Note that estimates of those critical parameters can not be easily obtained without the knowledge of the global network topology.

Micro-level prediction intends to answer questions like 'who should I pick as the next hop to pass the information so as to further propagate the information'. Likewise, the features-based methods can also be adopted, as done in [1,13,14, 16,17,19,23]. Since a micro-level prediction problem can also be formulated as a tweet recommendation problem, collaborative filtering can be utilized as in [6,27]. Nevertheless, the success of those algorithms still requires access to global information of the network. In contrast, our REPULSE framework described in Sect. 3.2 only relies on locally observable information extracted from the local neighborhood of a target node. For predicting whether a target user will retweet a message relayed by one of the target user's friends, REPULSE can achieve a reasonable prediction accuracy by using locally observable information only.

3   Problems Formulation

3.1   Terminologies and Definitions

In this paper, when a user posts a message onto his homepage, we also refer the message as a 'tweet'. When a user retweets a message, we use the word 'retweet', 'relay' and 'forward' interchangeably to describe this action and the resultant retweeted/ forwarded/ relayed message is referred as a 'retweet'.

Before proceeding to the problem formulation, the definition of 'ego network' needs to be clarified to avoid confusion. The ego network $G(e) = (V, E)$, as shown in Fig. 2a, contains the ego user and all his first-hop friends, as well as all the edges connecting those users. Notice that in Fig. 2a, those dotted edges are not part of the ego network. We study the prediction of retweet behavior using only locally observable information. Given an ego node e and its ego network $G(e)$, the locally observable information includes:

- The profile set $P_e = \{p_i \mid \forall i \in [1, |V|]\}$, where p_i is the profile information for user $v_i \in V$;
- The friend lists set $L_e = \{l_i \mid \forall i \in [1, |V|]\}$, where $l_i = \{f_1, f_2, f_3, ...\}$ is the friend list for user $v_i \in V$;
- The tweeting activity history set $TA_e = \{ta_i \mid \forall i \in [1, |V|]\}$, where $ta_i = \{tw_1, tw_2, tw_3, ...\}$ is a set of all the tweets posted by user $v_i \in V$. Notice that besides the tweet content, each tweet tw also contains meta information like the posting time and the tweet header which contains the retweet-path information.

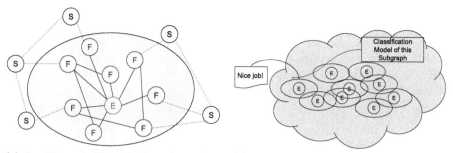

(a) Ego Network. Dotted edges do not be-   (b) Illustration of popular tweets predic-
long to the ego network.   tion scenario in a subgraph.

Fig. 2. Terminologies illustration

We use $I_e = \{P_e,\ L_e,\ TA_e\}$ to represent the locally observable information available to the ego node e.

In Renren, the retweet-path information is embedded in the tweet header and is locally observable to a downstream (ego) node. It is an important source of information used to boost the prediction performance. For each tweet tw relayed by the ego node e, tw contains the information indicating how the message is transmitted from the message author to the ego node e. This information is represented by a retweet-path $rp_{tw,e} = [v_1,\ v_2,\ v_3, ..., e]$. rp is a hop-by-hop ordered list, in which v_1 is the message author, and e is the ego node, and the rest are users who have retweeted the message before the ego node e did.

Since a lot of real world OSNs, e.g. Renren and Sina Weibo[1], actually embed the retweet-path of a tweet into its message header, it is possible for a downstream (ego) node to extract the retweet-path for every tweet that it receives.

3.2 REPULSE

With the increase volume of messages posted by each user's friends, it is necessary to effectively filter/prioritize incoming messages. Otherwise, the user will be overwhelmed by massive non-relevant messages and has very little chance to see messages he is interested in. To prioritize the incoming messages, we first need to identify messages that the user is interested in. Since the action of retweet is a strong indicator implying the user's interests in a particular message, we propose the REtweet Prediction Using Localized information SolEly (REPULSE) framework. The goal of REPULSE is to predict retweet behavior while using locally observable information only. Given an ego node e, when one of ego's friends retweets a message, we want to predict whether ego e will retweet that

---

[1] In Twitter, only a complete set of users who have retweeted the same message is shown, without disclosing the actual ordering. This set of forwarding users in Twitter aggregates information from different retweet-paths in the overall diffusion graph. Note that the set of forwarding users can serve the same purpose as retweet-paths do.

message while relying on information observable to e in its ego network only. We formulate the problem as follows:

Problem 1. Consider an ego node e and its corresponding ego network $G(e) = (V, E)$, as well as information I_e that is locally available to e in its ego network. We want to learn a prediction rule:

$$\psi_e : \mathcal{X} \to \mathcal{Y} \tag{1}$$

$\mathcal{X}$ is the feature space (the domain set), in which each sample from I_e is represented by a vector $\mathbf{x} = (f_1, \ f_2, \ ..., \ f_n)$ and n is the dimension of $\mathcal{X}$. $\mathcal{Y}$ is a two-element label set $\{-1, \ 1\}$, and $\exists$ friend $f_i \in V$ s.t. f_i has retweeted a message tw, then '1' means e also retweets tw after f_i did and '-1' means e does not retweet tw. The training set $S_{\psi_e} = \left((\mathbf{x}_1, y_1), (\mathbf{x}_2, y_2)... \right)$ is a list of pairs in $\mathcal{X} \times \mathcal{Y}$. We calculate the values of vector $\mathbf{x}$ using locally available information I_e only.

This localized retweet prediction can be treated as a supervised learning problem and tackled using classification algorithms. In particular, the features-based method is used and features are extracted from the locally available information I_e only. In Sect. 4, we discuss all the features used to construct the feature space $\mathcal{X}$.

3.3   HOTPIE

As we introduce a new set of community-related features, to demonstrate the power of those features, we also implement a popular message predictor using the framework named HOt Tweets Prediction by Individual Ego (HOTPIE). The popularity is measured by the number of retweets in the subgraph. Instead of predicting the exact number of retweets, we define several categories of popularity. This is because it is not essential to know the exact number of retweets if we want to identify the breaking news. Given a subgraph of OSN, when a tweet appears inside the subgraph for the first time via retweeting by an ego node, we intend to predict whether this tweet will become popular in the future by using information available in the ego network only. We formulate the problem as follows:

Problem 2. Consider a subgraph $G_{sub} = (V_G, E_G)$ in the social network, where V_G is the set of all the nodes and E_G is the set of all the edges in the subgraph. Given a tweet tw posted by someone outside G_{sub}, an ego node $e \in V_G$ is the first user to retweet tw in G_{sub}. Then by using $G(e)$ and I_e, we want to learn a prediction rule:

$$\rho : \mathcal{X} \to \mathcal{Y} \tag{2}$$

$\mathcal{X}$ is the feature space (the domain set), in which each sample is represented by a vector $\mathbf{x} = (f_1, \ f_2, \ ..., \ f_n)$ and n is the dimension of $\mathcal{X}$. $\mathcal{Y}$ is a three-element label set $\{0, \ 1, \ 2\}$, where '0' represents 'Normal', and '1' represents

'Popular', and '2' represents 'Super Popular'. Each category of popularity is measured by the number of retweets of tw in the subgraph G_{sub}. The training set $S_\rho = \left((\mathbf{x}_1, y_1), (\mathbf{x}_2, y_2)... \right)$ is a list of pairs in $\mathcal{X} \times \mathcal{Y}$. We calculate the value of vector $\mathbf{x}$ using only locally available information I_e.

We use Fig. 2b to illustrate the problem. The gray cloud is a subgraph of the whole social network, which consists of many ego networks. Then someone outside the subgraph posts a new tweet. Once an ego node e inside the subgraph retweets that tweet, by observing the behavior inside the ego network $G(e)$, we want to predict whether this tweet will become a popular tweet inside the subgraph. Notice that the classifier ρ is trained using all samples from the subgraph, but once the classifier is trained, each ego node can use this classifier with features extracted from the ego network to conduct the prediction. It is like a distributed monitor system, in which each ego node reports popular tweets once he identifies them. Compared to a centralized monitor which needs to collect network information throughout the subgraph, our prediction is distributed to each ego who is capable of predicting using the classifier independently.

Similarly, we formulate this prediction problem as a supervised learning problem and use all samples from the subgraph to train the classifier. However, features are extracted only using information available in the ego network. Namely we only use I_e to calculate the value of vector $\mathbf{x} = (f_1, f_2, ..., f_n)$. Once the classifier is trained, the prediction can be conducted by the ego node e using only features extracted from I_e.

3.4   Difference Between HOTPIE and REPULSE

Figure 3 shows the workflow of REPULSE. It has three modules: feature extraction, dataset rebalance and SVM learning. The workflow of HOTPIE is available in Appendix A. Similar to REPULSE, HOTPIE also consists of three modules and the techniques used in each module are also similar to those used in REPULSE. However, HOTPIE differs from REPULSE in two aspects:

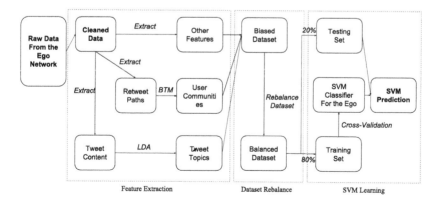

Fig. 3. Workflow of REPULSE

- Features are extracted solely from information available in the ego network, but then all samples from different ego networks are used to train a single classifier. However, REPULSE allows each ego to train its own classifier using features extracted using information available in the ego network.
- Only one single classifier for the entire subgraph is trained to predict popular tweets and each ego shares the same classifier. In contrast, the number of classifiers is equal to the number of egos when using REPULSE, and each ego uses its own classifier to predict local retweet behavior.

4   Extracting Features for the Classifiers

Selection of features is critical to the performance of classification algorithms for the supervised learning. Besides those features widely used by other work [18], we introduce the user community-related features by applying the topic modeling algorithm in retweet-paths. We consider five different families of features including: ego profile features, tweet metadata features, community-related features, tweet content-based features and friends feedback features. We provide the full list of features in Appendix C. For ego profile features and tweet metadata features, they are extracted directly from the ego profile and tweets correspondingly. Friends feedback features are extracted from friends feedback after the target user retweets a message, and are only used by HOTPIE. We apply the Latent Dirichlet Allocation algorithm to infer topics of each tweet and then extract tweet content-based features. For the community-related features, we recast the overlapping community detection problem to the topic modeling problem using retweet-paths. Details are provided next.

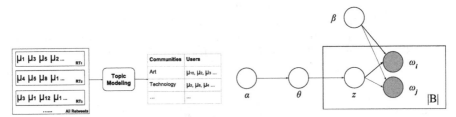

(a) Apply the topic modeling algorithm on retweet-paths corpus.   (b) Graphic representation of BTM.

Fig. 4. Use BTM to infer user's communities.

4.1   Community-Related Features

Different from Twitter and Facebook, Renren preserves a retweet-path inside the message header to record users who have retweeted this message. In other words, the downstream user knows how this message is relayed hop-by-hop from the message originator to him. The reason to exploit retweet-paths is that users who retweet the same tweet should share similar interests [5]. Although Twitter and Facebook have no such retweet-path, lists of users who have retweeted or liked a particular message can serve for the same purpose. The majority of online social network providers allow third parties to retrieve those lists by providing APIs (e.g. statuses/retweets/:id in Twitter). So it is straightforward to extract similar features for other platforms.

The retweet-path is not used to build the feature space in previous work. If two users always retweet the same set of tweets, then they must have lots of interests in common. Different from traditional social networks which are built upon personal friendships, in some 'open' OSNs, e.g. Sina Weibo, one's posting can be retweeted by a totally stranger who happens to visit the author's homepage in Sina Weibo. No explicit friendship needs to be established before one can retweet another's messages. As such, retweeting alone may not imply friendships. Two users several hops away in the friends network can be in the same retweet-path and could be classified into the same community if they co-occur frequently. Besides, we want to determine the community membership of each user using locally observable retweet-paths only. It becomes an overlapping community detection problem in OSNs using locally observable relationships solely. Towards this end, we recast the overlapping community detection problem to the topic modeling problem.

The topic modeling algorithms in the information retrieval field can be adopted here without changes. Figure 4a illustrates how we use the topic modeling algorithm to solve the community detection problem. RT represents the retweet-path and μ represents the user. When applying topic modeling on corpus with documents, the output is a probabilistic distribution over words for each topic. However, when applying topic modeling on retweet-paths, as shown in Fig. 4a, the output is a probabilistic distribution over users for each community. Each user is like a word, and each retweet-path with different users is like a bag of words document. Consequently topics discovered by the topic modeling algorithm are communities of users. For each community, we could have a probabilistic distribution over users, indicating the membership strength. In other words, we aim to learn a function

$$\sigma_e : \mathcal{R}_e \rightarrow \mathcal{C} \tag{3}$$

for the ego node e. $\mathcal{R}_e$ is the set of retweet-paths in tweet activity set TA_e. $\mathcal{C}$ is the community memberships for users in retweet-paths in $\mathcal{R}_e$.

However, there are some additional challenges to overcome due to the relatively short nature of each retweet-path. On average the length of a retweet-path in our dataset is 3 users. Most of existing topic modeling algorithms suffer from the sparsity of words co-occurrence patterns in the short length documents. We use Biterm Topic Modeling (BTM) [24] to detect communities due to its outstanding performance in short length documents. BTM is an extension of Latent Dirichlet Allocation [3] and overcomes the sparsity issue by maximizing the likelihood of the biterm set for the entire corpus. Figure 4b shows the graphic representation of BTM. α and β are the Dirichlet prior parameters. Topic z and words ω_i, ω_j are chosen from multinomial distributions. The probability of a biterm $b(\omega_i, \omega_j)$ is given by:

$$P(b) = \sum_z P(z)P(\omega_i|z)P(\omega_j|z) = \sum_z \theta_z \phi_{i|z} \phi_{j|z} \qquad (4)$$

And the likelihood of the entire biterm set B is:

$$P(B) = \prod_{(i,j)} \sum_z \theta_z \phi_{i|z} \phi_{j|z} \qquad (5)$$

BTM calculates the community memberships by maximizing the likelihood of the biterm set. Once each user is assigned with the community membership, we extract six community-related features.

5   Predictor Using Global Information

We also implement a baseline predictor using global information. In particular, compared to locally observable information, global information contains all the data we have collected, which is beyond the ego network. The baseline predictor differs from REPULSE/HOTPIE in two aspects.

First, values of community-related features and content-based features are different for the baseline predictor and REPULSE/HOTPIE. Using the baseline predictor, the corpus of BTM/LDA is all the tweets we have crawled, and the diffusion tree of each message serves as a document. However, the corpus is limited to all tweets inside the ego network when using REPULSE/HOTPIE and each retweet-path serves as a document. As a result, there is only one BTM/LDA model trained and shared by all users for the baseline predictor, but each user has its own BTM/LDA model when using REPULSE/HOTPIE.

Second, the number of classifiers is different. Using the baseline predictor, only one single classifier is trained using all samples from the dataset. However, for REPULSE, each ego node e has its own classifier ψ_e as defined in Eq. 1, which is trained using samples from I_e. Although HOTPIE has only one classifier, the prediction can be conducted using only locally observable information once the classifier is trained.

We refer the baseline predictor for REPULSE as RPuG (Retweet Prediction using Global information), and the baseline predictor for HOTPIE as PPuG (Popularity Prediction using Global information). The performance comparison is available in the next section.

6   Performance Evaluation

We use RESTful APIs provided by Renren to create the dataset. We conducted the crawling process from June 3, 2015 to September 26, 2015, and then again from December 21, 2015 to April 16, 2016. The dataset contains 4,840,843 user profiles and the detailed activity records of 21,499 users including their tweet history, content of each tweet as well as their friend lists. In terms of tweets, it contains 7,512,356 tweets and 1,807,152 retweets. Some users seldom retweet during the entire data collection period, so we decide to filter out those users with less than 5 retweets. This results in a dataset containing 15,699 egos with their first-hop neighborhood in Renren. We then use the dataset to evaluate the prediction performance of REPULSE and HOTPIE.

6.1   Localized Retweet Prediction

In this section, we study the performance of REPULSE in predicting whether a target user would retweet a message relayed by one of the target user's friends. Four feature families are used for this problem, including ego profile features, tweet metadata features, content-based features and community-related features. We use precision, recall, accuracy and F1 score as evaluation metrics.

Table 1. Retweet prediction performance with different feature sets.

Feature sets	Ego profile tweet metadata	Ego profile tweet metadata users communities	Ego profile tweet metadata tweet content	All features using RPuG	All features using REPULSE
Precision	0.7673	0.9240	0.7675	0.9229	0.8291
Recall	0.6750	0.8984	0.6757	0.9000	0.8288
Accuracy	0.7351	0.9123	0.7355	0.9124	0.8250
F1 score	0.7182	0.9110	0.7187	0.9113	0.8182

Impact of Different Feature Sets. Since REPULSE allows each user to train its own classifier, the performance difference may be due to user's own characteristics. To eliminate the bias introduced by each user, we use RPuG to measure the impact of different feature sets. As community-related features are first introduced in this paper, we also need to compare their effectiveness against other commonly used features such as tweet content-based features. Since ego profile features and tweet metadata features are widely studied, we focus on tweet content features and community-related features. Four experiments are conducted to assess the impact of different feature sets and the prediction results are shown in Table 1. By comparing the first column with the third one, we can find tweet content features do not help the prediction performance. In contrast, results in the first two columns demonstrate that by considering the community memberships of users involved in the retweet-path of a message, the accuracy of retweet prediction is boosted significantly. The insignificant difference between the prediction accuracy in the first column and the second column reconfirms the limited value of the tweet content-based features for classification. The lack of contribution of the content-based features may be explained as follows: On one hand, Chinese is different from English regarding the segmentation. English phrases and words can be separated by space, but that of Chinese will purely rely

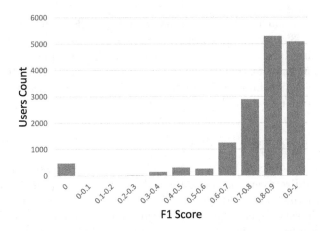

Fig. 5. Histogram of users with different F1 scores using REPULSE. The F1 score of RPuG is 0.91 and average F1 score of REPULSE is 0.82.

on the segmenter tools, which may not correctly retrieve all the words; On the other hand, the content of tweets may contain lots of noise since users have the complete flexibility to post any content, like emoji and notations. The influence of those noise on the topic modeling algorithm has not been well studied. As a result, we claim that by including community-related features, a reliable retweet prediction can be made even without analyzing the tweet content.

Performance Comparison of REPULSE and RPuG. As the F1 score is a weighted average of the precision and recall, we will only show the F1 score difference between RPuG and REPULSE. Values of other metrics are shown in the last two columns in Table 1. The average F1 score of all users using REPULSE is 0.818, compared to the F1 score of 0.911 achieved by RPuG. Some similar problems have been studied by others. Different from REPULSE, Petrovic *et al.* [17] only predicted if a tweet will be retweeted, without specifying which user will retweet. The highest F1 score achieved by Petrovic *et al.* is 0.466. Tang *et al.* [19] generated a list of target audience for promotion-oriented messages, but the F1 score is 0.8, which is less than that of RPuG and the average of REPULSE. Our work studies the retweet behavior for a particular user and each prediction involves the target user and an incoming tweet.

In this paper, we use a dummy 2-class random classifier, namely predicting to be either case with 0.5 probability, as the reference. As shown before, the F1 score of RPuG is 0.91, which is larger than the average F1 score of REPULSE by around 0.09. The confusion matrix of retweet prediction using REPULSE and RPuG is given in Table 2. We can see the prediction accuracy of the two classes is similar. For both classes, the prediction accuracy of REPULSE is lower than that of RPuG. Given that REPULSE only takes information available in the ego network to predict, the performance loss is still reasonable.

Figure 5 is the histogram showing the number of users with different F1 scores using REPULSE. The y-axis is the number of users and x-axis is the range of F1 scores. Notice that a 2-class random classifier can achieve an F1 score of 0.5. There is a reasonable amount of users with an F1 score less than that of a random classifier. Especially, there are 469 users with an F1 score of 0. Among those users, 163 users perform poorly due to the privacy settings of their friends. In those cases, the crawler can not access the homepages (which contain the tweets, activity logs as well as friend lists) of the friends of the target ego node. As a result, no negative sample is collected. Besides, 294 out of the remaining 306 users retweet less than 20 times within the observed data traces. Majority of those with an F1 score of 0 are not active users and the poor prediction performance on their retweet behavior is mainly due to the small number of training and testing samples available locally. However, there are still around 5,000 users whose F1 score is larger than 0.9, which is even better than that of using RPuG. Besides, there are 2,258 out of those 5,000 users with an F1 score of 1. By calculating the Pearson correlation of each ego profile feature and the F1 score, we find users associated with high F1 scores have a large number of retweets and retweet frequently from their friends. In other words, even with locally observable information, the prediction performance of REPULSE for some active users can be better than that of a classifier using global network information.

By now, our experiments have demonstrated that suffering from the absence of global network information, the retweet prediction performance will decline by around 0.09 measured by the F1 score when using REPULSE. However, if we purely rely on the locally observable information, prediction performance can still beat RPuG for some active users.

6.2   Localized Popular Tweets Prediction

We use the cascade size to measure tweets popularity. The cascade size of a tweet is defined as the number of retweets in the subgraph. The majority (96.5 %) have a cascade size less than 5, and only 0.07 % tweets have a cascade size larger than 49. We define three levels of popularity: 'Normal', with cascade size less than or

Table 2. Confusion matrices for REPULSE and RPuG

Ground Truth \ Prediction	User will retweet		User will not retweet		Accuracy Per Class	
	REPULSE	RPuG	REPULSE	RPuG	REPULSE	RPuG
User will retweet	296,100	334,737	47,012	26,694	86.30%	92.61%
User will not retweet	49,147	36,718	293,965	324,713	86.55%	89.84%

equal to 5; 'Popular', with cascade size larger than 5 and less than 50; 'Super Popular', with cascade size larger than or equal to 50.

We use four feature families to predict tweet popularity, including ego profile features, tweet metadata features, friends feedback features and community-related features.

Performance Comparison of HOTPIE and PPuG. In our experiment, the F1 score of HOTPIE is above 0.5 and the accuracy for the super popular tweets is 0.72. Some similar problems have been tackled, but HOTPIE differs from them in both the prediction performance and the information used to conduct the prediction. Zhao *et al.* [29] could predict 60 % of the top 100 popular tweets after observing 25 % of the total retweets. In contrast, HOTPIE identifies 72.31 % of the super popular tweets using locally observable information only. Although Hong *et al.* [11] could predict the most popular tweets with a high accuracy, the classifier is trained using all the retweets for each tweet. However, it is practically infeasible to collect all the retweets for each tweet in most OSNs.

Table 3. Performance of predicting popular tweets with different experiment settings.

	Random classifier	w/o community features	PPuG	HOTPIE	HOTPIE w/o friends feedback features
Micro F1	0.33	0.425	0.472	0.554	0.498
Macro F1	0.33	0.413	0.446	0.532	0.483

In this paper, a dummy random classifier is used as the comparison reference. The macro and micro F1 scores [20] are chosen as the evaluation metrics. We also conduct one experiment without community-related features to measure the impact of those features. The results are shown in Table 3, and corresponding confusion matrices are available in Appendix B. Different from REPULSE, HOTPIE outperforms PPuG by around 17 % in terms of the micro F1 score. It is probably due to the fact that global network information trains a classifier suitable to everyone, but locally observable information has customized information for each user, which is supposed to be able to characterize the user behavior more accurately. The last column in Table 3 also indicates by using friends feedback features, the F1 score is increased by around 10 %. Besides, the results also prove the power of community-related features, which have been shown to be able to improve the F1 score by 25 %.

7   Conclusion

In this paper, we systematically study the impact of global vs. locally observable information when predicting retweet behavior in OSNs. We propose a framework named REPULSE to predict whether a target user (the so-called ego node) will retweet a message relayed by one of his/her friends, while using only locally observable information. By comparing the performance with the classifier using global network information, we find that although the average performance of REPULSE is worse than that of the classifier using global network information, the prediction for some active users (around one third of users in the dataset) still outperforms the classifier using global network information. We also analyze the correlations between ego profile features and the prediction F1 score, and find that REPULSE can precisely predict the behavior of those who retweet from their friends frequently. Besides, we have identified a new locally available information source, namely the retweet-paths. By utilizing the topic modeling algorithm to solve the overlapping community detection problem, we can infer community membership for each user in the retweet-paths. Several features are extracted from the user community membership then. By adopting the newly proposed community-related features extracted from locally observable retweet-paths, the F1 score of retweet prediction is increased from 0.72 to 0.91. To demonstrate the power of aforementioned community-related features, we conduct another prediction experiment to identify popular tweets using the proposed framework HOTPIE. We find that the classifier trained using features extracted from locally available information outperforms the one trained using global network information. Besides, the use of community-related features improves the micro F1 score from 0.43 to 0.55 for the corresponding 3-class classification problem.

All the experiments in this paper are conducted using data crawled from Renren. Even though there is no retweet-path from Twitter or Facebook, it is relatively straightforward to implement REPULSE and HOTPIE for those platforms. The observation is that users in the same retweet-path share common interests in some topics. The list of users who have ever retweeted or liked a particular tweet, which is available in the majority of online social networks including Twitter and Facebook, can play the role of the retweet-path, albeit the lack of node ordering information as in the case of the retweet-paths in Renren. Deploying REPULSE and HOTPIE to other platforms is also part of our future work. Furthermore, both REPULSE and HOTPIE are still offline frameworks at present. However, due to their localized nature and the widespread use of the online SVM algorithm, it should be possible to implement the two frameworks to conduct real time prediction.

A   Workflow of HOTPIE

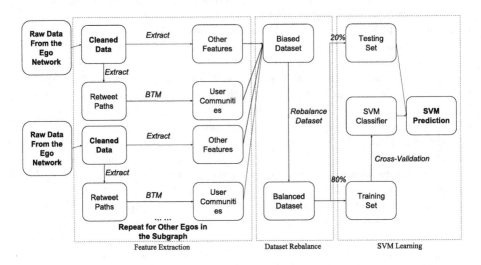

Fig. 6. Workflow of HOTPIE

B   Confusion Matrices of HOTPIE and PPuG

Table 4. Confusion matrix of using HOTPIE, with per class accuracy

Ground Truth \ Prediction	Normal	Popular	Super Popular	Accuracy Per Class
Normal	108	22	15	74.48%
Popular	70	37	38	25.52%
Super Popular	29	20	96	66.21%

Table 5. Confusion matrix of using PPuG, with per class accuracy

Ground Truth \ Prediction	Normal	Popular	Super Popular	Accuracy Per Class
Normal	85	22	28	62.96%
Popular	64	24	47	17.78%
Super Popular	30	23	82	60.74%

Table 6. Confusion matrix without community-related features, with per class accuracy

Prediction / Ground Truth	Normal	Popular	Super Popular	Accuracy Per Class
Normal	79	33	33	54.58%
Popular	59	31	55	21.38%
Super Popular	32	38	75	51.72%

C Full Feature List

Table 7. Feature names with feature IDs

Families of Features	Details of Each Feature Family	
Ego Profile Features	1. Number of tweets 2. Number of retweets 3. Number of friends 4. Number of retweets from friends 5. Maximum retweet-path length	6. Average retweet-path length 7-10. Number of tweets posted in the morning, afternoon, night and mid-night 11. Gender 12. Age
Metadata Features	13. Length of the retweet-path 14. Number of friends in the retweet-path 15. Posting time 16. Number of @	17. Number of friends being @ 18. Tweet type 19. Number of URLs 20. Tweet length
Community-related Features	21. Number of distinct comm-unities in the retweet-path 22. Number of shared commu-nities between the leaf user and the ego node 23. Number of users sharing at least one community with the ego in the retweet-path	24. Number of ego's communities 25. Number of shared communit-ies between each user in the path and the ego 26. Binary indicator showing if the ego is the first user to retweet
Content-based Features	27. Number of topics the tweet contains 28. Ids of the top-2 dominant topics	29. Strength of the dominant topic 30. Strength difference between the dominant topic and the most minor topic
Friends Feedback Features	31. Number of friends retweeting the same tweet 32. Degree of past interaction between friends and the ego 33. Average retweet time gap	

References

1. Artzi, Y., Pantel, P., Gamon, M.: Association for computational linguistics: human language technologies. In: Predicting responses to microblog posts. In: Proceedings of the 2012 Conference of the North American, pp. 602–606. Association for Computational Linguistics (2012)
2. Bandari, R., Asur, S., Huberman, B.A.: The pulse of news in social media: Forecasting popularity. arXiv preprint (2012). arXiv:1202.0332
3. Blei, D.M., Ng, A.Y., Jordan, M.I.: Latent dirichlet allocation. J. Mach. Learn. Res. **3**, 993–1022 (2003)

4. Bollen, J., Mao, H., Zeng, X.: Twitter mood predicts the stock market. J. Comput. Sci. **2**(1), 1–8 (2011)
5. Boyd, D., Golder, S., Lotan, G.: Tweet, tweet, retweet: conversational aspects of retweeting on twitter. In: 2010 43rd Hawaii International Conference on System Sciences (HICSS), pp. 1–10. IEEE (2010)
6. Chen, K., Chen, T., Zheng, G., Jin, O., Yao, E., Yu, Y.: Collaborative personalized tweet recommendation. In: Proceedings of the 35th International ACM SIGIR Conference on Research and Development in Information Retrieval, pp. 661–670. ACM (2012)
7. Cheng, J., Adamic, L., Dow, P.A., Kleinberg, J.M., Leskovec, J.: Can cascades be predicted? In: Proceedings of the 23rd International Conference on World Wide Web, pp. 925–936. ACM (2014)
8. Chesney, T.: Networked individuals predict a community wide outcome from their local information. Decis. Support Syst. **57**, 11–21 (2014)
9. Earle, P.S., Bowden, D.C., Guy, M.: Twitter earthquake detection: earthquake monitoring in a social world. Annal. Geophys. **54**(6), 708–715 (2012)
10. Guille, A., Hacid, H., Favre, C., Zighed, D.A.: Information diffusion in online social networks: a survey. ACM SIGMOD Rec. **42**(2), 17–28 (2013)
11. Hong, L., Dan, O., Davison, B.D.: Predicting popular messages in twitter. In: Proceedings of the 20th International Conference Companion on World Wide Web, pp. 57–58. ACM (2011)
12. Kupavskii, A., Ostroumova, L., Umnov, A., Usachev, S., Serdyukov, P., Gusev, G., Kustarev, A.: Prediction of retweet cascade size over time. In: Proceedings of the 21st ACM International Conference on Information and Knowledge Management, pp. 2335–2338. ACM (2012)
13. Lee, K., Mahmud, J., Chen, J., Zhou, M., Nichols, J.: Who will retweet this? Automatically identifying and engaging strangers on twitter to spread information. In: Proceedings of the 19th International Conference on Intelligent User Interfaces, pp. 247–256. ACM (2014)
14. Luo, Z., Osborne, M., Tang, J., Wang, T.: Who will retweet me? finding retweeters in twitter. In: Proceedings of the 36th International ACM SIGIR Conference on Research and Development in Information Retrieval, pp. 869–872. ACM (2013)
15. Ma, Z., Sun, A., Cong, G.: On predicting the popularity of newly emerging hashtags in twitter. J. Am. Soc. Inform. Sci. Technol. **64**(7), 1399–1410 (2013)
16. Naveed, N., Gottron, T., Kunegis, J., Alhadi, A.C.: Bad news travel fast: A content-based analysis of interestingness on twitter. In: Proceedings of the 3rd International Web Science Conference, p. 8. ACM (2011)
17. Petrovic, S., Osborne, M., Lavrenko, V.: Rt to win! predicting message propagation in twitter. In: ICWSM (2011)
18. Suh, B., Hong, L., Pirolli, P., Chi, E.H.: Want to be retweeted? large scale analytics on factors impacting retweet in twitter network. In: 2010 IEEE Second International Conference on Social Computing (socialcom), pp. 177–184. IEEE (2010)
19. Tang, L., Ni, Z., Xiong, H., Zhu, H.: Locating targets through mention in twitter. World Wide Web **18**(4), 1019–1049 (2015)
20. Tsoumakas, G., Katakis, I., Vlahavas, I.: Mining multi-label data. In: Data Mining and Knowledge Discovery Handbook, pp. 667–685. Springer (2009)
21. Tsur, O., Rappoport, A.: What's in a hashtag?: content based prediction of the spread of ideas in microblogging communities. In: Proceedings of the Fifth ACM International Conference on Web Search and Data Mining, pp. 643–652. ACM (2012)

22. Tumasjan, A., Sprenger, T.O., Sandner, P.G., Welpe, I.M.: Predicting elections with twitter: What 140 characters reveal about political sentiment. In: ICWSM 2010, pp. 178–185 (2010)
23. Uysal, I., Croft, W.B.: User oriented tweet ranking: a filtering approach to microblogs. In: Proceedings of the 20th ACM International Conference on Information and Knowledge Management, pp. 2261–2264. ACM (2011)
24. Yan, X., Guo, J., Lan, Y., Cheng, X.: A biterm topic model for short texts. In: Proceedings of the 22nd International Conference on World Wide Web, pp. 1445–1456. International World Wide Web Conferences Steering Committee (2013)
25. Yang, J., Counts, S.: Predicting the speed, scale, and range of information diffusion in twitter. In: ICWSM 2010, pp. 355–358 (2010)
26. Zaman, T., Fox, E.B., Bradlow, E.T., et al.: A bayesian approach for predicting the popularity of tweets. Annal. Appl. Stat. 8(3), 1583–1611 (2014)
27. Zaman, T.R., Herbrich, R., Van Gael, J., Stern, D.: Predicting information spreading in twitter. In: Workshop on Computational Social Science and the Wisdom of Crowds, Nips, vol. 104, pp. 17599–601. Citeseer (2010)
28. Zhang, X., Fuehres, H., Gloor, P.A.: Predicting stock market indicators through twitter i hope it is not as bad as i fear. Procedia-Soc. Behav. Sci. 26, 55–62 (2011)
29. Zhao, Q., Erdogdu, M.A., He, H.Y., Rajaraman, A., Leskovec, J.: Seismic: A self-exciting point process model for predicting tweet popularity. In: Proceedings of the 21th ACM SIGKDD International Conference on Knowledge Discovery and Data Mining, pp. 1513–1522. ACM (2015)

Social Influence: From Contagion to a Richer Causal Understanding

Dimitra Liotsiou$^{(\boxtimes)}$, Luc Moreau, and Susan Halford

University of Southampton, Southampton, UK
{dl1g13,l.moreau}@ecs.soton.ac.uk, susan.halford@soton.ac.uk

Abstract. A central problem in the analysis of observational data is inferring causal relationships - what are the underlying causes of the observed behaviors? With the recent proliferation of Big Data from online social networks, it has become important to determine to what extent social influence causes certain messages to 'go viral', and to what extent other causes also play a role. In this paper, we present a causal framework showing that social influence is confounded with personal similarity, traits of the focal item, and external circumstances. Combined with a set of qualitative considerations on the combination of these sources of causation, we show how this framework can enable investigators to systematically evaluate, strengthen and qualify causal claims about social influence, and we demonstrate its usefulness and versatility by applying it to a variety of common online social datasets.

Keywords: Social influence · Contagion · Causal inference · Graphical causal models · Confounding · Computational social science

1 Introduction: Social Influence and Confounded Causes Behind Observed Actions

Social influence has long been an important research topic in the social sciences. With the emergence of online social network platforms like Facebook and Twitter over the last decade, Big Data from social interactions has been produced at an unprecedented volume and detail, offering scientists new kinds of 'found' observational data through which to examine social processes. This has led to social influence becoming an increasingly prominent topic of study in the field of computer science, as well as to the birth of the interdisciplinary field of computational social science [33] for which methods need to be developed for systematically combining the social and the computational sciences [18,34,48].

Understanding social influence is pivotal since it has been claimed that social influence drives the spread of behaviors and attitudes as diverse as smoking, obesity, happiness, and political participation along social ties, in a process analogous to the contagious spread of viruses [2,5,17,29,31,36], to the extent that ensuring a select few trend-setting individuals (the so-called 'influentials') adopt a behavior would suffice to lead a large population to follow their example and

© Springer International Publishing AG 2016
E. Spiro and Y.-Y. Ahn (Eds.): SocInfo 2016, Part II, LNCS 10047, pp. 116–132, 2016.
DOI: 10.1007/978-3-319-47874-6_9

also adopt this behavior. If social influence does operate in this manner, then harnessing its power would bring immense benefits to marketing, public policy, and public health interventions.

This type of contagion-based paradigm for social influence has been extensively applied to theoretical and observational studies of online social networks like Twitter and Flickr [3,6,8,25,26]. Here, if user j's social connection i mentions the same entity as them (e.g. a URL or a hashtag), within a narrow time window, or if i re-shares or up-votes j's post, or chooses to follow j, or mentions j's username [14,24], then i's action is assumed to be due to social influence from j. One may say that such measures of online activity represent the levels of attention or interest that a given piece of content has generated [1,49]. However, beyond indicating some degree of attention, it is far from straightforward to infer the *meaning* or the *causes* behind such measures of observed actions, and indeed [3,6] recognize that this approach yields an overestimate of social influence. Moreover, it has been acknowledged that the ideal way to make causal claims in empirical settings is to use controlled experiments, but this can often be difficult or infeasible in practice [3,43,45,47].

The difficulty in estimating the extent of social influence from non experimental, observational data is that social influence is only one of many possible causes behind a pair of observed actions. Rather than social influence, there may be other unobserved common causes (called *confounders* [38]) behind two observed actions. These other, often unobserved, causes are commonly grouped into the classes of: similarity of personal traits, intrinsic properties of the focal item, and external circumstances [3,10,15,43,44]. The *focal item* might be a message, behavior, action, or some other item involved in the observable actions (*outcomes*) that the investigators want to study. Observationally determining that a cause of a given action is social influence rather than any one of the other causes, or a mix of many of these causes, is known to be a very difficult problem [3,4,6,43–45].

Therefore, we focus on the questions of why does a person (or a group of people) take a given observed action -what are the underlying causes and the mechanism that determine whether this person (or group) takes this action? If one were to intervene upon a causal factor, e.g. recruiting an 'influential' to endorse a product or healthy behavior on social media [6], what might be the reaction of people exposed to this? These are questions typical of *causal* inference [47].[1]

In this paper, we propose a causal framework for social influence, expanding on [43], and use it to show that social influence is confounded with causes related to personal similarity, traits of the focal item, and external circumstances. We then describe how this framework enables an investigator to systematically evaluate, improve and qualify causal claims on social influence versus each of the other types of possible causes, focusing on observational ('found') data from online social settings. This framework merges computational methods with causal assumptions rooted in social science findings, offering a promising way

---

[1] As opposed to inference based on *statistical* prediction methods [9,20–23,28,47], which have been used elsewhere in the literature (e.g. [11,16,40]).

to address the need for interdisciplinary common methodological ground in the nascent field of computational social science [18,33,34,48]. We limit our focus here to building this theoretical framework, and to performing an initial evaluation using previous studies of online datasets. A full empirical application to, and validation of the framework on, an online dataset that can adequately capture the confounding causes (typically left at least partly unobserved in online social datasets) is in our future work plans.

The rest of this paper is structured as follows: We first present the necessary background on social influence and the other three classes of possible causes. We then describe our methodology, which is based on graphical causal models, and in the following section apply it to the context of social influence, and show how graphical causal models both make the causal confounding visible and indicate how it can be removed to yield an unbiased estimate of social influence. Following this, we discuss some important qualitative and meaning-related aspects of social influence. We then demonstrate and evaluate how applying our causal framework to well known online social interaction settings enables one to assess the adequacy of the datasets and methods used, and to strengthen one's causal claims. We finally discuss possible directions for future work and present concluding remarks.

2   Social Influence and Other Classes of Causes

This section lays out the necessary background on social influence and the other possible causes behind observed actions, namely similarity of personal traits, intrinsic traits of the focal item and external circumstances. In all cases, we note that each factor may cause two people to exhibit the same observed behavior regardless of whether there is a social tie between them or not [10,30].

Social Influence. Social influence can be defined as the phenomenon where a person's behavior (action, opinion, or belief) is caused by another person's observed behavior: a person i may perform an action that person j performed earlier because j's performing of the action was so inspirational, persuasive, or impressive (e.g. j making a persuasive argument based on domain expertise) that i was convinced or became inclined to also perform it [30,43]. We only consider cases where i has free choice, i.e. j cannot force i to perform the given action. For instance, [35] defines influence as a form of causation, occurring in a possibly covert, unclear, or unintentional way, that does not involve force or coercion. Similarly in [39], a seminal work from the communications literature, the term social influence is used in the sense of a person causing another person to change their behavior, through the use of appropriate incentives.

Similarity of Personal Traits. Two people i and j may *independently* adopt the same behavior because they share one or more personal traits, such as interests, values, beliefs, opinions, needs, desires, personality profile, or demographic characteristics, like age, race, gender, social class [5,7,46]. For instance, two people may each independently post about political news on Twitter, because they each have an active interest in politics.

Intrinsic Properties of the Focal Item. In the social psychology, management, and marketing literature [10,32], it has been established that certain features can be 'engineered into' a *focal item* (e.g. a message, a product) that entice people to reshare it with others, making it 'go viral' and potentially increasing sales or adoption rates. An important type among them is features that invoke emotional arousal, specifically *activating* emotions such as excitement or anger, as these have been found to increase the chances that the viewer will then reshare, discuss or even adopt this message, behaviour or product. Hence, investigators should account for such relevant features, as well as other more general features (e.g. the price of a product; the effort or risk associated with a behavior [13]) that play a causal role in a person's reaction to a focal item.

External Circumstances. External circumstances may be the common cause why two people may *independently* take the same action, e.g. users i and j may post the same video or URL on social media because it relates to an important current news item, or a popular trend, that they both are aware of. External circumstances encompass factors from the external environment (e.g. a news item, a trend or a currently popular belief or attitude, a new law, a natural disaster), outside the personal traits of person i and j, and outside the traits of the focal item.

3   Methodology: Graphical Causal Models

It is an often-repeated cautionary phrase in statistics that 'correlation does not imply causation.' The field of causality theory, which saw rapid developments in the last thirty years, allows one to go beyond correlations and reason about causation in a rigorous, formal way, using tools like graphical causal models, which in turn are based on directed graphs and probability theory [38]. In this paper we will be using graphical causal models to reason about social influence versus the other possible causes of observed actions, expanding upon the work presented in [43]. We present the relevant theory here, based on [37,38,42].

A graphical causal model can be represented as a directed acyclic graph G, comprised of a set of nodes, N, and a set of directed edges, or arrows, E - that is, $G = \{N, E\}$. Nodes represent variables, and edges denote causal relationships. A directed edge from a node A to a node B denotes the *direct causal effect* of A on B, where A is a cause of B; A is called a *parent* or *ancestor* of B, and B a *child* or *descendant* of A. If a node has no arrow pointing to it, i.e. no parents, it is called *exogenous*, otherwise it is called *endogenous*. A *path* is a sequence of consecutive edges that do not all necessarily point in the same direction. Which nodes are connected to which depends on the modeller's causal assumptions, which should be well-justified and grounded in domain expertise [38]. The rules for manipulating graphical causal models then show what causal inferences can be made from these causal assumptions.

Graphical causal models are particularly useful for identifying latent (unobserved) variables that introduce *confounding bias* to the estimate of the causal effect of a variable X on another variable Y, and for then *adjusting for* those variables to obtain the unbiased causal effect of X on Y.

We illustrate this using the simple example causal model in Fig. 1, whose structure appears in our model of the causal effect of social influence and other factors on observed outcomes, as we shall see. Figure 1 represents a situation where the observed variable X is a cause of the observed variable Y, but variable U is a latent (unobserved) cause of both X and of Y. As causal graphs are governed by the Causal Markov Condition, whereby endogenous variables only depend on their parents [38], the joint probability distribution representing Fig. 1 is: $P(y, x, u) = P(u)P(x|u)P(y|x, u)$, where $P(w)$ is short for $P(W = w)$, since Y's parents are U and X, U is the parent of X, and U has no parents.

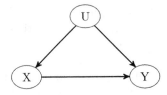

Fig. 1. Example graphical causal model

We want to estimate the causal effect of X on Y, which we write as $P(Y = y|do(X = x))$ in Pearl's do-notation, denoting the distribution of Y which would be generated, counterfactually, if X were set to the particular value x through experimental *manipulation* or *intervention*. In the causal graph this would mean deleting all arrows into X, setting X's value to x, and leaving the rest unchanged. This post-intervention distribution of Y is not in general the same as the ordinary conditional distribution $P(Y = y|X = x)$, as the latter represents taking the original, pre-intervention, population and *selecting* from it only the sub-population where $X = x$. The mechanisms that set X to that value may have also influenced Y through other channels, so the latter distribution would not typically really tell us what would happen if we externally manipulated X.

Figure 1 illustrates this point. If we consider the dependence of Y on X, in the form of the conditional $P(Y = y|X = x)$, we see there are two channels of information flow from cause to effect: one is the direct, causal path from X to Y, represented by $P(Y = y|do(X = x))$. However, there is also another, indirect path, between X and Y through their unobserved common cause U, where observing X gives information about its parent U, and U gives information about its child Y. If we just observe X and Y, we cannot distinguish the causal effect from the indirect inference -the causal effect is *confounded* with the indirect dependence between X and Y created by their common cause U. More generally, the effect of X on Y is confounded whenever $P(Y = y|do(X = x)) \neq P(Y = y|X = x)$. If there is a way to write $P(Y = y|do(X = x))$ in terms of distributions of observables, then we say that the confounding can be removed by an *identification, or deconfoundng, strategy*, which renders the causal effect *identifiable*.

Formally, to test whether there is confounding, we must first test whether some variables "*block*" (stop the flow of information or dependency along) all paths from X to Y, using the so-called *d-separation criterion* (as per [38]): A set of nodes Z *block* or *d-separate* a path p if and only if (i) p contains a *chain* $i \rightarrow m \rightarrow j$ or a *fork* $i \leftarrow m \rightarrow j$ such that the middle node m is in Z, or (ii) p contains a *collider* $i \rightarrow m \leftarrow j$ such that neither the middle node m, nor any of its descendants, are in Z. Then, a set Z d-separates X from Y if and only if Z blocks *every* path from X to Y. Further, a set of variables Z satisfies the *back-door criterion* (as per [38]) relative to X and Y if: (i) no node in Z is a descendant of X, and (ii) Z blocks every path between X and Y that contains an arrow into X. Then the set Z is called a *sufficient, admissible* or *deconfounding* set. Finding this deconfounding set permits the confounding bias to be removed, thus rendering the causal effect X on Y identifiable from non-experimental data, using the *back-door adjustment* formula [37,38]:

$$P(Y = y|do(X = x)) = \sum_z P(Y = y|X = x, Z = z)P(Z = z) \qquad (1)$$

Since the right-hand side of Eq. 1 contains only probabilities which are estimable (e.g. by regression) from our observational, non-experimental data, the causal effect of X on Y can be estimated from such data without bias.

In the example of Fig. 1, we see that variable U satisfies the back-door criterion, and hence, to obtain the direct causal effect of X on Y, one should simultaneously measure X, Y and U for every member of the randomly-selected sample under study, and then obtain the causal effect by using the back-door adjustment formula (Eq. 1) for $Z = \{U\}$.

In summary, to remove confounding and obtain the unbiased causal effect of X on Y, our *deconfounding strategy* is: (1) select a large random sample from the population of interest, (2) for every individual in the sample, measure X, Y, and all variables in Z, and (3) adjust for Z by partitioning the sample into groups that are homogeneous relative to Z, assess the effect of X on Y in each homogeneous group, and then average the results, as per Eq. 1.

4   Application to Social Influence: Confounding with Other Possible Causes

In this section, we use graphical causal models to reason about possible confounding of social influence with other causes, when working with observational data. We begin with the framework presented in [43], which we then simplify slightly without affecting its results with respect to confounding. We then adjust this framework such that it can model confounding even in the absence of a social tie. We next construct similar causal frameworks which show how social influence is confounded with personal traits, with intrinsic traits of the focal item, and with external circumstances. Finally, we put these separate models together into a single graphical causal model which shows the causal relations between causes and outcomes, and makes visible which variables should be measured and adjusted for to remove confounding.

4.1   Social Influence Is Confounded with Homophily

We begin by presenting the graphical causal model used in [43], which demonstrated that the phenomena of homophily (the tendency of people to form social ties with people similar to them) and of behaviour adoption due to social influence from friends are confounded in observational social network data. We follow their notation for continuity: Symbols X_k and Z_k denote sets of random variables representing, respectively, the unobserved and observed personal traits of person k. Each of those may be discrete or continuous, and is assumed to remain constant during the time period studied. $A_{k,l}$ is an observed variable, for simplicity in this case assumed to by binary, with value 1 if person k considers person l to be a 'friend', and with value 0 otherwise. $Y_{k,t}$ is an observed response variable, denoting whether person k performs action Y at a time t, and may be discrete or continuous. For simplicity, we assume time progresses in discrete steps (although this is not essential, as stated in [43]). It is also assumed that there is *latent* homophily in this system, hence whether two people are friends, i.e. whether $A_{i,j} = 1$, depends causally on their latent personality traits X_i and X_j. The model is shown in Fig. 2a.

We are interested in estimating social influence, i.e. the *direct causal effect* of person j's performing of action Y, $Y_{j,t-1}$, on person i's subsequent performing of the same action, $Y_{i,t}$, represented by the arrow $Y_{j,t-1} \rightarrow Y_{i,t}$: person i performs action Y because person j's example inspired them to do the same.[2] Homophily introduces a backdoor path between $Y_{i,t}$ and $Y_{i,t-1}$ through the latent X_i and X_j: $Y_{i,t} \leftarrow X_i \rightarrow A_{i,j} \leftarrow X_j \rightarrow Y_{j,t-1}$, i.e. the latent X_i and X_j are in the deconfounding set, thus social influence (the direct causal effect of $Y_{j,t-1}$ on $Y_{i,t}$) is confounded with homophily. So X_i and X_j should be measured and adjusted for, to retrieve the pure causal effect of $Y_{j,t-1}$ on $Y_{i,t}$.

Before we move on to apply this type of modeling to show how influence is confounded with other causes, we first simplify the model for ease of examination of paths and of manipulation. As [43] say, the assumption that $Y_{i,t-1}$ has a direct causal effect on $Y_{i,t}$ can be dropped without affecting the results of the investigation. Therefore, we remove $Y_{i,t-1}$, and, similarly for j, we remove $Y_{j,t}$. Since we are interested in examining the causes behind why i did Y at time t, $Y_{j,t}$ is not relevant.[3] In addition, since the observed personal traits Z_i and Z_j do

---

[2] In [43], it is assumed that one can be directly socially influenced only by those people she considers her 'friends' ($A_{i,j} = 1$), and not by anyone else.

[3] We note that $Y_{i,t-1}$ might represent a plausible and relevant kind of cause, e.g. that i does Y at time t because i did Y at $t-1$ and was happy with the results, or out of habit from having done it previously at time $t-1$. However, this previous happiness or habit may best be included in X_i as a variable representing an interest in Y.

not play a role in either introducing or removing confounding in this model or in our next models, we also remove those, and assume that all personality traits are unobserved, hence represented by X_i and X_j - indeed, usually there is no, or insufficient, data on users' personal traits in observational online social network studies. This simplification yields the model in Fig. 2b.

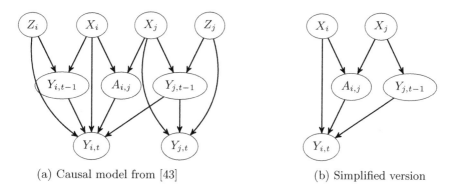

(a) Causal model from [43]   (b) Simplified version

Fig. 2. Graphical causal model from [43] (a), and simplified version (b)

4.2 Social Influence Is Confounded with Similarity in Personality Traits, Focal Item Traits, and External Circumstances

In this section, we present the graphical causal models that show how social influence is confounded with each of the following types of causes: similarity in personality traits, focal item traits, and external circumstances. We note that all confounding cases are due to structurally equivalent back-door paths of the form presented in Fig. 1 - each could essentially be regarded as a common cause: person-internal (personal traits), item-internal, or external.

Confounding with Similarity in Personality Traits. To show how a shared personality trait may be a cause behind i and j independently performing the same action Y, we now replace the previous latent personal trait variables X_i and X_j with W, representing the latent shared traits between i and j (i.e. W is the intersection of sets X_i and X_j), and W_i, i's remaining latent traits that j does not share, and respectively W_j for j's latent traits that i does not share. This produces the model of Fig. 3a, which shows that $Z = \{W\}$ is the deconfounding set on which to perform back-door adjustment.

Confounding with Traits of Focal Item. Similarly to Fig. 3a, b shows that variable F, representing the focal item traits, lies on a backdoor path $Y_{i,t} \leftarrow F \rightarrow Y_{j,t-1}$. Hence, the deconfounding set to be back-door adjusted is $Z = \{F\}$.

Confounding with External Circumstances. Similarly to Fig. 3b, in Fig. 3c variable U represents the external common cause (e.g. a shocking news item), and the back-door path $Y_{i,t} \leftarrow U \rightarrow Y_{j,t-1}$ introduces confounding. Hence $Z = \{U\}$ is the deconfounding set that should be back-door adjusted.

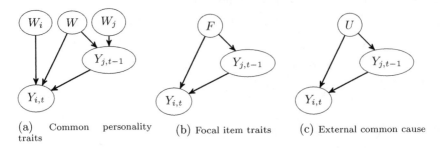

(a) Common personality traits

(b) Focal item traits

(c) External common cause

Fig. 3. Graphical causal models for social influence versus similarity in personality traits (a), focal item traits (b), and external circumstances (c)

4.3 Putting It All Together: Social Influence, Personal Similarity, Focal Item Traits, External Circumstances

We now put together all the above graphical causal models, to show the full picture of all causes that affect person i's decision to perform action Y at time t, and how these, if left unobserved and unadjusted for, introduce confounding bias into our estimate of social influence from person j's action Y at time $t-1$.

Keeping the same notation, we present two models, one without a social tie variable $A_{i,j}$ in Fig. 4a, and one with that tie in Fig. 4b. Given our split of personal traits into those that both people have in common (W) and those they do not (W_i, W_j), we assume that the decision to consider someone a 'friend' depends on having enough things in common (W), and also on not having too many differences in personality (e.g. to the extent that one cannot tolerate or is offended by the other's value system) - hence, besides W, we assume that W_i and W_j also causally affect whether a social tie will be fostered.

Therefore, the minimal deconfounding set for Fig. 4a is $Z = \{F, U, W\}$, and for Fig. 4b it is $Z' = \{F, U, W, W_j\}$[4]. Therefore, in order to retrieve the pure direct causal effect of $Y_{j,t-1}$ on $Y_{i,t}$, an investigator must implement our deconfounding strategy - crucially, all variables in the appropriate minimal deconfounding set must be measured for *every* individual in the random sample, and adjusted for as per Eq. 1.

We see that this full model presents a complex picture, with many factors playing a role in i's decision to take action Y. Indeed, as we shall discuss in the following two sections, it is known in the social sciences that social influence alone is seldom enough to ensure $Y_{i,t}$ - rather, a specific combination of social influence and all the other causal factors is needed.

---

[4] W_i could be in Z' but it is redundant, due to the assumed asymmetry of A_{ij}; if there was an edge $A_{ij} \rightarrow Y_{j,t-1}$ then W_i would have to be in the minimal confounding set.

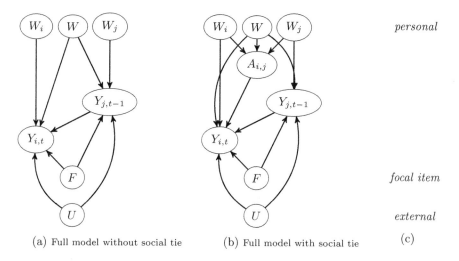

Fig. 4. Full graphical causal models for social influence versus other causes, without social ties (a), and with social ties (b), with the legend (c) on the right showing the context of each latent causal variable

5 The Impact of Causal Factor Characteristics on the Nature of Observed Outcomes

In this section, we aim to shed some further light on the question of what kinds of causal circumstances are needed for a person or group to take a given action. In the empirical and the theoretical literature [10,29,30,46,50] it has been widely acknowledged that no person is a clean slate, and no situation is 'neutral', therefore social influence does not operate in a vacuum, and on its own is rarely sufficient to ensure one or more individuals i take a specific action or commit to a new behavior $(Y_{i,t})$ (e.g. making some online content 'go viral', or a product sell out): a single well-connected person j alone is not enough to reliably influence others i to act a certain way; rather, a combination of compatible personal traits (W and W_j), a focal item with appropriate features F, and beneficial external conditions U are also needed.

Therefore, we next examine some important qualitative aspects of how different combinations of causal factors may lead to different qualities in the final observed outcomes. These qualitative aspects affect the extent and nature of claims one can make about social influence, and hence should be measured, e.g. by recording more details of the decision-making process than is common in observational social network datasets, or (e.g. to avoid making the process intrusive for participants) through interviews, or through a combination of methods.

Magnitude, Direction, and Duration. Instead of modeling the outcome Y_{it} as binary, it could instead have a magnitude, duration, and direction. The magnitude would represent the intensity of i's engagement with Y from time t onwards, whether this engagement is only superficial (small magnitude) or serious and incorporated into their value system (large magnitude), while duration would capture how short-lived or long-lasting this is [12,30]. The direction would capture whether i does the same as j with respect to Y (positive direction), or the opposite (negative), e.g. because j's way of engagement with Y was against i's values, or whether i does not take substantive action in relation to Y, e.g. out of loss of interest [2]. For instance, Facebook's addition of specific reaction buttons for love, anger, etc. to the Like button (which was previously used to express any type of reaction) [27], is one approach to capturing direction.

Normative versus Informational Social Influence. A person may change their behavior or take an action not because they find the traits of that behavior or action (F) inherently worthwhile, but rather because they want to please or feel accepted by someone they know (j, $A_{i,j}$) or by a wider social group (U). In [19], the former type is termed *informational influence*, and the latter *normative influence*, as discussed in [30]. Which type of social influence occurs in a given case depends on all the causal variables.

Generalizability of Observed Outcomes. Often, investigators use observational social media data capturing the levels of online interest in a product of behavior as proxies for estimating a different outcome like product sales or adoption of that behavior. However, it has been shown that the levels of interest on social media may not translate to actual purchases or behavior change [12] (e.g. the case of the popular Evian advert that did not increase sales, in [10]). That is because the causal factors in the two cases are very different: in the latter case, factors that do not apply in online discussions, like for instance the price, qualities, effort and/or risk associated with this product or behavior, F, and society's views of adopting it, U, come into play. Therefore, when using data from online social networks as proxies, the underlying causal factors should be adequately similar.

Changing Deep Rooted Behaviors: Identity, Effort and Risk. It has been claimed that social influence drives behaviors as diverse as sharing a message with friends, purchasing decisions, smoking habits, and happiness levels [10,17]. However, some behaviors (e.g. quitting or restarting smoking, or becoming happier) are much more deeply rooted in a person's identity, psychology or worldview (X_i plays a stronger role), are more difficult to change (F), and carry more risk in terms of social acceptance (U) [10,13], than other actions (e.g. re-sharing some information on social media, or choosing which brand of bottled water to buy).

6   Evaluation

In this section, we demonstrate how our causal framework and qualitative considerations might help investigators position their findings within the full causal picture for social influence, assess the extent and types of causal claims on influence their data allows them to make, and determine what causal variables should next be measured and adjusted for in order to make more robust causal claims. We examine examples of studies that actively try to capture causal effects of influence by reducing the effects of confounders, using quantitative and/or qualitative methodologies, in research settings involving one or more of the disciplines of sociology, social psychology, marketing, and computer science. We use our framework to examine how these studies lay out potential avenues, as well as expose caveats, for future attempts at measuring and adjusting for confounders and at capturing the qualitative aspects of social influence processes.

In [5], a controlled experiment on Facebook is performed, with the focal item being a Facebook app about films. It is randomized which friends i of j see messages $Y_{j,t-1}$ declaring j's use of this app, aiming to measure social influence versus susceptibility (i's tendencies to adapt to $Y_{j,t-1}$ by also downloading the focal item). It is assumed that randomly choosing the subjects i who will be exposed to $Y_{j,t-1}$ will suffice to control for homophily (similarity W among friends i and j linked through $A_{i,j}$) and for exposure to common external causes (U). Hence, it is assumed that whenever an exposed person i also downloads the app ($Y_{i,t}$) the only cause must be social influence ($Y_{j,t-1} \rightarrow Y_{i,t}$). However, we note that since the alternative causes have not been measured, they may continue to introduce confounding, despite the random selection - for instance, it might have been that all people who also downloaded the app did so because they themselves had an interest (W) in films, and all the people who did not download it did so because they had no interest in films. Therefore, the cause might rather have been a common personal trait W - we cannot know whether the cause was social influence or another cause, until we have measured and adjusted for the confounders for every person i in the sample.

Taking steps to observationally measure personality traits for each participant, [4] use an observational dataset containing many personal traits (X_i, X_j) for each pair of users, in an attempt to disentangle homophily from social influence. Still, as explained in [43], due to the methods used, there may still remain some latent personal similarity (W) which affects behavior adoption ($Y_{i,t}$). Moreover, we note that the confounders relating to the focal item traits (F), and to external common cause (U) remain latent. Still, this study shows a way to observationally measure X_i and X_j to some extent.

In an online randomized experiment, [41] manage to measure some confounders and obtain a relatively close estimate of the causal effect of aggregate social influence on users' choices of whether to download a song (focal item). It is randomized which users i are exposed to aggregate social influence (total number of downloads a song has received, $\sum_j Y_{j,t-1}$, where the identities of users j are not displayed). To reduce the effect of external common cause U, special care is taken (including conducting surveys) to ensure the displayed songs and artists are virtually unknown. The songs are kept the same (F constant) while some participant groups see the number of downloads for each song and other groups do not. However, as W has not been measured, and neither has F (e.g. song genre), a small possibility remains that the same song might have been dowloaded more in a social influence group than in a neutral one not because of social influence (from the displayed download count), but rather because that group contained more participants who were fans (W) of that song's genre (F). Therefore, some confounding due to latent W and F might remain, so these should be measured and adjusted for. Still, this study offers a good example of a significant and detailed effort to reduce U while experimentally controlling F.

In [45], observational data is used to study the causal effect of Amazon recommendations of the form 'Customers who bought this [product A] also bought [that product B]' on the views of product B (the focal item). Again, i cannot see the identities of customers j who bought both products. The investigators attempt to control for F to an extent, by studying many different product categories, and try to ensure that external causes U are held constant as much as possible. They also investigate the effect of the type of users i they have studied (X_i) on the causal effect of the recommendation. In qualitative terms, they recognize that a user's clicking on a recommendation might be due to convenience rather than the persuasive qualities of this particular recommendation. Overall, they caution that their results are still an upper bound for the causal effect of social influence, but a stricter one than under naive assumptions, and acknowledge that their results may not readily generalize to the average Amazon user, or to all Amazon product categories, or to other recommendation settings.

An example of how qualities of outcomes can be measured at a fine granularity and over time is presented in [2]. Here, the social influence from one participants' emotional state on another's (effect of $Y_{j,t-1}$ on $Y_{i,t}$), in the setting of face-to-face offline interactions, is measured using a mixed methodology of infrared sociometric sensors (badges) and questionnaires. The authors measure here many 'directions' of outcomes: not just mimetic (termed 'attraction'), but also neutral or negative (termed 'inertia, repulsion and push') at three points per day. They also measure participants' fixed personality traits X_i and X_j, but do not measure other confounders, and are careful to clarify that their social influence claims are correlational, not causal.

The offline controlled experiments in [30] offer useful examples of how to design experiments, control for some confounders, and use varied types of questionnaires, and how to measure the ways in which the combination of causal circumstances $(U, F, Y_{j,t-1})$ affect the nature of the resulting outcome $Y_{i,t}$. Here, the goal is to empirically evaluate how different combinations of causal circumstances (particularly $Y_{j,t-1}, U$) lead to different types of outcomes (termed 'compliance, identification and internalization'). Still, the broader external environment U (e.g. popular attitudes relevant to the topic of the focal message) and the participants' personal views (W) remain unmeasured and so may introduce confounding. Experimentally, the core of the argument (F) is kept the same, but the way it is framed $(Y_{j,t-1})$ is varied. To measure the 'magnitude' of the outcome, i.e. extent to which it was internalized and incorporated into i's worldview and value system, and its duration, questionnaires are used which include open-ended questions, both soon after exposure to $Y_{j,t-1}$ and some weeks after.

In summary, we have demonstrated how our causal framework and qualitative considerations can be used to help one position, assess and improve the claims they can make on social influence by ensuring they measure all relevant confounders as much as possible and adjust for them. To demonstrate how this might be achieved in practice, we have assessed the merits of practical attempts at reducing confounding and at accounting for qualitative aspects, both in observational and experimental settings, whether online or in mixed online-offline setups, covering quantitative and qualitative methods.

7   Conclusion and Future Work

Overall, we have proposed a methodological framework for assessing the causal effect of social influence, covering the space of other types of causes that may lead to an observed action (outcome), namely similarity of personal traits, traits of the focal item, and external circumstances. We have shown that social influence is confounded with each of these types of causes, using the formal rules of graphical causal models and based on robust causal assumptions about what types of causes might directly affect one's actions, which stem from well-established results from the social sciences literature. In merging computational rules with social science-based causal assumptions, this framework offers a promising interdisciplinary methodology of the type that is much-needed in computational social science. Drawing from social and computational disciplines, we then presented some important characteristics of the observed outcomes and the causal variables, which affect the nature, form and extent of the claims one can make on social influence. We then demonstrated how our causal framework and qualitative considerations may be applied in practice, by using them to evaluate the robustness of social influence estimates (how much confounding has been successfully adjusted for, how much still remains, and what qualitative aspects have been examined) from a set of diverse social influence studies from the social science and computer science literature that employed a varied range of practical methods.

As discussed, typical online social datasets do not adequately capture all relevant confounding causes. So, in future work, in order to make robust causal claims about social influence, we plan to apply our proposed framework to our own online dataset, taking care to obtain data that is detailed enough in capturing all relevant causes as much as possible. Further, it would be worth investigating how to harness social science expertise to devise systematic methods for identifying which specific causal variables for each type of cause are relevant in a given setting and should be measured, and how this may vary across different settings. Moreover, since the observed outcome (whose causes we aim to estimate) reflects a possibly subjective decision made by a specific person, we note that this person's choice and interpretation of relevant causes might differ from the investigator's, so it may be worth accounting for this potential difference using social science expertise (e.g. from social psychology).

References

1. Ackland, R.: Web social science: Concepts, data and tools for social scientists in the digital age. Sage, London (2013)
2. Alshamsi, A., Pianesi, F., Lepri, B., Pentland, A., Rahwan, I.: Beyond contagion: Reality mining reveals complex patterns of social influence. PloS One **10**(8), e0135740 (2015)
3. Anagnostopoulos, A., Kumar, R., Mahdian, M.: Influence and correlation in social networks. In: Proceedings of the 14th ACM SIGKDD International Conference on Knowledge Discovery and Data Mining, pp. 7–15. ACM (2008)
4. Aral, S., Muchnik, L., Sundararajan, A.: Distinguishing influence-based contagion from homophily-driven diffusion in dynamic networks. Proc. Nat. Acad. Sci. **106**(51), 21544–21549 (2009)
5. Aral, S., Walker, D.: Identifying influential and susceptible members of social networks. Science **337**(6092), 337–341 (2012)
6. Bakshy, E., Hofman, J.M., Mason, W.A., Watts, D.J.: Everyone's an influencer: quantifying influence on twitter. In: Proceedings of the fourth ACM international conference on Web search and data mining. pp. 65–74. ACM (2011)
7. Bakshy, E., Rosenn, I., Marlow, C., Adamic, L.: The role of social networks in information diffusion. In: Proceedings of the 21st International Conference on World Wide Web, pp. 519–528. ACM (2012)
8. Barbieri, N., Bonchi, F., Manco, G.: Influence-based network-oblivious community detection. In: 2013 IEEE 13th International Conference on Data Mining (ICDM), pp. 955–960. IEEE (2013)
9. Barnett, L., Barrett, A.B., Seth, A.K.: Granger causality and transfer entropy are equivalent for gaussian variables. Phys. Rev. Lett. **103**(23), 238701 (2009)
10. Berger, J.: Contagious: Why Things catch on. Simon and Schuster, New York (2013)
11. Borge-Holthoefer, J., Perra, N., Gonçalves, B., González-Bailón, S., Arenas, A., Moreno, Y., Vespignani, A.: The dynamics of information-driven coordination phenomena: A transfer entropy analysis. Sci. Adv. **2**(4), e1501158 (2016)
12. Cebrian, M., Rahwan, I., Pentland, A.S.: Beyond viral. Commun. ACM **59**(4), 36–39 (2016)
13. Centola, D., Macy, M.: Complex contagions and the weakness of long ties1. Am. J. Soc. **113**(3), 702–734 (2007)
14. Cha, M., Haddadi, H., Benevenuto, F., Gummadi, P.K.: Measuring user influence in twitter: The million follower fallacy. ICWSM **10**, 10–17 (2010)
15. Cheng, J., Adamic, L., Dow, P.A., Kleinberg, J.M., Leskovec, J.: Can cascades be predicted? In: Proceedings of the 23rd International Conference on World Wide Web, pp. 925–936. International World Wide Web Conferences Steering Committee (2014)
16. Chikhaoui, B., Chiazzaro, M., Wang, S.: A new granger causal model for influence evolution in dynamic social networks: The case of dblp. In: Twenty-Ninth AAAI Conference on Artificial Intelligence (2015)
17. Christakis, N.A., Fowler, J.H.: Social contagion theory: examining dynamic social networks and human behavior. Statist. Med. **32**(4), 556–577 (2013)
18. Counts, S., De Choudhury, M., Diesner, J., Gilbert, E., Gonzalez, M., Keegan, B., Naaman, M., Wallach, H.: Computational social science: Cscw in the social media Era. In: Proceedings of the Companion Publication of the 17th ACM Conference on Computer Supported Cooperative Work and Social Computing, pp. 105–108. ACM (2014)

19. Deutsch, M., Gerard, H.B.: A study of normative and informational social influences upon individual judgment. J. Abnorm. Soc. Psychol. **51**(3), 629 (1955)
20. Diebold, F.X.: Elements of forecasting. Citeseer, Ohio (1998)
21. Diebold, F.X.: Forecasting. Department of Economics, University of Pennsylvania (2015). http://www.ssc.upenn.edu/~fdiebold/Textbooks.html
22. Eichler, M.: Graphical modelling of multivariate time series. Probab. Theor. Relat. Fields **153**(1–2), 233–268 (2012)
23. Eichler, M.: Causal inference with multiple time series: principles and problems. Philos. Trans. Royal Soc. London A Math. Phys. Eng. Sci. **371**(1997), 20110613 (2013)
24. Ghosh, R., Lerman, K.: Predicting influential users in online social networks. In: Proceedings of KDD Workshop on Social Network Analysis (SNA-KDD), July 2010
25. González-Bailón, S., Borge-Holthoefer, J., Rivero, A., Moreno, Y.: The dynamics of protest recruitment through an online network. Sci. Rep. **1**, 197 (2011)
26. Goyal, A., Bonchi, F., Lakshmanan, L.V.: Learning influence probabilities in social networks. In: Proceedings of the Third ACM International Conference on Web Search and Data Mining, pp. 241–250. ACM (2010)
27. Greenberg, J.: Advertisers don't like facebook's reactions. They love them. WIRED (2016). http://www.wired.com/2016/02/advertisers-feel-facebooks-new-reactions-%F0%9F%98%8D/
28. Hlaváčková-Schindler, K., Paluš, M., Vejmelka, M., Bhattacharya, J.: Causality detection based on information-theoretic approaches in time series analysis. Phys. Rep. **441**(1), 1–46 (2007)
29. Katz, E., Lazarsfeld, P.F.: Personal Influence, The Part Played by People in the Flow of Mass Communications. The Free Press, New York (1955)
30. Kelman, H.C.: Processes of opinion change. Public Opin. Q. **25**(1), 57–78 (1961)
31. Kempe, David, Kleinberg, Jon, Tardos, Éva: Influential nodes in a diffusion model for social networks. In: Caires, Luís, Italiano, Giuseppe, F., Monteiro, Luís, Palamidessi, Catuscia, Yung, Moti (eds.) ICALP 2005. LNCS, vol. 3580, pp. 1127–1138. Springer, Heidelberg (2005). doi:10.1007/11523468_91
32. Kilduff, M., Chiaburu, D.S., Menges, J.I.: Strategic use of emotional intelligence in organizational settings: Exploring the dark side. Res. Organ. Behav. **30**, 129–152 (2010)
33. Lazer, D., Pentland, A., Adamic, L., Aral, S., Barabsi, A.L., Brewer, D., Christakis, N., Contractor, N., Fowler, J., Gutmann, M., Jebara, T., King, G., Macy, M., Roy, D., Van Alstyne, M.: Computational social science. Science **323**(5915), 721–723 (2009). http://www.sciencemag.org/content/323/5915/721.short
34. Mason, W., Vaughan, J.W., Wallach, H.: Computational social science and social computing. Mach. Learn. **95**(3), 257 (2014)
35. Morriss, P.: Power: A Philosophical Analysis. Manchester University Press, Manchester (1987)
36. Nickerson, D.W.: Is voting contagious? evidence from two field experiments. Am. Polit. Sci. Rev. **102**(01), 49–57 (2008)
37. Pearl, J.: Causal inference in statistics: An overview. Stat. Surv. **3**, 96–146 (2009)
38. Pearl, J.: Causality. Cambridge University Press, Cambridge (2009)
39. Rogers, E.M.: Diffusion of Innovations. Simon and Schuster, New York (2003)
40. Runge, J.: Quantifying information transfer and mediation along causal pathways in complex systems. Phys. Rev. E **92**(6), 62829 (2015)
41. Salganik, M.J., Dodds, P.S., Watts, D.J.: Experimental study of inequality and unpredictability in an artificial cultural market. Science **311**(5762), 854–856 (2006)

42. Shalizi, C.: Advanced Data Analysis from an Elementary Point of View. Cambridge University Press, New York (2013)
43. Shalizi, C.R., Thomas, A.C.: Homophily and contagion are generically confounded in observational social network studies. Sociol. Methods Res. **40**(2), 211–239 (2011)
44. Sharma, A., Cosley, D.: Distinguishing between personal preferences and social influence in online activity feeds. In: Proceedings of the 19th ACM Conference on Computer-Supported Cooperative Work and Social Computing, pp. 1091–1103. CSCW 2016, NY, USA (2016). http://doi.acm.org/10.1145/2818048.2819982
45. Sharma, A., Hofman, J.M., Watts, D.J.: Estimating the causal impact of recommendation systems from observational data. In: Proceedings of the Sixteenth ACM Conference on Economics and Computation, pp. 453–470. ACM (2015)
46. Sperber, D.: Explaining culture: A naturalistic approach. Cambridge University Press, New York (1996)
47. Spirtes, P.: Introduction to causal inference. J. Mach. Learn. Res. **11**(May), 1643–1662 (2010)
48. Wallach, H.: Computational social science: Toward a collaborative future. In: Computational Social Science: Discovery and Prediction (2016)
49. Watts, D.: Challenging the influentials hypothesis. WOMMA Measuring Word Mouth **3**(4), 201–211 (2007)
50. Watts, D.J.: Everything is obvious: * Once you know the answer. Crown Business (2011)

Influence Maximization on Complex Networks with Intrinsic Nodal Activation

Arun V. Sathanur[(✉)] and Mahantesh Halappanavar

Pacific Northwest National Laboratory, Richland, WA 99352, USA
arun.sathanur@pnnl.gov

Abstract. In many complex networked systems such as online social networks, at any given time, activity originates at certain nodes and subsequently spreads on the network through influence. Under such scenarios, influencer mining does not involve explicit seeding as in the case of viral marketing. Being an influencer necessitates creating content and disseminating the same to active followers who can then spread the same on the network. In this work, we present a simple probabilistic formulation that models such self-evolving systems where information diffusion occurs primarily because of the intrinsic activity of users and the spread of activity occurs due to influence. We provide an algorithm to mine for the influential seeds in such a scenario by modifying the well-known influence maximization framework with the independent cascade diffusion model. A small example is provided to illustrate how the incorporation of intrinsic and influenced activation mechanisms help us better model the influence dynamics in social networks. Following that, for a larger dataset, we compare the lists of influential users identified by the given formulation with a computationally efficient centrality metric derived from a linear probabilistic model that incorporates self activation.

Keywords: Complex networks · Influence maximization · Social influence · Self-activation · Centrality · Spectral methods

1 Introduction and Related Work

Diffusion of information in complex networks has been the subject of intense scrutiny for researchers and practitioners in many fields. A particular problem that has captured the attention is the identification of central or influential nodes on the network. One rigorous approach to finding influential users with motivations originating in viral marketing is the approach based on influence maximization. We can define the influence maximization problem as follows. Consider a directed graph $G = (V, E)$ that abstracts a complex network, where V is the set of vertices $V = \{v_1, v_2, v_3 \dots\}$ and E is the set of directed edges $\{(v_u, v_w) | v_u, v_w \in V\}$. Further, the vertices are labeled as either *Passive* or *Active* denoting the state of the vertex and a necessary but not sufficient condition for an active vertex v_u to activate a passive vertex v_w is that $(v_u, v_w) \in E$. The other conditions come from the nature of the local diffusion model that is also provided.

© Springer International Publishing AG 2016
E. Spiro and Y.-Y. Ahn (Eds.): SocInfo 2016, Part II, LNCS 10047, pp. 133–141, 2016.
DOI: 10.1007/978-3-319-47874-6_10

Given that there is budget to activate k vertices, the influence maximization problem aims to find that particular set of k *seed vertices* called the *seed set S* , that when activated results in maximal activations on the network amongst all possible such sets of k vertices.

Starting with the landmark paper by Kempe, Kleinberg and Tardos [7], several works have explored newer diffusion models and variations to the ones studied in the work by Kempe *et al.*, namely the independent cascade model and the linear threshold model. These models explicitly address one or more *sociological* aspects of influence. Li *et al.* in [11] consider influence dynamics and influence maximization under a general voter model with positive and negative edges. In a follow-up work Kempe *et al.* [8] discuss a diverse set of models including the so called decreasing cascade model where attempts by multiple neighbors to activate a node results in decreasing probabilities for activation, as the size of the set of neighbors trying to activate the node increases. The authors in ref. [14] propose a general diffusion model that takes into account different granularities of influence, namely pair-wise, local neighborhood etc. The authors in [2], consider influence maximization under the scenario where negative opinions may emerge and propagate. In [5], the authors consider the problem of identifying the individuals whose strong positive opinion about a product will maximize the overall positive opinion about the product. In the process, the authors leverage the social influence model proposed by Friedkin and Johnson [4].

Next we consider the models that address two different types of activation namely intrinsic and influenced. For example, in an online social network (OSN) these can refer to users posting content on their own and users retweeting or liking the posts respectively. Myers, Zhu and Leskovec investigate the diffusion of information, with origins external to that of a social network, through the internal social influence mechanism [12]. In a recent work [3] the authors recognize that the events on social media can be categorized as exogenous and endogenous and model the overall diffusion through a multivariate Hawke's process. While being similar in spirit to these works, our work is more geared towards mining influential nodes in scenarios with intrinsic and influenced activation.

We make the following contributions in this work. Our approach provides for probabilistically modeling the intrinsic and extrinsic activation mechanisms. We then examine these mechanisms in the context of influencer mining from *two different perspectives*, namely the well-known combinatorial influence maximization perspective and a generalized PageRank-type centrality perspective. Carefully chosen experiments on real-world-like and real-world graphs are used to illustrate the two perspectives.

2   Modified Influence Maximization Approach

Considering nodal activation to originate from two specific mechanisms, namely, *Intrinsic* and *Influenced*, allows us to effectively model the so-called *self-evolving* systems such as OSNs that are comprised of content creators (higher probability of getting activated intrinsically) and content consumers (activated via social influence and spreading the content). Recognizing that most of the users are in some sense content creators and content consumers at the same time, we introduce a real-valued parameter $\alpha \in [0, 1]$ that models the probability of self activation. The total probability for activation for a given node (user) i is a weighted sum of the probability for activation from the two different mechanisms. The parameters $\alpha(i)$ and $\beta(i)$ denote the probability of activation intrinsically and through influence respectively and we have $\alpha(i) \geq 0$, $\beta(i) \geq 0$. The influenced part of the probability for activation is then expressed as a weighted sum of the activation probabilities of the connected neighbors of the user under scrutiny. The w_{ij}s denote the probability of user j activating user i, given that user j is activated by either of the above means. These mechanisms and the associated coefficients are described in Fig. 1. The above probabilistic formulation is similar to the Friedkin-Johnson social influence model for opinion change [4] where the authors recognize that the dynamics of opinion change are governed by two mechanisms - the intrinsic opinion and the influenced opinion.

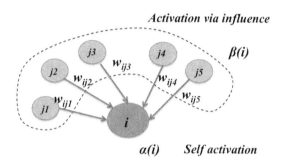

Fig. 1. A concise representation of the self and influenced mechanisms of activation of a node i

Note that all the model parameters discussed above can be efficiently determined as maximum likelihood estimates by observing the activity on the desired portion of the network. For example the proportion of tweets by a user i that are intrinsic in nature can quantify $\alpha(i)$ while a particular weight w_{ij} can be determined by the proportion of user i's retweets (or influenced activity) having their origin in the activity of user j that user i follows. While these *local influence models* can be determined in alternate ways, our focus is to find the overall influencers once these model parameters are estimated.

We propose a simple modification to the classic influence maximization framework using the greedy hill-climbing optimizer [7] working with the independent cascade (IC) model, to incorporate the self-activation mechanism. Let us assume that we are seeking k influential nodes out of a total of N nodes on the network. Let S^p be the set of influential nodes at step $p \leq k$. The greedy hill-climbing optimizer expands the set to size $(p + 1)$ by polling each of the vertices not in S^p and augmenting those vertices, one at a time to form the set

$S^p \cup v$ and looking for the best marginal gain in terms of the activations. At each such step p, instead of setting each of the nodes in $S^p \cup v$ to be activated and then computing the activations according to the IC model, we probabilistically activate each node in $S^p \cup v$ with a probability given by the corresponding α values to simulate the intrinsic activation process. This modification is depicted in Algorithm 1. Given the probabilistic nature of the algorithms, the overall activation numbers are obtained by running the diffusion model in a Monte Carlo fashion by invoking n independent trials involving randomized graphs with corresponding edge weights. We further accelerate the process of finding the influential nodes by parallelizing the Monte Carlo runs by leveraging *multi-threaded platforms*.

Algorithm 1. Selects a set of k influential nodes that cause maximal activations on a network, following the independent cascade(IC) model with self-activation (SA). The inputs are a directed graph $(G = (V, E))$, set of edge weights $(P = \{p_{uv} : (uv) \in E\})$, vector of alpha values $(\alpha = \{\alpha_v : v \in V\})$, number of samples (n), and number of seeds to be identified (k).

1: **procedure** IC-SA(G, P, α, k, n)
2:     Generate n random numbers $r_{uv}^1 \ldots r_{uv}^n$ for each edge in E and generate a set SG containing n subgraphs such that in subgraph i, $p_{uv} \geq r_{uv}^i$
3:     $S \leftarrow \emptyset$ ▷ Set of influential nodes to be mined
4:     **while** $|S| < k$ **do**
5:         $v_{best} \leftarrow \emptyset, a_{best} \leftarrow 0$
6:         **for** each node v in $V \setminus S$ **do**
7:             $a \leftarrow 0$
8:             **for** each $G_i \in SG$ in **parallel do**
9:                 $\hat{S} \leftarrow$ active nodes in $S \cup \{v\}$ based on α
10:                Compute number of nodes, $\hat{a}$, in $V \setminus \hat{S}$ that are reachable from the $\hat{S}$
11:                $a \leftarrow a + \hat{a}$ ▷ Synchronized update
12:            **if** $a \geq a_{best}$ **then**
13:                $v_{best} \leftarrow v$
14:                $a_{best} \leftarrow a$
15:        **if** $v_{best} \neq \emptyset$ **then**
16:            $S \leftarrow S \cup \{v_{best}\}$
17:     **return** S

We also adopt the weighted-cascade method for normalizing the edge probabilities [7]. Thus if $\boldsymbol{W}$ denotes the sparse weight matrix that characterizes the IC edge probabilities, we require that $\boldsymbol{W}$ be row-stochastic. That is $\sum_{j,(j,i)\in\mathcal{E}} w_{ij} = 1$. Further by assuming that the nodes are not *lazy* and are activated by either of the two mechanisms that we outline, we set $\beta(i) = (1 - \alpha(i))$. This will render the overall IC probability between nodes j and i to be $(1 - \alpha(i))w_{ij}$. The assumption that $(\alpha(i) + \beta(i)) = 1$ is being relaxed in the ongoing work where we allow $(\alpha(i) + \beta(i)) \leq 1$ thereby modeling the slack with a *lazinesss* factor.

3   Experiments

3.1   Small Organization Tree

We first consider a small directed and weighted network with 23 nodes, organized in a tree-like fashion. The graph is depicted on the left side of Fig. 2. In this experiment, we consider the tree-like network to depict a small organization with a Director (Node D), two Managers (M1 and M2) and twenty Employess (E1-E20), with 10 employees each working under the two managers. We set $\alpha^0(D) = 0.95$ signifying that the Director almost exclusively acts intrinsically. We also set $\alpha^0(M1) = \alpha^0(M2) = 0.25$. All the employees have an α of 0.25 as well. As for the weights, the edges ending at node D receive weights of 0.5 each (when the Director chooses to be influenced, the director gets influenced equally by the two managers). As for the managers, they have a weight of 0.5 each on the edges that are incoming from D and the remaining 0.5 is split equally among the edges originating at the 10 employees each. The employees carry a weight of 1.0 on the edges originating from their managers. We then perturb this baseline case to mimic a situation where the director starts becoming more susceptible to influence while the manager $M1$ starts becoming inflexible. This is done by setting $\alpha(D) = \alpha^0(D) - \delta$ and $\alpha(M1) = \alpha^0(M1) + \delta$. We then sweep δ from 0.05 to 0.45. The results are shown in the right panel of Fig. 2 where we can see that D starts out as the most influential node as expected, but then M1 becomes more influential than D at a certain value of δ and will eventually have reach over all the employees on the network. Note that the activation numbers plotted on the y-axis are the cumulative activations over all the $n = 3200$ Monte Carlo samples. This simple experiment shows that influence on social networks is sensitive to the extent of intrinsic activation and clearly such scenarios cannot be easily captured by the traditional influence maximization framework.

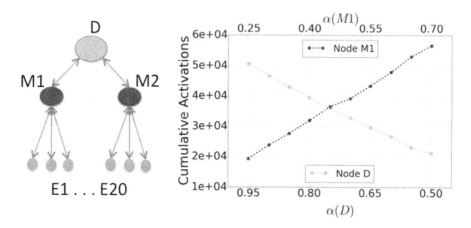

Fig. 2. The small organizational tree network (left) and the behavior of the influence functions with the various α values.

3.2   Larger Graphs and the Influence Function

Our dataset of larger graphs consists of

- LFR-1000 graph with 1000 vertices and 11433 edges is a synthetic network that follows the generative LFR model that mimics real-worlsd graphs [10] .
- The PBlogs graph [1] that represents a real-world blogs network and has 1095 vertices and 12597 edges.

The details of these graphs are discussed in [6]. Visualizations of the two larger graphs are shown in the inset in Fig. 3.

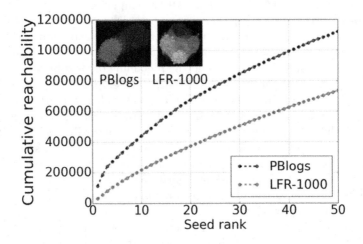

Fig. 3. Submodular nature of the influence function under self-activation. *Inset: The PBlogs (left) and LFR-1000 (right) networks visualized in Gephi*

For the classic influence maximization problem with the independent-cascade model, the greedy hill-climbing optimizer is shown to be optimal in the sense that it provides a $(1 - \frac{1}{e} - \epsilon)$ approximation guarantee because the expected influence spread is a sub-modular function. For the case of intrinsic nodal activation, we have not been able to prove the sub-modularity of the influence function. While leaving the formal proof as an open problem, we conjecture that the sub-modularity holds because for each seed l selected, we can introduce an edge pointing from a dummy node to the seed l with an activation probability equal to $\alpha(l)$ and seed the dummy node. This process does not interfere with the normal operation of the network except activating the seed l with a probability $\alpha(l)$, exactly as required and therefore preserves the sub-modularity of the influence function. When we applied the modified influence maximization approach given by Algorithm 1 to the LFR-1000 and the PBlogs graphs and requested for 50 seeds, we observed from Fig. 3 that the cumulative influence spread (total number of activations from all the samples considered) showed a sub-modular character as evidenced by the diminishing gains [9] in the total number of activations for each seed added.

4   Comparison with an Equivalent PageRank-Type Influence Measure Based on a Linear Model

We examine the influencer mining on networks with intrinsic and influenced nodal activations from a slightly different perspective. By collecting the various probabilities together and recognizing the recursive nature of influence spread on a social network, we arrive at a generalized, computationally efficient, PageRank type spectral influence measure that incorporates the two activation mechanisms [13]. As demonstrated in [13], when considering activity on an OSN, this approach is a better measure of influence spread than a purely topological metric such as PageRank.

For a given node i, from Fig. 1, the total probability of activation $p_A^T(i)$ can be written as

$$p_A^T(i) = \alpha(i) + \beta(i) \left(1 - \prod_{j,(j,i) \in \mathcal{E}} \left(1 - w_{ij} p_A^T(j) \right) \right) \tag{1}$$

By retaining the leading-order terms, we get a linearized version of Eq. 1 as

$$p_A^T(i) = \alpha(i) + \beta(i) \sum_{j,(j,i) \in \mathcal{E}} w_{ij} p_A^T(j) \tag{2}$$

$p_A^T(i)$ denotes the total probability of activation for node i (intrinsic and influenced). The parameters $\alpha(i)$ and $\beta(i)$ denote the weights for activation intrinsically and through influence respectively as before and we set $\beta(i) = (1 - \alpha(i))$ as before. We now extend Eq. 2 to the entire network with N nodes to obtain a matrix-vector equation.

$$\boldsymbol{p_A^T} = \boldsymbol{\alpha}\mathbf{1} + ((\boldsymbol{I} - \boldsymbol{\alpha})\, \boldsymbol{W})\, \boldsymbol{p_A^T} \tag{3}$$

Here $\boldsymbol{p_A^T}$ is a vector of size $N \times 1$ denoting respectively the total probability of activation for all the nodes on the network. $\boldsymbol{I}$ denotes the identity matrix of size $N \times N$. $\boldsymbol{\alpha}$ denotes the diagonal matrix with entries corresponding to the intrinsic activation probability for all the nodes on the network. $\boldsymbol{W}$ denotes the sparse, stochastic weight matrix with entires given by the w_{ij}s discussed earlier. $\mathbf{1}$ is the all-ones vector of size $N \times 1$

We can then express the total activation probabilities as

$$\boldsymbol{p_A^T} = \mathbf{1}^T \boldsymbol{G}; \boldsymbol{G} = (\boldsymbol{I} - (\boldsymbol{I} - \boldsymbol{\alpha})\boldsymbol{W}))^{-1} \boldsymbol{\alpha} \tag{4}$$

Here $\mathbf{1}^T$ is utilized to give us the column-sum of $\boldsymbol{G}$. We also note that because the matrix $\boldsymbol{W}$ is a row-stochastic matrix, the matrix $\boldsymbol{G}$ is also row-stochastic. The quantity $C_A(i) = \left(\sum_{j=1}^N G_{ji} \right)$ which corresponds to the sum of the entries in column i of $\boldsymbol{G}$, represents the expected number of hosts activated by node i and is a measure of influence. We term this *amplification factor* as *activation centrality* and it can be directly computed as a linear solve.

Table 1. Correlations, two ways, between the proposed approaches for the two inputs PBlogs and LFR1000 for different sizes of seed sets (10, 20, 30, and 50). Closer the metric to one the better.

Correlation type	Input	$k = 10$	$k = 20$	$k = 30$	$k = 50$
Jaccard	PBlogs	0.538	0.818	0.875	0.818
RBO	PBlogs	0.817	0.846	0.851	0.868
Jaccard	LFR1000	0.818	0.905	0.765	0.818
RBO	LFR1000	0.979	0.963	0.947	0.937

In our experiments, the α and W entries were randomized with entries drawn from the uniform distribution over $[0, 1]$ and W was converted to a row-stochastic matrix. We then compare the sets of top-k influencers identified by both the methods on two larger graphs in our dataset. The comparison was carried out with respect to two measures - Jaccard similarity and the rank-biased overlap (RBO) that also considers ordering with higher weights given to matches that happen at the top [15]. These results are presented in Table 1 where we see excellent agreement between the sets of influential nodes obtained by both the methods. Thus the activation centrality metric which includes the *intrinsic activation* mechanism represents a computationally more viable alternative to the full-scale influence maximization framework, retaining the essence of the model and being amenable to a sparse matrix based linear solve.

5   Conclusions and Ongoing Work

In this short paper we introduced the notion of vertices on a social graph getting activated by two mechanisms, namely, intrinsically and through social influence as commonly observed in online social networks. Utilizing a modified version of the influence maximization framework that combines the self-activation probability with the independent cascade model for influenced activation, we were able to find the influential nodes on such a network. We then introduced a spectral centrality measure of influence that takes into account, the intrinsic and influenced activation mechanisms and demonstrated that the sets of influential users identified by the two mechanisms agree very well. Building on this preliminary work, ongoing work is exploring multiple facets of this problem including the exploration of how a social network can be successful in the long run by balancing the two modes of activation. We are also extending these methods other complex systems such as for attack modeling in cyber networks.

Acknowledgement. This research was supported by the High Performance Data Analytics program at the Pacific Northwest National Laboratory (PNNL). PNNL is a multi- program national laboratory operated by Battelle Memorial Institute for the US Department of Energy under DE-AC06-76RLO1830.

References

1. Adamic, L.A., Glance, N.: The political blogosphere and the 2004 us election: divided they blog. In: Proceedings of the 3rd International Workshop on Link Discovery, pp. 36–43. ACM (2005)
2. Chen, W., Collins, A., Cummings, R., Ke, T., Liu, Z., Rincon, D., Sun, X., Wang, Y., Wei, W., Yuan, Y.: Influence maximization in social networks when negative opinions may emerge and propagate. In: SIAM Data Mining, pp. 379–390 (2011)
3. Farajtabar, M., Du, N., Gomez-Rodriguez, M., Valera, I., Zha, H., Song, L.: Shaping social activity by incentivizing users. In: Advances in Neural Information Processing Systems, pp. 2474–2482 (2014)
4. Friedkin, N.E., Johnsen, E.C.: Social influence networks and opinion change. Ad. Group Proces. **16**(1), 1–29 (1999)
5. Gionis, A., Terzi, E., Tsaparas, P.: Opinion maximization in social networks. In: SIAM Data Mining Conference, pp. 387–395. SIAM (2013)
6. Halappanavar, M., Sathanur, A., Nandi, A.: Accelerating the mining of influential nodes in complex networks through community detection. In: Proceedings of the 13th ACM International Conference on Computing Frontiers, CF 2016, Como, Italy, May 16–18, 2016
7. Kempe, D., Kleinberg, J., Tardos, E.: Maximizing the spread of influence through a social network. In: Proceedings of ACM SIGKDD, pp. 137–146. ACM, New York (2003)
8. Kempe, D., Kleinberg, J., Tardos, É.: Influential nodes in a diffusion model for social networks. In: Caires, L., Italiano, G.F., Monteiro, L., Palamidessi, C., Yung, M. (eds.) ICALP 2005. LNCS, vol. 3580, pp. 1127–1138. Springer, Heidelberg (2005). doi:10.1007/11523468_91
9. Krause, A., Golovin, D.: Submodular function maximization. Tractability Pract. Approaches Hard Probl. **3**(19), 8 (2012)
10. Lancichinetti, A., Fortunato, S.: Benchmarks for testing community detection algorithms on directed and weighted graphs with overlapping communities. Phys. Rev. E **80**(1), 016118 (2009)
11. Li, Y., Chen, W., Wang, Y., Zhang, Z.L.: Influence diffusion dynamics and influence maximization in social networks with friend and foe relationships. In: Proceedings of the Sixth ACM International Conference on Web Search and Data Mining, pp. 657–666. ACM (2013)
12. Myers, S.A., Zhu, C., Leskovec, J.: Information diffusion and external influence in networks. In: ACM SIGKDD, pp. 33–41. ACM (2012)
13. Sathanur, A.V., Jandhyala, V., Xing, C.: Physense: Scalable sociological interaction models for influence estimation on online social networks. In: IEEE International Conference on Intelligence and Security Informatics, pp. 358–363. IEEE (2013)
14. Srivastava, A., Chelmis, C., Prasanna, V.K.: Influence in social networks: A unified model? In: 2014 IEEE/ACM International Conference on Advances in Social Networks Analysis and Mining (ASONAM), pp. 451–454. IEEE (2014)
15. Webber, W., Moffat, A., Zobel, J.: A similarity measure for indefinite rankings. ACM Trans. Inf. Syst. (TOIS) **28**(4), 20 (2010)

Applicability of Sequence Analysis Methods in Analyzing Peer-Production Systems: A Case Study in Wikidata

To Tu Cuong[(✉)] and Claudia Müller-Birn

Freie Universität Berlin, Berlin, Germany
to.cuong@fu-berlin.de, clmb@inf.fu-berlin.de

Abstract. Building a shared understanding of a specific area of interest is of increasing importance in today's information-centric world. A shared understanding of a domain can be realized by building a structured knowledge base about it collaboratively. Our research is driven by the goal to understand participation patterns over time in collaborative knowledge building efforts. Consequently, we focus our study on one representative project – Wikidata. Wikidata is a free, structured knowledge base that provides structured data to Wikipedia and other Wikimedia projects. This paper builds upon previous research, where we identified six common participation patterns, i.e. roles, in Wikidata. In the research presented here, we study the applicability of sequence analysis methods by analyzing the dynamics in users' participation patterns. The sequence analysis is judged by its ability to answer three questions: (i) "Are there any preferable role transitions in Wikidata?"; (ii) "What are the dominant dynamic participation patterns?"; (iii) "Are users who join earlier more turbulent contributors?" Our data set includes participation patterns of about 20,000 users in each month from October 2012 to October 2014. We show that sequence analysis methods are able to infer interesting role transitions in Wikidata, find dominant dynamic participation patterns, and make statistical inferences. Finally, we also discuss the significance of these results with respect to the understanding of the participation process in Wikidata.

Keywords: Sequence analysis · Peer-production system · User behavior · Wikidata

1 Introduction

With the explosion of information on the Web, the effective access to and use of this information enables the development of intelligent applications. One approach for a more effective use of available information is to make it understandable to humans and machines alike. A prerequisite for enabling this understanding is to create a shared conceptualization in a domain of interest, i.e. ranging from a structured vocabulary to an ontology, that is commonly agreed to by all participants.

© Springer International Publishing AG 2016
E. Spiro and Y.-Y. Ahn (Eds.): SocInfo 2016, Part II, LNCS 10047, pp. 142–156, 2016.
DOI: 10.1007/978-3-319-47874-6_11

In order to create such a shared conceptualization, different approaches exist that are translated into software. Collaborative Protegé [12], for example, extends existing features of a more top-down approach for knowledge modeling as opposed to Semantic Media Wiki [8], which extends existing features of the Wiki software. In our study, we focus on Wikidata[1]. This is an open community that allows easier curating of structured data that can be presented as raw facts within the Wikipedia ecosystem [14]. This leads to a higher consistency and data quality across the various Wikipedia language versions. Currently, Wikidata consists of more than 19 million data *items*. Each of them contains multiple *statements* about their characteristics (cf. example in Fig. 1). *Sitelinks* connect each item to Wikipedia articles in the different language versions.

Users can contribute to Wikidata by adding, editing, or removing *statements, properties*, etc. Thus, an online community emerged around Wikidata. Participation in online communities is not evenly distributed – people often start with a peripheral level of participation. As their experience grows, it appears that their contribution and experience will often rise in tandem [16]. A layered model describes this development. The different layers represent distinct roles in the community. Ye and Kishida [17] describe the traveling through these layers as "role transformation". Preece and Shneiderman [11] generalize this line of research into the *Reader-to-Leader Framework*. The connecting element in all these approaches is that roles allow some conclusions to be drawn from users' activities. These activities represent users' participation patterns.

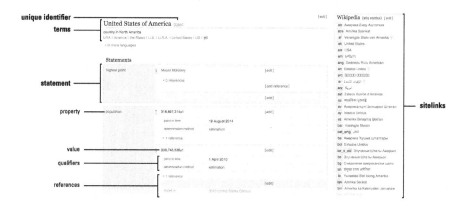

Fig. 1. Item page of Wikidata (Q30) "United States of America" showing the different concepts.

Such participation patterns constitute the participatory architecture of a community. If community members change the regularity and type of their involvement, they also change the social dynamics within the community. Kittur et al. [7], for example, show that both the amount and type of participation in

---

[1] http://wikidata.org.

Wikipedia appear to evolve over time. At the beginning of the Wikipedia project, for example, most of the editing work was carried out in the main namespace, which contains all encyclopedic articles. Over the course of the project, the focus shifted to other namespaces, for example the community namespace.

In our research we wanted to understand how users actually participate in Wikidata. Since knowledge representation has been carried out mainly by experts in the past, we wanted to learn about existing participation patterns in this community and how the modeling of knowledge is organized within the community. In previous research, we determined existing participation patterns in Wikidata [10]. Based on Wikidata's edit history, we described the contributing behavior of users by identifying overlapping and varying areas of activity, for example, editing a sitelink or adding an statement. Based on a k-means method we identified six mutually exclusive clusters, i.e. participation patterns. We called them *Reference Editor, Item Creator, Item Editor, Item Expert, Property Editor,* and *Property Engineer.* The ordering of these activity patterns corresponds to their proximity to the knowledge modeling task. Adding a sitelink, as *Reference Editors* do, does not, for example, require any knowledge about describing knowledge. As opposed to *Property Engineers*, who are mainly involved in the creation of new properties[2] that are used to describe items by statements. *Item Creators* mostly focus on creating items. These items are derived from existing Wikipedia pages. The *Item Editors* primarily edit labels and descriptions of items. *Item Experts* focus on describing characteristics of items by defining statements. Finally, *Property Editors* again edit labels and description on properties. By comparing these participation patterns on one page on Wikidata (cp. Fig. 1), it can be seen that participation is quite focussed on one specific area of the knowledge modeling process.

However, our previous research allows us only to describe users' behavior only from a static perspective. It omits the transition between participation patterns and, therefore, the community dynamics. This motivates us to study the dynamics in the community participation processes. A dynamic perspective allows us to understand how a user's behavior change over time and how they might get a better understanding of knowledge representation by taking on a more complex tasks. The overarching goal of this research is to learn from these transition processes and support their activities by means of softwares. The analysis of the dynamics of user participation is guided by the following questions: (i) "Are there any preferable role transitions in Wikidata?"; (ii) "What are the dominant dynamic participation patterns?"; and (iii) "Are users who join earlier more turbulent contributors?"

Our paper is organized as follows. Section 2 surveys selected researches from the area of online communities. Section 3 describes our research methodology. We notice that the sequential nature of users' participation patterns lends itself naturally to sequence analysis methods, which are popular with social scientist

---

[2] Wikidata is organized into namespaces. Each namespace contains a specific kind of artifacts. Property namespace, for example, groups all Wikidata's pages about properties whereas the main namespace comprises all the items.

studying life trajectories [1]. We present our results in Sect. 4. We then discuss their significance in Sect. 5. Finally, Sect. 6 concludes our study together with discussion about future works.

2   Related Work

Existing studies about online communities have focused traditionally on using techniques from visualization together with statistical methods to analyze event logs. Most of them ignore the temporal dynamics of events within online communities, for example, the order of edits on a Wiki page. Welser et al. [15] use visualization to identify potential social roles in online discussion groups and confirm them by regression techniques. Once again, the temporal dimension of discussion is ignored. Later, Gleave et al. [5] define the concept of "social role" that begins with the structural foundation in commonalities of behaviors. The role of a mother, for example, starts with playing, protecting, and caring for her children. Moreover, authors argue that one needs to combine structural analysis and content analysis to identify social roles accurately in online communities. Again, the authors do not take into account the temporal dynamics of social interactions.

Viegas et al. [13] use history flow to track the evolution of Wikipedia pages. By providing a quick overview of edit history for each page, one can, for example, track who contributes the most or detect edit patterns such as edit wars, i.e. users keep deleting each other's contribution.

Recently, Keegan et al. [6] argued that sequence analysis methods developed from biology and sociology [1,9] lend themselves naturally to problems in computer-supported cooperative work (CSCW) due to the temporal-ordered nature of event logs, which are popular in CSCW. Their approach is similar to ours. The difference is one of granularity – we apply sequence analysis methods not to event logs, but to social roles of users within Wikidata.

3   Methodology

In our study, we observed about 20,000 active users on Wikidata from October 2012 to October 2014 on Wikidata. We recorded their participation pattern in each month. This results in a data set of about 20,000 participation pattern sequences. We borrowed quantitative techniques for life trajectories analysis from social scientists to gain further insights.

Social scientists are interested in life trajectories, such as occupational histories or professional careers [1,3]. Their typical research questions include "Do people's life stages obey social norms?", "How do a group of people develop in terms of social advancement over time?", or "Why do some people tend to have more chaotic life trajectories?". We transfered this idea into the Wikidata context.

Life trajectories are represented using state sequences, where each position in a sequence is in a state. Exemplarily, a life trajectory could include attending school, doing an apprenticeship, or having a full-time job. Life trajectories are also called categorical time series. By viewing participation patterns as states, we converted users' temporal-ordered participation patterns into a data set of state sequences. This enabled us to apply sequence analysis methods to the study of Wikidata users' dynamic behaviors. We use the two terms *participation pattern* and *state* interchangeably from now on.

3.1 Modeling Dynamic Participation Patterns as State Sequences

In our study, we modeled participation patterns, i.e. user roles, as sequence states. These roles were derived from the monthly editing patterns of users[3]. They are *Item Creator, Item Expert, Item Editor, Reference Editor, Property Editor*, and *Property Engineer*.

There were also two decisions that we made when converting participation pattern sequences into categorical time series. Firstly, there was an issue of misalignment of time spans. Users start and end contributing edits at different time. Given a sequence that starts after October 2012, we treated the previous months's participation patterns as "missing" ones, i.e. unobserved ones. This resulted in a set of state sequences that had each of them starting at the same time, i.e., October 2012, and possibly ending in different months.

Secondly, some users contribute edits for a few months, disappear for months, and then become active again. We treated this gap of inactive months as missing patterns.

3.2 State Sequence and Its Characteristic Measurements

In this section, we define formally what a state sequence is and some of its characteristic measurements, which are useful in explaining our results in Sect. 4.

A *state sequence* is a temporal-ordered list of states. The time unit could be one hour, one month, or another length of time. The set of possible states S, or the alphabet, are predefined.

The *entropy of a state distribution*, at a position in time, is defined as

$$h(p_1 \ldots p_{|S|}) = -\sum_{i=1}^{|S|} p_i \log p_i$$

where p_i is the proportion of the ith state, i.e., participation patterns. This entropy can be seen as a measure of the diversity of states at a given time. The lower the entropy, the lower the state diversity is. While the entropy of state distribution is concerned with state uncertainty at a given position over all sequences, *within-sequence entropy* measures the same value, but with respect to individual sequences.

---

[3] The monthly period of time has been chosen to determine stable patterns of participation with fewer fluctuations than seen in weekly periods of time.

The *within-sequence entropy* is defined for each state sequence as

$$h(\pi_i \ldots \pi_{|S|}) = -\sum_{i=1}^{|S|} \pi_i \log \pi_i$$

where π_i is the proportion of occurrences of the ith state in a particular sequence.

The *transition rate* from state s_i to state s_j is computed as the ratio between the number of its instances over the total number of transitions.

There are several measurements of dissimilarity (distance) between two state sequences. Among those, *Optimal Matching* (OM) [3], also known as the Levenshtein distance, is a popular choice. OM is defined as the minimal cost in terms of transformation operations to transform one state sequence into another. The transformation operations include substitution, deletion, and insertion. The cost per operation depends on the application domains. Since OM allows deletion and insertion, we use OM so as to measure the distance between two state sequences of unequal length.

4   Results

In the following, we present our results from applying sequence analysis methods to investigate existing role transitions, determine dominant participation patterns, and the stability of these patterns depending on their joining time. We will discuss our results in more details in the subsequent section.

4.1   Are There Any Preferable Role Transitions in Wikidata?

We computed transition rates among states to understand how users' participation patterns change between consecutive months. Knowing the transition rates allowed us to identify participation patterns that have a strong relationship with each other.

We showed the initial states on y-axis in Fig. 2. These states represent the six participation patterns with the seventh state ("*") representing users who showed no participation in a month. The x-axis shows in each case the subsequent states. The color scale is from white (low value) to red (high value). One can notice that the diagonal transition rates are often relatively high compared to others. A high value of the diagonal elements describes users who tended to maintain their participation patterns over two consecutive months. The most stable pattern in this regard is, independently from the unobserved participation pattern, the *Reference Editor* where almost 70 percent of the users stayed in this pattern. Except for *Property Engineers* and the *Property Editors*, users in the other roles are likely to become inactive in the next months. Moreover, *Item Creators* also have the tendency to become *Reference Editor*. This make sense since when a user creates an item on Wikidata, it is highly possible that he/she also adds a reference link back to where the information of the item is provided,

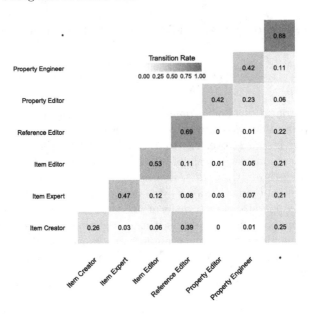

Fig. 2. The heat map of transition rates among participation atterns. The color scale is from white (low value) to red (hight value). "*" represents an unobserved participation pattern, i.e. no participation. (Color figure online)

for example, a Wikipedia article. Additionally, there is a 23 % chance that *Property Editors* become *Property Experts* in the following months indicating that these two patterns are closely connected.

While transition rates provides a good overview of the strength of participation patterns' relationship, it has one limit. It only provides information about two consecutive time steps, i.e. months in our study. We would also like to study the composition of participation patterns over time.

We plotted the traversal state distribution (Fig. 3), which is the visualization of state distribution at each time step, to address the limitation of transition rates. We can see that in the first five months *Item Editors* dominate the community while *Reference Editors* dominate in the following months. Moreover, the proportions of *Item Editors*, *Property Engineers*, and *Item Experts* remain stable over time. This can be explained by the fact that users focus on creating the classes at the beginning. Later on, new users concentrated more on populating classes' properties.

4.2   What Are the Dominant Dynamic Participation Patterns?

We next investigated whether there were any dominant dynamic participation patterns. In other words, is there any typical development of participation patterns over time? We computed the set of distinct representative participation patterns whose coverage is greater than a threshold θ, i.e. the percentage of

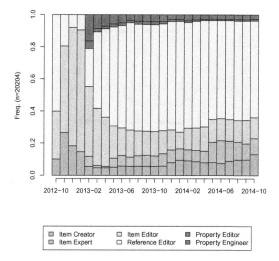

Fig. 3. Transversal state distribution of Wikidata categorical sequence data. The *Item Editor* role was popular at the beginning before taken over by being the *Reference Editor* role afterwards.

state sequences that are in their neighborhood. A participation pattern is in a neighborhood of a set of representative participation patterns if its distance to any of them is less than a preset threshold θ [4].

There is a trade-off between θ and the percentage of participation patterns that are assigned to these sets. Ideally we would like to minimize the number of representative sets (i.e. the number of clusters) while maximizing θ. We experimented with different θ values and found that $\theta = 0.6$ is a suitable threshold. Particularly, 63.7 % of all participation patterns are assigned to representative sets. We also computed representative sets with $\theta = 0.8$ for comparison.

The results are shown in Fig. 4. The upper graphic (Fig. 4a) shows the sets of representative participation patterns with a θ of 0.6 and the lower graphic (Fig. 4b) shows the sets with $\theta = 0.8$, respectively. Each graphic consists of two charts. The distance axis is in the top chart. Here, the theoretical maximum between two participation patterns in our data set is 50. The four signs, $\triangle$, $\diamond$, $\square$, $\circ$, represents the four representative participation patterns. In the bottom chart, the representative participation patterns are visualized as horizontal bars. They are plotted bottom-up according to their representativeness score with their width proportional to the number of participation patterns assigned to them. In Fig. 4a, for example, the bar, with a $\diamond$ sign next to it, represents a 25-month long participation pattern whose first 23 months are in unobserved states and the last two months are in the *Item Editor* and *Item Expert* pattern, respectively. Finally, on the left-hand side, we have the information regarding the number of representative participation patterns and the total number of participation patterns (users). In Fig. 4b, these numbers would be 4 and 20,204, respectively.

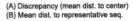

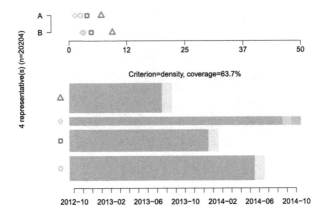

(a) 63.7% of users participation patterns are similar to one of the 4 representative ones: △, ◇, □, ○.

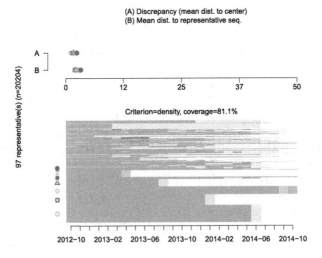

(b) 81.1% of users participation patterns are similar to one of the 97 representative ones: △, ◇, □, ○, etc.

Fig. 4. Representative participation patterns selected by neighborhood density criterion, i.e. the number of sequences assigned to them. They are represented by horizontal bars with width proportional to their corresponding density.

We noticed two facts from Fig. 4. Firstly, it takes only 4 representative participation patterns to represent 63.7% users. This shows that more than half of the users exhibited simple behaviors. They were active for only one month and never came back. On the other hand, the rest of the users exhibited erratic behavior. Figure 4b shows that we need 93 more representative patterns just to cover an extra of 17.4% users.

We call users whose participation patterns represented by representative participation pattern sets in Fig. 4a as *covered* users. The rest of them are called, naturally, *uncovered* ones. To understand the difference in behavior of these two groups better, we visualized their participation patterns side by side (Fig. 5). Each participation pattern is represented by a horizontal bar. There are 6,991 uncovered users and 13,213 covered ones. From a bird's-eye view, one can see that uncovered users seems to be more active compared to ones represented by covered ones, i.e. there are many more unobserved states in a gray color in Fig. 5a than in Fig. 5b. Figure 5a is also rather yellowish, which means participation patterns such as *Item Editor* and *Reference Editor* seems to be dominant in the uncovered sequences.

4.3   Are Users Who Join Earlier More Turbulent Contributors?

Looking at Fig. 5, we noticed that uncovered users are not only more active at making edits, but also switch their participation pattern more often than covered ones. We also saw that the participation patterns in Fig. 5a tend to start earlier than ones in Fig. 5b. Could it be that earlier adopters have more erratic yet lively contribution behavior than later ones?

We use *turbulence* measure proposed by Elzinga et al. [2] to quantify the activeness and variability in contribution behaviors. This measure takes into account how many distinct states and state changes there are in a given sequence. The higher the turbulence, the more distinct states and state changes a sequence has.

We then performed a regression analysis, with the response variable being *turbulence* and the predictor variables being *joining time*, *ending time*, and *within-sequence entropy* (cf. Sect. 3) of a dynamic participation pattern. We selected *joining time* as a predictor variable because we wanted to quantify the relationship between the time somebody joins Wikidata and their behavior over time. *ending time* was added as a control variable. Finally *within-sequence entropy* was chosen so as to take into account its effect on *turbulence*, i.e. it is a confounding variable.

Table 1 shows the results of the regression analysis mentioned previously. The *Std. Error* column shows the average amount that an estimate, shown in the *Estimate* column, differs from the actual values. The t-values measure the number of standard deviations, or how far an estimate is away from 0. They are used in hypothesis testing on estimated coefficients. If there is no relationship between the response variable and predictor variable, i.e., the estimated coefficient is equal to 0, then the corresponding t-value will be small. All of the t-values in our result are big compared to the corresponding estimates. And the

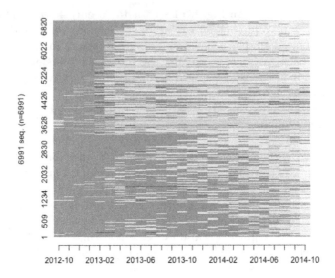

(a) Uncovered participation patterns.

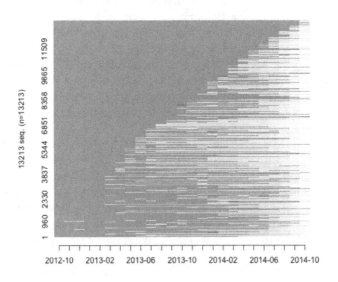

(b) Covered participation patterns.

Fig. 5. User participation patterns are classified into two sets: uncovered and covered ones, with respect to $\theta = 0.6$. Uncovered users seems to be more active than covered ones.

chance this happens due to purely random circumstances is very small, see *p-value* column of Table 1. In other words, the regression analysis shows a reliable correlation between our predictor variables and response variable.

We can see from Table 1 that *joining time* has a negative influence on the turbulence of a dynamic participation pattern. Its coefficient is -0.336. In other words, the earlier an user joins Wikidata, the more turbulent his/her dynamic participation pattern is.

Table 1. Results from Linear Regression of Effects on Turbulence of Dynamic Participation Patterns.

	Estimate	Std. Error	t value	p-value
(Intercept)	2.449	0.031	79.98	<0.0001
Joining time	−0.336	0.002	−196.68	<0.0001
Ending time	0.308	0.002	173.00	<0.0001
Entropy	4.817	0.056	86.68	<0.0001

5 Discussion

In this paper, we apply sequence analysis techniques to infer preferable role transitions, find out the dominant dynamic participation patterns, and perform a regression analysis that find a strong correlation between users' joining time and their contribution behavior.

While finding preferable role transitions, in addition to the known fact that most users tend to stay within their role over time, we also notice that an *Item Creator* likely becomes a *Reference Editor*. This represents a typical workflow in this community. An article in Wikipedia is created as an item on Wikidata and then this item is connected to the article on Wikipedia via a sitelink. A *Reference Editor* would rather stay in the same role or leave the project than switch to another role. It makes sense, since this role represents people who set the language links between Wikipedia articles. They might not even be aware that setting a link – a sitelink – would create an edit on Wikidata.

Moreover, once an user is inactive during one month, it is highly possible that she/he remains so in the subsequent month. We also notice that *Property Engineer*s and *Property Editor*s are more likely to remain active in the next month. Their transition rates to become inactive (*"*"*) are relatively low, 0.11 and 0.06, respectively. This corresponds with previous findings where users belonging to these roles are core member of the Wikidata community in terms of their understanding of structured (semantic) vocabulary [10].

In order to find dominant dynamic participation patterns, we applied the method of finding the representative participation patterns set. This set consists of participation patterns that have distance from others in our data set less than a given threshold. The percentage of dynamic participation patterns are represented by the representative set is said to be its coverage. Some participation patterns are just far away from all representative patterns, and, hence, are not covered by the representative set.

A good representative set is computed by maximizing its coverage, while minimizing its size. We experimented with different percentage values and found out the best coverage is 63.7 % which yields 4 representative dynamic participation patterns. These patterns are very simple and they reflect our insights from the first analysis. Looking over the complete period of time, a large percentage of users are rather inactive and if these users become active they are very likely to become *Reference Editor*. As has already been discussed, these users are rather Wikipedia's and not an active part of the Wikidata community. It seems to be necessary to differentiate between users that handle properties, i.e. *Property Editors* and *Property Engineers*, and those who do not. The community of Wikidata seems quite different to other's, since there is probably a strong separation between the people who belong to the core – the semantic aware users – and the users who contribute but quite irregularly. It might be very insightful to repeat the analysis of the contribution patterns as carried out by Müller-Birn et al. [10] but omit users who change sitelinks. The contribution patterns derived from such analysis might be more representative of the Wikidata community. By visualizing the patterns in the representative set's neighborhood and ones that are not, we notice that users, whose participation patterns that are not represented, make edits to Wikidata more regularly compared to the others. A conclusion could be drawn from this is that only a small number of users make most of contributions to Wikidata.

This hypothesis is supported by our last analysis. We performed regression analysis to confirm the strong negative correlation between users' joining time, i.e. the first month users become active, and their turbulence in editing behavior. We can show that users who joined earlier are persistent contributors even though they take part in different roles, whereas users who join late are quite stable in their behavior.

6 Conclusion and Future Works

In this paper, we study the efficacy of sequence analysis methods in understanding dynamic participation patterns in a peer-production system such as Wikidata. We have shown that these methods are able to identify and quantify the strength of relationship among participation patterns. The typology of dynamic participation patterns is also uncovered and visualized effectively. The strong correlation between contribution consistency and early contributors is also confirmed.

We also note that there is a limitation in applying sequence analysis methods. They require a deterministic set of states or participation patterns in our study. However the latter were generated in our study using the k-means clustering technique [10], which is not completely deterministic. Nevertheless, we believe that sequence analysis methods are suitable for analyzing dynamic user behaviors as pointed out by Keegan et al. [6].

There are a number of possible future studies. Firstly, people coordinate when they edit Wikidata items and properties and, thus, the relationship between social network and content network, i.e. Wikidata ontology, needs to be investigated. We are pursuing this line of research. Secondly, a set of appropriate states of user behaviors need to be determined to apply effectively sequence analysis techniques in studying users behaviors within peer-production communities.

References

1. Abbott, A., Tsay, A.: Sequence analysis and optimal matching methods in sociology review and prospect. Sociol. Methods Res. **29**(1), 3–33 (2000)
2. Elzinga, C.H., Liefbroer, A.C.: De-standardization of family-life trajectories of young adults: a cross-national comparison using sequence analysis. European Journal of Population/Revue européenne de Démographie **23**(3), 225–250 (2007). http://dx.doi.org/10.1007/s10680-007-9133-7
3. Gabadinho, A., Ritschard, G., Mueller, N.S., Studer, M.: Analyzing and visualizing state sequences in R with traminer. J. Stat. Softw. **40**(4), 1–37 (2011)
4. Gabadinho, A., Ritschard, G., Studer, M., Müller, N.S.: Extracting and rendering representative sequences. In: Fred, A., Dietz, J.L.G., Liu, K., Filipe, J. (eds.) IC3K 2009. CCIS, vol. 128, pp. 94–106. Springer, Heidelberg (2011). doi:10.1007/978-3-642-19032-2_7
5. Gleave, E., Welser, H.T., Lento, T.M., Smith, M.A.: A conceptual and operational definition of'social role in online community. In: 42nd Hawaii International Conference on System Sciences, HICSS 2009, pp. 1–11. IEEE (2009)
6. Keegan, B.C., Lev, S., Arazy, O.: Analyzing organizational routines in online knowledge collaborations: a case for sequence analysis in CSCW. arXiv preprint arXiv:1508.04819 (2015)
7. Kittur, A., Suh, B., Pendleton, B.A., Chi, E.H.: He says, she says: conflict and coordination in wikipedia. In: Proceedings of the SIGCHI Conference on Human Factors in Computing Systems, pp. 453–462. ACM (2007)
8. Krötzsch, M., Vrandečić, D., Völkel, M., Haller, H., Studer, R.: Semantic wikipedia. Web Seman. Sci. Serv. Agents World Wide Web **5**(4), 251–261 (2007)
9. McVicar, D., Anyadike-Danes, M.: Predicting successful and unsuccessful transitions from school to work by using sequence methods. J. Roy. Stat. Soc.: Ser. A (Appl. Stat.) **165**(2), 317–334 (2002)
10. Müller-Birn, C., Karran, B., Luczak-Rösch, M., Lehmann, J.: Peer-production system or collaborative ontology development effort: what is wikidata? (2015)
11. Preece, J., Shneiderman, B.: The reader-to-leader framework: motivating technology-mediated social participation. AIS Trans. Hum. Comput. Interact. **1**(1), 13–32 (2009)
12. Tudorache, T., Noy, N.F., Tu, S., Musen, M.A.: Supporting collaborative ontology development in Protégé. In: Sheth, A., Staab, S., Dean, M., Paolucci, M., Maynard, D., Finin, T., Thirunarayan, K. (eds.) ISWC 2008. LNCS, vol. 5318, pp. 17–32. Springer, Heidelberg (2008). doi:10.1007/978-3-540-88564-1_2
13. Viegas, F.B., Wattenberg, M., Kriss, J., Van Ham, F.: Talk before you type: coordination in wikipedia. In: 40th Annual Hawaii International Conference on System Sciences, HICSS 2007, pp. 78–87. IEEE (2007)

14. Vrandečić, D., Krötzsch, M.: Wikidata: a free collaborative knowledge base. Commun. ACM **57**(10), 78–85 (2014)
15. Welser, H.T., Gleave, E., Fisher, D., Smith, M.: Visualizing the signatures of social roles in online discussion groups. J. Soc. Struct. **8**(2), 1–32 (2007)
16. Wenger, E.: Communities of practice: learning as a social system. Syst. Thinker **9**(5), 2–3 (1998)
17. Ye, Y., Kishida, K.: Toward an understanding of the motivation of open source software developers. In: 25th International Conference on Software Engineering, Proceedings, pp. 419–429. IEEE (2003)

Network-Oriented Modeling
and Its Conceptual Foundations

Jan Treur[✉]

Behavioural Informatics Group, Vrije Universiteit Amsterdam,
Amsterdam, The Netherlands
j.treur@vu.nl
https://www.researchgate.net/profile/Jan_Treur,
http://www.few.vu.nl/~treur

Abstract. To address complexity of modeling the world's processes, over the years in different scientific disciplines separation assumptions have been made to isolate parts of processes, and in some disciplines they have turned out quite useful. It can be questioned whether such assumptions are adequate to address complexity of integrated human mental and social processes and their interactions. In this paper it is discussed that a Network-Oriented Modeling perspective can be considered an alternative way to address complexity for modeling human and social processes.

Keywords: Network-oriented modeling · Separation · Interaction · Conceptual foundations

1   Introduction

To address complexity of modeling the world's processes, over the years different strategies have been used. From these strategies isolation and separation assumptions are quite common in all scientific disciplines and have often turned out very useful. They traditionally serve as means to address the complexity of processes by some strong form of decomposition.

This also holds for classical disciplines such as Physics, where, for example, for mechanical modeling for building construction only forces from objects on earth are taken into account and not forces from all other objects in the universe, that still do have some effects as well. It is recognized that these distant effects from sun, moon, planets and other objects do exist, but it assumed that they can be neglected.

For such cases within Physics such an isolation assumption may be a reasonable choice, but in how far is it equally reasonable to address complexity of human mental and social processes? Over the years within the Behavioural and Social Sciences also a number of assumptions have been made in the sense that some processes can be studied by considering them as separate or isolated phenomena. However, within these human-directed sciences serious debates or disputes have occurred time and time again on such a kind of assumptions. They essentially have the form of arguments pro or con an assumption that some processes can be studied by considering them as separate or

E. Spiro and Y.-Y. Ahn (Eds.): SocInfo 2016, Part II, LNCS 10047, pp. 157–175, 2016.
DOI: 10.1007/978-3-319-47874-6_12

isolated phenomena. Examples of such separation assumptions to address human complexity of mental and social processes concern:

- mind versus body
- cognition versus emotion
- individual processes versus collective processes
- non-adaptive processes versus adaptive processes
- earlier versus later: temporal separation

It can be questioned whether, for example, mind can be studied while ignoring body, or cognition while ignoring emotion, or sensory processing in isolation from action preparation. Or, put more general, in how far are these traditional means to address complexity by separation still applicable if the complexity of human mental and social processes has to be addressed? Do we need to break with such traditions to be able to make more substantial scientific progress in this area addressing human processes? And, not unimportant, are there adequate alternative strategies to address human complexity? This is discussed in this paper, and it is pointed out that network-oriented modeling can be considered an alternative way to address complexity.

In this paper, first in Sect. 2 the five separation assumptions mentioned above are discussed in some more detail. Next, in Sect. 3 it is discussed how as an alternative, interaction in networks can be used to address complexity. In Sect. 4 the development of a Network-Oriented Modeling perspective is discussed. In Sect. 5 the Network-Oriented Modeling approach based on temporal-causal networks presented in [104, 105] is briefly pointed out. Finally, Sect. 6 is a discussion.

2   Addressing Human Complexity by Separation Assumptions

The position taken in this paper is that indeed a number of the traditional separation and isolation habits followed in order to address human complexity have to be broken to achieve more progress in scientific development. Partly due to the strong development of Cognitive, Affective and Social Neuroscience, in recent years for many of the issues mentioned above, a perspective in which dynamics, interaction and integration are key elements has become more dominant: a perspective with interaction as a point of departure instead of separation. Given this background, for each of the separation issues listed above this will be discussed below in more detail. It will be pointed out how in many cases separation assumptions as mentioned lead to some types of discrepancies or paradoxes.

2.1   Mind Versus Body

A first isolation assumption that has a long tradition is the assumption that the mind can be studied in separation from the body. There has been debate about this since long ago. Aristotle [3] considered properties of 'mind and desire' as the source of motion of a living being: he discusses how the occurrence of certain internal (mental) state properties (desires) within a living being entails or causes the occurrence of an action in the external world; see also [78]. Such internal state properties are sometimes called by him 'things in the soul', 'states of character', or 'moral states'. In that time such 'things' were not considered part of the physical world but of a ghost-like world indicated in this case by 'soul'. So, in this context the explanation that such a creature's position gets changed is that there is a state of the soul driving it. This assumes a separation between the soul on the one hand, and the body within the physical world on the other hand. How such nonphysical states can affect physical states remains unanswered.

Over time, within Philosophy of Mind this has been felt as a more and more pressing problem. The idea that mental states can cause actions in the physical world is called *mental causation* (e.g., [61, 62]). The problem with this is how exactly nonphysical mental states can cause effects in the physical world, without any mechanism known for such an effect. Within Philosophy of Mind a solution for this has been proposed in the form of a tight relation between mental states and brain states. Then it is in fact not the mental state causing the action, but the corresponding (physical) brain state. Due to this the separation is not between the soul or mind, and the body, but between the brain and the body [10, 61, 62].

However, this separation between brain and body also has been debated. More literature on this from a wider perspective can be found, for example, in [19, 66, 112]. It is claimed that mind essentially is embodied: it cannot be isolated from the body. One specific case illustrating how brain and body intensely work together and form what is called an embodied mind is the causal path concerning feelings and emotional responses. A classical view is that, based on some sensory input, due to internal processing emotions are felt, and based on this they are expressed in some emotional response in the form of a body state, such as a face expression. However, James [58] claimed a different order in the causal chain in which expressing an emotion comes before feeling the emotion (see also [24], pp. 114–116):

stimulus → sensory representation → preparation for a body state
→ expressed emotion in body state → sensed body state
→ representation of body state → felt emotion

The perspective of James assumes that a *body loop* via the expressed emotion is used to generate a felt emotion by sensing the own body state. So, the body plays a crucial role in the emergence of states of the brain and mind concerning emotions and feelings. Damasio made a further step by introducing the possibility of an *as-if body loop* bypassing actually expressed bodily changes by assuming a direct causal connection from preparation state to representation of the body (e.g., [21], pp. 155–158; see also [22], pp. 79–80; [24]). A brief survey of Damasio's ideas about emotion and feeling can be found in [24], pp. 108–129. According to this perspective emotions relate to actions, whereas feelings relate to perceptions of body states caused by these actions. It takes a cyclic process involving both mind and body that (for a constant environment) can lead to equilibrium states for both emotional response (preparation) and feeling.

2.2 Cognition Versus Emotion

Another assumption made traditionally is that cognitive processes can be described independently, leaving affective states aside. The latter types of states are considered as being part of a separate line of (affective) processes that produce their own output, for example, in the sense of emotions and expressions of them. However, this assumed separation between cognitive and affective processes has been questioned more and more. Specific examples of questions about interactions between affective and cognitive states are: how does desiring relate to feeling, and in how far do sensing and believing relate to feeling? To assume that desiring can be described without involving emotion already seems a kind of paradox, or at least a discrepancy with what humans experience as desiring. Recent neurological findings suggest that this separation of cognitive and affective processes indeed may not be a valid and fruitful way to go. For example, Phelps [85] states:

> 'The mechanisms of emotion and cognition appear to be intertwined at all stages of stimulus processing and their distinction can be difficult. (..) Adding the complexity of emotion to the study of cognition can be daunting, but investigations of the neural mechanisms underlying these behaviors can help clarify the structure and mechanisms.' ([85], pp. 46–47).

Here it is recognized that an assumption on isolating cognition from emotion is not realistic, as far as the brain is concerned. Therefore models based on such an assumption cannot be biologically plausible and may simply be not valid. Moreover, it is also acknowledged that taking into account the intense interaction between emotion and cognition 'can be daunting'; to avoid this problem was a main reason for the isolation assumption as a way to address complexity. However, Phelps [85] also points at a way out of this: use knowledge about the underlying neural mechanisms. In the past when there was limited knowledge about the neural mechanisms this escape route was not available, and therefore the isolation assumption may have made sense, although the validity of the models based on that can be questioned. But, now Neuroscience has shown a strong development, this provides new ways to get rid of this isolation assumption.

Similar claims about the intense interaction between emotion and cognition have been made by Pessoa [83]. In experimental contexts different types of effects of affective states on cognitive states have indeed been found; see, for example, [31, 34, 38, 113]. Moreover, more specifically in the rapidly developing area of Cognitive Neuroscience (e.g., [41, 89]) knowledge has been contributed on mechanisms for the interaction and intertwining of affective and cognitive states and processes (for example, involving emotion, mood, beliefs or memory); see, for example, [28, 65, 83, 85, 98].

Not only for desiring and believing the isolation assumption for cognition vs emotion is questioned, but also for rational decision making. Traditionally, rationality and emotions often have been considered each other's enemies: decision making has often been considered as a rational cognitive process in which emotions can only play a disturbing role. In more recent times this has been questioned as well. For example, in [69], p. 619, and [59, 80, 90] it is pointed at the positive functions served by emotions. In particular, in decision making it may be questioned whether you can make an adequate decision without feeling good about it. Decisions with bad feelings associated to them may lack robustness. This indicates another paradox or discrepancy between the isolation assumption and how real life is experienced: emotions can be considered a vehicle for rationality; for more details, see [106] or [105], Chapt. 6.

2.3 Individual Versus Collective

Yet another isolation assumption concerns the distinction between mental processes within an individual and social processes. The former are usually referred to the territory of Psychology, whereas the latter are referred to the territory of Social Science. From both sides only modest attempts are made to also involve elements of the other territory, for example, Social Psychology. The general idea is to study social processes as patterns emerging from interactions between individuals thereby abstracting from the processes within each of the individuals. This easily leads to some kind of paradoxes. For example, as persons in a group are autonomous individuals with their own neurological structures and patterns, carrying, for example, their own emotions, beliefs, desires and intentions, it would be reasonable to expect that it is very difficult or even impossible to achieve forms of sharedness and collectiveness. However, it can be observed that often groups develop coherent views and decisions, and, even more surprisingly, the group members seem to share a positive feeling about it. In recent years by developments in Neuroscience new light has been shed on this seeming paradox of individuality versus sharedness and collectiveness. This has led to the new discipline called Social Neuroscience; e.g., [17, 18, 25, 26, 47]. Two interrelated core concepts in this discipline are mirror neurons and internal simulation of another person's mental processes. Mirror neurons are neurons that not only have the function to prepare for a certain action or body change, but are also activated upon observing somebody else who is performing this action or body change; e.g., [57, 86, 91]. Internal simulation is internal mental processing that copies processes that may take place externally, for example, in mental processes in another individual; e.g., [21–24, 40, 44, 49–51]. Mechanisms involving these core concepts have been described that provide an explanation of the emergence of sharedness and collectiveness from a biological perspective. This new perspective breaks the originally assumed separation between processes within individuals and processes of social interaction.

2.4 Adaptive Versus Nonadaptive Processes

Another assumption that sometimes is debated is that mental and social processes are modelled as if they are not adaptive. In reality processes usually have adaptive elements incorporated, but often these elements are neglected and sometimes studied as separate phenomena. One example in a social context is the following. Often a contagion principle based on social interaction is studied, describing how connected states affect each other by these interactions, whereas the interactions themselves are assumed not to change over time (for example, qua strength, frequency or intensity). But in reality the interactions also change, for example based on what is called the homophily principal: the more you are alike, the more you like (each other); for example, see [16, 72, 73]. Another way of formulating this principal is: birds of a feather flock together. It can often be observed that persons that have close relationships or friendships are alike in some respects. For example, they go to the same clubs, watch the same movies or TV programs, take the same drinks, have the same opinions, vote for the same or similar parties. Such observations might be considered support for the contagion principle: they were together and due to that they affected each other's states by social contagion, and therefore they became alike. However, also a different explanation is possible based on the homophily principle: in the past they already were alike before meeting each other, and due to this they were attracted to each other. So, the cyclic relation between *the states of the members* and *the strength of their connection* leads to two possible causal explanations of being alike and being connected:

being connected → being alike (contagion principle)

being alike → being connected (homophily principle)

Such circular causal relations make it difficult to determine what came first. It may be a state just emerging from a cyclic process without a single cause. For more discussion on this issue, for example, see [2, 76, 94, 97].

Another example illustrating how adaptivity occurs fully integrated with the other processes, concerns the function of dreaming. From a naïve perspective, dreaming might be considered as just playing some movie, thereby triggering some emotions, and that's all. But in recent research, the idea has become common that dreaming is a form of internal simulation of real-life-like processes serving as training in order to learn or adapt regulation of fear emotions; see, for example, [27, 46, 67, 81, 96]. To this end in dreams adequate exercising material is needed: sensory representations of emotion-loaden situations are activated, built on memory elements suitable for high levels of arousal. The basis of what is called 'fear extinction learning' is that emotion regulation mechanisms are available which are adaptive: they are strengthened over time when they are intensively used. This is learning of fear inhibition connections in order to counterbalance the fear associations which themselves remain intact (e.g., [67], p. 507); this can be based on a Hebbian learning principle [48].

2.5 Earlier Versus Later: Temporal Separation

Another traditionally made separation assumption is that processes in the brain are separated in time. For example, sensing, sensory processing, preparation for action and action execution are assumed to occur in linearly ordered sequential processes. For the case of emotions it was already discussed that such linear temporal patterns are not applicable. But also more in general it can be argued that such linear patterns are too much of a simplification, as in reality such processes occur simultaneously, in parallel; often a form of internal simulation takes place, as put forward, among others, by [5, 21, 22, 44, 49–51, 70, 84]. The general idea of internal simulation that was also mentioned above in the specific context of emotions and bodily processes, is that sensory representation states are activated (e.g., mental images), which in response trigger associated preparation states for actions, which, by prediction links, in turn activate other sensory representation states for the predicted effects of the prepared actions. The latter states represent the effects of the prepared actions or bodily changes, without actually having executed them. Being inherently cyclic, the simulation process can go on indefinitely, as the latter sensory representations can again trigger preparations for actions, and so on, and everything simultaneously, in parallel, as in the world no process is freezing to wait for another process to finish first. Internal simulation has been used, for example, to describe (imagined) processes in the external world, e.g., prediction of effects of own actions [6], or processes in another person's mind, e.g., emotion recognition or mind-reading [44] or (as discussed above) processes in a person's own body by as-if body loops [21]. This breaks with the tradition that there is a temporal separation of processes such as sensing – sensory processing – preparing for action – executing action.

3   Addressing Complexity by Interaction in Networks

The separation assumptions to address complexity as discussed in Sect. 2 are strongly debated, as they all come with shortcomings. In this section it is discussed that in fact the problem is not so much in the specific separation assumptions, but in the general idea of separation itself. In social contexts it is clear that the intense interaction between persons based on their mutual and often interrelated cyclic relationships, makes them not very well suitable for any separation assumptions: all these interactions take place all the time simultaneously, in parallel. And this does not only apply to social processes but also to individual mental processes, as will be discussed in some more detail here.

In the domain of Neuroscience the structures and mechanisms found suggest that many parts in the brain are connected by connections that often are part of cyclic paths, and such cycles are assumed to play an important role in many mental processes (e.g., [9, 20, 88]). As an example also put forward above, there is a growing awareness, fed by findings in Neuroscience that emotions play an important mediating role in most human processes, and this role often provides a constructive contribution, and not a disturbing contribution as was sometimes assumed. Usually mental states trigger emotions and these emotions in turn affect these and other mental states. To address this type of circular effects, different views on causality and modeling are required, compared to the traditional views in modeling of mental processes. For example, Scherer [93] states:

> 'What is the role of causality in the mechanisms suggested here? Because of the constant recur-
> sivity of the process, the widespread notion of linear causality (a single cause for a single effect)
> cannot be applied to these mechanisms. Appraisal is a process with constantly changing results
> over very short periods of time and, in turn, constantly changing driving effects on subsystem
> synchronization (and, consequently, on the type of emotion). (...) Thus, as is generally the case
> in self-organizing systems, there is no simple, unidirectional sense of causality (see also [68]).'
> ([93], p. 3470).

Also in the domain of Philosophy of Mind this issue of cyclic causal connections is recognized, for example, by Kim [61]. The idea is that a mental state is characterized by the way it mediates between the input it receives from other states and the output it provides to other states; this is also called the *functional* or *causal role* of the mental state. As a simplified example on the input side a mental state of being in pain is typically caused by tissue damage and in turn on the output side it typically causes winces, groans and escape behavior; e.g., [61], p. 104. So, in this perspective the question what exactly is pain can be answered as the state that forms a causal bridge (or causally mediates) from tissue damage to winces, groans, and escape behavior. Kim describes the overall picture as follows:

'Mental events are conceived as nodes in a complex causal network that engages in causal transactions with the outside world by receiving sensory inputs and emitting behavioral outputs' ([61], p. 104)

As input not only sensory input can play a role but also input from other mental states such as in the pain example 'being alert'. Similarly, as output not only behavioral output can play a role but also other mental states can be affected, such as in the pain example feeling distress and a desire to be relieved of it. Within Philosophy of Mind this is often considered challenging:

'But this seems to involve us in a regress or circularity: to explain what a given mental state is, we need to refer to other mental states, and explaining these can only be expected to require reference to further mental states, on so on – a process that can go on in an unending regress, or loop back in a circle' ([61], pp. 104–105)

In Fig. 1 an example of such a cyclic causal path is depicted. Here mental state S_1 has a causal impact on mental state S_2, but one of the states on which S_2 has an effect, in turn affects one of the input states for S_1.

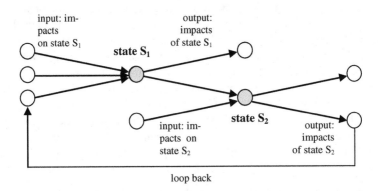

Fig. 1. Mental states with their causal relations conceived as nodes in a complex, often cyclic causal network (see also [61], p. 104)

This view from Philosophy of Mind is another indication that a modeling approach will have to address causal relations with cycles well. To obtain an adequate understanding of such cycles and their dynamics and timing it is inevitable to take into account the temporal dimension of the dynamics of the processes effectuated by the causal relations. In principle, this situation makes that an endless cyclic process over time emerges, which in principle works simultaneously, in parallel, and in interaction with other processes. In such a graph at each point in time activity takes place in every state simultaneously (it is not that one state waits for the other). The notion of state at some point in time used here refers to a specific part or aspect of the overall state of a model at this point in time. Such an overall state can include, for example, at the same time a 'being in pain' state, a 'desire to get relieved' state, and an 'intention to escape' state. The overall state at some point in time is the collection of all states at that point in time. All the time the processes in the brain occur in parallel, in principle involving all specific states within the overall state, mostly in an unconscious manner. In this sense the brain is not different from any other part of the universe where everywhere processes take place simultaneously, in parallel. During all this parallel processing, any change in state S_1 in principle will lead to a change in state S_2, which in turn will lead to another change in state S_1, which leads to another change in state S_2, and so on and on. The state changes in such a process may become smaller and smaller over time, and the cyclic process eventually may converge to an equilibrium state in which no further changes occur anymore; but also other patterns are possible, such as limit cycles in which the changes eventually end up in a regular, periodic pattern of changes.

In the sense described above, mental processes can show patterns similar to patterns occurring in social interactions, where cycles of connections are natural and quite common. An example from the context of modeling social systems or societies can be found in [77]:

'Instead of rigid rural-urban dichotomies and other absolute, "container" views of space, there is a need to recognise spatial overlaps and complexities such the pervasiveness of so-called translocal livelihood systems. Accordingly, much more *relational, network-oriented modelling approaches* are needed.' ([77], p. 1)

Also here it is claimed that separation in the form of what they call 'container' views of space falls short in addressing the complexities involved, and as an alternative a Network-Oriented Modeling approach is suggested to address human social complexity. Similar claims are made from the area of organization modeling by Elzas [32]:

The study of the process-type of organization, which still - at this moment because its relative novelty - requires modelling to evaluate, can benefit from certain network-oriented modelling formalisms because of the very nature of the organizational concept. (...) in addressing (...) the specific coordinating problems of the adaptively interrelated distributed-action organizational units as they are found in process-based organizational models ([32], p. 162)

So, both from the area of the analysis of mental processes and from the area of analysis of social processes, the notion of network is suggested as a basis. In next section the notion of Network-Oriented Modeling is discussed in some more detail.

4   Network-Oriented Modeling

This paper started in Sect. 2 by some reflection on traditional means to address complexity by assuming separation and isolation of processes, and the shortcomings, discrepancies and paradoxes entailed by these assumptions. In Sect. 3 the circular or cyclic, interactive and distributed character of many processes (involving interacting sub-processes running simultaneously, in parallel) was identified as an important challenge to be addressed, and it was recognized that a perspective based on interactions in networks is more suitable for this. In this section a Network-Oriented Modeling perspective is proposed as an alternative way to address complexity. This perspective takes the concept of network and the interactions within a network as a basis for conceptualization and structuring of any complex processes. Network-Oriented Modeling is *not* considered here as modeling *of* (given) networks, but modeling any (complex) processes *by* networks. It is useful to keep in mind that the concept network is a just a mental concept and this is used as a conceptual structuring tool to conceptualize any processes that exist in reality.

The concept of network is easy to visualize on paper, on a screen or mentally and as such provides a good support for intuition behind a model. Moreover, as the Network-Oriented Modeling approach presented here (see Sect. 6) also incorporates a temporal dimension enabling interpretation of connections as temporal-causal connections, the mental concept of network also provides support for the intuition behind the dynamics of the modelled processes.

The scientific area of networks has already a longer tradition within different disciplines of more than 80 years. But it has developed further and within many other disciplines, such as Biology, Neuroscience, Mathematics, Physics, Economics, Informatics or Computer Science, Artificial Intelligence, and Web Science; see, for example [12, 43, 107]. These developments already show how processes in quite different domains can be conceptualized as networks. Historically the use of the concept network in different domains can be traced back roughly to the years 1930 to 1950, or even earlier, for studying processes such as brain processes in Neuroscience by neural networks (e.g., [71, 92]), metabolic processes in Cell Biology by metabolic networks (e.g., [79, 110]), social interactions within Social Science by social networks; (e.g., [1, 15, 75]), processes in Human Physiology (e.g., [56, 111]), processes in engineering in Physics (e.g., [13, 55], and processes in engineering in Chemistry (e.g., [33, 101]). Within such literature often graphical representations of networks are used as an important means of presentation. For a historical overview of the development of social network analysis, for example, see [37].

After getting accustomed to such conceptualizations as networks of processes that exist in the real world, a belief may occur that these networks actually exist in reality (as neural networks, or as computer networks, or as social networks, for example), and *modeling by networks* happens sometimes to be phrased alternatively as *modeling networks*. However, it still has to be kept in mind that the concept 'network' is a mental concept used as a tool to conceptualize any type of processes. To make this distinction more clear linguistically, the phrase Network-Oriented Modeling is used as indication for modeling by networks. Within this book the preferred use of the word 'network' is to indicate a model or conceptualization of some process, not to indicate the process in the real world itself. For example, social media such as Facebook, Twitter, WhatsApp, Instagram,... do not form or create social networks in reality, but they create social interactions in reality that can be described (conceptualized, modelled) by (social) networks or by network models.

Network-Oriented Modeling offers a conceptual tool to model complex processes in a structured, intuitive and easily visualizable manner, but the approach described here also incorporates the dynamics of the processes in these models. Using this approach, different parts of a process can be distinguished, but in contrast to the separation and isolation strategy to address complexity, a network-oriented approach does not separate or isolate these parts, but emphasizes and explicitly models the way how they are connected and interact. Moreover, by adding a temporal dimension to incorporate a dynamic perspective, it is explicitly modelled how they can have intense and circular causal interaction, and how the timing of the processes is. For more background information on the dynamic perspective to cognitive modeling, see [104] pp. 134–136, or [105], Chap. 1.

5 Network-Oriented Modeling by Temporal-Causal Networks

As discussed above, both internal mental processing and social processing due to social interactions often involve multiple cyclic processes and adaptive elements. This has implications for the type of modeling approach to be used. Within Network-Oriented Modeling, the network models considered have to integrate such cycles, and also allow adaptive processes by which individuals can change their connections. To model such dynamics, a dynamical modeling perspective is needed that can handle such combinations of cycles and the adaptation of connections over time. Therefore, within the Network-Oriented Modeling approach as discussed here, the dynamic perspective has to be incorporated as well: a temporal dimension is indispensable. This is what is achieved in the Network-Oriented Modeling approach based on temporal-causal networks described in [104, 105].

The Network-Oriented Modeling approach based on temporal-causal networks is a generic and declarative dynamic AI modeling approach based on networks of causal relations (e.g., [63, 64, 82]), that incorporates a continuous time dimension to model dynamics. As discussed above, this temporal dimension enables causal reasoning and simulation for cyclic causal graphs or networks that usually inherently contain cycles, such as networks modeling mental or brain processes, or social interaction processes, and also the timing of such processes. States in such a network are characterised by the connections they have to other states, comparable to the way in which in Philosophy of Mind mental states are characterised by their causal roles, as discussed in Sect. 3. Moreover, adaptive elements can be fully integrated. The modeling approach can incorporate ingredients from different modeling approaches, for example, ingredients that are sometimes used in specific types of (continuous time, recurrent) neural network models, and ingredients that are sometimes used in probabilistic or possibilistic modeling. It is more generic than such methods in the sense that a much wider variety of modeling elements are provided, enabling the modeling of many types of dynamical systems, as described in [104, 105].

As discussed in detail in [104, 105] temporal-causal network models can be represented at two levels: by a conceptual representation and by a numerical representation. These model representations can be used not only to display interesting graphical network pictures, but also for numerical simulation. Furthermore, they can be analyzed mathematically and validated by comparing their simulation results to empirical data.

Moreover, they usually include a number of parameters for domain, person, or social context-specific characteristics. To estimate values for such parameters, a number of parameter tuning methods are available.

A conceptual representation of a temporal-causal network model in the first place involves representing in a declarative manner states and connections between them that represent (causal) impacts of states on each other, as assumed to hold for the application domain addressed. The states are assumed to have (activation) levels that vary over time. What else is needed to describe processes in which causal relations play their role? In reality not all causal relations are equally strong, so some notion of *strength of a connection* is needed. Furthermore, when more than one causal relation affects a state, in which manner do these causal effects combine? So, some way to *aggregate multiple causal impacts* on a state is needed. Moreover, not every state has the same extent of flexibility; some states may be able to change fast, and other states may be more rigid and may change more slowly. Therefore, a notion of *speed of change* of a state is used for timing of processes. These three notions are covered by elements in the Network-Oriented Modeling approach based on temporal-causal networks, and are part of a conceptual representation of a temporal-causal network model:

- **Strength of a connection** $\omega_{X,Y}$: each connection from a state X to a state Y has a *connection weight value* $\omega_{X,Y}$ for the strength of the connection.
- **Combining multiple impacts on a state** $c_Y(..)$: for each state (a reference to) a *combination function* $c_Y(..)$ is chosen to combine the causal impacts on Y.
- **Speed of change of a state** η_Y: for each state Y a *speed factor* η_Y is used to represent how fast a state is changing upon causal impact.

Combination functions in general are similar to the functions used in a static manner in the (deterministic) Structural Causal Model perspective described, for example, in [74, 82, 114], but in the Network-Oriented Modeling approach described here they are used in a dynamic manner, as will be pointed out below briefly, and in more detail in [104, 105].

Combination functions can have different forms. How exactly does one impact on a given state add to another impact on the same state? In other words, what types of combination functions can be considered? The more general issue of how to combine multiple impacts or multiple sources of knowledge occurs in various forms in different areas, such as the areas addressing imperfect reasoning or reasoning with uncertainty or vagueness. For example, in a probabilistic setting, for modeling multiple causal impacts on a state often independence of these impacts is assumed, and a product rule is used for the combined effect; e.g., [30]. In practical applications, this assumption is often questionable or difficult to validate. In the areas addressing modeling of uncertainty also other combination rules are used, for example, in possibilistic approaches minimum- or maximum-based combination rules are used; e.g., [30]. In another different area, addressing modeling based on neural networks yet another way of combining effects is used often. In that area, for combination of the impacts of multiple neurons on a given neuron usually a logistic sum function is used: adding the multiple impacts and then applying a logistic function; e.g., [7, 45, 52–54].

So, there are many different approaches possible to address the issue of combining multiple impacts. The applicability of a specific combination rule for this may depend much on the type of application addressed, and even on the type of states within an application. Therefore the Network-Oriented Modeling approach based on temporal-causal networks incorporates for each state, as a kind of parameter, a way to specify how multiple causal impacts on this state are aggregated. For this aggregation a number of standard combination functions are made available as options and a number of desirable properties of such combination functions have been identified; e.g., see [104], Table 10, or [105], Chap. 2, Table 2.10. Some of these standard combination functions are scaled sum, product, complementary product, max, min, and simple and advanced logistic sum functions. These options cover elements from different existing approaches, varying from approaches considered for reasoning with uncertainty, probability, possibility or vagueness, to approaches based on recurrent neural networks; e.g., [7, 29, 30, 42, 45, 52–54, 115]. In addition, there is still the option to specify any other (non-standard) combination function, preferably taking into account the desired properties.

The above three concepts (connection weight, speed factor, combination function) can be considered as parameters representing characteristics in a network model. In a non-adaptive network model these parameters are fixed over time. But to model processes by *adaptive networks*, not only the state levels, but also these parameters can change over time. For example, the connection weights can change over time to model evolving connections in network models. A conceptual representation of temporal-causal network model can be transformed in a systematic or even automated manner into a numerical representation of the model as described in [104, 105], thus obtaining the following *difference* and *differential equation* for states Y:

$$Y(t + \Delta t) = Y(t) + \eta_Y \left[c_Y \left(\omega_{X_1,Y} X_1(t), \ldots, \omega_{X_k,Y} X_k(t) \right) - Y(t) \right] \Delta t$$

$$dY(t)/dt = \eta_Y \left[c_Y \left(\omega_{X_1,Y} X_1(t), \ldots, \omega_{X_k,Y} X_k(t) \right) - Y(t) \right]$$

For modeling processes as adaptive networks, some of parameters (such as connection weights) are handled in a similar manner, as if they are states. For more detailed explanation, see [104, 105].

Summarizing, the Network-Oriented Modeling approach based on temporal-causal networks described here provides a complex systems modeling approach that enables a modeler to design conceptual model representations in the form of networks described as cyclic graphs (or connection matrices), which can be systematically transformed into executable numerical representations that can be used to perform simulation experiments. The modeling approach makes it easy to take into account on the one hand theories and findings from any domain from, for example, biological, psychological, neurological or social sciences, as such theories and findings are often formulated in terms of causal relations. This applies, among others, to mental processes based on complex brain processes, which, for example, often involve dynamics based on interrelating and adaptive cycles, but equally well it applies to social interaction processes and their adaptive dynamics. This enables to address complex adaptive phenomena such as the integration of emotions within all kinds of cognitive processes, of internal simulation and mirroring of mental processes of others, and dynamic social interaction patterns, as shown in [105] by a number of example models; see also [11, 14, 95, 100, 102, 103].

6 Discussion

Concerning the scope of applicability, it has been shown [104, 105] that any smooth continuous state-determined system (any smooth dynamical system described as a state-determined system or by a set of first order differential equations) can also be modeled by temporal-causal networks, by choosing suitable parameters such as connection weights, speed factors and combination functions. In this sense it is as general as modeling approaches put forward, for example, in [4, 7, 8, 35, 36, 60, 87, 99, 108, 109], and approaches such as described, for example in [39, 45, 52–54].

To facilitate applications, dedicated software is available supporting the design of models in a conceptual manner, automatically transforming them into an executable format and performing simulation experiments. A variety of example models that have been designed illustrates the applicability of the approach in more detail.

The topics addressed have a number of possible applications. An example of such an application is to analyse the spread of a healthy or unhealthy lifestyle in society. Another example is to analyse crowd behaviour in emergency situations. A wider area of application addresses socio-technical systems that consist of humans and devices, such as smartphones, and use of social media. For such mixed groups, in addition to analysis of what patterns may emerge, also for the support side the design of these devices and media can be an important aim, in order to create a situation that the right types of patterns emerge. This may concern, for example, safe evacuation in an emergency situation or strengthening development of a healthy lifestyle. Other application areas may address, for example, support and mediation in collective decision making and avoiding or resolving conflicts that may develop.

References

1. Aldous, J., Straus, M.A.: Social networks and conjugal roles: a test of bott's hypothesis. Soc. Forces **44**(576–580), 965–966 (1966)
2. Aral, S., Muchnik, L., Sundararajan, A.: Distinguishing influence based contagion from homophily driven diffusion in dynamic networks. Proc. Nat. Acad. Sci. (USA) **106**, 1544–1549 (2009)
3. Aristotle: Physica (translated by R.P. Hardie and R.K. Gaye) (350 BC)
4. Ashby, W.R.: Design for a Brain, Chapman and Hall, London (second extended edition). 5th edn., 1952 (1960)
5. Barsalou, W.: Simulation, situated conceptualization, and prediction Lawrence. Phil. Trans. R. Soc. B **364**, 1281–1289 (2009)
6. Becker, W., Fuchs, A.F.: Prediction in the oculomotor system: smooth pursuit during transient disappearance of a visual target. Exp. Brain Res. **57**, 562–575 (1985)
7. Beer, R.D.: On the dynamics of small continuous-time recurrent neural networks. Adapt. Behav. **3**, 469–509 (1995)
8. Beer, R.D.: Dynamical approaches to cognitive science. Trends Cogn. Sci. **4**, 91–99 (2000)

9. Bell, A.: Levels and loops: the future of artificial intelligence and neuroscience. Phil. Trans. R. Soc. Lond. B **354**, 2013–2020 (1999)
10. Bickle, J.: Psychoneural Reduction: The New Wave. MIT Press, Cambridge (1998)
11. Blankendaal, R., Parinussa, S., Treur, J.: A temporal-causal modelling approach to integrated contagion and network change in social networks. In: Proceedings of the 22nd European Conference on Artificial Intelligence, ECAI 2016. IOS Press (2016)
12. Boccalettia, S., Latorab, V., Morenod, Y., Chavez, M., Hwanga, D.-U.: Complex networks: structure and dynamics. Phys. Rep. **424**, 175–308 (2006)
13. Bode, H.W.: Network Analysis and Feedback Amplifier Design. Van Nostrand, Princeton (1945)
14. Bosse, T., Hoogendoorn, M., Klein, M.C.A., Treur, J., van der Wal, C.N., van Wissen, A.: Modelling collective decision making in groups and crowds: integrating social contagion and interacting emotions, beliefs and intentions. Auton. Agent. Multi-Agent Syst. **27**, 52–84 (2013)
15. Bott, E.: Family and Social Network: Roles, Norms and External Relationships in Ordinary Urban Families London. Tavistock Publications, London (1957)
16. Byrne, D.: The attraction hypothesis: do similar attitudes affect anything? J. Pers. Soc. Psychol. **51**, 1167–1170 (1986)
17. Cacioppo, J.T., Berntson, G.G.: Social neuroscience. Psychology Press, New York (2005)
18. Cacioppo, J.T., Visser, P.S., Pickett, C.L.: Social Neuroscience: People Thinking About Thinking People. MIT Press, Cambridge (2006)
19. Clark, A.: Being There: Putting Brain, Body, and World Together Again. MIT Press, Cambridge (1998)
20. Crick, F., Koch, C.: Constraints on cortical and thalamic projections: the no-strong-loops hypothesis. Nature **391**, 245–250 (1998)
21. Damasio, A.R.: Descartes' Error: Emotion. Reason and the Human Brain. Papermac, London (1994)
22. Damasio, A.R.: The Feeling of What Happens, Body and Emotion in the Making of Consciousness. Harcourt Brace, New York (1999)
23. Damasio, A.R.: Looking for Spinoza. Vintage books, London (2003)
24. Damasio, A.R.: Self Comes to Mind: Constructing the Conscious Brain. Pantheon Books, NY (2010)
25. Decety, J., Cacioppo, J.T. (eds.): Handbook of Social Neuroscience. Oxford University Press, New York (2010)
26. Decety, J., Ickes, W.: The Social Neuroscience of Empathy. MIT Press, Cambridge (2009)
27. Deliens, G., Gilson, M., Peigneux, P.: Sleep and the processing of emotions. Exp. Brain Res. **232**, 1403–1414 (2014)
28. Dolan, R.J.: Emotion, cognition, and behavior. Science **298**, 1191–1194 (2002)
29. Dubois, D., Lang, J., Prade, H.: Fuzzy sets in approximate reasoning, Part 2: logical approaches. Fuzzy Sets Syst. **40**, 203–244 (1991). North-Holland
30. Dubois, D., Prade, H.: Possibility theory, probability theory and multiple-valued logics: a clarification. Ann. Math. Artif. Intell. **32**, 35–66 (2002)
31. Eich, E., Kihlstrom, J.F., Bower, G.H., Forgas, J.P., Niedenthal, P.M.: Cognition and Emotion. Oxford University Press, New York (2000)
32. Elzas, M.S.: Organizational structures for facilitating process innovation. In: Real Time Control of Large Scale Systems, pp. 151–163. Springer, Heidelberg (1985)
33. Flory, P.J.: Network structure and the elastic properties of vulcanized rubber. Chem. Rev. **35**, 51–75 (1944)

34. Forgas, J.P., Goldenberg, L., Unkelbach, C.: Can bad weather improve your memory? An unobtrusive field study of natural mood effects on real-life memory. J. Exp. Soc. Psychol. **45**, 254–257 (2009)

35. Forrester, J.W.: Lessons from system dynamics modeling. Syst. Dyn. Rev. **3**, 136–149 (1987)

36. Forrester, J.W.: World Dynamics, 2nd edn. Pegasus Communications, Waltham (1973)

37. Freeman, L.C.: The Development of Social Network Analysis: A Study in the Sociology of Science. BookSurge Publishing, Vancouver (2004)

38. Frijda, N.H., Manstead, A.S.R., Bem, S.: The influence of emotions on beliefs. In: Frijda, N.H., et al. (eds.) Emotions and Beliefs: How Feelings Influence Thoughts, pp. 1–9. Cambridge University Press (2000)

39. Funahashi, K., Nakamura, Y.: Approximation of dynamical systems by continuous time recurrent neural networks. Neural Netw. **6**, 801–806 (1993)

40. Gallese, V., Goldman, A.: Mirror neurons and the simulation theory of mind-reading. Trends Cogn. Sci. **2**, 493–501 (1998)

41. Gazzaniga, M.S. (ed.): The Cognitive Neurosciences. MIT Press, Cambridge (2009)

42. Giangiacomo, G.: Fuzzy logic: Mathematical Tools for Approximate Reasoning. Kluwer Academic Publishers, Dordrecht (2001)

43. Giles, J.: Computational social science: making the links. Nature **488**, 448–450 (2012)

44. Goldman, A.I.: Simulating Minds: The Philosophy, Psychology, and Neuroscience of Mindreading. Oxford Univ. Press, New York (2006)

45. Grossberg, S.: On learning and energy-entropy dependence in recurrent and nonrecurrent signed networks. J. Stat. Phys. **1**, 319–350 (1969)

46. Gujar, N., McDonald, S.A., Nishida, M., Walker, M.P.: A role for REM sleep in recalibrating the sensitivity of the human brain to specific emotions. Cereb. Cortex **21**, 115–123 (2011)

47. Harmon-Jones, E., Winkielman, P. (eds.): Social Neuroscience: Integrating Biological and Psychological Explanations of Social Behavior. Guilford, New York (2007)

48. Hebb, D.: The Organisation of Behavior. Wiley, New York (1949)

49. Hesslow, G.: Will neuroscience explain consciousness? J. Theoret. Biol. **171**, 29–39 (1994)

50. Hesslow, G.: Conscious thought as simulation of behaviour and perception. Trends Cogn. Sci. **6**, 242–247 (2002)

51. Hesslow, G.: The current status of the simulation theory of cognition. Brain Res. **1428**, 71–79 (2012)

52. Hirsch, M.: Convergent activation dynamics in continuous-time networks. Neural Netw. **2**, 331–349 (1989)

53. Hopfield, J.J.: Neural networks and physical systems with emergent collective computational properties. Proc. Nat. Acad. Sci. (USA) **79**, 2554–2558 (1982)

54. Hopfield, J.J.: Neurons with graded response have collective computational properties like those of two-state neurons. Proc. Nat. Acad. Sci. (USA) **81**, 3088–3092 (1984)

55. Hubbard, J.C.: The Acoustic Resonator Interferometer: I. The Acoustic System and its Equivalent Electric Network. Phys. Rev. 38, 1011 (1931); Erratum. Phys. Rev. **46**, 525 (1934)

56. Huber, J.F.: The arterial network supplying the dorsum of the foot. Anat. Rec. **80**, 373 (1941)

57. Iacoboni, M.: Mirroring People: the New Science of How We Connect with Others. Farrar, Straus & Giroux, New York (2008)

58. James, W.: What is an emotion. Mind **9**, 188–205 (1884)

59. Jenison, R.L., Rangel, A., Oya, H., Kawasaki, H., Howard, M.A.: Value encoding in single neurons in the human amygdala during decision making. J. Neurosci. **31**, 331–338 (2011)

60. Kelso, J.A.S.: Dynamic Patterns: The Self-Organization of Brain and Behavior. MIT Press, Cambridge (1995)
61. Kim, J.: Philosophy of Mind. Westview Press, Colorado (1996)
62. Kim, J.: Mind in a Physical world: an Essay on the Mind-Body Problem and Mental Causation. MIT Press, Cambridge (1998)
63. Kuipers, B.J.: Commonsense reasoning about causality: deriving behavior from structure. Artif. Intell. **24**, 169–203 (1984)
64. Kuipers, B.J., Kassirer, J.P.: How to discover a knowledge representation for causal reasoning by studying an expert physician. In: Proceedings Eighth International Joint Conference on Artificial Intelligence, IJCAI 1983. William Kaufman, Los Altos (1983)
65. LaBar, K.S., Cabeza, R.: Cognitive neuroscience of emotional memory. Nat. Rev. Neurosci. **7**, 54–64 (2006)
66. Lakoff, G., Johnson, M.: Philosophy in the flesh: The embodied mind and its challenge to western thought. Basic Books (1999)
67. Levin, R., Nielsen, T.A.: Disturbed dreaming, posttraumatic stress disorder, and affect distress: a review and neurocognitive model. Psychol. Bull. **133**, 482–528 (2007)
68. Lewis, M.D.: Self-organizing cognitive appraisals. Cogn. Emot. **10**, 1–25 (1996)
69. Loewenstein, G., Lerner, J.: The role of emotion in decision making. In: Davidson, R.J., Goldsmith, H.H., Scherer, K.R. (eds.) The Handbook of Affective Science, pp. 619–642. Oxford University Press, Oxford (2003)
70. Marques, H.G., Holland, O.: Architectures for functional imagination. Neurocomputing **72**, 743–759 (2009)
71. McCulloch, W.S., Pitts, W.: A logical calculus of the ideas immanent in nervous activity. Bull. Math. Biophysics **5**, 115–133 (1943)
72. McPherson, M., Smith-Lovin, L., Cook, J.M.: Birds of a feather: homophily in social networks. Annu. Rev. Sociol. **27**, 415–444 (2001)
73. Mislove, A., Viswanath, B., Gummadi, K.P., Druschel, P.: You Are Who You Know: Inferring User Profiles in Online Social Networks. Proc. WSDM 2010, 4–6 February 2010, New York City, New York, USA, pp. 251–260 (2010)
74. Mooij, J.M., Janzing, D., Schölkopf, B.: From differential equations to structural causal models: the deterministic case. In: Nicholson, A., Smyth, P. (eds.) Proceedings of the 29th Annual Conference on Uncertainty in Artificial Intelligence (UAI-13), pp. 440–448. AUAI Press (2013)
75. Moreno, J.L., Jennings: H.H.: Statistics of social configurations. Sociometry **1**, 342–374 (1938)
76. Mundt, M.P., Mercken, L., Zakletskaia, L.I.: Peer selection and influence effects on adolescent alcohol use: a stochastic actor-based model. BMC Pediatr. **12**, 115 (2012)
77. Naudé, A., Le Maitre, D., de Jong, T., Mans, G.F.G. Hugo, W.: Modelling of spatially complex human-ecosystem, rural-urban and rich-poor interactions (2008). https://www.researchgate.net/profile/Tom_De_jong/publication/30511313_Modelling_of_spatially_complex_human-ecosystem_rural-urban_and_rich-poor_interactions/links/02e7e534d3e9a47836000000.pdf
78. Nussbaum, M. (ed.): Aristotle's De Motu Animalium. Princeton University Press, Princeton (1978)
79. Ouellet, C., Benson, A.A.: The path of carbon in photosynthesis. J. Exp. Bot. **3**, 237–245 (1951)
80. Ousdal, O.T., Specht, K., Server, A., Andreassen, O.A., Dolan, R.J., Jensen, J.: The human amygdala encodes value and space during decision making. Neuroimage **101**, 712–719 (2014)

81. Pace-Schott, E.F., Germain, A., Milad, M.R.: Effects of sleep on memory for conditioned fear and fear extinction. Psychol. Bull. **141**, 835–857 (2015)
82. Pearl, J.: Causality. Cambridge University Press (2000)
83. Pessoa, L.: On the relationship between emotion and cognition. Nat. Rev. Neurosci. **9**, 148–158 (2008)
84. Pezzulo, G., Candidi, M., Dindo, H., Barca, L.: Action simulation in the human brain: Twelve questions. New Ideas Psychol. **31**, 270–290 (2013)
85. Phelps, E.A.: Emotion and cognition: insights from studies of the human amygdala. Annu. Rev. Psychol. **57**, 27–53 (2006)
86. Pineda, J.A. (ed.): Mirror Neuron Systems: The Role of Mirroring Processes in Social Cognition. Humana Press Inc. (2009)
87. Port, R.F., van Gelder, T.: Mind as Motion: Explorations in the Dynamics of Cognition. MIT Press, Cambridge (1995)
88. Potter, S.M.: What can artificial intelligence get from neuroscience? In: Lungarella, M., Bongard, J., Pfeifer, R. (eds.) Artificial Intelligence Festschrift: The Next 50 Years. Springer, Berlin (2007)
89. Purves, D., Brannon, E.M., Cabeza, R., Huettel, S.A., LaBar, K.S., Platt, M.L., Woldorff, M.G.: Principles of Cognitive Neuroscience. Sinauer Associates Inc., Sunderland (2008)
90. Rangel, A., Camerer, C., Montague, P.R.: A framework for studying the neurobiology of value-based decision making. Nat. Rev. Neurosci. **9**, 545–556 (2008)
91. Rizzolatti, G., Sinigaglia, C.: Mirrors in the Brain: How Our Minds Share Actions and Emotions. Oxford University Press (2008)
92. Rosenblatt, F.: The perceptron: a probabilistic model for information storage and organisation in the brain. Psych. Rev. **65**, 386–408 (1958)
93. Scherer, K.R.: Emotions are emergent processes: they require a dynamic computational architecture. Phil. Trans. R. Soc. B **364**, 3459–3474 (2009)
94. Shalizi, C.R., Thomas, A.C.: Homophily and Contagion are Generically Confounded in Observational Social Network Studies. Sociol. Methods Res. **40**, 211–239 (2011)
95. Sharpanskykh, A., Treur, J.: Modelling and analysis of social contagion in dynamic networks. Neurocomputing **146**, 140–150 (2014)
96. Sotres-Bayon, F., Bush, D.E., LeDoux, J.E.: Emotional perseveration: an update on prefrontal-amygdala interactions in fear extinction. Learn. Mem. **11**, 525–535 (2004)
97. Steglich, C.E.G., Snijders, T.A.B., Pearson, M.: Dynamic networks and behavior: separating selection from influence. Sociol. Methodol. **40**, 329–393 (2010)
98. Storbeck, J., Clore, G.L.: On the interdependence of cognition and emotion. Cogn. Emot. **21**, 1212–1237 (2007)
99. Thelen, E., Smith, L.: A dynamic Systems Approach to the Development of Cognition and Action. MIT Press, Cambridge (1994)
100. Thilakarathne, D.J., Treur, J.: Computational cognitive modelling of action awareness: prior and retrospective. Brain Informatics **2**, 77–106 (2015)
101. Treloar, L.R.G.: The elasticity of a network of longchain molecules. I. Trans. Faraday Soc. **39**, 241–246 (1943)
102. Treur, J.: A Cognitive Agent Model Incorporating Prior and Retrospective Ownership States for Actions. In: Walsh, T. (ed.), Proceedings of the Twenty-Second International Joint Conference on Artificial Intelligence, IJCAI 2011, pp. 1743–1749 (2011)
103. Treur, J.: A cognitive agent model displaying and regulating different social response patterns. In: Walsh, T. (ed.) Proceedings of the Twenty-Second International Joint Conference on Artificial Intelligence, IJCAI 2011, pp. 1735–1742 (2011)

104. Treur, J.: Dynamic modelling based on a temporal-causal network modelling approach. Biologically Inspired Cogn. Architectures **16**, 131–168 (2016)
105. Treur, J.: Network-Oriented Modeling: Addressing Complexity of Cognitive. Affective and Social Interactions. Understanding Complex Systems Series. Springer Publishing (2016)
106. Treur, J., Umair, M.: Emotions as a vehicle for rationality: rational decision making models based on emotion-related valuing and hebbian learning. Biologically Inspired Cogn. Architectures **14**, 40–56 (2015)
107. Valente, T.W.: Social Networks and Health: Models, Methods, and Applications. Oxford University Press, New York (2010)
108. van Gelder, T.: The dynamical hypothesis in cognitive science. Behav. Brain Sci. **21**, 615–665 (1998)
109. van Gelder, T., Port, R.: It's about time: an overview of the dynamical approach to cognition. In: Port, R.F., van Gelder, T.: Mind as Motion: Explorations in the Dynamics of Cognition, pp. 1–43. MIT Press, Cambridge (1995)
110. Westerhoff, H.V., Groen, A.K., Wanders, R.J.A.: Modern theories of metabolic control and their applications. Biosci. Rep. **4**, 1–22 (1984)
111. Wiener, N., Rosenblueth, A.: The mathematical formulation of the problem of conduction of impulses in a network of connected excitable elements, specifically in cardiac muscle. Arch. Inst. Cardiol. Mexico. **16**, 202 (1946)
112. Wilson, M.: Six views of embodied cognition. Psychon. Bull. Rev. **9**, 625–636 (2002)
113. Winkielman, P., Niedenthal, P.M., Oberman, L.M.: embodied perspective on emotion-cognition interactions. In: Pineda, J.A. (ed.) Mirror Neuron Systems: the Role of Mirroring Processes in Social Cognition. Humana Press/Springer Science, pp. 235–257 (2009)
114. Wright, S.: Correlation and Causation. J. Agric. Res. **20**, 557–585 (1921)
115. Zadeh, L.: Fuzzy sets as the basis for a theory of possibility. Fuzzy Sets Syst. **1**, 3–28 (1978). Reprinted in Fuzzy Sets and Systems 100 (Suppl.), 9–34 (1999)

Poster Papers: Politics, News, and Events

Social Contribution Settings and Newcomer Retention in Humanitarian Crowd Mapping

Martin Dittus[1]([✉]), Giovanni Quattrone[2], and Licia Capra[2]

[1] ICRI Cities, UCL, London, UK
martin.dittus.11@ucl.ac.uk
[2] Department of Computer Science, UCL, London, UK

Abstract. Organisers of crowd mapping initiatives seek to identify practices that foster an active contributor community. Theory suggests that social contribution settings can provide important support functions for newcomers, yet to date there are no empirical studies of such an effect. We present the first study that evaluates the relationship between colocated practice and newcomer retention in a crowd mapping community, involving hundreds of first-time participants. We find that certain settings are associated with a significant increase in newcomer retention, as are regular meetings, and a greater mix of experiences among attendees. Factors relating to the setting such as food breaks and technical disruptions have comparatively little impact. We posit that successful social contribution settings serve as an attractor: they provide opportunities to meet enthusiastic contributors, and can capture prospective contributors who have a latent interest in the practice.

Keywords: Humanitarian OpenStreetMap Team · Crowdsourcing · Crowd mapping · Volunteering · Mapathon · Mapping party

1 Introduction

Since its inception in 2010, the Humanitarian OpenStreetMap Team (HOT) has coordinated thousands of volunteers in the creation of maps for humanitarian purposes. All maps are published on the online mapping platform OpenStreetMap (OSM), free to use under a liberal license. Contributors have traced satellite images and digitised data collected in the field in response to Typhoon Haiyan in the Philippines, the earthquake in Nepal in early 2015, and other disasters where humanitarian aid teams required updated maps to coordinate their work. Despite these efforts, vast regions of the inhabited world remain unmapped. In November 2014, HOT and partnering aid organisations launched Missing Maps (MM), a proactive effort to produce new maps before they are needed in times of crisis. However, while emergency response initiatives regularly benefit from new contributor influx resulting from widespread media coverage, MM has to learn how to build mapper capacity in the absence of urgent causes.

Since its inception, MM organisers have refined practices that support the collective effort. Among them are so-called mapathons, social event settings which

ⓒ Springer International Publishing AG 2016
E. Spiro and Y.-Y. Ahn (Eds.): SocInfo 2016, Part II, LNCS 10047, pp. 179–193, 2016.
DOI: 10.1007/978-3-319-47874-6_13

allow regional community groups to come together in person, to learn the practice and to socialise. Organisers of these events pursue several outcomes: to initiate newcomers to the practice, to have them produce maps over the course of the evening, but importantly also to then retain these new contributors for future activities. The volume of mapping data produced at mapathons can easily be measured with existing tools, however it is not currently clear how many attendees remain active afterwards. Do these events have a measurable impact on contributor retention? There is some evidence that communal event settings can play an important role in fostering sustained contributor activity. In a recent study of HOT contributor engagement, it was found that mapping initiatives which organised mapathons had higher newcomer retention rates [7], however it is not yet known whether this can be attributed to the mapathon format itself.

The present study places a focus on the group experience of HOT mappers: mapathons as social contribution environments, and their impact on newcomer retention. The research addresses two primary concerns, to produce new empirical evidence for the effects of colocated practice in online crowd mapping, and to identify some of the contributing factors. We identify three groups of first-time contributors who physically meet in different social contribution settings, and compare their retention to two groups of online contributors who likely never met in person. We find that participation in mapathons can be associated with a significant increase in newcomer retention. In particular, retention was highest for cohorts that meet regularly, compared to cohorts that only met once, or that likely never attended mapathons. A comparative analysis of different aspects of the setting (such as food breaks and technical disruptions) revealed that these had comparatively little impact on longer-term engagement. The results suggest that organisers may be able to increase newcomer retention by offering regular opportunities for social encounter and peer learning.

2   Related Work

In the study of computer-supported cooperative work, work contexts are often considered along two dimensions: whether participants are colocated (they work in the same place), and whether they operate synchronously (they work at the same time) [1,12]. According to this model, HOT online practice is asynchronous and remote, and contributors can act entirely independently of each other. HOT mapathons however are synchronous and colocated. In the context of a global online community, colocation may appear an artificial and needless constraint. Yet a range of literature suggests that it can have important benefits for the experience of first-time contributors.

In distance learning and online education, it was found that **colocated practice** can augment online settings in important ways. The proximity of real-world social interactions can have important benefits for the learning experience of participants, in part by allowing for different forms of knowledge exchange [10]. Similarly, studies of communal software development settings found that social encounters within a community of practice create opportunities for mentoring and learning, provided there is a mix of experiences among attendees [8,18]. Such events can allow newcomers to become expert contributors through situated learning, or so-called peripheral participation, and this can become an important motivation to continue participating in the community [13]. However it was also found that there are tradeoffs between the mix of experts and novices, and task interdependence: events at which experts contribute to independent tasks may yield outputs, but contribute less to community growth. Conversely, events with a larger share of newcomers may contribute to community growth [18]. For open source development groups it was suggested that project attractiveness and individual motivations play an important role in the decision of a newcomer to join a project, however that there can be many hindering factors that lead to an aborted onboarding process [17]. In a study of sustained open source participation it was found that newcomers can particularly benefit if the nature of an early task fosters interactions between participants [8].

In **online practice**, there is evidence that the socialisation experience of first-time contributors can increase their contributions and long-term retention: in an evaluation of Wikipedia socialisation tactics, it is observed that early user retention was increased by the use of welcome messages, assistance, and constructive criticism [4]. On the other hand, invitations to join yielded a steeper decline in contributions by new editors. A further study confirms that a successful early socialisation experience among Wikipedia contributors is associated with and can sometimes predict increased contributor engagement [5]. However the authors also observe that the causal structure between socialisation, motivation, and participation is not entirely clear. Further studies identify similar effects [3,4,9].

To our knowledge there are no quantitative studies of colocated practice in a global **crowd mapping** project, and of its effects on newcomer retention. There are early studies of OSM mapping parties, these are similarly structured around social mapping experiences, however they typically aim to map the local area and involve colocated practice by necessity [11,15]. It is not clear how contemporary HOT mapathons compare to these earlier settings. Among early OSM contributors, it was found that an individual's local geographic knowledge was the most significant driver to contribute [2]. On the other hand, it is conceivable that the global scope and perceived social benefit of HOT mapathons attracts different audiences, and fosters a different form of long-term engagement compared to OSM mapping parties.

3   Research Questions

RQ1: Does mapathon attendance improve newcomer retention?

According to existing research, colocated practice in social contribution environments can be associated with improved newcomer retention rates. Qualitative literature in social psychology, online community studies, and related domains provide support for such an association [6,14,16,19]. However, there is no empirical evidence available to confirm such an effect in a crowd mapping context. We seek to establish such evidence by analysing contributor activities of the HOT crowd mapping community.

RQ2: Which specific factors contribute to increased retention?

Organisers of HOT mapathons have some influence on the setting and format of the events, however to date they have no basis to justify certain choices. In particular, it is not currently known which factors of the setting may affect subsequent newcomer retention. We seek to identify specific aspects of social contribution settings that have an impact on newcomer retention, based on the observation of HOT mapathons and an empirical analysis of their outcomes.

4   Methodology

All our analyses are based on a public record of HOT contributions. We captured the contribution activity of first-time mappers belonging to two separate mapathon cohorts in London, observing a total of 14 events. Subsequent newcomer retention is compared with that of an online control group, a set of first-time online contributors who did not attend the London mapathons. We further compared these cohorts to a second control group of participants of the Arup "Mappy Hour", at an employee-initiated regular mapping event. Furthermore, we co-developed a set of mapathon features in a workshop with MM organisers. Participation outcomes of the 14 mapathons were then compared in relation to these features. The following sections explain these steps in more detail.

4.1   HOT Contribution History

A primary data source for the research is the HOT Tasking Manager, a website which helps coordinate the work of thousands of online contributors while reducing edit conflicts.[1] It presents a list of currently active HOT projects, along with contextual information and mapping instructions. Within each project, work is divided into smaller tasks. Contributors start by selecting a specific project and task, and then contribute to the map using OSM tools. The Tasking Manager also serves as a public record of HOT participation: every project records a list of its past contributors. A further data source for the research is the full OSM edit history, a large public data set which captures OSM map contributions over time. All HOT mapping activity takes place on OSM, and the map contributions by HOT volunteers are contained in this edit history. The full data set is freely available for download.[2]

In a preparatory stage, we identified the map contributions for every Tasking Manager project. Since summer 2015, OSM editing tools automatically annotate

---

[1] http://tasks.hotosm.org.
[2] http://planet.osm.org/planet/full-history/.

changesets with a HOT project identifier, which makes such an identification straightforward. In cases where this was not provided, edits were instead identified based on their location, date, and contributor. This provided us with the full set of HOT mappers, their contributions to HOT, and any further contributions they made to OSM which were not linked to HOT activities.

4.2   Study Period

From November 2014, MM organisers in London started hosting regular mapathons that were open to the public. These provided us with an opportunity to observe contributor retention over time. The first MM mapathon marks the start of our study period. We seek to study participants after their first attendance for a subsequent period of up to 90 days. Our evaluation is based on a snapshot of the OSM edit history that was published on 11th of January 2016, which means the cutoff date for the inclusion of an event is 13th of October 2015. The study considers newcomer activity between the date of the first mapathon on 24th of November 2014, and the last mapathon held on 6th of October 2015. In this period, a total of 14 mapathons took place.

4.3   Study Cohorts

Mapathon Cohorts: Monthly and Corporate. The organisers of Missing Maps mapathons in London are affiliated with the British Red Cross and Médecins Sans Frontières. Throughout the study period, two types of mapathons were organised by this team. Ongoing *monthly mapathons* are open to the wider public, and hosted at a different venue every month. Event sizes are limited by venue capacity rather than interest, and typically vary between 50 and 100 people. Events start in the early evening on a weekday, and typically last three hours. From early 2015, MM further organised a number of *corporate mapathons* for staff members at large corporations, these are one-off events that are not open to the public. The setting and format is comparable to monthly mapathons, however the attendee mix differs in some important ways. Typically all attendees are first-time contributors, and training is limited to basic mapping techniques. According to organisers, participants tend to be office workers and highly computer literate.

These form our mapathon cohorts:

– 11 monthly mapathons between 24th November 2014 and 6th October 2015.
– 3 corporate mapathons on 12th February, 15th May, and 6th October 2015.

For our analysis we seek to identify newcomers who attended these events, and then observe their activity in the subsequent days and weeks. However, there is no public register of HOT mapathon attendance. Instead, we estimate event attendance based on a limited set of information that is readily available: event dates and times, and the list of HOT projects which were worked on during each event. Event dates are generally made public, for example on the

MM homepage[3]. Project lists were collected by participating in the events, or by consulting with organisers after the fact, and in total comprise 19 HOT projects across the 14 mapathons. Since MM mapathons involve proactive HOT mapping initiatives rather than urgent crises, their projects are not listed in a prominent position on the Tasking Manager homepage. Instead remote mappers need to make a conscious effort to find them, either by paging through the listing of active projects, or by searching for them by name. As a result, mapathon activity is clearly visible in the contribution timelines of these projects, and there are only few contributors in the hours before or after a mapathon.

We further identified HOT newcomers among these attendees, first-time mappers with at most one prior day of OSM contributions. The threshold of one day was chosen because attendees are generally asked to sign up to OSM and make some test contributions before the event. As a result, it is plausible that many newcomers may already have made some minor contributions to the map before they attend their first event.

Table 1. Estimated attendance at the 14 mapathons under study, including the number and share of first-time attendees.

Date	Cohort	Attendees	Newcomers	% newcomers
2014-11-24	Monthly	64	37	57.8 %
2014-12-15	Monthly	58	24	41.4 %
2015-01-27	Monthly	52	16	30.8 %
2015-02-12	Corporate	50	44	88.0 %
2015-02-24	Monthly	49	25	51.0 %
2015-03-31	Monthly	62	29	46.8 %
2015-04-28	Monthly	51	19	37.3 %
2015-05-15	Corporate	191	174	91.1 %
2015-06-02	Monthly	27	6	22.2 %
2015-07-07	Monthly	51	15	29.4 %
2015-08-04	Monthly	87	49	56.3 %
2015-09-01	Monthly	41	15	36.6 %
2015-10-06	Corporate	30	28	93.3 %
2015-10-06	Monthly	69	24	34.8 %

Table 1 summarises our attendee estimates per mapathon, accounting for both newcomers and more experienced participants. In total, more than 600 distinct attendees participated across the 14 events. Among these, we identified 505 newcomers, approximately evenly split between the two cohorts. They represent 82 % of all attendees across the 14 events. The data shows that the share

---

[3] http://www.missingmaps.org/#events.

of newcomers differs significantly between the event cohorts: corporate events on average are attended by 90 % newcomers, while the monthly mapathons are attended by 40 % first-time attendees.

Online Control Groups: Matched MM and Nepal. As first online control group we identified HOT contributors who engaged in comparable work, but likely never attended a mapathon. We identified new HOT contributors in the study period who started with one of the same 19 MM projects used for mapathons, but who were not among the attendees identified for these events. In total, 550 first-time HOT contributors matched these criteria. Some of these may have attended mapathons in other cities, however the large sample size and long study period makes it likely that a significant share of this group started as online contributors. This group comprises our "matched" online cohort: contributors who started out doing the same work as MM mapathon attendees, but who were unlikely to have done so at a mapathon. That is, they likely joined the crowd mapping platform online.

As a further control group we added a second online cohort of newcomers who started mapping during an urgent disaster event, contributing to a different set of projects than the other groups. This group was included so we could compare the previous settings to a different kind of stimulus which may feasibly attract new engaged mappers. We selected HOT newcomers who joined to help with the Nepal emergency response in April 2015, their initial contributions were to urgent projects that were focused on emergency response mapping. None of these volunteers attended a MM mapathon in London when they first started mapping. In total, 4,518 first-time HOT contributors fall into this group, they comprise the "Nepal" online cohort.

Arup Mappy Hour. As a final point of reference we chose Arup "Mappy Hour" participants, these are staff members of the multinational engineering consultancy Arup who regularly meet to contribute to HOT. Their office setting may be comparable to that of corporate mapathons, although their events are peer-organised by staff members, not external organisers.[4] According to organisers, Arup Mappy Hour emerged in early 2015 out of the independent activities of multiple staff members. Mappy Hour groups are comparatively small (under 20 attendees). Mappy Hour attendees can be identified in the HOT contribution history because they annotate their HOT contributions with an #Arup tag. Based on these annotations, we found that 135 HOT newcomers had attended a Arup Mappy Hour session in the observation period. These are contributors who had at most one day of prior OSM contribution experience before they attended their first Mappy Hour.

---

[4] http://doggerel.arup.com/mapping-the-worlds-most-vulnerable-regions/.

4.4   Mapathon Features

In order to address RQ2, we sought to identify specific aspects of the mapathon format that may have an impact on participant engagement and retention, with a focus on aspects that are under organiser control. This analysis is restricted to monthly and corporate mapathons only. For these cohorts, it was possible to observe hundreds of participants over the course of the observation period. In comparison, Arup Mappy Hour events are held in a non-public setting. Online cohorts are excluded from the analysis because the contribution setting of participants is not known.

We organised a workshop with MM organisers to identify aspects of a mapathon that may plausibly encourage or discourage continued engagement. Workshop participants developed a set of hypotheses of potential mapathon aspects that may affect participant engagement. Based on these we developed a set of event features which are easily observed, comparable across events, and were identified as potentially important factors because they can affect the actions of and interactions between attendees.

These features are summarised in Table 2. In the following sections we will describe each of the features in turn, discuss our motivation to include it in the study, and describe the associated data collection process.

Table 2. Mapathon features collected per event.

Aspect	Variable	Description
Cohort	*cohort*	Monthly or corporate mapathon?
Attendees	*hot_mappers*	% with prior HOT contributions
	home_mappers	% who mapped at home
	osm_experts	% with >50 days of OSM activity
	prev_attendees	% repeat attendees
Setting	*social_food*	Food served in separate area?
	tech_issues	Larger technical disruptions?
Tool use	*josm_learners*	% newcomers learning JOSM?

Attendee Mix. The attendee mix was considered an important aspect of the attendee experience: according to organisers, attendees who are experienced in mapping can provide important peer support, and the presence of an existing community of practice may affect the motivation of newcomers to keep coming back. Mapathon features relating to the attendee mix were derived from the OSM edit history. We identified the share of attendees with prior contributions to HOT, and separately those who contributed to HOT outside of a mapathon ("at home"). We further identified OSM experts with more than 50 days of prior OSM contributions, this approximately captures the 10% most experienced attendees across all events. Finally, we identified repeat attendees: the share of attendees who have been to at least one previous mapathon. This share of repeat attendees can be regarded as an indicator of the presence of a community of participation.

Food Served in Separate Area. Organisers further debated the role of food as social catalyst. Attendees are always provided with free food. At most events, food is served at the desks, and attendees can resume work while they eat. At some events, however, the setting is more conducive to social interactions between attendees. Organiser experience showed that food that is served in a separate room may disrupt the work, but it also tends to encourage mingling. At three of the 14 mapathons, food was served away from desks, for example as a buffet in a separate room, introducing opportunities to socialise.

Larger Technical Disruptions. Technical issues at mapathons can have a negative impact on the overall event experience when they disrupt the contribution process of many attendees. In some cases this merely interrupts the contribution process for a short time. However more severe disruptions can lead to frustrating experiences for both organisers and attendees, for example when earlier work is lost as a result, and has to be repeated. Four events were disrupted by technical issues that affected all attendees.

Share of Newcomers Learning JOSM. There was further debate among organisers on the topic of tool use. Mapathon organisers train most newcomers in the use of iD first. This editor is web-based, simple to learn, and does not require the installation of software. However, some attendees start by learning JOSM, this editor is more complex but also more powerful, and it allows for faster mapping. It needs to be installed on the attendee's laptop, and this process can take some time. In conversation, organisers stated that contributors with professional GIS and IT backgrounds tended to prefer JOSM over iD, however there is a concern that some newcomers may be discouraged by the more complex interface. Annotations in the OSM edit history allow us to determine which editor was used for a particular contribution. Based on this data we computed the share of JOSM learners at each event.

4.5   Approach

In order to address RQ1, we compare newcomer retention across three event cohorts and two online cohorts, involving the study of hundreds of first-time mappers during their initial activity period of 45 days. We distinguish three aspects of participation: the initial learning of the contribution practice during the event (*initiation*), subsequent mapping at home over the following days (*activation*), optional repeat attendance at a mapathon (*revisit*). Measures of initiation capture whether the participant started with the JOSM editor or iD, whether they abandoned their session within the first 30 min, and whether they completed at least one task. A contributor is considered activated if they contribute to any HOT project in the subsequent 7 days following the mapathon event. A revisit takes place when a contributor attends a subsequent mapathon in the 45 days following their first attendance. The full set of newcomer features is listed in Table 3.

We further compare longer-term *retention* across all five cohorts by means of a survival analysis. We observed contribution activity by HOT newcomers after their first attendance for a period of 90 days to identify the last known moment of contribution. We considered contributors 'dead' if they had been inactive for at least 45 days by the end of this survival period. The last known date of contribution before that point marks their 'death event'.

Our analysis of potential causal factors for RQ2 makes additional use of the observational data collected at monthly and corporate mapathons as described in Sect. 4.4. It seeks to explain the newcomer activation and retention measures listed in Table 3 by considering the event setting, attendee mix, and attendee tool choice as shown in Table 2. All analyses were computed on a per-user basis, first with a pairwise Spearman correlation, and finally as a logistic regression model to explain the particular outcomes for all first-time attendees. In regression models we further included aggregate outcome measures as control variables: the share of attendees who have been successfully initiated, activated, or retained at each event (*initiation_rate*, *activation_rate*, *retention_rate*). Before analysis we standardised numerical features using z-scores, so that all variables have a mean of 0 and a standard deviation of 1.

Table 3. Attendee features computed per mapathon newcomer.

Phase	Variable	Description
Initiation	*josm*	Started with JOSM?
	abandoned	Active for less than 30 min?
	completed	Submitted at least one task?
Activation	*active_{7d}*	Active in the first week?
Revisit	*revisit*	Repeat mapathon attendance?

5   Results

5.1   RQ1: Newcomer Retention

Activation and revisit rates across the 14 mapathons are shown in Table 4. On average, at monthly mapathons 11.9 % of newcomers are activated in the following 7 days, and 4.6 % attend a subsequent mapathon. In comparison, at corporate events on average only 3.6 % newcomers are activated and 2.0 % revisit. These numbers indicate that the two mapathon cohorts have markedly different retention profiles.

To confirm this we computed survival functions for each cohort based on a Kaplan-Meier estimate with a 98 % confidence interval. A corresponding survival plot is shown in Fig. 1, this also includes the two online cohorts. The monthly mapathon cohort had the highest predicted retention rates, with a 20 % chance of newcomer survival after 28 days. In contrast to this, the corporate mapathon cohort had the lowest retention rates, with a near-zero likelihood of survival in the same amount of time. In comparison, the matched online cohort had a survival rate of 6 %, and the Nepal online cohort 2 % after 28 days. A pairwise logrank test confirmed that the four cohorts have distinct survival distributions ($p < 0.001$).

Compared to online cohorts, the Arup Mappy Hour cohort has a significantly higher retention rate, as the survival plot in Fig. 2 shows. Retention of this group is comparable to the highly engaged monthly mapathon group: after 28 days, almost 25 % of first-time contributors are still actively contributing to HOT.

Table 4. Activation and repeat attendance rates among first-time mapathon attendees.

Date	Type of event	% activated	% revisits
2014-11-24	Monthly	13.5 %	0.0 %
2014-12-15	Monthly	8.3 %	8.3 %
2015-01-27	Monthly	12.5 %	6.3 %
2015-02-12	Corporate	6.8 %	0.0 %
2015-02-24	Monthly	16.0 %	4.0 %
2015-03-31	Monthly	3.4 %	3.4 %
2015-04-28	Monthly	5.3 %	0.0 %
2015-05-15	Corporate	1.7 %	2.3 %
2015-06-02	Monthly	50.0 %	16.7 %
2015-07-07	Monthly	6.7 %	6.7 %
2015-08-04	Monthly	14.3 %	6.1 %
2015-09-01	Monthly	20.0 %	6.7 %
2015-10-06	Corporate	10.7 %	3.6 %
2015-10-06	Monthly	8.3 %	4.2 %

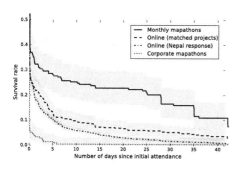

Fig. 1. Newcomer survival rate for mapathon and online cohorts, with 98 % confidence intervals.

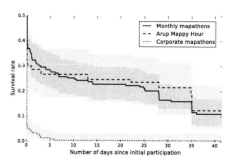

Fig. 2. Newcomer survival rate for mapathon and Arup Mappy Hour cohorts.

5.2   RQ2: Newcomer Retention Factors

Table 5 shows significant associations between mapathon features and event outcomes, as identified with a pairwise correlation analysis. According to these results, only two of the hypothesised mapathon features have a significant link to participant activity during the event. First-time mappers at monthly mapathons with a higher share of first-time JOSM users were more likely to complete at least one task ($\rho_S = 0.18$, $p < 0.01$). On the other hand, technical problems during a monthly mapathon negatively affected newcomer task submission ($\rho_S = -0.16$, $p < 0.02$). In neither case was there a significant association to subsequent activation or retention outcomes. In other words, during monthly mapathons, technical problems affected performance during the event, but they did not necessarily carry any longer-term engagement effects. Only one feature was significantly associated with a subsequent activation outcome: newcomers who contributed at least one task were more likely to contribute on at least one more day in the following week ($\rho_S = 0.16$, $p < 0.02$). Pairwise correlation found no significant associations for newcomer retention.

However it is not certain that these associations are indicators of real effects. We sought to assess the relationship of the effects by means of logistic regression models. Our models included the observations for both monthly and corporate events, taking into account all mapathon features and engagement outcomes. A model to explain activation of individual newcomers confirmed the relationship between task completion and activation, however had a bad model fit ($pseudoR^2 = 0.146$), and none of the other parameters were found significant. A further logistic regression model to explain newcomer repeat attendance had bad model fit, with no regression parameters found significant.

Table 5. Significant correlations between mapathon features and event outcomes, for monthly mapathons only.

Aspect	Variable	Outcome	ρ_S	p
Setting	$tech\_issues$	$completed$	-0.16	<0.02
Tool use	$josm\_learners$	$completed$	0.18	<0.01
Initiation	$completed$	$active_{7d}$	0.16	<0.02

6   Discussion and Implications

The monthly mapathon format appears to be working well, many attendees were retained for longer periods. On average, 50 % of monthly mapathon attendees were repeat visitors. Around 10 % of first-time attendees subsequently mapped at home in the first week, and a similar proportion returned to a future event. Retention rates of Arup Mappy Hour participants were similarly high. However there are clear differences across the remaining cohorts. Newcomers at corporate mapathons contributed to the same projects, but were rarely retained. Only few attendees at these events were activated in the first week. Similarly, online participants who contributed to the same projects were approximately three times less likely to be retained for future work than the monthly cohort.

A potentially important difference between the settings is the frequency at which social events are held. The two cohorts with the highest retention rates hosted regular events, the remaining cohorts were one-offs or online cohorts. The survival curves of the top cohorts show drops in retention after 28 and 35 days, visible as steps in Figs. 1 and 2, which would correspond to a monthly event frequency. This suggests that such future events may foster continued newcomer participation. However our observational data does not allow us to isolate event frequency from other factors, such as the attendee mix.

Furthermore, event frequency alone does not account for the fact that a difference in newcomer activation rates is immediately apparent within the first days after each event: activity levels for corporate and online cohorts already drop within the first few days. The reasons for this consistent difference across cohorts are currently unclear.

In addition to these effects, our results provide empirical evidence that sustained engagement relates to factors of the individual. People who managed to complete at least one task during their first mapathon were more likely to remain engaged in future activities. This effect was found regardless of the setting.

6.1 Implications

Based on this research we make the following recommendations to organisers:

– The cohort that held regular monthly events has the highest newcomer retention. Based on this, we suggest that organisers of other HOT groups provide regular opportunities for existing and latent enthusiast contributors to come together. This is difficult to achieve at large scale, however there may be opportunities to provide similar social experiences in an online setting.
– In particular, we suggest to experiment with forms of online support that imitate the mapathon experience: expert guidance, peer support, the presence of a community of practice, and other aspects.
– The most engaged mapathon cohort also had the most diverse mix of attendees in terms of prior experience, from newcomers to highly experienced mappers. It is feasible that newcomers can be swayed by others' enthusiasm, even if they are selected from a cohort that is unlikely to be retained. We recommend to experiment with the attendee mix, for example by organising events where corporate and monthly mapathon groups come together.
– Corporate mapathons has the lowest newcomer retention, although their event format was comparable to other mapathons. This difference in outcome may be related to the recruiting strategy: some participant groups may be more likely to become engaged mappers that others. We recommend to carefully evaluate the outcomes of such recruiting efforts, and to focus on groups that are more likely to be interested in sustained participation.

7   Conclusion and Future Work

A growing body of evidence suggests that sustained contributor engagement also has to do with aspects of the participation context and attendee selection, rather than just specific details of the setting and contribution process. In particular, there is evidence for the importance of regular social events and the presence of an existing community of practice.

Findings from this study can form a basis on which organisers can design interventions. Introducing newcomers to an existing and active community of practice may have a longer lasting effect than just the demonstration of the contribution process alone. Furthermore, instead of aspiring for indiscriminate growth, it may be advisable to identify prospective contributors who already have a propensity for the practice, and who may already be embedded with existing contributor communities.

Our study captures the presence of potential socialisation triggers at each mapathon, but we do not determine whether people actually made use of them. Attendee surveys could identify which factors were involved in the decision to remain engaged. Studies could further test for a match between participant motivations and project needs, prior experience with the practice outside of an OSM context (such as GIS experience), the presence of social ties, and related motivational factors.

Further research is needed to identify other unobserved factors which can help improve our model fit. For example, the work could be augmented with a comparative study of mapathons in other cities, and organised by different teams. Such a study could seek to confirm the observed relationships between attendee selection, event setting, and subsequent newcomer retention. Additionally it could seek to observe differences in recruiting and organising practices that were not captured by our London-focused study. At the time of writing, HOT enthusiasts are organising mapathons in a growing number of cities around the world, and there is similar growth in informal mapping groups at universities and other institutions.

References

1. Baecker, R.M.: Readings in Human-Computer Interaction: Toward the Year 2000. Morgan Kaufmann, San Francisco (2014)
2. Budhathoki, N.R.: Participants' motivations to contribute geographic information in an online community. Ph.D. thesis, University of Illinois (2010)
3. Burke, M., Marlow, C., Lento, T.: Feed me: motivating newcomer contribution in social network sites. In: Proceedings of SIGCHI 2009, pp. 945–954. ACM (2009)
4. Choi, B., Alexander, K., Kraut, R.E., Levine, J.M.: Socialization tactics in Wikipedia and their effects. In: Proceedings of CSCW 2010, pp. 107–116. ACM (2010)
5. Ciampaglia, G.L., Taraborelli, D., Francisco, S.: MoodBar: increasing new user retention in Wikipedia through lightweight socialization. In: Proceedings of CSCW 2015, pp. 734–742 (2015)

6. Clary, E.G., Snyder, M., Ridge, R.D., Copeland, J., Stukas, A.A., Haugen, J., Miene, P.: Understanding and assessing the motivations of volunteers: a functional approach. J. Pers. Soc. Psychol. **74**(6), 1516 (1998)
7. Dittus, M., Quattrone, G., Capra, L.: Analysing volunteer engagement in humanitarian mapping: building contributor communities at large scale. In: Proceedings of CSCW 2016, pp. 107–116. ACM (2016)
8. Fang, Y., Neufeld, D.: Understanding sustained participation in open source software projects. J. Manage. Inf. Syst. **25**(4), 9–50 (2009)
9. Farzan, R., Kraut, R., Pal, A., Konstan, J.: Socializing volunteers in an online community: a field experiment. In: Proceedings of CSCW 2012, pp. 325–334. ACM (2012)
10. Haythornthwaite, C., Kazmer, M.M., Robins, J., Shoemaker, S.: Community development among distance learners: temporal and technological dimensions. J. Comput. Mediated Commun. **6**(1) (2000). Blackwell Publishing Ltd. ISSN: 1083-6101. http://dx.doi.org/10.1111/j.1083-6101.2000.tb00114.x
11. Hristova, D., Quattrone, G., Mashhadi, A., Capra, L.: The life of the party: impact of social mapping in OpenStreetMap. In: Proceedings of ICWSM 2013, pp. 234–243 (2013)
12. Johansen, R.: Groupware: Computer Support for Business Teams. The Free Press, New York (1988)
13. Lave, J., Wenger, E.: Situated Learning: Legitimate Peripheral Participation. Cambridge University Press, Cambridge (1991)
14. Nov, O.: What motivates Wikipedians? Commun. ACM **50**(11), 60–64 (2007)
15. Perkins, C., Dodge, M.: The potential of user-generated cartography: a case study of the OpenStreetMap project and Mapchester mapping party. North West Geogr. **8**(1), 19–32 (2008)
16. Schervish, P.G., Havens, J.J.: Social participation and charitable giving: a multivariate analysis. Voluntas Int. J. Voluntary Nonprofit Organ. **8**(3), 235–260 (1997)
17. Steinmacher, I., Gerosa, M.A., Redmiles, D.: Attracting, onboarding, and retaining newcomer developers in open source software projects. In: Workshop on Global Software Development in a CSCW Perspective (2014)
18. Trainer, E.H., Chaihirunkarn, C., Kalyanasundaram, A., Herbsleb, J.D.: Community code engagements: summer of code & hackathons for community building in scientific software. In: Proceedings of the 18th International Conference on Supporting Group Work, pp. 111–121. ACM (2014)
19. Wilson, J.: Volunteering. Ann. Rev. Sociol. **26**, 215–240 (2000)

A Relevant Content Filtering Based Framework for Data Stream Summarization

Cailing Dong[1]([⊠]) and Arvind Agarwal[2]

[1] Department of Information Systems, University of Maryland, Baltimore County,
Baltimore, USA
cailing.dong@umbc.edu
[2] Palo Alto Research Center (PARC), Palo Alto, USA

Abstract. Social media platforms are a rich source of information these days, however, of all the available information, only a small fraction is of users' interest. To help users catch up with the latest topics of their interests from the large amount of information available in social media, we present a relevant content filtering based framework for data stream summarization. More specifically, given the topic or event of interest, this framework can dynamically discover and filter out relevant information from irrelevant information in the stream of text provided by social media platforms. It then captures the most representative and up-to-date information to generate a sequential summary or event story line along with the evolution of the topic or event. This framework does not depend on any labeled data, it instead uses the weak supervision provided by the user, which matches the real scenarios of users searching for information about an ongoing event. The experiments on two real events traced by Twitter verified the effectiveness of the proposed framework. The robustness of using the most easy-to-obtain weak supervision, i.e., trending topic or hashtag indicates that the framework can be easily integrated into social media platforms such as Twitter to generate sequential summaries for the events of interest. We also make the manually generated gold-standard sequential summaries of the two test events publicly available (https://drive.google.com/open?id=15jRw13i0xARUW3HqBn3BdR45IXk7P2Qj-HO_OFmMW0) for future use in the community.

Keywords: Social media · Data stream · Content filtering · Summarization · Microblog · Twitter

1 Introduction

In the last few years, social media, in particular micro-blogging websites has seen a steep rise in their popularity with increasing number of users contributing the content in terms of short text messages. These short text messages are status updates consists of various and ever-changing topics, ranging from simple status updates about personal life such as *going to visit a friend*, to text messages

E. Spiro and Y.-Y. Ahn (Eds.): SocInfo 2016, Part II, LNCS 10047, pp. 194–209, 2016.
DOI: 10.1007/978-3-319-47874-6_14

about ongoing real-life events such as *FIFA world cup*, to more involved conversations about the topics of general interests such as *global warming, terrorism* etc. Because of the nature of streaming data i.e. large volume with high velocity and variety, users are constantly struggling to keep up with the latest information. Thus, it has become a necessity these days to filter the relevant content from irrelevant one and summarize it according to the topic of interests.

Most of the work on micro-blogging summarization has not put as much effort on relevant content filtering before performing summarization, despite the fact that most of the content is unrelated to the topics of mainstream news. As an example, in Twitter only 3.6 % of the posts are related to the topics of mainstream news [7]. Thus, the performance of summarization largely depends on the quality of the relevant content with regard to the given topic or event.

In this paper, we present a relevant content filtering based framework for data stream summarization, namely **W**eakly **S**upervised **S**tream **F**ilter and **S**ummarizer (WS²FS), which is suitable for both *relevant content filtering* and *sequential summarization*. Unlike classical supervised method that relies on the availability of labeled data, the proposed framework does *not* use any manually labeled data, it instead uses *weak supervision* from users, which can be as simple as topical keywords, or in the form of any rule that can provide global *feature-level* information. When using the most easy-to-obtain supervision, i.e., hashtags, our framework can be treated as an almost unsupervised method, making it practical to be used for summarizing both *personalized events* (i.e. events of personal interests) and *general events* (events of general interests). Another important strength of the framework is its ability to handle streaming data. The framework is modeled as an online classification framework, which evolves i.e., learns from the new data as it becomes available. In general, our contribution in this paper is as follows:

- We propose a relevant content filtering based framework for data stream summarization. It couples the two important tasks in social media, i.e., *relevant content filtering* and *event summarization* in one integrated framework. This framework is not only able to capture the event evolution and dynamically filter out relevant content, but also generates sequential summary or event story line effectively.
- The proposed framework is almost unsupervised since it does not use any manual labeled data[1]. The *weak supervision* it uses can either be done automatically or be provided by information seeker.
- It best simulates the real scenarios of users searching for relevant information with self-defined search queries from any social stream websites. Thus, it can be integrated into any social stream websites such as Twitter readily to generate the summaries of the event of interest in a timely fashion.
- Our experimental results showed that the hashtag-based *weak supervision* produces the best results. As such supervision is easy to obtain, our framework

---

[1] It is not explicitly labeled for the classification task, rather than obtained from the data itself.

can be easily extended to generate both personal event summary and global event summary.

- We make the manually generated summaries of the two test events publicly available, which can be used readily in the community.

The remainder of the paper is structured as follows. Section 2 provides some related work. In Sect. 3, we describe the general structure, major components and detailed implementation of the proposed framework. Section 4 demonstrates the comprehensive experiments on two real events delivered in Twitter, and examines its performance using different types of weak supervisions. Finally we discuss and conclude the paper in Sect. 5.

2   Related Work

2.1   Relevant Content Filtering

Most of the work in content filtering has been based on the following two types of methods: (1) Information Retrieval (IR) based methods, (2) Machine Learning (ML) based method. In IR based methods, a query is formed based on the information that is being sought, and then, the query is executed to find the relevant content. Although in theory, any traditional IR based method can be used for this, streaming nature of the data on micro-blogging website makes it hard to implement. For online streaming social media content, using IR based method that employs pre-built indexes is not feasible. In ML based methods, a typical approach is to build topic specific *supervised* classifier [8]. However, these supervised classification methods have various limitations which makes them less appropriate for streaming data. First of all, supervised classification methods need labeled data. Getting labeled data is both expensive and time consuming. Secondly, supervised classification methods are not easily extensible to new topics. Every time a new topic comes, one has to create new labeled data and then build a new classifier. Since the topics keep evolving in the data stream, it is not reliable to use a fixed labeled dataset to capture the whole event. To the best of our knowledge, our previous work [3] is the first study specifically focusing on filtering relevant content in streaming data.

2.2   Micro-blogging Summarization

Previous work on micro-blogging summarization can be divided into three categories, i.e., frequency-based methods [14,18], graph-based methods [15,19] and context-based methods [2,20]. Frequency-based methods are based on the assumption that if a word or a set of words in a data instance (such as a tweet) has a high frequency of being repeated, the instances containing the set of high-frequency words must be good candidates for generating summary. Based on the similar assumption, graph-based methods build a word graph to capture common sequences of words about the given topic. The path with the highest total weight is regarded as a candidate summary instance. Typical graph-based methods include TextRank [12], LexRank [4] and Phrase Reinforcement

(PR) algorithm [19]. Context-based approaches rate the importance of a data instance not only based on its textual importance, but also based upon other non-textual features, such as user influence, data instance popularity and temporal signals [2]. Although verified to be very effective in generating single-sentence summary, none of these algorithms were specifically designed for or have been used on streaming data. Furthermore, these methods are pure summarization methods assuming the relevant content is ready to use. Putting them into the streaming environment, the effectiveness of these summarization methods would not be guaranteed as they could fail to capture the evolution of the given topic based on *static* keywords. In contrast, our proposed framework integrates both *relevant content filtering* and *summarization*. It dynamically changes its behavior according to the arriving data from the stream. We emphasize here that these summarization methods are not competitors to the proposed framework, they are rather complementary, i.e., any of these summarization methods can be integrated with the *relevant content filter* of our framework.

2.3 Event Tracking and Summarization

Lately, event tracking and summarization has raised lots of attention, where one key task is to detect the relevant content about the event. One major application domain is summarizing scheduled events [1,13]. For instance, Chakrabarti and Punera [1] employed a modified Hidden Markov Model (HMM) to learn the structure and vocabulary of multiple American football games, in order to detect relevant content with regard to future games. Nichols et al. [13] used an unsupervised algorithms to generate summary for sporting events, in which relevant content were extracted by detecting spikes in volume of status updates. Compared with scheduled events which usually have specific "moments" and terminology, tracking unscheduled events are more challenging, but of general applicability. In [16], Osborne et al. classified a tweet as relevant or not based on the score distribution within a set of tweets, and further generated summary by removing redundancy among the selected tweets. In general, most of the previous work focus on detecting data volume changes, extracting sub-topics and further clustering them into the same events [10,11]. Different from these methods, our framework employs more sophisticated techniques grounded in machine learning (i.e. online learning), and it is designed for both scheduled and unscheduled events.

3 Weakly Supervised Stream Filter and Summarizer (WS²FS)

To filter out the relevant content with regard to a given topic or event from a data stream, we need to classify each instance (e.g., each tweet) into "relevant" or "non-relevant" categories, which is a classic binary classification problem. A good classifier for streaming data needs:

- *Reliable training datasets.* For a text stream containing almost infinite set of topics, creating such training datasets through manual annotations is impractical. It calls for an automatic approach to creating reliable training datasets.
- *Maintain the "main thread" and capture the "evolution" of the event.* A good classifier for streaming data should be continuously learning. It should capture not only the main "theme" of the event, but also the content as the event unfolds, which is also an important feature of building a good summarizer on streaming data.

Motivated by the above two important tasks, we propose a **W**eakly **S**upervised **S**tream **F**ilter and **S**ummarizer (**WS^{2}FS**) to filter out relevant content and further generate sequential summary from data stream. The general framework is shown in Fig. 1.

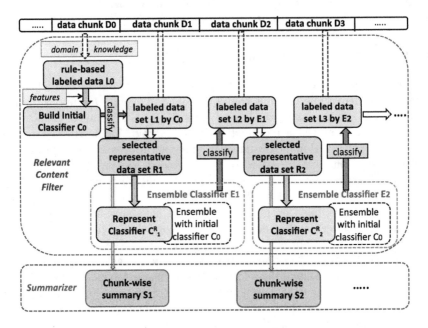

Fig. 1. Framework of WS^{2}FS

3.1   Relevant Content Filter of WS^{2}FS

The general structure of the *relevant content filter* in WS^{2}FS is demonstrated at the top part of Fig. 1. The first step is to define the appropriate size of a stream chunk and thereafter split the data stream into different data stream chunks. The size of the chunk represents the granularity of filtering. It is flexible to define, e.g., according to timestamp, data volume, etc. The next step is choosing an appropriate starting point. Our framework can start from any time or any stream chunk. A good starting point is when the given event starts to emerge (similar to the spike in data volume), which is a good timing to capture the "main thread" of the event. In general, *relevant content filter* consists of three main components: (1) a one-time INITIAL CLASSIFIER C_0 builder which builds C_0 using the first stream chunk D_0, (2) a representative dataset R_i builder and (3) an ENSEMBLE CLASSIFIER E_i builder, for each chunk D_i ($i \geq 1$). In the following, we describe each of these components in detail.

First of all, build the initial classifier C_0. Once the stream is split into chunks, an initial classifier C_0 is built on first chunk D_0. Since D_0 does not have labels, the very first task is to get labels for D_0. One major contribution of the proposed framework is that it does not use any manually labeled data. Instead, the labels of instances in D_0 are created automatically based on *weak supervision* provided by the information seeker. We emphasize here that the labels are not obtained in a classical way, i.e., by asking label for each instance, rather, information seekers provide rules that operate on the whole corpora which in turn produce each label. This component results in a rule-based labeled dataset L_0, which is used to construct the INITIAL CLASSIFIER C_0.

Second and third components correspond to building representative training datasets and classifiers for each of the following chunks. These two components function alternatively, i.e., classifier in chunk i i.e., C_i is used to classify the data in the next chunk i.e., D_{i+1} (in other words get L_{i+1}) which is then further processed to build the classifier for chunk $i+1$, i.e., C_{i+1}. The training dataset for *each* chunk should be such that it captures the dynamic nature of the event in that chunk. In other words, it should contain any new content that appeared in that chunk. We build training dataset for chunk $i + 1$ using classifier from chunk i, followed by some further processing. This further processing consists of using *weak supervision* along with other available information in the dataset such as the number of followers in Twitter (detailed information will be elaborated in the following part). We call such dataset as *chunk representative dataset* and denote it by R_i. The positive instances in R_i are selected in a way that they are highly reliable, and contain subtopics of the given event in chunk i. Thus they capture the "evolution" of the event and set up the up-to-date criteria for filtering relevant data from the next chunk.

In order to build a content filtering classifier for chunk i, we first build a classifier called REPRESENT CLASSIFIER C_i^R upon each R_i. As this classifier is built upon R_i, which mainly captures the localized subtopics that appeared in chunk i, it will fail to capture the main theme of the event. While, the main theme of the event is often captured by the first classifier C_0. In order to capture both the main theme of the event and the updated subtopics, for each chunk i, an ensemble classifier is built using two base classifiers, i.e., chunk-specific REPRESENT CLASSIFIER C_i^R and the INITIAL CLASSIFIER C_0. We call this classifier ENSEMBLE CLASSIFIER E_i. For each instance, its confidence of being a relevant instance is calculated by combining the confidence values produced by both of the two base classifiers, as given by:

$$Conf_{E_i}(x) = (1 - \alpha) * Conf_{C_0}(x) + \alpha * Conf_{C_i^R}(x). \tag{1}$$

The weight α is flexible to set. If the provided *weak supervision* is not strong, we may need to put more weight on C_0 to better maintain the main "theme". On the contrary, if we are more interested in the evolution of the given event, we can put more weight on C_i^R to better capture the sub-topics.

Chunk Representative Dataset R_i builder. How to create *chunk representative dataset R_i* to build REPRESENT CLASSIFIER C_i^R is the key component of *relevant content filter* in WS^{2}FS. As mentioned before, the positive instances R_i^+ in R_i should capture the evolution of the given event. This evolution is usually captured by the words that most frequently co-occurred with the provided *weak supervision* or its induced rules. We call such words as *companion words*. It is worth noting that the purpose of the *weak supervision* goes beyond getting initial labels for D_0. They are used all along the text stream to obtain *companion words*.

Procedure 1. Qualification Criteria $\mathcal{Q}$ of Selecting R_i

---

Input: Labeled data L_i from D_i and corresponding confidence value $Conf_{E_i}$ $(i \geq 0)$
Input: Given topic of interest T
Input: #candidate instances $= p$; #companion words $= m$; #representative instances $= n$
Output: *chunk representative dataset R_i sorted by relevancy*

$\qquad\qquad\qquad\qquad\qquad\qquad\qquad\qquad\qquad$ ▷ **Select companion words**

1: $Avg^+(Conf_{E_i})$: average *confidence* value of positive instances L_i^+
2: $L_i^{+'} \leftarrow \{x \in L_i^+ : Conf_{E_i}(x) \geq Avg^+(Conf_{E_i}(x))\}$
3: $L_i^{C+} \leftarrow$ sort $L_i^{+'}$ by *correlation* descendingly
4: $L_i^{R+} \leftarrow$ sort $L_i^{+'}$ by *credibility* descendingly
5: $L_i^{cand+} \leftarrow$ (top-p (L_i^{C+}) $\cap$ top-p (L_i^{R+})) $\cup$ [top (L_i^{C+})]
6: $Comp_i \leftarrow$ top-m most frequent words in L_i^{cand+}

$\qquad\qquad\qquad\qquad\qquad\qquad\qquad\qquad\qquad\qquad$ ▷ **Select distant words**

7: $Avg^-(Conf_{E_i})$: average *confidence* value of negative instances L_i^-
8: $L_i^{-'} \leftarrow \{x \in L_i^- : (Conf_{E_i}(x) \leq Avg^-(Conf_{E_i}(x))\}$
9: $L_i^{C-} \leftarrow$ sort $L_i^{-'}$ by *correlation* ascendingly
10: $L_i^{R-} \leftarrow$ sort $L_i^{-'}$ by *credibility* ascendingly
11: $L_i^{cand-} \leftarrow$ (top-p (L_i^{C-}) $\cap$ top-p (L_i^{R-})) $\cup$ [top (L_i^{C-})]
12: $Distant_i \leftarrow$ top-m most frequent words in L_i^{cand-}

$\qquad\qquad\qquad\qquad$ ▷ **Select representative instances and build R_i**
$\qquad\qquad\qquad\qquad\qquad\qquad\qquad\qquad\qquad\qquad\qquad\qquad$ ▷ **check diversity**

13: $Common_i = Comp_i \cap Distant_i$
14: **if** ($\frac{|Common_i|}{|Comp_i|} < 0.5$)&($\frac{|Common_i|}{|Distant_i|} < 0.5$) **then**
15: $\qquad Comp_i \leftarrow Comp_i \cup T$
16: $\qquad S_i^+ \leftarrow$ top-$n(L_i^{+'})$ on frequency of $Comp_i$
17: $\qquad S_i^- \leftarrow$ top-$n(L_i^{-'})$ on frequency of $Distant_i$
18: $\qquad R_i \leftarrow S_i^+ \cup S_i^-$
19: **end if**

The **qualification criteria** $\mathcal{Q}$ for selecting representative dataset R_i is given by the pseudocode in Procedure 1. The selection of *companion words* (line 1–6) depends on the "3C" factors, i.e., *Confidence, Correlation* and *Credibility*. *Confidence* means the confidence of being relevant as judged by the ENSEMBLE CLASSIFIER E_i. *Correlation* measures the semantic relatedness to the given event. And *credibility* measures the reliability of the "source" of the instance. Specifically, line 1–2 make sure the candidate positive instances have high *confidence*. Line 3 and line 4 create two sorted list of instances according to *correlation* and *credibility* in descending order. The qualified instances are selected as the intersection of the top-p instances in the two sorted lists, guaranteeing both high *correlation* and *credibility*. If the number of qualified instances is smaller than p, the top instances in the sorted list based on *correlation* are added as supplements (line 5). Finally, the top-m most frequent words in the qualified instances are selected as *companion words* (line 6). In order to build REPRESENT CLASSIFIER, we also need to select representative negative instances. The procedure is described in line 7–12 in Procedure 1, which is based on low values of *confidence, correlation*, and *credibility*. We call the words selected from the these negative instances as *distant* words. Finally, the instances that have highest frequency of *companion words* and *distant words* are selected as representative positive and negative instances, respectively (line 16–17). Before doing this, we need to check the *diversity* of the two sets of words (line 14). If the ratio of the size of the common words to the size of either set is low, it means the two sets of words have large diversity and therefore are reliable enough to be selected as representative instances. Otherwise, the representative dataset building process will be skipped from this given stream chunk.

3.2   Summarizer of WS²FS

The *summarizer* of WS²FS is seen at the bottom of Fig. 1. For each chunk, the *relevant content filter* in WS²FS has produced a list of representative positive instances R_i^+, sorted based on *confidence, correlation* and *credibility*. The final ranking signifies the importance and representativeness of the instances. Thus, the top-ranked instances in each stream chunk can be regarded as good candidates to generate the *chunk-wise summary* of the given event. Later, the *summarizer* of WS²FS combines all the chunk-wise summaries in chronological order, and further process it to generate the final *sequential summary*.

   A key point here is how to select a candidate data instance as a final chunk-wise or sequential summary instance. As the resultant summary should cover as many aspects of the event as possible, we use *diversity* as the selection measurement. *Diversity* refers to the opposite of redundancy here. That is, all the instances should minimally overlap with each other within the final *chunk-wise summary* and *sequential summary*. In our work, we employ ROUGE-L [9] to calculate the degree of overlapping between a candidate instance and each of the already chosen summary instances. ROUGE-L calculates the statistics (average recall, precision and F1 values) about the longest common subsequence between two string. Obviously, the higher the value is, the less diversity the candidate instance will bring. A threshold $\theta_{diversity}$ can be set flexibly based on the desired level of diversity.

4   Experiments

Twitter is a representative source of data stream. In our experiment, we apply WS^{2}FS on a Tweet stream dataset TWEETS2011 from TREC 2011 Microblog Track[2]. The dataset contains 16 million tweets sampled between January 23rd and February 8th, 2011. It also provides a set of manually labeled relevant tweets for 50 topics. Two important events happened during the two weeks of sampled tweets, *Moscow airport bombing* and *Egyptian revolution*. In this dataset, the first event is descried by one topic *Moscow airport bombing*, while the second event covers three topics, *Egyptian curfew*, *Egyptian evacuation* and *Egyptian protesters attack museum*. We combine all the tweets related to the three topics and associate them with event *Egyptian revolution*. As the original annotation is based on both the tweet text and the content of the URL [6], we asked two annotators to re-annotate the test dataset only based on tweet text. The inter-annotator agreement is 0.952 measured by Cohen's kappa coefficient.

4.1   Experimental Settings

(1) Stream chunks. We split the data stream into different chunks by the creation "date" of the tweets.

(2) Weak supervision. On Twitter, the related content is searched by simply applying keyword match on tweets. Thus, we define the *weak supervision* as: *"A tweet is a relevant instance if any topical words (case insensitive) appear in the tweet"*. The results of applying the *weak supervision* on the testing dataset are treated as our *Baseline* method.

In reality, different users may provide different types/levels of *weak supervision* about the event. As event *Egyptian revolution* covers three aspects in the original datasets, we simulate the scenarios of users having the following three types of *weak supervision* about this event:

- *Type-a: general concept* – when a user cares about the event in general, described by topical words "Egyptian, revolution";
- *Type-b: trending topic/hashtag* – when a user is interested in the trending topic or hashtag "#Jan25";
- *Type-c: specific aspects* – when a user is interested in specific aspects of the event, described by the keywords "Egyptian, protesters, attack, museum, curfew, evacuation" (these keywords are chosen based on the topical words in the three topics in the original dataset).

(3) Features. WS^{2}FS is designed as a general framework that does not rely on any specific features, thus we only use the following three type of features:

- *Keywords-feq:* the total frequency of topical key words in the tweet text. It is used to measure one of the "3C" factors, i.e., *correlation*.

---

[2] http://trec.nist.gov/data/tweets/.

- *Status*: We define status as the normalized ratio between the number of followers and followings of the user who wrote or retweeted the current tweet. Intuitively, the higher the value of *status*, the more reliable is the tweet. We use it to measure the *credibility* in qualification criteria $\mathcal{Q}$.
- *Content-words*: the words in tweet text that carry the content. We use Twitter NLP toolkit[3] [5] to get part-of-speech (POS) tag for each word in tweet text. Only the content words with specific POS tags ("N", "^" , "S", "Z", "V", "A", "R" , "D") are kept.

(4) Classifier. As *Naïve Bayes* has been verified to be very effective in many text mining related tasks, we use it to build our classifiers. In addition, we set α in Eq. 1 to be 0.5 and 0.8 for events *Moscow airport bombing* and *Egyptian revolution* respectively, as the later involves many sub-events we want to capture.

(5) Other parameters. By testing different values on the parameters in Procedure 1, we finally settled on selecting around 10 % of the instances as candidate instances. In order to avoid topic drifting, only 10 % of the candidate instances are chosen as representative instances.

4.2 Performance of Relevant Content Filter

To demonstrate the performance of *relevant content filter* of WS^{2}FS, we focus on its fundamentals, i.e., *companion words*, and its performance on the fixed testing dataset. In Table 1, we showed the companion words in event *Moscow airport bombing* where the words in bold are event-relevant ones. We can find that at the early stage of the framework, the companion words are closely related to this event. As the topic/event starts to die out or submerge by new topics, the

Table 1. Companion words in event *Moscow airport bombing*

ChunkID	Companion words
Jan-24	**Injured, killed, blast, explosion, 31, dead, bombing, suicide**
Jan-25	**Blast, news, 35, killed, terrorist, attack, bombing, suicide, dead**
Jan-26	**Modern, revenge, russian**, heathrow, girlfriend, video, **bombing**
Jan-27	**Russian**, news, **bombing**, call, **police**, ap, sings
Jan-28	International, san, gatwick, blvd, **domodedovo, terror, attack, bombing**
Jan-29	Malaga, **blast**, shut, **investigators**, orlando, latest, **bombing**
Jan-30	International, passengers, stream, watch, **security,** people, live
Jan-31	London, news, subject, **bomber**, introduce, luton, #egypt
Feb-02	Source, **russian, victim**, dfw, international
Feb-03	Cairns, open, townsville, richmond, international, opening, **security**
Feb-07	**Islamist**, san, **umarov, ordered**, guardian, **doku**, operators, **rebel**
Feb-08	**Claims, umarov, ordered, leader, doku, rebel, bombing, chechen**

---

[3] http://www.ark.cs.cmu.edu/TweetNLP/.

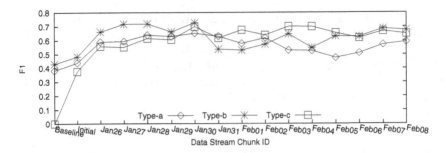

Fig. 2. Performance of *relevant content filter* on event *Egyptian revolution* with different types of *weak supervision*.

selected companion words starts to slightly drift from the main theme. However, when there are updated information, the given topic/event comes alive again. Overall, the companion words selected by *relevant content filter* can capture the evolution of the events to a large extent.

In Fig. 2, we show the F1 score of *content relevant filter* on the testing dataset of event *Egyptian revolution*, with three types of *weak supervision* defined earlier. As we can see, based on *Type-a weak supervision*, it retrieves the relevant content quite well, mainly because general concept is universally acknowledged and widely used to discuss about this event. However, the performance begins to decrease as the event evolves, which is probably because people tend to discuss more detailed aspects of the event along its evolution. Armed with the hashtag based *Type-b weak supervision*, the *relevant content filter* filters out the relevant content with high accuracy in the first few chunks. Later on, as the event evolves, more dynamic "labels" are created to describe the event, so the performance simply based on the hashtag slightly decreases in some chunks. Overall, the trending topic/hashtag born with the event can capture the main theme of the event along with its lifetime. As shown in this figure, the F1 score generated by *Baseline* method with *Type-c weak supervision* is around 0. This is because it simply classifies all the tweets containing any of those keywords as relevant content. But later on, our *content relevant filter* produces much more accurate relevant content as the event evolves. In general, the *relevant content filter* of WS^2FS can effectively capture the dynamically changing relevant content of an event, with different levels of *weak supervision*.

4.3   Performance of Summarizer

Comparison Summarization Methods. We choose the following commonly used classical summarization methods as baseline methods:

(1) *Centroid:* the centroid instance in the dataset is chosen to be the candidate summary instance [17].

(2) *LexRank:* each instance is modeled as a vertex and the edges are created based on the cosine similarity of the TF-IDF vectors of the two vertices. Graph ranking method (e.g. PageRank) is used to select candidate summary instances [4].

(3) *Query-based method:* from the perspective of information retrieval, the summary instance is chosen from the most relevant documents to the given query. It contains the following three different kinds of similarity measures:

 – *QueryCosine (Q1):* uses cosine similarity of the TF-IDF vectors.
 – *QueryCosineNoIDF (Q2):* uses cosine similarity on TF vectors.
 – *QueryWordOverlap (Q3):* uses the overlapping of uni-grams in both document and query to measure the similarity.

Evaluation Criteria for Generated Summaries. We compare these baseline methods with the *summarizer* of WS^{2}FS on both *chunk-wise summary* and *sequential summary*. To guarantee the quality of the gold-standard summaries, we ask the annotators to grasp the "big moments" along the timeline of the events. Specifically, we extract some facts happened on different dates about the event *Moscow airport bombing*, based on its Wikipedia page[4]. For event *Egyptian revolution*, the annotations should capture the key facts following the timelines provided by both its Wikipedia page[5] and the report from Al Jazeera English[6]. The two annotators manually choose the top-3 (if available) most important tweets in each stream chunk as the gold-standard *chunk-wise summary*. Accordingly, using the selection criterion *diversity* described in Sect. 3.2, the top-3 tweets with highest score generated by *summarizer* of WS^{2}FS and all the baseline methods are regarded as their chunk-wise summaries. Each of the two annotators also manually creates a sequential summary along with all the chunks. Using the same selection criterion *diversity*, the *summarizer* of WS^{2}FS and each of the baseline methods also generate their own sequential summaries for both events.

Quality of Chunk-Wise Summary. In Table 2, we show the quality of *chunk-wise summary* generated by different summarization methods on event *Moscow airport bombing*. Due to the space limitations, we only list the average values across all chunks. As we can see, our summarizer of WS^{2}FS produces the best results in terms of *precision, recall* and *F1*. That is, the chunk-wise summary generated by WS^{2}FS are most similar to the manually generated summary.

In Table 3, we demonstrate the best average ROUGE-L scores of chunk-wise summarization produced by corresponding *weak supervision* type (indicated in the parenthesis) on event *Egyptian revolution*. On the whole, our summarizer produces the best results using *Type-b* supervision. For the baseline methods, the best results are usually generated using *Type-b* and *Type-c weak supervision*. This is because they largely rely on topical words, and can not fully explore the general concept (*Type-a*).

---

[4] http://en.wikipedia.org/wiki/Domodedovo_International_Airport_bombing.

[5] http://en.wikipedia.org/wiki/Timeline_of_the_Egyptian_Revolution_of_2011.

[6] http://www.aljazeera.com/news/middleeast/2011/01/201112515334871490.html.

Table 2. Average value of *chunk-wise summary* on event *Moscow airport bombing*.

	Centroid	LexRank	Q1	Q2	Q3	WS^{2}FS
Precision	0.176	0.247	0.518	0.607	0.586	**0.706**
Recall	0.235	0.167	0.385	0.374	0.316	**0.726**
F1	0.200	0.195	0.434	0.449	0.394	**0.714**

Table 3. Average value of *chunk-wise summary* on event *Egyptian revolution*.

	Centroid	LexRank	Q1	Q2	Q3	WS^{2}FS
Precision	0.272 (b)	0.232 (c)	0.192 (c)	0.203 (c)	0.205 (c)	**0.597** (b)
Recall	0.277 (b)	0.287 (c)	0.359 (b)	0.274 (c)	0.336 (b)	**0.576** (b)
F1	0.277 (b)	0.252 (c)	0.242 (c)	0.229 (c)	0.245 (c)	**0.585** (b)

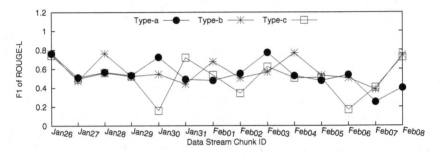

Fig. 3. *Chunk-wise summary* on event *Egyptian revolution* with different types of *weak supervision*.

Furthermore, we are interested in investigating how different types of *weak supervision* affect the performance of our *summarizer* in producing summary in each chunk along with the event evolution. The F-1 values on each chunk are shown in Fig. 3. In general, the performance fluctuates more with *Type-c weak supervision*, probably due to the matching "degree" between its keywords and the corresponding big moments of the event. In the real scenario, what keywords will be associated to an unscheduled event is nearly unable to predict. Thus, the method which can generate good summary based on hashtag (*Type-b*) or general information (*Type-a*) is more useful than the method which rely on specific topical words (*Type-c*), which again verified the practical usefulness of our proposed framework which makes better use of *Type-b* supervision.

Results of Sequential Summarization. Table 4 shows the quality of sequential summaries generated by different methods on event *Moscow airport bombing*, compared with the the gold-standard sequential summary. From this table, we can see that *query-based* methods tend to produce higher precision but lower recall. *Centroid* and *LexRank* produce poor results overall. Our *summarizer* produces the best results. The evaluation of sequential summarization on event *Egyptian revolution* are shows in Fig. 4, with different methods under different types of *weak supervision*. It shows the similar results as of the chunk-wise summaries in Table 3. That is, WS^2FS produces the best results. In terms of *weak supervision*, those baseline methods benefit more from *Type-c weak supervision* as they rely more on topical words. WS^2FS produces the best results with hashtag-based *Type-b weak supervision*, and generates good enough results based on *weak supervision* coming from general concept (*Type-a*) as well. The resultant sequential summary produced by the *summarizer* on the two events are publicly available along with the gold standard summaries[7].

Table 4. Quality of *sequential summary* on event *Moscow airport bombing*.

	Centroid	LexRank	Q1	Q2	Q3	WS^2FS
Precision	0.267	0.439	0.544	**0.682**	0.649	0.631
Recall	0.338	0.177	0.416	0.457	0.374	**0.662**
F1	0.299	0.252	0.471	0.547	0.474	**0.646**

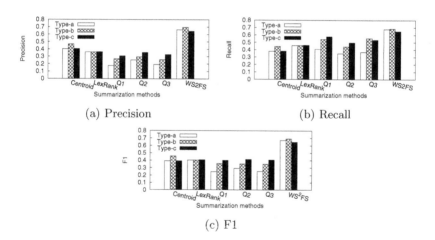

(a) Precision   (b) Recall

(c) F1

Fig. 4. Quality of *sequential summary* on event *Egyptian revolution* with different types of *weak supervision*.

---

[7] (https://drive.google.com/open?id=15jRw13i0xARUW3HqBn3BdR45IXk7P2Qj-HO_OFmMW0).

5   Discussion and Conclusion

In this paper, we have presented a relevant content filtering based framework for data stream summarization in social media platforms. One of the important inputs to our system is *weak supervision*, this weak supervision should not only be easy to obtain for a variety of events but also be effective in generating meaningful summaries. In our experiments, we have shown that it is possible to achieve both goals simultaneously. More specifically, we have experimented and shown results for both *relevant content filtering* and *summarization* based on three different types of *weak supervision* that are relatively easier to obtain. Among these three types of weak supervision, *type-b* weak supervision (i.e. trending-topic/hashtag based) has performed the best. This hashtag-based weak supervision also happen to be the one that can be most easily obtained. Other than its easy availability, hashtag-based weak supervision also has an immediate practical advantage. For the users who tweet with a particular hashtag, the proposed summarization method can be used to provide them an up-to-date summary of the events related to their hashtags, which is a very useful feature to be integrated in social media. In addition to the personalized event summary, the framework can also be used to provide global event summary based on the hashtags trending on the site. Such advantage demonstrates the practicality of our method and strengthens the argument for its adaptation into real world applications.

References

1. Chakrabarti, D., Punera, K.: Event summarization using tweets. In: ICWSM (2011)
2. Chang, Y., Wang, X., Mei, Q., Liu, Y.: Towards Twitter context summarization with user influence models. In: Proceedings of the Sixth ACM International Conference on Web Search and Data Mining, pp. 527–536. ACM (2013)
3. Dong, C., Agarwal, A.: WS^2F: a weakly supervised framework for data stream filtering. In: 2014 IEEE International Conference on Big Data, Big Data 2014, Washington, DC, USA, October 27–30, pp. 50–57 (2014)
4. Erkan, G., Radev, D.R.: Lexrank: graph-based lexical centrality as salience in text summarization. J. Artif. Intell. Res. (JAIR) **22**(1), 457–479 (2004)
5. Gimpel, K., Schneider, N., O'Connor, B., Das, D., Mills, D., Eisenstein, J., Heilman, M., Yogatama, D., Flanigan, J., Smith, N.A.: Part-of-speech tagging for Twitter: annotation, features, and experiments. In: Proceedings of the 49th Annual Meeting of the Association for Computational Linguistics: Human Language Technologies, pp. 42–47 (2011)
6. Ounis, I., Craig Macdonald, J.L., Soboroff, I.: Overview of the TREC-2011 microblog track. In: Proceedings of the 20th Text REtrieval Conference (TREC 2011) (2011)
7. Kelly, R.: Twitter study - august 2009 (August 2009). http://pearanalytics.com/wp-content/uploads/2012/12/Twitter-Study-August-2009.pdf
8. Khan, M.A.H., Iwai, M., Sezaki, K.: An improved classification strategy for filtering relevant tweets using bag-of-word classifiers. J. Inf. Process. **21**(3), 507–516 (2013)
9. Lin, C.Y.: Rouge: a package for automatic evaluation of summaries. In: Text Summarization Branches Out: Proceedings of the ACL-2004 Workshop, pp. 74–81 (2004)

10. Long, R., Wang, H., Chen, Y., Jin, O., Yu, Y.: Towards effective event detection, tracking and summarization on microblog data. In: Wang, H., Li, S., Oyama, S., Hu, X., Qian, T. (eds.) WAIM 2011. LNCS, vol. 6897, pp. 652–663. Springer, Heidelberg (2011)

11. Marcus, A., Bernstein, M.S., Badar, O., Karger, D.R., Madden, S., Miller, R.C.: Twitinfo: aggregating and visualizing microblogs for event exploration. In: Proceedings of the SIGCHI Conference on Human Factors in Computing Systems, pp. 227–236. ACM (2011)

12. Mihalcea, R., Tarau, P.: Textrank: bringing order into texts. Association for Computational Linguistics (2004)

13. Nichols, J., Mahmud, J., Drews, C.: Summarizing sporting events using Twitter. In: Proceedings of the 2012 ACM International Conference on Intelligent User Interfaces, pp. 189–198. ACM (2012)

14. Olariu, A.: Hierarchical clustering in improving microblog stream summarization. In: Gelbukh, A. (ed.) CICLing 2013. LNCS, vol. 7817, pp. 424–435. Springer, Heidelberg (2013)

15. Olariu, A.: Efficient online summarization of microblogging streams. In: EACL 2014, p. 236 (2014)

16. Osborne, M., Moran, S., McCreadie, R., Von Lunen, A., Sykora, M., Cano, E., Ireson, N., Macdonald, C., Ounis, I., He, Y., et al.: Real-time detection, tracking, and monitoring of automatically discovered events in social media. In: Proceedings of the 52nd Annual Meeting of the Association for Computational Linguistics: System Demonstrations, pp. 37–42 (2014)

17. Radev, D.R., Jing, H., Styś, M., Tam, D.: Centroid-based summarization of multiple documents. Inf. Process. Manage. **40**(6), 919–938 (2004)

18. Sharifi, B., Hutton, M.A., Kalita, J.K.: Experiments in microblog summarization. In: 2010 IEEE Second International Conference on Social Computing, pp. 49–56 (2010)

19. Sharifi, B., Hutton, M.A., Kalita, J.: Summarizing microblogs automatically. In: Human Language Technologies: The 2010 Annual Conference of the North American Chapter of the Association for Computational Linguistics, pp. 685–688 (2010)

20. Yang, X., Ghoting, A., Ruan, Y., Parthasarathy, S.: A framework for summarizing and analyzing twitter feeds. In: Proceedings of the 18th ACM SIGKDD International Conference on Knowledge Discovery and Data Mining, pp. 370–378. ACM (2012)

Relevancer: Finding and Labeling Relevant Information in Tweet Collections

Ali Hürriyetoğlu[1,2]([⊠]), Christian Gudehus[3], Nelleke Oostdijk[2],
and Antal van den Bosch[2]

[1] Statistics Netherlands, P.O. Box 4481, 6401 Heerlen, CZ, The Netherlands
a.hurriyetoglu@cbs.nl
[2] Centre for Language Studies, Radboud University, P.O. Box 9103, 6500 Nijmegen,
HD, The Netherlands
{a.hurriyetoglu,n.oostdijk,a.vandenbosch}@let.ru.nl
[3] Faculty of Social Science, Ruhr University Bochum,
150 Building GB 04/148, 44801 Bochum, Germany
christian.gudehus@ruhr-uni-bochum.de

Abstract. We introduce a tool that supports knowledge workers who want to gain insights from a tweet collection, but due to time constraints cannot go over all tweets. Our system first pre-processes, de-duplicates, and clusters the tweets. The detected clusters are presented to the expert as so-called information threads. Subsequently, based on the information thread labels provided by the expert, a classifier is trained that can be used to classify additional tweets. As a case study, the tool is evaluated on a tweet collection based on the key terms 'genocide' and 'Rohingya'. The average precision and recall of the classifier on six classes is 0.83 and 0.82 respectively. At this level of performance, experts can use the tool to manage tweet collections efficiently without missing much information.

Keywords: Social media analysis · Event analysis · Data mining · Text mining · Machine learning · Social signal processing · Decision support systems · Genocide · Rohingya

1 Introduction

Keyword-based collections of tweets tend to be overly rich in the sense that not all tweets are relevant for the task at hand. Tweets can be irrelevant for a particular task, for instance because they are posted by non-human accounts, contain spam, refer to irrelevant events, or point to an irrelevant sense of an ambiguous keyword used in collecting the data. This richness has a number of dynamic characteristics, which can be present in a static or continuously updated collection, as well. There are no guarantees that tweet collections will have similar characteristics across different periods of time.

With the aim of managing tweet collections, we introduce the term *information thread* and our tool, Relevancer. An information thread characterizes a

© Springer International Publishing AG 2016
E. Spiro and Y.-Y. Ahn (Eds.): SocInfo 2016, Part II, LNCS 10047, pp. 210–224, 2016.
DOI: 10.1007/978-3-319-47874-6_15

specific informationally coherent set of tweets. Relevancer is the tool we developed to analyze a tweet collection in terms of information threads.

Related sets of tweets (information threads) are initially detected using unsupervised machine learning. They are then confirmed by a human expert, and are used as training data in order to classify any remaining or new tweets using supervised machine learning. An expert can be anybody who is able to make knowledgeable decisions about how to annotate tweet clusters in order to understand a tweet collection in a certain context.

Relevancer enables an expert to analyze a tweet collection, i.e. any set of tweets that has been collected by using keywords. The tool requires expert feedback in terms of cluster annotation in order to complete the analysis. Experts can repeat the annotation process in case they collect new data with the same keywords. Alternatively they can decide to do another type of annotation once they understand the collection better after evaluating the first clusters. Our method advances the state of art in terms of efficient and complete understanding and management of a non-standard, rich, and dynamic data source.

This paper illustrates the functionality of Relevancer with a use case based on a particular tweet collection that we processed. It serves to illustrate the different steps in the description of our approach and the way that the Relevancer tool we developed supports it. The strength of our approach is the ability to scale to a large collection without sacrificing the precision or the recall by understanding intrinsic characteristics of the used key terms on social media. Finally, sharing the responsibility for completeness and precision with the users of the tool ensures they will achieve and preserve the target performance they require.

The remainder of the paper is structured as follows. In Sect. 2, we first describe related research. Section 3 introduces the concept 'information thread'. Then, in Sects. 4 and 5, we describe the structure of the tool and give information about the tweet collection used for the case study we did based on two key terms ('genocide' and 'Rohingya'). The results are presented in Sect. 6. Finally, Sect. 7 concludes this paper.

2   Related Studies

Identifying different uses of a word in different contexts is a word sense induction task [8]. This task is especially challenging for tweets, as they have a limited context [5]. Moreover, the diversity of the content on Twitter [3] and specific information need of an expert require a more flexible definition than a sense or topic of the content for tweet collections. We introduce the term information thread, which can be seen as contextualization of a sense.

Popular approaches of word sense induction on social media data are Latent Dirichlet allocation (LDA) [6,7], and user graph analysis [11]. The success of the former method depends on the given number of topics, which is challenging to determine, and the latter assumes the availability of user communities. Both methods provide one-fit-all solutions that are not flexible enough to allow users to customize the output based on a particular collection and the needs that an expert or a researcher has. Therefore, we designed our tool Relevancer in such a fashion that it is not restricted by these assumptions. Relevancer makes it possible to discover information threads without any a priori restrictions.

Social science researchers have been seeking ways of utilizing social media data [4] and have developed various tools [1] to this end. Although these tools have many functions, they do not focus on identifying the uses of key terms. A researcher should continue to navigate the collection by specific key term combinations. Our study aims to enable researchers to discover specific or new uses, information threads, of the already used key terms and focus on the related tweets of a particular information thread.

A tweet collection management tool must take into account the characteristics of the social media data and achieve certain tasks. Available tools[1] have been designed for specific domains, languages, and use cases. Each of these suffers from one or more of the following restrictions: (a) they are restricted to analyzing tweets that contain at least a certain number of well-formed key terms or content words in specific languages; (b) they do not take into account tweet characteristics such as emoticons and personal language use; (c) they rarely use other attributes of a tweet text, such as mentions and URLs; and (d) they assume the availability of a group of annotators that are willing to label a sufficient number of tweets. We designed Relevancer[2,3] in such a way that it does not suffer from any of the aforementioned restrictions.

Enabling human intervention is crucial to ensuring high level performance in text analytics approaches [9]. Therefore we need to build a flexible pipeline that can facilitate human input in order to yield customized results [2]. Our approach responds to this challenge and need by providing the adaptive steps of analysis.

3   Information Threads

The task we address here is a specific case for collecting data using key terms from Twitter. The use and the interpretation of the key terms depends partly on the social context of a Twitter user and the point in time this word is used. Often, the senses and nuances or aspects that a word may have on social media cannot all be found in a dictionary. Therefore, we focus on the automatic identification of sets of tweets that contain the same contextualization of a sense, namely tweets that convey the same meaning and nuance (aspect). We name this kind of tweet groups as information threads.

For example, the word 'flood' has multiple senses; including being covered with water and filling somewhere with large amount of something[4]. A researcher

---

[1] For example, https://wiki.ushahidi.com/display/WIKI/SwiftRiver, https://github.com/qcri-social/AIDR/wiki/AIDR-Overview, and https://github.com/JakobRogstadius/CrisisTracker.

[2] https://bitbucket.org/hurrial/relevancer.

[3] http://relevancer.science.ru.nl.

[4] http://www.oed.com/view/Entry/71808.

working in the field of water management will want to focus on only the water-related sense. At the same time, he will want to discriminate between different nuances or aspects of this sense: past, current, up-coming events, effects, measures taken, etc. By incrementally clustering and labeling these, the collection is analyzed into different information threads.

The information thread concept allows a fine-grained management of all uses of a key word. In the case of this study, this approach enables the user of the tool to focus on uses of a key term at any level of granularity. For instance, tweets about a certain event, which takes place at a certain time and place, and tweets about a type of event, without a particular place or time, can be processed at the same level of complexity.

4   Tweet Collection Analysis with Relevancer

We retrieved a tweet collection from the public Twitter API[5] with the key terms 'genocide' and 'Rohingya' between May 25 and July 7 2015. We use this tweet collection to illustrate how Relevancer is used.

Relevancer begins by cleaning and exploring the tweet collection: it detects duplicates, extracts detailed features, and divides the collection into coherent subsets to which an expert can attach labels. Labeled tweets are then used to train an automatic classifier. This classifier can be used to analyze the remaining tweets or a new collection.

The analysis steps after duplicate and near-duplicate elimination of the tool are presented in Fig. 1. The details are explained in the following subsections.

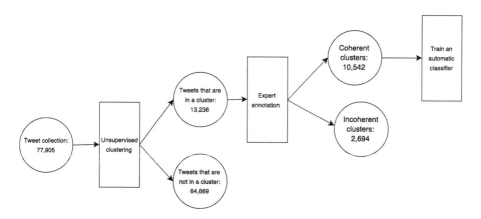

Fig. 1. Phases in the analysis process with the number of tweets at each step starting after duplicate and near-duplicate elimination

---

[5] https://dev.twitter.com/rest/public.

4.1 Pre-processing

Any tweet that has an indication of being a retweet is excluded. We use two types of retweet detection in a tweet: (a) the retweet identifier of the tweet's JSON file, (b) when the tweet starts with "RT @".

We proceed with normalizing the user names and URLs to "usrusrusr" and "urlurlurl" respectively. The normalization eliminates the noise introduced by huge number of different user names and URLs and preserves the abstract information that a user name or a URL is present.

Finally, we detect and exclude exact duplicate tweets. The importance of this step can be appreciated when we identify tweets such as exemplified in (1) and (2) below were posted 5,111 and 2,819 times respectively. However, we have to note that short tweets that signal the same information in different contexts are eliminated as well.

1. *usrusrusr The 2nd GENOCIDE against #Biafrans as promised by #Buhari has begun,3days of unreported aerial Bombardment in #Biafraland*
2. *usrusrusr New must read by usrusrusr usrusrusr A genocide & Human trafficking at a library in Sweden urlurlurl*

4.2 Feature Extraction

Any token that occurs in a tweet text is used as a feature in similarity calculations and in machine learning. Tokens are any space-delimited sequences of alphanumeric characters, emoticons, and sequences of punctuation marks. Features can be words, hashtags, letters, numbers, letter and number combinations, and emoticons. Punctuation mark sequences are treated differently. Sequences of punctuation marks comprising two, three or four items are considered as one feature. If the punctuation marks sequence is longer than four, we split them from left to right in tokens of length 4 by ignoring the last part if it is a single punctuation mark. The limit of length 4 is used for punctuation mark combinations, since longer combinations can be rare, which may cause them not to be used at all.

Detected tokens are used as unigram and bigram features. We apply a dynamically calculated threshold to the features. The threshold is half of the log of the number of tweets in a collection.

The motivation for using all aforementioned features is that this makes it possible to capture any nuance that may be present in groups of tweets. As a result, we can detect and handle many stylistic and textual characteristics properly. Moreover, using just the space for determining the features enables the tool to operate on any language that uses the blank space as token separator. Since none of the steps contain any language-specific optimization and the language and task related information is provided by a human user, we consider our method to be language-independent.

4.3   Near-Duplicate Detection

Most near-duplicate tweets occur because the same quote is used, the same message is tweeted, or the same news article is shared with other people. Since duplication does not add information and may harm efficiency and performance of the basic algorithms, we remove near-duplicate tweets by keeping only one of them in the collection.

The near-duplicate tweet removal algorithm is based on cosine similarity of the tweets. If the pairwise cosine similarity of two tweets is larger than 0.8, we assume that these tweets are near-duplicates of each other. In case the available memory does not allow to process all tweets at once, we apply a recursive bucketing and removal procedure. The recursive removal starts with splitting the collection into buckets of tweets of a size that can be handled efficiently. After removing near-duplicates from each bucket, we merge the buckets and shuffle the remaining tweets, which are unique in their respective bucket. We repeat the removal step until we can no longer find duplicates in any bucket.

For instance, the near-duplicate detection method recognizes the tweets (1) and (2) below as near-duplicates and leaves just one of them in the collection.

1. *Actor Matt Dillon puts rare celebrity spotlight on Rohingya urlurlurl#news*
2. *urlurlurl Matt Dillon puts rare celebrity spotlight on Rohingya urlurlurl*

4.4   Information Thread Detection

The information thread detection step aims at finding available information threads related to the key terms used to collect a tweet collection. In case the tweets in the collection do not share a certain key term, this step will still find related groups of tweets that are about certain uses of the words and word combinations in this collection. This groups will represent information threads.

We think that any approach that tries to find a relation between all tweets will fail to some extent. Because twitter data is extremely rich in a sense that it is not realistic to think that every tweet can be related to particular tweets. Therefore, our method aims at detecting only related group of tweets and ignoring tweets that do not present any clear relation to other tweets. These outlier tweets will be available to be analyzed by the classifier or manually at the end of the process.

The clustering aims to identify coherent clusters of tweets in order to understand the collection in terms of clusters, which constitute an information thread. We run a basic algorithm, K-Means for small collections and MiniBatch K-Means for large ones[6]. We repeat the clustering and cluster selection steps in iterations until we reach the requested number of coherent clusters, which is provided by the expert. Clusters that fit our coherence criteria are picked for annotation and the tweets in them are not included in the following iteration of clustering.

---

[6] We used scikit-learn v0.17.1 for all machine learning tasks in this study http:// scikit-learn.org.

4.5   Relevancer Parameters

Formulas for parameter value assignment were identified empirically with several principles in mind. Since tweet collections do not demonstrate a clear separation of tweet clusters for the whole set, we assume that optimizing the parameters is not feasible. Moreover, having expert feedback for the selected coherent clusters allows us not to spend relatively excessive amounts of time on optimizing the clustering parameters for yielding a better automatic clustering. This generic approach aims at finding only main coherent clusters of tweets.

Relevancer facilitates two levels of parameters: clustering and coherence parameters. The clustering parameters depend on the collection size (n) and the time spent on searching for clusters. The cluster coherence parameters control the selection of clusters for annotation. These parameters are updated at each iteration (i) based on the requested number of clusters (r). A requested number of clusters is given as a stopping criterion for the exploration and as an indicator of the adaptation step for the value of the other parameters at each cycle.

Clustering parameters. There are two clustering parameters. K is the number of expected clusters and t is the number of initializations of the clustering algorithm before it delivers its result. These parameters are adjusted at each iteration.

Coherence parameters. The second layer of the parameters contains the number of clusters that should be generated for the expert (r) and the cluster coherence criteria. Although these parameters have default values, they can be set by the expert as well. Adaptation of the cluster coherence criteria steps is small if we are close to our target.

The value of k at each iteration is assigned based on Eq. 1. The parameter k is equal to half of the square root of the tweet collection size at that iteration plus the number of previous iterations times the difference between the requested number of clusters and the detected number of clusters (a). This adaptive behavior ensures that if we do not have many clusters after several iterations, we will be searching for smaller clusters at each iteration.

The result of the clustering, which is evaluated by the coherence criteria, is the best score in terms of inertia after initializing the clustering process (t) times in an iteration. As provided in Eqs. 1 and 2, (t) is log of the tweet collection at the current iteration plus the number of iterations performed until that point times (t), Eq. 2. This formula ensures that the more time it takes to find coherent clusters, the more the clustering will be initialized before it delivers any result.

$$k = \frac{\sqrt{n_i}}{2} + (i \times (r_i - a_i)) \tag{1}$$

$$t = \log_{10} n_i + i \tag{2}$$

If the requested number of clusters is too high for the collection, the adaptive relaxation of the coherence parameters stops at a level any cluster may qualify as a coherent. We think it is unrealistic to expect relatively good clusters in such a situation. In such a case, the available clusters are returned before they reach the number requested by the expert. On the other hand, in case the algorithm exceeds the number of requested clusters in an iteration, it completes the search for coherent clusters and yields all detected clusters.

4.6 Cluster Annotation

Automatically selected clusters are presented to an expert for identifying clusters that present certain information threads[7]. After the annotation, each thread may consist of several clusters. In other words, similar clusters should be labeled with the same label. Clusters that are not clear and fall out outside the focus of a study should be labeled as incoherent and irrelevant respectively. This decision is taken by a human expert. The tweets that are in an incoherent labeled cluster are treated as the tweets that are not placed in any coherent cluster.

Sample tweets, the closest and the farthest to the cluster center, from coherent and incoherent clusters are presented in Table 1. A coherent cluster (CH) tweets have a clear relation that allows us to treat this group of tweets as pertaining the same information thread. Incoherent clusters are summarized under IN1 and IN2. In the former group, tweets do not have any relation between the first and last tweet[8]. The latter group contains a meta-relation that can be considered as an indication of being about a video. However, if the expert is interested in the content, this cluster should be annotated as incoherent.

Table 1. Tweets that are the closest and the farthest to the cluster center for coherent (CH), incoherent type 1 (IN1) and type 2 (IN2) clusters

CH	— myanmar rejects'unbalanced' rohingya remarks in oslo (from usrusrusr urlurlurl — shining a spotlight on #myanmar's #rohingya crisis: usrusrusr remarks at oslo conf on persecution of rohingyas urlurlurl
IN1	— un statement on #burma shockingly tepid and misleading, and falls shortin helping #rohingya says usrusrusr usrusrusr urlurlurl — usrusrusr will they release statement on bengali genocide 10 months preceding'71 ?
IN2	— i liked a usrusrusr video urlurlurl rwanda genocide 1994 — i liked a usrusrusr video urlurlurl fukushima news genocide; all genocide paths lead to vienna and out of vienna

For instance, clusters that contain tweets like "plain and simple: genocide urlurlurl" and "it's genocide out there right now" can be gathered under the

---

[7] The annotation is designed to be done or coordinated by a single person in our setting.

[8] The expert may prefer to tolerate a few different tweets at the end of the group in case majority of the tweets are coherent and treat the cluster as coherent.

same label, e.g., actual ongoing events. If a cluster of tweets is about a recent event, a label can be created for that particular event as well. For instance, the tweet "the plight of burma's rohingya people urlurlurl" is about a particular event related to the Rohingya people. If we want to focus on this event related to Rohingya, we should provide a specific label for this cluster and use it to specify this particular context. We can use 'plight' as a label as well. In such a case, the context should specify cases relevant to this term.

In case the expert is not interested in or does not want to provide a specific label to a CH, the expert should attach the label irrelevant, which behaves as an umbrella label for all tweet groups that are not the present focus of the expert.

We developed a web-based user interface for presenting the clusters and assigning a label to the presented cluster. We present all tweets in a cluster if the number of tweets is smaller than 20. Otherwise, we present the first and the last 10 tweets in terms of the distance to the cluster center. This setting provides an overview and enables spotting incoherent clusters effectively. A cluster can be skipped without providing a label for it. In such a case, that cluster is not used in the following steps.

At the end of this process, an expert is expected to have defined the information threads for this collection and have identified the clusters that are related to these threads. Tweets that are part of a certain information thread can be used to understand the related thread or to create an automatic classifier that can classify new tweets, e.g., ones that were not included in any selected cluster at the clustering step, in classes based on detected and labeled information threads.

4.7   Creating a Classifier

Creating classifiers for tweet collections is a challenge that is mainly affected by label selection of the expert annotator, nature of the key word, and time of the collection. Consequently, the classes are or prone to be imbalanced.

The labeled tweet groups are used as training data for building automatic classifiers that can be applied 'in the wild' to any previous or new tweets, particularly if they are gathered while using the same query.

Relevancer facilitates the Naive Bayes algorithm for creating a classifier, which is one of the most basic supervised machine learning algorithms. We prefer this classifier due to its short training time and comparable performance to sophisticated machine learning algorithms [10] for text classification. We need time efficiency in order to be able to re-train a classifier in case an expert prefers to update the current classifier with new data or create another classifier after observing the results of a particular classifier.

Parameters of the Naive Bayes algorithm are optimized by using a grid search on the training data. The performance of the classifier is based on 15 % of the training data, which was held out and not used at the training step.

4.8   Scalability

Relevancer applies various methods in order to remain scalable independent of the number of tweets and features used. High number of tweets and features force this tool to have the scalability at the center of its design.

Large amount of tweets and a wide variety of features are processed by using basic and fast algorithms. Depending on the size of the collection, K-means or MiniBatch K-Means algorithms are employed in order to rapidly identify candidate clusters. The main parameter k, number of clusters parameter for K-Means, in these algorithms is increased at each iteration in order to target finding more and smaller clusters than previous iteration. Targeting more and smaller clusters increase chance of identifying coherent clusters.

Tweets in coherent clusters are excluded from the part of the collection that enter the subsequent clustering iteration. This approach shrinks the collection size at each iteration. Moreover, the criteria for coherent cluster detection is relaxed at each step until certain thresholds are reached in order to prevent the clustering from being repeated.

The Naive Bayes classifier was chosen in order to create and evaluate a classifier at a reasonable time. The speed of this step enables users to decide whether they will use a particular classifier or need to generate and annotate additional coherent clusters immediately.

This optimized cycle enables experts to be able to provide feedback frequently without having to wait too long. As a result, the quality of the results increases with minimal input and time for a particular task.

5   Tweet Collection

We collected 363,959 tweets with the key terms 'genocide' and/or 'Rohingya' between May 25 and July 7 2015. The number of tweets that contain only 'genocide' and only 'Rohingya' are 269,131 and 137,319 respectively; 12,889 tweets contain both terms.

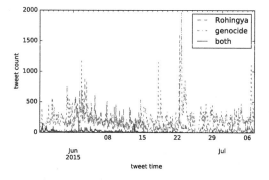

Fig. 2. Tweet distribution per key term

Figure 2 shows the distribution of the tweets for each subset. We can already observe that there are many peaks and a constant tweet ratio for each key term. Our analysis of the collection aims at understanding the constantly mentioned content and the peaks separately.

6   Results

We start the analysis by preprocessing the tweets in the collection. As a first step, we excluded 198,049 tweets, which are all retweets, of the 363,959 tweets, leaving 165,910 tweets in the collection. Then the exact duplicates were excluded from the collection, leaving 103,987 tweets in the collection. Finally 26,082 near-duplicate tweets were removed.

Figure 3 illustrates the final tweet distribution. We observe that the distribution was changed after we eliminated the repetitive information. Large peaks in the 'genocide' data were drastically eliminated and some of the small peaks disappeared. Thus, only relatively important peaks in the 'genocide' data remain and the peaks in tweets containing the key term 'Rohingya' becomes apparent.

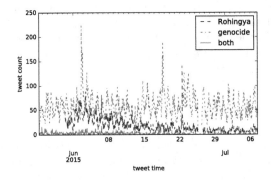

Fig. 3. Tweet distribution per key term after removing retweets, duplicates, and near-duplicates

The summary of the evolution of the size of the tweet collection is presented in Fig. 4. At each step, a large portion of the data is detected as repetitive and excluded. This cleaning phase shows us how size of the collection depends on the preprocessing. Decreasing the size of the data without loosing information[9] enables us to apply sophisticated analysis techniques to social media data.

After removing the duplicates and near-duplicates, we clustered the remaining 77,905 tweets. Although the requested number of clusters was 100, the number of generated clusters was 145, which contain 13,236 tweets. The cluster parameters were set to begin with the following values: (i) the distances of the closest

---

[9] We note that the repetition pattern analysis is valuable in its own right. However, this information is not within the scope of the present study.

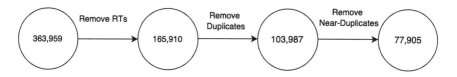

Fig. 4. Tweet number change at each step of the preprocessing in the collection

and farthest tweets to the cluster center have to be less than 0.725 and 0.875 respectively; and the difference of the distance to the cluster center between the closest and the farthest tweets in a cluster should not exceed 0.4.

The annotation of the 145 clusters by a domain expert yielded the results in Table 2. This process yielded 8 labels: Actual cases (AC), Cultural genocide (CG), Historical cases (HC), Incoherent (IN), Indigenous people genocide (IPG), Irrelevant (IR), Jokes (JO), and Mass migration (MM).

This step enabled the domain expert to understand the data by annotating only 17 % of the preprocessed tweets, which is 0.03 % of the complete collection, without the need of having to go over the whole set. Furthermore, annotating tweets in groups as suggested by the clustering algorithm improved the time efficiency of this process.

Table 2. Number of labeled clusters and total number of tweets for each of the labels

	# of clusters	# of tweets
AC	48	4,937
CG	7	375
HC	22	1,530
IN	32	2,694
IPG	1	109
IR	30	3,365
JO	1	30
MM	4	226
Total	145	13,236

We used the Relevancer to create an automatic classifier by using the annotated tweets. We merged the tweets that are under the JO (Jokes) label with the tweets under the Irrelevant label, since their number is relatively small compared to other tweet groups for generating a classifier for this class. Moreover, the incoherent clusters were not included in the training set. This leaves 10,552 tweets in the training set.

We used the same token characteristics explained in the feature extraction step to create the features used by the classifier. We added token trigrams to the unigrams and bigrams as features. The parameter optimization and the training

222   A. Hürriyetoğlu et al.

was done on 85 % of the labeled data and the test was performed on the remaining
15 %. The only parameter of the Naive Bayes classifier, α, was optimized on the
training set to be 0.105 by testing equally separated 20 values between 0 and 2.

The performance of the classifier is summarized in Tables 3 and 4 as a confusion matrix and evaluation summary. We observe that classes that have a clear
topic, e.g., HC and CG, perform reasonably well. However, classes that are potentially a combination of many sub-topics, such as AC and IR, which contain JO
labeled tweets as well, perform relatively worse. Detailed analysis yielded that
the HC thread contains only a handful past events that were referred to directly.
On the other hand, there is a lot of discussions in addition to clear event reference in the AC thread. As a result, labels that contain specific language use
work better than the labels contain diverse language use.

Table 3. The rows and the columns represent the actual and the predicted labels of
the test tweets. The diagonal provides the correct number of predictions.

	AC	CG	HC	IPG	IR	MM
AC	586	3	5	1	158	1
CG	1	42	0	0	3	0
HC	26	0	198	0	7	1
IPG	2	1	0	9	3	0
IR	62	1	3	0	441	1
MM	8	0	0	0	0	19

Table 4. Precision, recall, and F1-score of the classifier on the test collection. The
recall is based only on the test set.

	Precision	Recall	F1	Support
AC	.86	.78	.81	754
CG	.89	.91	.90	46
HC	.96	.85	.90	232
IPG	.90	.60	.72	15
IR	.72	.87	.79	508
MM	.86	.70	.78	27
Avg/Total	.83	.82	.82	1,582

The result of this step is an automatic classifier that can be used to classify
any tweet in the aforementioned six classes. Although the performance is relatively low for a class that is potentially a mix of various subtopics, the average
scores, which are 0.83, 0.82, and 0.82 for the precision, recall, and F1 respectively,
are sufficient for using this classifier in a real scenario.

7   Conclusion and Future Research

We presented Relevancer, our tweet collection management tool, and its performance on a collection collected with the key terms 'genocide' and 'Rohingya'. Each step of the analysis process was explained in quite some detail in order to shed light both on the tool and the characteristics of this collection.

The results show that the requested number of clusters should not be too small. Otherwise, missing classes may cause the classifier not to perform well on tweets related to information threads not represented in the final set of annotated clusters. Second, the reported performance was measured on the held out subset of the training set. This means that the test data has the same distribution with the subset of the training data that is used for the actual training. Therefore, generalization of those results to the actual social media context should be a concern of a further study.

At the end of the analysis steps, the expert successfully identified repetitive information, which is 79 % of the whole collection, analyzed coherent clusters, defined the main information threads, annotated the clusters, and created an automatic classifier that has 0.82 F1 score.

We plan to improve the feature extraction by including skip-grams and posting user dimensions and to introduce more sophisticated cluster evaluation metrics. Moreover, we can get feedback from the experts about which users or hashtags should best be ignored or included. This information can be used to update and continuously evaluate the classifiers. The posting user and hashtags have the potential to provide information about certain information threads. This information can enable the annotation step to be expanded to the tweets that are not in any cluster but contain information thread specific hashtags or users.

We do not carry out any post processing on the clusters. But we can identify outlier tweets in a cluster in terms of the distance to the cluster center in order to refine the clusters. This will enable the tool to have cleaner clusters and to find required clusters in fewer iterations.

Finally, the temporal distribution of the tweets in an information thread can guide us in defining its scope. An information thread over a short period will be treated in a manner different from a more persistent one.

Acknowledgements. This research was funded by the Dutch national research programme COMMIT. We gratefully acknowledge the contribution by Statistics Netherlands.

References

1. Borra, E., Rieder, B.: Programmed method: developing a toolset for capturing and analyzing tweets. Aslib J. Inf. Manage. **66**(3), 262–278 (2014). http://dx.doi.org/10.1108/AJIM-09-2013-0094
2. Chau, D.H.: Data mining meets hci: Making sense of large graphs. Technical report, DTIC Document (2012)

3. Choudhury, M.D., Counts, S., Czerwinski, M.: Find Me the Right Content! Diversity-based sampling of social media spaces for topic-centric search. ICWSM, pp. 129–136 (2011). http://www.aaai.org/ocs/index.php/ICWSM/ICWSM11/paper/viewPDFInterstitial/2792/3290
4. Felt, M.: Social media and the social sciences: How researchers employ Big Data analytics. Big Data Soc. **3**(1), 1–15 (2016). http://bds.sagepub.com/lookup/doi/10.1177/2053951716645828
5. Gella, S., Cook, P., Baldwin, T.: One sense per tweeter... and other lexical semantic tales of twitter. In: Proceedings of the 14th Conference of the European Chapter of the Association for Computational Linguistics, pp. 215–220 (2014). http://www.aclweb.org/anthology/E14-4042
6. Gella, S., Cook, P., Han, B.: Unsupervised word usage similarity in social media texts. In: Second Joint Conference on Lexical and Computational Semantics (*SEM), Proceedings of the Main Conference and the Shared Task: Semantic Textual Similarity, vol. 1, pp. 248–253 (2013). http://www.aclweb.org/anthology/S13-1036
7. Lau, J.H., Cook, P., McCarthy, D., Newman, D., Baldwin, T., Computing, L.: Word sense induction for novel sense detection. In: Proceedings of the 13th Conference of the European Chapter of the Association for Computational Linguistics (EACL 2012), pp. 591–601 (2012)
8. Mccarthy, D., Apidianaki, M., Erk, K.: Word sense clustering and clusterability. Comput. Linguist. **42**(2), 4943 (2016)
9. Tanguy, L., Tulechki, N., Urieli, A., Hermann, E., Raynal, C.: Natural language processing for aviation safety reports: from classification to interactive analysis. Comput. Ind. **78**, 80–95 (2016)
10. Wang, S., Manning, C.: Baselines and Bigrams: simple, good sentiment and topic classification. In: Proceedings of the 50th Annual Meeting of the Association for Computational Linguistics vol. 94305(1), pp. 90–94 (2012)
11. Yang, Y., Eisenstein, J.: Putting Things in Context: Community-specific embedding projections for sentiment analysis. CoRR abs/1511.0 (2015). http://arxiv.org/abs/1511.06052

Analyzing Large-Scale Public Campaigns on Twitter

Julia Proskurnia[1]([✉]), Ruslan Mavlyutov[2], Roman Prokofyev[2], Karl Aberer[1], and Philippe Cudré-Mauroux[2]

[1] École Polytechnique Fédérale de Lausanne, Lausanne, Switzerland
{iuliia.proskurnia,karl.aberer}@epfl.ch
[2] EXascale Infolab, University of Fribourg, Fribourg, Switzerland
{r.mavlyutov,r.prokofyev,c.cudremauroux}@unifr.ch

Abstract. Social media has become an important instrument for running various types of public campaigns and mobilizing people. Yet, the dynamics of public campaigns on social networking platforms still remain largely unexplored. In this paper, we present an in-depth analysis of over one hundred large-scale campaigns on social media platforms covering more than 6 years. In particular, we focus on campaigns related to *climate change* on Twitter, which promote online activism to encourage, educate, and motivate people to react to the various issues raised by climate change. We propose a generic framework to identify both the type of a given campaign as well as the various actions undertaken throughout its lifespan: official meetings, physical actions, calls for action, publications on climate related research, etc. We study whether the type of a campaign is correlated to the actions undertaken and how these actions influence the flow of the campaign. Leveraging more than one hundred different campaigns, we build a model capable of accurately predicting the presence of individual actions in tweets. Finally, we explore the influence of active users on the overall campaign flow.

1 Introduction

Social media have become central to our digital lives, as they allow individuals to share news, photos, or opinions, as well as to have online discussions in real-time. One particularly interesting phenomenon is social media marketing, which can be defined as the process of drawing attention to some specific issue or product via social media platforms. Such endeavors often take the form of extensive campaigns, whose aim is to raise the awareness of the public on a particular topic and potentially to engage it into concrete actions.

Social media platforms provide tools to effectively conduct these campaigns; On Twitter, for example, people use so-called *hashtags* to associate their messages to a certain topic. Many campaigns, therefore, have their own hashtags that uniquely identify them. Moreover, many tweets associated with a campaign convey some specific messages to the audience, such as requests for signing a petition, asking for a certain action, attending a demonstration, etc. These messages can be considered as certain *actions*, and their effect on the dynamics of

© Springer International Publishing AG 2016
E. Spiro and Y.-Y. Ahn (Eds.): SocInfo 2016, Part II, LNCS 10047, pp. 225–243, 2016.
DOI: 10.1007/978-3-319-47874-6_16

the campaigns remains largely unexplored in the scientific literature. Identifying and categorizing such messages within the context of a campaign would enable us to answer questions such as what drives attention to a particular topic or how to reach a certain target audience. In this work, we propose a number of categories to classify the actions from the perspective of the goals of the campaigns as well as a methodology to identify them. We build a classifier for the action types based on the tweets content and study the distribution of these action types for different types of campaigns.

In the second part of this work, we analyze the resulting user involvement patterns in order to explore the dynamics of the campaigns. Analyzing such patterns is key to understand how attractive the campaigns are and who are the main contributors to the information dissemination. We perform a comparative analysis of the campaigns and their contents, through which we identify noticeable differences between the various types of campaigns. We observe that campaigns where only a tiny fraction of users create the major part of the content are less likely to attract users on social media. Finally, we cluster the involvement patterns and study their correlations with the campaign types.

This work focuses on campaigns related to climate change and animal welfare. Those two topics recently gained increased attention and have the advantage of gathering a high number of users for relatively long periods of time, thus are well suited for our study. Moreover, these topics are mainly of interest for non-profit and governmental organizations, and this work might help them to better understand the impact of their actions on the audience.

In summary, the main contribution of this paper is a large-scale study on the dynamics of campaigns on social media. This study focuses on the following research questions:

- How to identify and compare various types of public campaigns and their corresponding actions? (Sect. 4)
- How is the initial goal and contents of a campaign correlated to the user engagement pattern of a campaign? (Sect. 4.2);
- Is there a relationship between the type of a campaign and the actions undertaken through the campaign lifespan? (Sect. 5);

The rest of this paper is structured as follows. We start with an overview of related work in the areas of Twitter analytics and social media analysis below in Sect. 2. Section 3 describes our data collection, aggregation, and cleansing processes. We analyze the collected data in Sect. 4 by extracting different types of campaigns and clustering them by their user engagement patterns. Section 5 extends our analysis by focusing on various types of tweets and on their distributions in campaigns, as well as building a classifier that is able to predict the type of a tweet. Finally, we discuss our results in Sect. 6 and draw conclusions in Sect. 6.

2   Related Work

Social media platforms quickly came to the attention of the research community, since they allow to conduct large-scale studies on various aspects of social network dynamics, such as popularity prediction. Many studies have recently focused on micro-blogging platforms such as Twitter[1], which provides an access to a small (compared to the overall data) sample of its data based on keyword queries.

In this work, we study the communication patterns and message type pre-eminence for various campaigns on climate change. A number of studies have focused on Twitter communication patterns, including studies on hashtag life-cycles [17,19], event detection and their analysis [12,20], food consumption patterns [1], and usage across different languages [10].

Climate change discourse. Climate change issues are receiving increased attention as they lead to a number of global challenges [3,11]. Many studies recently examined how the climate change debate is covered on social media channels [20,22,26]. However, coverage of debates does not reveal how campaigns develop, and how popular they are based on the messages they contain. As users tend to increasingly rely on their social entourage to filter information [9], we examine in this paper how different message types and techniques engage people in different ways throughout the campaign.

Campaign analytics. Social media is a very influential tool for widening public awareness on various issues as noted by [25]. Previous work on campaigns on social media mostly focused on political and protest campaigns. [13] used a bispace model based on a Poisson process to capture the propagation of information in both Twitter and non-Twitter environments. Additionally, [8] explores how social networks are used to spread protest information. In our work we focus not on the information dissemination but rather how the campaigns were conducted and what are the main actions that were taken to reach the goal. An in-depth study on the theoretical principles underpinning public communication campaigns is given in [4]. Finally, one of the most recent works on campaign analysis [6] focuses on the behavioural stage sequences of the users during the COP21 and EH2015 forums and proposes a framework to identify a user stage by her tweets. On the contrary, we focus on the campaign actions and the corresponding users' engagement rather than user behavioural stages. Moreover, our analysis is carried out on over a hundred public campaigns.

Tweet topic identification. In the context of topic identification, recent works focused on classifying and clustering tweets based on their topics [5,16,21]. Those techniques produce different sets of topics for different datasets. [2] performs an extensive evaluation of different tweet topic detection methods, including methods based on the combinations of syntactic and linguistic techniques. In our case, however, such approaches did not result in valid clusters of message types. To the best of our knowledge, this work is the first on tweet action type classification in campaigns. The work of [23] was an important motivation for the definition of further types of tweets, such as official meetings, calls for action and physical actions. Given our objective of comparing campaign agendas, we

---

[1] https://twitter.com.

look into a number of types of campaigns and actions in this paper in order to identify the correlation between the types of campaigns and the different actions.

3   Data Collection and Cleansing

In this section, we first describe the process through which we collected tweets related to the domains of climate change and animal welfare (Sect. 3.1). We then introduce the strategy we took for identifying campaigns in those domains. We describe the process of identifying the retweets and duplicated tweets in detail in Appendix B. The resulting dataset, consisting of more than 8.5M tweets, is available online for future research[2].

3.1   Twitter Data Collection

We developed a data collection pipeline (see Fig. 1) to gather a broad range of Twitter campaigns related to climate change and animal welfare. Those two domains are usually tightly connected [24]. For example, there are multiple articles [15] on the connection between the number of farm animals and the amount of methane released to the atmosphere and thus causing climate change.

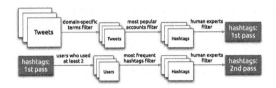

Fig. 1. Data collection pipeline

First pass. We proceeded in two phases in order to identify the campaign. In the first pass, we extracted all available tweets from Topsy[3] for two very prominent accounts that are related to climate change and animal welfare-related: @AlGore and @GreenPeace (2.77 M and 1.33 M followers respectively). This first pass resulted in 27 K tweets comprising 1250 unique hashtags. To select valid campaign hashtags out of the initial 1250 hashtags, we decided to rely on the annotations made by three authors of this paper. To determine whether a particular hashtag belongs to a given campaign, the authors were asked the following questions: (a) do tweets with the hashtag contain calls for actions, mentions of the campaign or an URL to the campaign website; (b) is there a Twitter account with an identical name and a description that corresponds to a given campaign. A hashtag was considered to be related to a campaign iff it was selected by

---

[2] https://github.com/toluolll/CampaignsDataRelease.
[3] Topsy (http://topsy.com/) is a partner of Twitter delivering search and analytic services and claiming to index all public tweets.

all of the annotators[4]. This manual annotation produced a set of 52 campaign hashtags.

Second pass. To increase the recall of our process, we ran a second pass. We identified further accounts (users) that mentioned at least two campaign tags (out of the first 52) in their messages. In that way, we identified 80 additional accounts for a total of 34 K unique hashtags. We filtered out hashtags that appeared in less than 50 tweets, which accounted for 75 % of the tweets. Similarly to the first pass, three authors of this paper annotated each resulting hashtag and 56 additional hashtags were identified as relevant. Overall, our process resulted in a dataset of 108 climate and animal welfare-related campaigns[5], each represented by a distinct Twitter hashtag. The total number of unique tweets (without retweets) in the resulting dataset amounts to 4M.

4   Campaign Analysis

Given that our research question connects two domains—climate change/animal welfare campaigns and social media content analysis—the framework we propose for campaign analysis is composed of two parts. First, we annotate the campaigns according to their primary goals. Next, we cluster them by examining user engagement patterns and by mining active users for the campaigns (i.e., users who tweet most often for a particular campaign). When organizing our data and defining the annotation process, we turned to dimensions considered in the theory of public communication campaigns [4,14,23]. For each campaign, we consider the major goal of the campaign (increase awareness, mobilize people), user engagement over time (ever-growing, regular, one-day, inactive), as well as user activity.

4.1   Types of Campaigns

Following the theoretical analysis of public communication campaigns by [4], we separate the campaigns into two classes based on their primary goals:

– *Mobilization campaigns* refer to the campaigns whose primary goal is to engage and motivate a wide range of partners, allies and individual at the national and local levels towards a particular problem or issue.
– *Awareness campaigns* refer to the campaigns whose primary goal is to raise people's awareness regarding a particular subject, issue, or situation. As discussed in Sect. 2, environmental awareness campaigns usually make a large use of mass media, and in particular, of Twitter.

---

[4] inter-rater agreement is ~95 %.
[5] It is worth noticing that many of the hashtags (around 20 each) in our campaign dataset are created using the morphological filters. For example, we collected hashtags that contain words such as save, protect, call, lead, act, 4, forthe, etc. (e.g. #savethedolphins, #call4action).

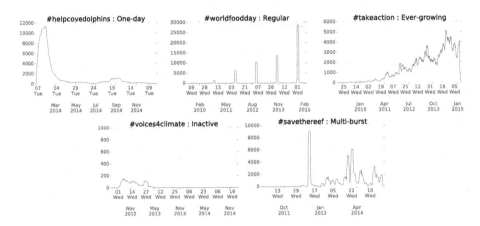

Fig. 2. Different user engagement patterns observed in campaigns.

These campaign types represent very different endeavors, which affects both the type of contents used in such campaigns as well as their user involvement pattern over time, which we analyze further.

Three authors of this paper manually annotated the campaigns as either mobilization or awareness campaigns. The category was considered as valid only when all experts agreed on it. This way, 50 awareness and 58 mobilization campaigns were identified. A few sample hashtags are #savesolar, #climateaction for **mobilization** and #cleanair4kids, #worldfoodday for **awareness** campaigns.

4.2 User Engagement Patterns

In the following, we present an analysis of user engagement in Twitter campaigns. We identify two main axes for analyzing user engagement: the first one focuses on user engagement patterns over time, while the second one analyzes the behavior of the most active users throughout the campaign.

User Engagement Patterns over Time. Subsequently, we cluster the campaigns by engagement patterns of their users to detect whether the engagement correlates with the campaign types. In order to do this, we first extract the number of unique daily users for each campaign hashtag and aggregate these numbers with a 30-day time window. Then, we compute the similarities between the resulting time series using Dynamic Time Warping [7] and cluster them using K-means by varying the K and chose the setup with the smallest in-cluster distance. This resulted in five major clearly distinguishable clusters. Sample campaigns with the above described types are shown on Fig. 2.

From our data collection through this process we have identified several major types of user involvement patterns. Following their overarching distribution, we name them:

- *one-day campaign*, a campaign that is organized over a short period of time to tackle some urgent issue;
- *regular campaign*, a campaign that happens on a regular basis, e.g., annually;
- *ever-growing campaign*, a campaign that gains traction over time;
- *multi-burst campaign*, a campaign that have multiple peaks of activity;
- *inactive campaign*, a campaign that shows a constantly low user engagement throughout its timespan[6]

Finally, we compare the representations of two major classes of campaigns with their user involvement patterns. The campaigns are distributed across the aforementioned engagement groups as 36 %, 10 %, 21 % 22 %, 11 %. As can be observed on Fig. 3, most of "regular" and "inactive" campaigns fall in the awareness category, while both "one-day" and "ever-growing" campaigns are dominated by the mobilization one. The main reason for the dominance of mobilization campaigns for the "one-day" type is the urgency of their issues and the need for immediate action. On the other hand, "regular"

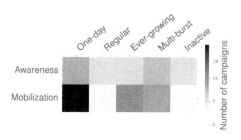

Fig. 3. Distribution of user engagement patterns for the different types of campaigns.

campaigns, that are organized on a periodic basis and pursue long-term goals consist of awareness campaigns mostly. "Ever-growing" campaigns also dominated by mobilization campaigns and focus on global issues, challenges, e.g., #saveanimals, #animalwelfare. "Multi-burst" campaigns are almost equally represented by the both types.

User Engagement Patterns by Volume. We observe that in many campaigns, there is a distinct subset of users who are authors of the majority of the campaign tweets. We call such a set of users a *campaign kernel*. A kernel identifies users with the most tweets and retweets in the campaign. In order to study the influence of a kernel, we propose the following technique: (1) for each user, compute the total number of original tweets and retweets posted in the campaign; (2) rank all users relatively to the volume of content produced for the campaign; (3) compute the Gini coefficient[7] based on the normalized per-user impact relative to the volume of messages in the campaign.

Figure 4(a) shows sample distributions of the relative amount of content generated by users participating in campaigns. We observe a clear distinction between campaigns where users are contributing the content almost equally

---

[6] "Inactive" category might be orthogonal to the other ones, however, it gives valuable insights regarding campaigns that have less traction on Twitter.

[7] http://wikipedia.org/wiki/Gini_coefficient.

(blue curve) and campaigns where only a tiny fraction of users create the major part of the content (red curve). Figure 4(b) shows campaigns with the lowest and the highest Gini coefficient values. High values denote campaigns where the majority of the contents is created by few users only. Values that are close to zero, on the other hand, characterize campaigns where users contribute almost equally. Characterists of the Gini coefficient play an important role in understanding how an information about campaigns is spread on social media, e.g., if actions of several active people can lead to more participation.

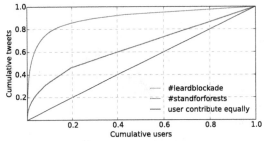

Hashtag	Gini
#protectthearcticrefuge	0.26
#savetheplanet	0.28
#standforforests	0.30
#connect4climate	0.33
...	
#saveafricananimals	0.88
#opkillingbay	0.88
#jpolyboycott	0.91
#unity4malaysia	0.99

(a) Example of the active and inactive kernel involvement. (b) Higher values are associated with more inequality of users contributions.

Fig. 4. Campaign kernel contributions presented as gini coefficient. (Color figure online)

Interestingly, we found a direct correlation between the total number of followers of the kernel users and the total amount of users participating in a campaign. The value of Pearson correlation coefficient for these variables is more than 0.85. This behavior is observed for various kernel sizes (that correspond to various proportions of total tweets they generated). The distribution of Pearson correlation coefficient between the number of kernel followers and the number of users engaged in the campaign for different kernel sizes has the shape of a bell curve ⌒, where thresholds for proportions of tweets are displayed on the x axes (50 % – 99 %), and Pearson correlation is on the y axes (0.70 – 0.858). The maximum correlation is reached for the kernels that produced 75 % of the content and on average this corresponds to the 2.5 % of the campaign users. Thus, we use this percentage of users further as a kernel of each campaign. Interestingly, we found no clear distinction between awareness and mobilization campaigns with regard to their kernels. However, the activity of kernel users differs with respect to the user engagement patterns described in Sect. 4.2. We observe that the majority of content in inactive campaigns is produced by a tiny fraction of users, while regular and ever-growing campaigns accumulate tweets from a much larger subset of users. Similarly, the kernels of inactive campaigns are 10 % smaller than one-day campaigns, while ever-growing campaigns have both small kernels and high participation of the involved users.

5   Tweet Type Identification and Classification

This section presents an in-depth analysis of the tweets from our dataset, focusing on the types of actions they contain (Sect. 5.1), and their correlations with the campaign types (Sect. 5.2). Generally, public campaigns perform a specific set of actions to reach their goals. While this information is not always publicly available, we propose a method to retrospectively identify campaign's actions based on their presence in Twitter. Such findings can be of great interest to less experienced campaign stakeholders, who could use this knowledge to adjust their agendas according to the most successful practices given their particular type of a campaign. We collected additional information about the tweets through a large-scale crowdsourcing experiment (Sect. 5.1), in order to collect enough annotations to build a supervised model capable of accurately predicting the type of action contained in a given tweet (Sect. 5.1).

5.1   Types of Tweets

As discussed in Sect. 2, the campaigns that are the most effective at influencing users are typically related to either promoting some positive behavior or preventing some negative actions [4]. In our context, prevention campaigns typically focus their attention on negative consequences rather than on positive alternatives. This introduces our first class of protest-related actions: physical actions [23]. Next, awareness campaigns that promote positive behaviors try to actively connect with either informational or instructional resources [14,22]. This motivates the definitions of two further types of actions: publications and calls for actions. Since most campaigns have some sort of supporters or base community, when running a campaign it is important to focus not only on the general public but also on specific stakeholders, e.g., to empower important communities, activate voluntary associations, or collaborate with governmental agencies. This often prompts the campaigns to organize official meetings, conferences, and debates [14,22,23]. Taking the above information into account, we consider five different classes of Tweets for our study.

- *Calls for action* correspond to tweets that contain a clear message calling for action, including actions to sign a petition, prevent events from happening, etc.
- *Publications* correspond to tweets that contain a reference to publication, news or some information related to the campaign, including videos, articles or background information on the campaign.
- *Official meetings* correspond to tweets that contain information about an official meeting, a conference, a convention or a debate related to the campaign.
- *Physical actions* correspond to tweets that contain information about past, current or upcoming actions organized by an individual, a group of people, or an organization that is related to the campaign. This includes proposals to participate to challenges, contests or to dedicate some time to a specific issue, e.g., cleaning streets or repairing homes.

– *Others*, finally, correspond to tweets that do not belong to the four categories above, such as content that is indirectly related to climate change or animal welfare domains, as well as personal opinions and experiences, or tweets in other languages.

Tweet Filtering and Annotation. Next, we explain how we classified the tweets from our dataset based on the classes introduced above. Since manually annotating our whole dataset is unrealistic, given the high number of tweets involved, we introduce a two-step process, where we first use micro-task crowdsourcing to annotate parts of the dataset and then leverage the resulting annotations in order to build an effective classifier.

The aim of the first step, i.e., crowdsourcing, is to collect as many high-quality annotations as possible pertaining to the types of tweets while limiting the involvement of the crowd. In order to do this, we first design a set of rules to preselect the tweets given our types. Those rules were created using simple regular expressions based on the analysis of a sample of the tweets, and are presented in Table 3 of Appendix A. In total, we created approximately 40 rules for each message type[8]. These rules were geared towards high recall based on the message types, rather than high precision. Nonetheless, they allowed us to significantly narrow down the number of tweets that would be presented to the crowd by focusing on subcategories early in the process. The resulting counts of tweets obtained from this process are given in Table 3 of the Appendix A.

We then crowdsource the action type annotation using the CrowdFlower platform[9]. The three authors of the paper manually labeled 5 % of the tweets beforehand to create a set of test questions for the crowd. Crowd workers could only work on our tasks if they correctly answered at least 7 out of 10 test questions. We additionally selected workers from English-speaking countries only and collected three independent judgments for every tweet. Agreement was obtained through the majority voting. We also made sure to identify and block malicious crowdworkers by leveraging a series of unambiguous test questions, following standard recommendations from CrowdFlower.

For each type of action, we considered a random sample of 2100 tweets. For more exploration, only half of these tweets is randomly selected from the collection complying with the regular expressions, while the other half is randomly selected from the rest of the campaign tweets. The results obtained through this process were consistent, with an agreement rate of 87.5 %. In general, human annotators applied our definitions for the types of actions very strictly. However, this sometimes narrowed the results; For instance, human annotators did not always correctly annotate the tweets related to the attendance of a conference or a meeting when obvious keywords or the acronym of the event were missing, e.g., "conference". As before, the annotated tweet collection is available online.

---

[8] https://github.com/toluolll/CampaignsDataRelease.
[9] http://www.crowdflower.com/.

Action Classification. At this stage, we use the results of the crowdsourced annotation campaign as a training set to create an effective type classifier for the tweets. For this task, we consider the following features:

- *Semantic features.* Having a large textual corpus of 10Gb, we trained a Word2Vec model [18] using the implementation from the Gensim library[10] with 200 word vector dimensions. To train the model, we preprocessed each tweet as follows: (a) deleted all punctuation excluding hashtag (#) and handler (@), (b) lowercased the tweets, (c) tokenized the tweets into words. Furthermore, we interpreted each tweet as a bag of word vectors and calculated an averaged vector for every tweet. The main motivation behind the choice of semantic features is their ability to capture the semantic similarity between words and phrases using contextual information [18].
- *Syntactic features.* In addition to the above features, we added manual rules based on the regular expressions from Sect. 5.1. This resulted in 46, 42, 38, 20 additional features for meetings, actions, calls for action and publications respectively.
- *Contextual features.* Finally, we added a feature whether a particular domain name is contained inside a tweet. We selected the most frequent domain names and used them as binary features for the classifier. The frequency threshold was chosen at one sigma.

In order to predict the type of a tweet, we trained an individual binary classifier for each of our action types. As a classification method, we used a state-of-the-art approach based on Decision Tree Ensembles[11], which effectively deals with diverse features. Appendix's A Table 2 shows its precision and recall results for the four types of actions using 10-fold cross-validation. We observe that the physical action type has the lowest precision and recall among all types. We connect this result to the relative subjectivity in the definition of physical actions and to the high linguistic variety of the tweets of this type. The prominence of physical actions is hard to determine in general, since they can encompass anything from territory cleanups and protests to film-making competitions and tweet-a-thons.

Further, the introduction of semantic features extracted from the tweet word vectors leads to a loss in precision and to some improvement in recall. This is due to the semantic representation of the tweets, which allows to identify semantically related tweets and words. For example, in the vector space representation produced by the Word2Vec model, the word "debate" is most similar to the words "politics, issue, discuss, policy, conversation". Overall, due to the very nature of the tweets (i.e., very limited length, use of slang, pictures, videos, or emoticons), recall is relatively low across all the categories.

As expected, we found that manually constructed syntactic rules result in better precision as compared to the Word2Vec features only. This is caused by

---

[10] https://github.com/piskvorky/gensim.
[11] http://scikit-learn.org/stable/modules/generated/sklearn.ensemble. ExtraTreesClassifier.html.

the fact that the rules are highly representative of the classes they are built for. Additionally, we observed that domain names play a more important role for meetings, calls for actions and publications, which is explained by the presence of conference websites and specialized websites to gather petition signatures.

5.2 Data Analysis

In order to detail content of the campaigns, we ran the tweet type classifiers over all tweets from all campaigns. We relied on the classifiers that were trained on all features from the previous section as they achieved the best F1-scores for all message types.

We applied the models on each campaign to identify the amount of contribution of a particular action to the overall contents of the campaign.

A visual summary of the outcome for the two main classes of campaigns is shown on Fig. 5(a). We observe major differences in terms of contents; in particular, we see that mobilization campaigns favor calls for actions that motivate the audience to react on the climate change issues, while, having relatively low physical actions. Interestingly, awareness campaigns encourage more physical actions and publication releases, while mobilization campaigns focus more on calls for actions and official meetings. Mobilization campaigns make a high use of official meetings, probably because they tend to raise more attention from the governments or particular stakeholders. To conclude, we see that mobilization and awareness campaigns get organized in very different ways, thus confirming the initial distinction we make between each other.

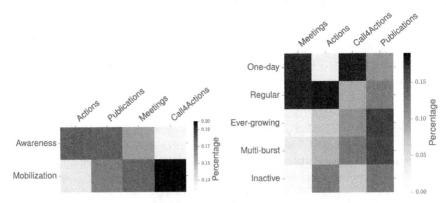

(a) Comparison of the distributions of actions for the types of climate change campaigns defined in Section 4.1. Dark indicate greater importance of a particular actions in the campaign type.

(b) Comparison of the distributions of actions for the two main categories of climate change campaigns: awareness and mobilization. Dark indicates a greater importance of a particular actions in the campaign type.

Fig. 5. Action distribution accross varius campaign types

Following the analysis given in Sect. 4, we performed a study on user engagement patterns. As shown on Fig. 5(b), "one-day" campaigns[12], focus on call for actions tweets which are mainly duplicated rather than retweeted. On the other hand, "regular" campaigns[13] are mostly represented either by regular meetings or physical actions, e.g., annual conferences, campings, etc. Interestingly, "ever-growing" and "multi-burst" campaigns[14], make larger use of publication and call for actions types, which significantly differs from the awareness campaign strategies in general. This can be explained by the targeted audience and by the issues tackled by those campaigns, such as global poverty, international divestments, dependence on fossil fuels, etc. All of these campaigns share global values and target international audiences around the globe.

Duplicate tweets. As described in Appendix B, some tweets from our dataset shared the same contents but were not strictly speaking retweets. This is due to some users trying to promote a tweet into a trending topic on Twitter. We decided to compute the proportions of such duplicated messages to see how they are distributed across different campaign types. To select the threshold at which a message should be treated as a duplicate, we considered the distribution of number of similar messages to the total amount of messages with these number as a half-normal distribution. In such way, the tweet was considered to be a duplicate if the number of such tweets exceeded three standard deviations.

Figure 6 illustrates the distribution of duplicate content for the different campaign types. As can be observed, duplicate content is especially significant for the mobilization campaigns, which can be explained by their spontaneous nature and the need to mobilize people in shorter periods of time. Awareness campaigns differ in the sense that they typically operate on longer-terms goals. From a user engagement perspective, both regular and inactive campaigns do not contain much duplicated content, while increasing, multi-burst, one-day campaigns make heavy use of it.

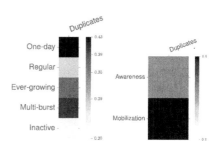

Fig. 6. Comparison of duplicate content between campaigns

Domain usage distribution. Users in the climate change community tend to make great use of links to images, facebook pages, youtube videos and petition sites. We explored the general distribution of the top domain names across the

---

[12] #helpcovedolphins, #savebucky, #freethearctic30, etc.
[13] #climatecamp, #climateweek, #worldenvironmentday, etc.
[14] #talkpoverty, #saveanimals, #saveenergy, #actonclimate, #divestment, #fossil-free.

campaigns and found that all types of campaigns extensively use visual content (youtube, facebook, photos, etc.). Nevertheless, both ever-growing and regular types of campaigns use such content more parsimoniously comparing to one-day and inactive on average. A similar trend was discovered between awareness and mobilization campaigns respectively. Interestingly, the tendency to overuse visual resources clearly does not affect the popularity of the posted content [27]. Among major domain names whose tweets gained the most retweets, we primarily observe contents related to the campaigns, i.e., the site of the campaigns, news and information about related issues.

6 Discussion

In the following, we take a step back, discuss the results we obtained and also make a series of recommendations in the context of public campaigns on social media. First, we proposed a framework for collecting campaigns and identifying their types. As explained in Sect. 3, we collected over 100 campaigns that were annotated with types, i.e., awareness and mobilization, as well as clustered by their user engagement patterns (Sect. 4). This resulted in a large collection of tweets that were partially annotated with action types using crowdsourcing and further generalized based on an annotated corpus using a machine learning classifier. Overall, our tweet action type detection technique showed high precision (~90%) and recall (~60%). This allowed us to automate action identification in tweets and to understand the overall campaigns' contents.

Subsequently, we focused on the analysis of campaigns classified by their initial goal and their user engagement pattern. The goals of awareness and mobilization campaigns differ significantly, and so do their contents. While awareness campaigns often involve physical actions and promote scientific publications, mobilization campaigns make great use of official meetings and calls for actions; For the mobilization campaigns, the more official meetings are organized the more leverage can be obtained from governmental organizations. The analysis of user involvement patterns also showed noticeable differences between campaign types and their agendas. "One-day" campaigns were dominated by calls for actions, while "regular" and "ever-growing" ones contained more physical meetings and publications on climate. This insight represents an important foundation on which specific campaigns studies and their contents can be built.

With the various techniques we leveraged for campaign analysis, we noted major differences in the way users duplicate messages. "One-day" and "ever-growing" campaigns in general contain 20% more duplicated content as compared to the "inactive" campaigns. In the "one-day" campaigns, this phenomenon can be explained by the spontaneous nature of particular tweets and the need to mobilize people in a short period of time. On the other hand, "awareness" campaigns typically operate on a longer basis, therefore their communities do not actively use duplicated tweets, i.e., on average 15% less duplicates than mobilization. This can be explained by the actions required during shorter periods of time for the mobilization campaigns. In a context of user involvement patterns, both "regular" and "inactive" campaigns do not contain as much duplicated content, while "ever-growing" and "one-day" campaigns make a heavy use of them.

Regarding the effects that drive user engagement, we observe that first-degree neighbors are essential for getting higher numbers of retweets (about half or the retweets from popular tweets originate from direct neighbors), while duplicated content attracts less retweets in general. Finally and most interestingly, the less diverse the main contributors of the campaign, the less likely it is to gain bigger audiences (as shown in Sect. 4.2).

Overall, this work has a potential to empower governmental and non-profit organizations by facilitating campaign analysis. The analysis of the collected campaigns combined with the analysis of individual tweets provides foundation for many applications, e.g., detecting public campaigns, or identifying means to boost user engagement.

Even though we expanded the campaign coverage by performing several iterations of the data collection, our methodology is focused on English-speaking tweets. The @AlGore and @GreenPeace accounts we used are biased towards the US, and so are the English terms and hashtags that were used for the climate change topic. Therefore, it would be beneficial to further expand the data collection by reiterating over the steps from Sect. 3.1 to sample more campaign hashtags over other languages and countries. At the same time, a given campaign may leverage multiple hashtags, which can affect the results of the analysis.

7 Conclusions

In this work, we analyzed large-scale social media campaigns related to climate change and animal welfare from various perspectives, including analyses on their primary goals, the types of messages they relay, as well as their user involvement patterns. In the context of climate change and animal welfare, we showed that public campaigns are represented by two main narratives: awareness and mobilization. Our subsequent analysis of user participation revealed that campaigns significantly differ in terms of their user involvement patterns. Finally, we presented a study on the best ways towards increasing user involvement for public campaigns by combining core users, followers, and actions. The high-level patterns that were found in our study lay a solid foundation for future work on specific campaigns and their fine-grained segmentation. As a possible extension, a more fine-grained classification of campaigns and campaign actions could reveal more sophisticated patterns and correlations that appear during the campaign life-span. For example, political or non-profit events might exhibit different user involvement patterns. Similarly, characteristics of geographically focused or global campaigns may differ.

Acknowledgements. The authors would like to thank Alexandra Olteanu for suggestions and feedback. The work was supported by the Sinergia Grant (SNF 147609).

A   Action Detection

Examples of tweet actions are shown in Table 1.

Table 1. Sample tweets for each type of action considered.

Type of action	Sample tweets
Official meeting	Monday Dec 1, U.N. COP climate talks begin Lima Peru @Yeb-Sano Just witnessed a sign of hope at the climate talks in #Cancun - ... #UNFCCC #tcktcktck
Physical action	#WorldEnvironmentDay #treeplanting is taking place around 09:00 at Tsarogaphoka in #Soshanguve We came. We swooped. We're camping!!! #climatecamp
Call for action	The #GreatBarrierReef is not a dump! Protect our World Heritage. #UNESCO #FightfortheReef Take Action: Stand with me and support clean #energy and a safer #climate future! #CleanAir4Kids
Publication	660 million Indians could lose 2.1 billion years as a result of air pollution... #gofossilfree Water Fact: Fact: At 1 drip per second, a faucet can leak 3,000 gallons per year. #savewater

Results of the action detection based on Decision Tree Ensembles is shown in Table 2. It shows the precision and recall results for the four types of actions using 10-fold cross-validation.

Table 2. Precision, Recall and F1-score values for classification of different types of actions with different sets of features.

	Meetings			Actions			Call for actions			Publications		
	P	R	F1	P	R	F1	P	R	F1	P	R	F1
All features	0.896	**0.616**	**0.730**	0.771	**0.577**	**0.660**	0.902	**0.664**	**0.765**	0.897	**0.528**	**0.665**
Sem	0.723	0.605	0.659	0.707	0.510	0.592	0.751	0.441	0.556	0.857	0.461	0.599
Sem + Cont	0.788	0.531	0.635	0.703	0.518	0.597	0.792	0.439	0.565	0.865	0.487	0.623
Sem + Synt	**0.912**	0.590	0.717	0.754	0.480	0.587	0.920	0.563	0.699	0.862	0.464	0.603
Synt	0.895	0.375	0.529	**0.816**	0.276	0.412	0.920	0.472	0.624	0.890	0.134	0.232
Synt + Cont	0.901	0.384	0.538	0.812	0.300	0.438	**0.921**	0.643	0.758	**0.911**	0.325	0.479

Below we show examples of manually created rules to preselect the tweets given specified action types.

Table 3. Examples of rules and number of tweets for each type of action.

Type of action	Sample rules	N# of tweets
Official meeting	$/speakingat(demo|the)/$	113989
Physical action	$/actionat(the)?park/$	154874
Call for action	$/tell(the)?to(keep|protect)/$	328603
Publication	$/greatnews/$	2559063

B Unique Tweets Identification and Retweets Count

One of the main issues with the data collected from Topsy is that the tool does not provide information about retweets. Therefore, we had to create heuristics to make sure that we could properly identify all retweeted messages. Taking into account that all tweets returned by the tool are sorted by timestamp, we can easily figure out the origin of all the tweets using a simple regex pattern (`(RT|MT) @author tweet_prefix`). This approach has a number of limitations, however. It does not identify complex retweet structures, such as where a tweet text is cited using quotes. We found that such cases are quite rare on Twitter and amount for ~0.5 % of all tweets.

In order to compute the complex retweet cases, we aggregated the tweets with at most 5 characters edit distance. Further, we discarded explicit retweets (`(RT|MT) @author`) and exact duplicates. However, certain retweets can be missing when a hashtag does not fit into the message due to the tweet length limit. To solve this problem, we leveraged the Topsy API, by returning and analyzing related tweets for each requested tweet in order to identify all further retweets. Finally, we note that we apply this process recursively—searching for retweets of retweets iteratively—in order to capture potentially complex retweet patterns. When no new retweets can be identified, we identify content that was not retweeted but duplicated. The practice of duplicating tweets gained traction on the platform as it can help promote topics into *Twitter Trends*. We consider a tweet to be a duplicate whenever at least 80 % of its contents exactly matches an original tweet excluding punctuation.

References

1. Abbar, S., Mejova, Y., Weber, I.: You tweet what you eat: studying food consumption through twitter. arXiv preprint arXiv:1412.4361 (2014)
2. Aiello, L.M., Petkos, G., Martin, C., Corney, D., Papadopoulos, S., Skraba, R., Goker, A., Kompatsiaris, I., Jaimes, A.: Sensing trending topics in twitter. Trans. Multi. **15**(6), 1268–1282 (2013). http://dx.doi.org/10.1109/TMM.2013.2265080
3. Alley, R.B., Marotzke, J., Nordhaus, W.D., Overpeck, J.T., Peteet, D.M., Pielke, R.A., Pierrehumbert, R.T., Rhines, P.B., Stocker, T.F., Talley, L.D., Wallace, J.M.: Abrupt climate change. Science (New York, N.Y.) **299**(5615), 2005–2010 (2003). http://www.sciencemag.org/content/299/5615/2005.short

4. Atkin, C., Rice, R.: Theory and principles of public communication campaigns. Public Communication Campaigns (2012)
5. Becker, H., Naaman, M., Gravano, L.: Beyond trending topics: Real-world event identification on twitter. In: ICWSM 2011, pp. 438–441 (2011)
6. Fernandez, M., Piccolo, L.S.G., Maynard, D., Wippoo, M., Meili, C., Alani, H.: Talking climate change via social media: Communication, engagement and behaviour. In: Proceedings of the 8th ACM Conference on Web Science, WebSci 2016, pp. 85–94. ACM, New York (2016). http://doi.acm.org/10.1145/2908131.2908167
7. Giorgino, T.: Computing and visualizing dynamic time warping alignments in R: The DTW package. J. Stat. Softw. **31**(7), 1–24 (2009)
8. Gonzalez-Bailon, S., Wang, N.: Networked discontent: the anatomy of protest campaigns in social media (2013). SSRN 2268165
9. Hermida, A., Fletcher, F., Korell, D., Logan, D.: Share, like, recommend: decoding the social media news consumer. Journalism Stud. **13**(5–6), 815–824 (2012)
10. Hong, L., Convertino, G., Chi, E.H.: Language matters in twitter: a large scale study. In: ICWSM 2011 (2011)
11. Hoornweg, D.: Cities and Climate Change: Responding to an Urgent Agenda. World Bank Publications, Washington DC (2011)
12. Jackoway, A., Samet, H., Sankaranarayanan, J.: Identification of live news events using twitter. In: Proceedings of the 3rd ACM SIGSPATIAL International Workshop on Location-Based Social Networks, pp. 25–32. ACM (2011)
13. Jin, F., Khandpur, R.P., Self, N., Dougherty, E., Guo, S., Chen, F., Prakash, B.A., Ramakrishnan, N.: Modeling mass protest adoption in social network communities using geometric brownian motion. In: KDD 2014, pp. 1660–1669. ACM, New York (2014). http://doi.acm.org/10.1145/2623330.2623376
14. Kirilenko, A.P., Stepchenkova, S.O.: Public microblogging on climate change: one year of Twitter worldwide. Global Environ. Change **26**, 171–182 (2014). http://www.sciencedirect.com/science/article/pii/S0959378014000375
15. Koneswaran, G., Nierenberg, D.: Global farm animal production and global warming: impacting and mitigating climate change. Environmental Health Perspectives (2008)
16. Lee, K., Palsetia, D., Narayanan, R., Patwary, M.M.A., Agrawal, A., Choudhary, A.: Twitter trending topic classification. In: ICDMW 2011, pp. 251–258. IEEE Computer Society, Washington, DC (2011). http://dx.doi.org/10.1109/ICDMW.2011.171
17. Lehmann, J., Gonçalves, B., Ramasco, J.J., Cattuto, C.: Dynamical classes of collective attention in twitter. In: WWW 2012, pp. 251–260. ACM, New York (2012). http://doi.acm.org/10.1145/2187836.2187871
18. Mikolov, T., Corrado, G., Chen, K., Dean, J.: Efficient estimation of word representations in vector space. In: Proceedings of the International Conference on Learning Representations (ICLR 2013), pp. 1–12 (2013)
19. Norris, P.: The handbook of comparative communication research. The ICA Handbook Series (2012)
20. Olteanu, A., Castillo, C., Diakopoulos, N., Aberer, K.: Comparing events coverage in online news and social media: the case of climate change. In: ICWSM 2015 (2015)
21. Quercia, D., Askham, H., Crowcroft, J.: TweetLDA. In: WebSci 2012, pp. 247–250. ACM, New York, June 2012. http://dl.acm.org/citation.cfm?id=2380718.2380750
22. Schmidt, A., Ivanova, A., Schäfer, M.S.: Media attention for climate change around the world: a comparative analysis of newspaper coverage in 27 countries. Global Environ. Change **23**(5), 1233–1248 (2013)

23. Segerberg, A., Bennett, W.L.: Social media and the organization of collective action: using twitter to explore the ecologies of two climate change protests. Commun. Rev. **14**(3), 197–215 (2011)
24. Shields, S., Orme-Evans, G.: The impacts of climate change mitigation strategies on animal welfare. Animals **5**(2), 361–394 (2015)
25. Slovic, P.: Informing and educating the public about risk. In: Slovic, P. (ed.) The Perception of Risk, pp. 182–198. Earthscan, London (2000). cited by 0000
26. Williams, H.T., McMurray, J.R., Kurz, T., Lambert, F.H.: Network analysis reveals open forums and echo chambers in social media discussions of climate change. Global Environ. Change **32**, 126–138 (2015). http://www.sciencedirect.com/science/article/pii/S0959378015000369
27. Xu, Z., Zhang, Y., Wu, Y., Yang, Q.: Modeling user posting behavior on social media. In: SIGIR 2012, pp. 545–554. ACM, New York (2012). http://doi.acm.org/10.1145/2348283.2348358

Colombian Regulations for the Implementation of Cognitive Radio in Smart Grids

Julián Giraldo Torres[1], Brayan S. Reyes Daza[2], and Octavio J. Salcedo Parra[1,2(✉)]

[1] Universidad Nacional de Colombia, Bogotá D.C., Colombia
jgiraldo@unal.edu.co
[2] Internet Inteligente Research Group, Universidad Distrital Francisco José de Caldas,
Bogotá D.C., Colombia
bsreyesd@correo.udistrital.edu.co, osalcedo@udistrital.edu.co

Abstract. The implementation of smart grids is the future in the way of the modernization of the electrical system as we know, this modernization brings lots of challenges to be tackled, one of them is the deployment of an effective, robust and reliable communications network that support the requirements of Smart Grids. Wireless communications become a viable solution to complete this task using the benefits considered in the proposed technology under the concept of Cognitive Radio, it uses techniques such as dynamic spectrum access to optimize communication way and improve characteristics in terms of information channels performance, however its implementation depends on the proper legislation regarding the use of radio spectrum in each country, in our case the legislative gap under Colombian regulations is wide, where dynamic access of this resource has not been take in count, so it's necessary to make an analysis from the point of view of legislation for the use of this emerging technology and its practical application concerning SG communications among others.

Keywords: Modernization · Smart grids · Cognitive radio · Radio electric spectrum

1 Introduction

It is a fact that the current power grid will change in the near future for the purpose of modernization, this job is specified under the concept of Smart Grids. Understand the general concept of the called Smart Grids will give us an overview of the challenge of set a communications infrastructure relevant that allows fulfill the requirements of the system being that communications are the pillar that supports the responsibility to make this work possible. Because of the electrical system is extensive and covers wide geographic areas, proposing a robust solution of communications covering criteria as security, reliability, latency, scalability, availability, maintenance and others it's a hard task, but the concept of cognitive radio will be discussed agree its characteristics of design and performance which promise be a viable alternative for supply the proposed requirements.

© Springer International Publishing AG 2016
E. Spiro and Y.-Y. Ahn (Eds.): SocInfo 2016, Part II, LNCS 10047, pp. 244–258, 2016.
DOI: 10.1007/978-3-319-47874-6_17

In Sect. 2 the most representative elements of the smart grids will be named and will be looking show the reader in a wide way what is intended with this concept, then in Sect. 3 the division of the radio spectrum will be specified in order to understand the medium of communication wireless technology, therefore the resource of work and object of optimization of Cognitive Radio, explained in Sect. 5 where further its viability as a solution will be explained in the larger concept of Smart Grids, in Sect. 4 are named some aspects of Colombian law regarding the use of radio spectrum and finally in Sect. 6 the conclusions will be expose and some issues which must be clearly referenced when taking the decision to realize the implementation of a communication infrastructure based on Cognitive Radio.

2 Smart Grids

The actual electrical grid is based on a model proposed in the last third of the XIX century, in 1887 the polyphase system model is provided by the physicist and engineer Nikola Tesla who planted the bases of the alternate current. The centralized model of generation, transmission, distribution and consumption of electrical energy that we know and use actually began implementing in the early XX century and its fundament remains intact over a century (Chun-Hao and Nirwan 2012) approximately, making improvements in elements of the system without changing its philosophy.

Smart grids want to change this paradigm by detecting system faults and automate it using a two-way flow in electricity and information and obtaining a distributed and automated system, implementing communication nets that allow interaction between the producer and the user, in other words, integrates an intelligent communications net for power net (Kumar et al. 2011) allowing the monitoring and control of the different elements connected to the network. SG is the modernization of the entire electrical system, want to integrate the technologies of information and communication complement infrastructure that take advantage of integration of renewable energy and allow two-way communication between the consumer and provider of electric service to improve efficiency considering concepts such as demand response, smart measure and power quality.

The benefits to be obtained of intelligent networks are among others (Fig. 1):

- Automation and remote reading of energy meters. (AMI - AMR, Advanced metering infrastructure - Advanced meter reading).
- Improving the quality of power.
- Communication from the generation, transmission, distribution and end user.
- Two-way flow of electricity and information
- Support for distributed generation.
- Flexible system of rates.
- Easy maintenance and operation.
- Integration of all types of generation in a single network.

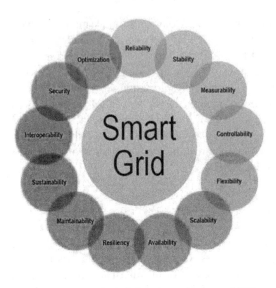

Fig. 1. Smart grids requirements. (Behnam 2012)

When we present these items, we realize that telecommunications play a leading role in the implementation of smart grids, thus making the deployment of a communications network will be a consuming and delicate task because they must ensure the principles and essential requirements to consider the system as an intelligent network.

So we got that SG is a wide proportions system in which large number of interconnected components interact producing massively information, these data should be transmitted, transported, administered and analyzed under schemes and architectures of communication that solve the loading and management of information generated in the system. It also means that design a communication architecture that meets all specifications of SG is not an easy task, many services, components and applications have different requirements in terms of bandwidth, capacity, latency, security, and others, so, realize that the current ICT infrastructure and standards used are unable to reach the requirements set in this concept, opens a start from here to propose viable alternatives to achieve the intention to establish the basis of will be smart grids of future, where choose the right architecture becomes critically important to consider ensuring an efficient, safe and reliable model delivering information between network components. In this way it is shown that the communication infrastructure becomes the pillar that supports the load of the smart grid in the perspective view (Yan et al. 2013), the Fig. 2 illustrates this premise.

Multiple topologies and network architectures have been proposed with the solid intention of providing the services and functional requirements in this complex system, it's the case of Ettus research, (Krattenmaker 1995); (Mitola 1999) and others ones, there we can find common concepts as the network architecture required. With the will to unify the criteria found, it can be deducted three servings network Access segment, Distribution segment and Main segment o backbone.

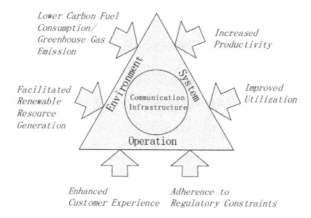

Lower Carbon Fuel
Consumption/
Greenhouse Gas
Emission

Increased
Productivity

Facilitated
Renewable
Resource
Generation

Communication
Infrastructure

Improved
Utilization

Operation

Enhanced
Customer Experience

Adherence to
Regulatory Constraints

Fig. 2. Smart grid pillar. (Yan et al. 2013)

Once it's understood the importance of the telecommunications network in the context of SG, the idea is to analyze the technologies available to reach the challenge posed to implement smart grids in Colombia, due to ease and cost of deployment to deploy a telecommunications network according to requirements of smart grids, we discuss an emerging technology that is part of wireless solutions because they offer advantages and facilities compared wired technologies, is also widely known that the transmission rates have improved greatly in recent years, a feature that makes it look a very attractive option. Also, it have been planted optimization techniques of the electromagnetic spectrum to use this resource in a right manner, this concept is called cognitive radio (Vehbi and Dilan, 2012) and will be explained later.

The electromagnetic spectrum is the medium diffusion of wireless communications, for this reason it is important to know this resource and its regulations to make wise decisions based on solid and consistent arguments from the point of view of engineering.

With the will to unify the criteria found, it can be deducted three servings network as the follows:

- Access segment.
- Distribution segment.
- Main segment o backbone.

The previous corresponds to the geographical location of the elements that make part of the power grid, they will be distributed throughout the infrastructure all telecommunications devices, sensors and meters that will be reporting information to the analysis and management systems.

To see graphically the previously, we observed Fig. 3, where portions of net are specified according to their distribution.

Also, it can clearly see that as the network portion, exist significant names to divide or segment the focus of work within the larger smart grid system, highlighting the following concepts that will not be extended in this article, but they will be delve in the references cited.

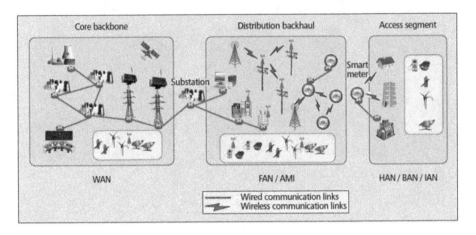

Fig. 3. Network architecture for SG. (Krattenmaker 1995)

The concepts that should be expanded are:

- HAN: Home Area Network
- BAN: Building Area Network
- NAN: Neighborhood Area Network
- IAN: Industrial Area Network
- FAN: Field Area Network
- WAN: Wide Area Network

Once it's understood the importance of the telecommunications network in the context of SG, the idea is to analyze the technologies available to reach the challenge posed to implement smart grids in Colombia, due to ease and cost of deployment to deploy a telecommunications network according to requirements of smart grids, we discuss an emerging technology that is part of wireless solutions because they offer advantages and facilities compared wired technologies, is also widely known that the transmission rates have improved greatly in recent years, a feature that makes it look a very attractive option. Also, it have been planted optimization techniques of the electromagnetic spectrum to use this resource in a right manner, this concept is called cognitive radio (2012) and will be explained later.

The electromagnetic spectrum is the medium diffusion of wireless communications, for this reason it is important to know this resource and its regulations to make wise decisions based on solid and consistent arguments from the point of view of engineering.

3   Spectrum Division

We know consequently use wireless technology is a feasible tendency since it's trying to implement an efficient telecommunications network and affordable cost to provide the necessary services to reach the requirements of SG. To think of using the radio electric spectrum with the exposed goal, we must have awareness of how this resource is used, for it is primordial to know which is divided into 9 frequency bands (ITU 2012) designated by integers as shown in Table 1.

Table 1. Spectrum division (ITU 2012).

Band number	Symbol	Frequency ranges	Example of service or application
4	VLF	3 to 30 kHz	Meteorology
5	LF	30 to 300 kHz	Radio navigation
6	MF	300 to 3000 kHz	Time Signals
7	HF	3 to 30 MHz	AM Radio
8	VHF	30 to 300 MHz	FM Radio
9	UHF	300 to 3000 MHz	Wi-Fi, TV, Cell
10	SHF	3 to 30 GHz	Aeronautical mobile
11	EHF	30 to 300 GHz	Communication between satellites
12		300 to 3000 GHz	

In this way and according to the characteristics of each band, commercial use and standards widely known, we can get an idea of which technologies use to implement a suitable telecommunications network within SG.

3.1 Practical Use of the Spectrum

In our daily life we use many applications or services based on the transmission of information across the radio electric spectrum, is the case of radio (AM 500–1500 kHz band 6, 80–108 MHz FM band 8), television (VHF UHF bands 8 and 9 respectively), mobile phone (800–2500 MHz using the band 9), Bluetooth (2.4 GHz, lane 9), Wi-Fi (depending on the implemented standard could be 2.4 GHz or 5 GHz bands 9 and 10 respectively) and others. It also should not be ignored that there is a strict regulation in each country for the use of spectrum, there it can be found based on the technical specifications of the sector, frequency bands licensed and unlicensed (900 MHz, 2.4 GHz, 5 GHz).

4 Colombian Regulations for the Spectrum Use

In the Colombian legal framework, the electromagnetic spectrum is an inalienable and an endless lapsed-time property, in this way the control and management are only responsibility of the state as it says in Article 75 of the 1991 constitution: "The electromagnetic spectrum is an inalienable and endless lapsed-time public property attached to the management and control by the State. Equal opportunities is guaranteed in access of use that terms of law established.

To guarantee media informative pluralism and competition, the State will intervene as mandated by the law to prevent monopolistic practices in the use of the electromagnetic spectrum." (Asamblea Nacional Constituyente 1991).

In most regimens, the state is responsible for controlling the use of the electromagnetic spectrum, in our case, Article 4 of the law 72 of 1989 provided: "The radioelectric channels and other mediums of transmission that Colombia uses or may use in the field of telecommunications are the exclusive property of the State." (Guerra de la Espriella and Oviedo 2011). Later in 1990 it's stated in Article 18 of Decree Law 1900 that the electromagnetic spectrum "is the exclusive property of the State and such as a public domain, inalienable and endless lapsed-time domain, the management, administration and control correspond the Ministry of communications ". (Guerra de la Espriella and Oviedo 2011).

In the same way, in that norm set in Article 20, it's established that the radioelectric frequency use requires prior permission from the then Ministry of Communications, actually the Ministry of Information Technologies and Communications, also for this reason must pay the rights that apply to the use of this resource. (Guerra de la Espriella and Oviedo 2011).

Article 19 of Decree Law 1900 of 1990, the management and administration of the spectrum is also named prioritizing the activities of planning, coordination, fixing the frequency table, allocation and verification frequencies, granting permission for use, protection and defense of the radioelectric spectrum, radio emissions testing technique, establishment of technical conditions of terminal equipment, vigilance for the irregularities control that contribute to maintain the correct use of the resource in question. (Guerra de la Espriella and Oviedo 2011). Also in Act 142 of 1994, Article 8 refers as follows: "It is the responsibility of the Nation: 8.1. In privative form plan, allocate, manage and control the use of the electromagnetic spectrum." (Congreso de Colombia 1994).

The Article 25 defines that: "Concessions, and environmental and health permits. Those who provide public services require concession contracts, with the competent authorities according to law, to use the water; to use the electromagnetic spectrum in the provision of public services will require a license or concession contract." (Congreso de Colombia 1994).

Continuing with this idea, Decree 4392 of 2010 specifies that the radioelectric spectrum, as defined by the International Telecommunication Union-ITU, is "the set of electromagnetic waves, which frequencies conventionally lower than 3000 GHz, which propagated in space without artificial guide." (Information and communication Technologies Ministery 2010). Just there it's named that the organization is reflected in the national table frequency attribution bands (NTFAB) in accordance with national and international standards and technological developments.

The order of NTFAB (Guerra et al. 2010) is in charge by Ministry of Information Technology and Communication on recommendations of the National Spectrum Agency (NSA), this one has as its main objective to plan, monitor and control the Radioelectric Spectrum in Colombia, as well as provide technical advice for the efficient management and promote their knowledge (ANE, 2011), among others, the functions of advising the Ministry of Information Technologies and Communications in design and formulation of policies, plans and programs related to radioelectric spectrum, designing and formulating policies, plans and programs related to the prevention and control of spectrum, in accordance with national and sectorial policies and proposals by the competent international organizations, when it'll be necessary, study and propose, in line with sector trends and technological developments, optimal schemes of surveillance and control of the radioelectric spectrum, including satellite ones, except as provided in Article 76 of the Constitution and in accordance with current regulations, exercise supervision and control of the radioelectric spectrum, except for the provisions of Article 76 of the Constitution, on the other hand it is fine to know that the control entities responsible for managing is the procurement, comptroller and congress.

Meanwhile in Law 1341 of 2009, Articles 11 and 72 indicate, that the use of radioelectric spectrum requires previous permission granted by the Ministry of Information Technologies and Communications, who wants to ensure transparent processes and maximizing resources for the state, for that reason prior to granting permission to use radioelectric spectrum allocation or service concession is determined if there is indeed a plural number of applicants in the corresponding frequency band so as to apply objective selection procedures. In this same decree objective selection procedures are established, the determination of multiple applicants, content of applications, evaluation and award of spectrum, performance guarantees, compensations, penalties, service continuity, temporality among others (MINTIC 2009).

In this way we realize that this wireless technology offers benefits to think about deploying a communications network focused to meet the requirements of SG, it should board the regulation of our own nation and consequent to have clarity in the application of technology designs, prototypes or wish to submit proposals, the previous reach the objective of informing and consider legal aspects of the use of the most important resource in the transmission of information via wireless.

5 Cognitive Radio

In 2003, the FCC (Federal Communications Commission) of the United States, claiming that he had to rethink current wireless network architectures giving way to the concept of cognitive radio (CR = Cognitive Radio) on which the fundamental principle for the design would be new networks (Aguilar and Navarro, 2011). Here CR is defined as the radio frequency system capable of varying its transmission parameters based on their interaction with the environment in which it operates.

According to the working principle of cognitive radio, we have the following main features: Cognition capacity: Functionality that allows census the environment for capture information concerning the radioelectric spectrum and identify sub channels used or blanks spaces. Auto-configuration: Permits a device change its transmission and reception parameters dynamically according to the census information (modulation, frequency, power). With the goal to optimize the spectrum, the FCC promoted the opening bands used in television broadcasting for use by unlicensed users, called secondary users. The principle of this proposal is based on techniques to share spectrum, this technique is called OSS (Opportunistic Spectrum Sharing).

Other organizations have advanced in techniques of cognitive radio, in the case of DARPA (Defense Advanced Research Projects Agency) who developed a technology that allows multiple telecommunication systems share the same spectrum using an adaptive mechanism called DSA (Dynamic Spectrum Access). On the other side we find the military forces of the United States (US Army) who also based his research developed the technique ASE (Adaptive Spectrum Exploitation). For its part, the IEEE has worked in the specification of the standard 802.22 WRAN (Wireless Regional Area Network) which aims to access the Internet using the frequency band allocated to television system (54 MHz – 862 MHz), a control of cognitive medium access is included, point-to-multipoint topology in the regional network composed of fixed base stations and portable devices based on the bands 8 and 9 (IEEE 802.11 Working Group, 2014).

Due to the growing demand for Radio frequency (RF) communications, strict policies of portions allocation of spectrum and its inefficient use (Vehbi and Dilan, 2012), it is necessary to use emerging technologies in this area as it is timely access to spectrum by taking advantage of the benefits offered technical dynamic management. The DSA technique for its part, proposes a logical strategy to enable the efficient use of this resource, this way has two actors, the primary user (PU) and secondary user (SU). The primary user is the owner of the licensed channel, for that simply reason he will have priority in the use of the medium, on the other hand we find the secondary user, who must find a appropriate time to use the resource for that reason he will be responsible to census the environment and identify sub channels used to make use of them in the absence of primary users. With an overall description of this functionality, we can realize that the cognitive radio technology has the potential to integrate sampling, communications and computer systems with the aim of serving the real life applications, including SG.

The RC offers viable solutions applicable to SG because of their adaptive capacity in existing spectrum conditions, seeking to improve the performance of the communication network in key areas such as reliability, delay, multipath, noise, changing characteristics of the spectrum in accordance with location, weather and environmental conditions. In summary, the motivation to think in cognitive radio as a viable solution in the implementation of SG are named below: A substantially improve the problem of interference associated with the collision of packets in wireless channels for the characteristic of timely access to the spectrum. It can handle very low latencies for the use of sub-channels used in the TV broadcast band. Wide geographic areas covering. Covering for multiple networks according to the principle of sharing the spectrum. It is very common finding in studies of RC that is appointed the television broadcast spectrum or that development approach to work in this band, however there are not established criteria why focus on this portion of the spectrum being the technology to adapted to the operation of any frequency band. This is the case (Oh et al. 2010); (Peha 2008), where are named approaches and development studies of cognitive radio using the white spaces in the TV band.

On the other hand it is important to note that certain policies of working have been established to address this issue, in (Oh et al. 2010) are highlighted initiatives such as the protection of the rights of the licensed user, rules for sharing the 5 GHz frequency with a dynamic selection in the range 5230–5350 MHz and 5470–5725 issued by the American organization FCC (Federal Communication Commission) which is involved in aspects such as the availability of the channel time-based, movement between channels, non-occupancy period, likewise are named selection criteria channel based on PSD

(Power Spectral Density), signal power limitations, antenna characteristics and emission patterns. For his part in (Peha 2008) we find political cooperation and coexistence between users of the same level and Primary-Secondary spectrum users, finding the right information presented in Fig. 2 of such reference, which seeks to find a working relationship based on the premise of spectrum share, and also valid signal interference percentages. For his part in (Behnam et al. 2012) a device called "Policy Reasoning" which makes decisions on dynamic spectrum access policies based on previously established as mentioned before and languages for information exchange for the sync devices which belong to a RC network.

5.1   Architecture, Topology and Application

We have seen that the RC is a new technology with features well defined that may be applicable to intelligent networks being that can be improve aspects of communication such as interference, coverage area and latency speaking in terms of wireless solutions and being aware of the use of spectrum according to Colombian regulations.

Now it will show realistic applications using cognitive radio in SG will. The immediate and most obvious example is AMI-AMR. Measuring points in the electrical system are strategically placed to measure loads, the information from the measuring points must be collected and analyzed for different purposes, pricing, power quality, TOU (Time Of Use), reports, events etc.

The management and data collection system requires a communications infrastructure to do its work in a reliable, secure, scalable and effective. In a normal wireless communication within AMI-AMR environment it is noisy and overloaded with signals from different devices operating in unlicensed bands in turn this results in many lost packets in themselves collisions medium means that reflected in high latency in communication channels, on the other hand, in the licensed channels have high prices translate into higher budgets to maintain the communication infrastructure. This leads to consider the RC technology to provide dynamic and timely allowing media access and use licensed channels without interfering with undergraduate information systems using common bands within the spectrum to improve the performance of the communication of information.

Within the framework proposed by SG it takes in count the ability to support distributed generation allowing the use of clean and renewable energy that will bring benefits how efficiency and reliability of electrical supply. Maintaining stability and proper balance between generation and demand is possible through exchange of information, as if it were a diagram of closed loop control. The wireless communication channels based on cognitive radio have advantages in this scheme for its wide range of coverage, little interference and dynamic spectrum access. Work monitoring network are easier with an extensive coverage with features such as detection of power outages or overheating problems or other types of damage to transmission lines or distribution, threats or warning alarm generation for specific situations. Noting the Fig. 4, we realize the tree network topology based on a hierarchical RC proposed where the full scope of the network is looking to satisfy given architecture in Sect. 2, Fig. 3. Likewise, HGW (HAN gateway cognitive) are identified in the access segment, the NGW (NAN cognitive gateway) in the distribution segment, finally the control center as well the cognitive radio base stations in the main segment or backbone network.

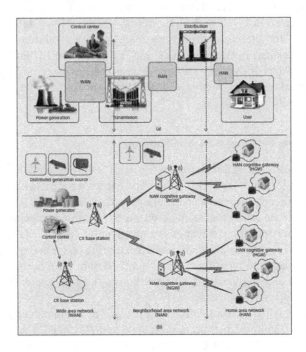

Fig. 4. Hierarchy communication architecture. (Yang, Kim, and Zhang, 2012).

According to this proposal, the access segment work with unlicensed bands, while other segments will be working in licensed bands for a chance to access the medium dynamically. HAN provide meter information (read data as energy, power, instrumentation variables, load profiles etc.) in real time. Taking advantage of the cognitive characteristics of the gateway (HGW), the operating parameters are dynamically configured to deliver the information to the next level, it means, to NGW. In other words, the HGW devices allow you to connect local networks to larger networks, for example Internet. NGW is responsible for transport the information from NAN to control centers and vice versa, also they are responsible for synchronizing the cognitive parameters to radios HGW either own census or sync level, it means control center and stations RC base. This synchronization of parameters is done through fixed channels shared by the radios in the network, this way we find devices in charge of monitoring the environment to make the decisions of the frequencies used in the transmission and reception of information, it is notable that this sampling should be performed in distant geographic areas to prevent interference between the same radio and thus respect the coverage area of each network segment.

Now that we know that the idea is to understand the technology based on RC and the Colombian regulatory framework, it is important to name a case study conducted in Bogotá DC (Paz et al. 2014), this paper analyzes the migration of the cellular communications to the UHF TV band, throwing as conclusions that are necessary flexible radio platforms and according to the available platforms today, the implementation is not adequate from the point of view of flexibility, cost, and energy expenditure, however notes that overcoming this barrier, the RC technology will bring significant advantages over current wireless communication models that allow optimal way to exploit the spectrum giving the ability to have shared services using the same frequency range.

6   Conclusions

It is understood that the technology explained under the concept of RC is viably applicable to SG for optimization that pretends to make the communication media on which it is based by providing wide coverage to use different bands for the transmission of information while avoiding packet collision at offering low latency to avoid retransmission of information important to meet the requirements of SG if it takes in count infrastructure and telecommunications features.

On the other hand, there is a legislative gap in the dynamic spectrum access, as the band allocation rules and rights that must be paid for the use of information channels distributed in the spectrum on the frequency bands set in the CNABF and assigned to the DSA technique called PU as RC. Focusing on this point is where the lacks are detected in terms of regulation refers, a logical fact because it is an emerging technology which has no networks in production but with a great potentially for the solutions that allows to implement, on the other hand highlights the advance work by different agencies such as the FCC and NTIA both American institutions who are in charge of the commercial spectrum users control and federal government users respectively, being ahead of regulation works policies as the mentioned in the 5 GHz frequency (Oh et al. 2010). Furthermore Ofcom who is the communications regulatory body in the UK, as well as the communications regulatory Commission of Ireland (ComReg) agree with the FCC to impose economic incentives (Oh et al. 2010) to standardize policies that assist developers and support developed works by RC FCC and IEEE.

This contrasts with the mentioned legislative lack, because until the day of writing this paper no rules with force of law have been found in the searches done over technical and scientific texts databases nor in web search engines with chains focused on investigating by legislation in different countries in Europe and north America who take the lead in this and many other technologies.

On the other hand it was observed that the band Television is a central focus in the study and development of the RC, although no solid criteria to support this trend has been found, it can understand the commercial use of this band how a striking aspect of the exploitation of this portion of the spectrum. Because of the transmit frequency is high, the advantage of being able to carry a lot of information because with more data more frequently but less than wavelength, and that turns into more power to achieve more coverage area.

By the simple fact of having live and direct broadcasts reaching multiple regions in the same country and guided by the normal operation of the TV we can deduce that it has low latency in the signal and areas of wide coverage, but this it is because our TV is a passive agent, that means, only receives the signal and the operator (responsible for the diffusion channel) is responsible for radiating the signal power to ensure coverage of the population. In the case of an RC network in this band, it should be noted that the teams will be active, so it is receiving and sending information, and for that reason energy requirements should be high if you want to reach a wide area of coverage, also a task of processing information related will be done with the means for determining the characteristics of transmission and reception of information-based policies as mentioned in Sect. 5 which must be clearly defined by the control entities related as ANTV, ANE, MinTIC, Congress and related participants such as TV channels, radio channels and different spectrum users.

On the other hand it a primary point of attention needs to clear rules of appropriate use of the resource for SU respecting the priority of the PU. According with the exposed in Sect. 4, we knew the procedure that must be one PU for the right channel on a spectrum, starting from this basis ¿how the scheme should be to use the same channel for SU?, ¿under which figure should apply for a permit to use the resource?, ¿such permission must be processed before the PU who won the right and license to use the resource or to the State in accordance with the previously stated is solely responsible for the management and protected by article 75 of the constitution of 1991 and other laws established?. If a permits scheme from the PU is followed, it could rise to malicious market behaviors that favor special interests, contrary to the provisions as pluralism and competition is sought in the objective selection of users. Also as an immediate effect is a decrease of the operating cost of the spectrum under the premise of RC should not be managed as a license to use, opening the discussion of suppression licenses for spectrum exploitation to reach the RC massif use. Because the previous reason, it is confirmed that the RC seen from the point of view of engineering is feasible and applicable, but for been a complex theme, a legislative scheme is needed to regulate this technology covering aspects such as limitations of power, features of signal emission types (analog and digital), antennas, channel occupancy, occupation times, regulations to prevent illegal trade of spectrum signaling occupation, mechanisms to prevent a user unlicensed indefinite use of a channel, interference limits, control and surveillance agencies.

Finally, according to the specification of Decree 4392 of 2010, the CNABF is named in accordance with national and international standards and technological developments, so it is understood that this framework is subject, between others, to technological developments which the RC fits perfectly, so it should be consider adjustments, modifications or additions to the regulations shown in national law for the implementation or use of RC in the communication networks that must be implemented for different purposes.

References

Aguilar Renteria, J., Navarro Cadavid, A.: Radio cognitiva – Estado del art. Sistemas y Telemática **9**(16), 31–53 (2011)

Asamblea Nacional Constituyente: Constitución política de Colombia (1991). http://www.alcaldiabogota.gov.co/sisjur/normas/Norma1.jsp?i=4125

Behnam, B., Amol, D., "Jerry" Park, J.-M.: Spectrum access policy reasoning for policy-based cognitive radios. Comput. Netw. **56**, 2649–2663 (2012). Elsevier www.arias.ece.vt.edu/pdfs/PR-Compnet.pdf

Congreso de Colombia: Ley 142 online (1994). http://www.alcaldiabogota.gov.co/sisjur/normas/Norma1.jsp?i=2752

Guerra de la Espriella, M., Medina Velandia, D., Pedraza Pineda, W., Roldán Perea, J.: National framework of band atribution frequency appendix d of the national management of the radioelectric spectrum, republic of Colombia. Inf. Commun. Technol. Ministery (2010). http://www.ane.gov.co/cnabf/index.php/conozca-el-cnabf-mnu

Vehbi, G., Dilan, C.S.: Cognitive radio networks for smart grid applications a promising technology to overcome spectrum inefficiency. In: IEEE Vehicular Technology Magazine, Raleigh, NC, USA (2012). http://ieeexplore.ieee.org/xpl/articleDetails.jsp?reload=true&arnumber=6184366

Information and communication Technologies Ministery: Decreto número 4392 (2010)

ITU: Regulations of radio communications articles (2012). www.itu.int/dms_pub/itu-s/oth/02/02/S02020000244501PDFE.pdf

Peha, J.M.: Sharing spectrum through spectrum policy reform and cognitive radio. In: Proceedings of the IEEE Special Issue on Cognitive Radio, Pittsburgh, PA, USA (2008)

Mitola III., J.: Cognitive radio for flexible multimedia communications. In: 1999 IEEE International Workshop on Mobile Multimedia Communications (MoMuC 1999), McLean, VA, USA, pp. 3–10 (1999)

Kumar, P., Ganesh, R., Sankar, S., Kumar, U.P., Shaiju, G.: Smart grid communication - deployment scenarios in distribution network. In: 2011 IEEE PES on Innovative Smart Grid Technologies - India (ISGT India), Trivandrum, India (2011). http://ieeexplore.ieee.org/xpl/articleDetails.jsp?arnumber=6145406

Krattenmaker, T.G.: Telecommunications Law and Policy. Carolina Academic (1995). Praw.duke.edu/fac/benjamin/.../opening_pages.pdf

Yang, L., Kim, H., Zhang, J.: Pricing-based decentralized spectrum access control in cognitive radio networks. IEEE Netw. **25**(5) 41–66 (2012)

Chun-Hao, L., Nirwan, A.: The progressive smart grid system from both power and communications aspects. IEEE Commun. Surv. Tutorials **14**(3), 1–23 (2012)

Guerra de la Espriella, M., Oviedo, J.: De las telecomunicaciones a las TIC: Ley de TIC de Colombia (L1341/09), CEPAL estudios y perspectivas, Bogotá, Colombia (2011)

MINTIC: Ley 1341 (2009). http://www.alcaldiabogota.gov.co/sisjur/normas/Norma1.jsp?i=36913

Paz, H., Bohórquez, M., Rodríguez, D.: Radio cognitive technology over ultra-high frequency band (UHF). Tecnura. **18**(39), 138–151 (2014)

Oh, T., Young, B., Guthrie, M., Kristi, H., Copeland, D., Kim, T.: Reeling in cognitive radio: the issues of regulations and policies affecting spectrum management. In: Ślęzak, D., Kim, T., Chang, A.C., Vasilakos, T., Li, M., Sakurai, K. (eds.) Communications and Networking, pp. 360–369. Springer, Heidelberg (2010)

Objetivos, Funciones y Deberes, Agencia Nacional del Espectro (2011). http://www.ane.gov.co/index.php/conozca-la-ane/objetivos-y-funciones-news.html

Mody A.: Working group on wireless regional area networks, IEEE 802.22 (2014). http://www.ieee802.org/22/

Yan, Y., Qian, Y., Sharif, H., Tipper, D.: A survey on smart grid communication infrastructures: motivations, Requirements Challenges. IEEE Commun. Surv. Tutorials **15**(1), 5–13 (2013)

Using Demographics in Predicting Election Results with Twitter

Eric Sanders[(✉)], Michelle de Gier, and Antal van den Bosch

CLS/CLST, Radboud University, Nijmegen, The Netherlands
e.sanders@let.ru.nl

Abstract. The results of two Dutch elections are predicted by counting political party mentions from tweets. In an attempt to improve the predictions, gender and age information from the Twitter users is automatically derived and used to adapt the party counts to the demographics in the election turnout. The prediction improves only slightly in one of the elections where the correlation between election outcome and Twitter-based prediction was relatively lower to begin with (0.86 versus 0.97). The relatively inaccurate estimation of Twitter user age may hinder a larger improvement.

Keywords: Twitter · Political election prediction · Demographics

1 Introduction

1.1 Twitter as Predictor of Political Election Outcome

The social media web platform Twitter[1] has been used as a basis for predicting many different types of events and outcomes of processes, among which the outcome of political elections. Traditionally the result of elections are predicted by questioning a representative part of society, which is costly and time consuming. If these polls could be replaced by predictions based on tweets, that are available anyway, costs could be largely reduced.

Tumasjan et al. were one of the first to report on election predictions based on tweets [17]. Although some flaws in their study were exposed [8], other researchers have used their procedure of predicting by simply counting tweets that mention political parties, with varying success. There are reports on this procedure for elections in a number of countries [7]. For the Netherlands, this method has been conducted by Tjong Kim Sang and Bos [14] and by Sanders and Van den Bosch [13]. In this latter study we compared the mentions of political parties in tweets of ten days before the Dutch parliamentary elections of 2012 with the polls and the election results. We found a very high correlation with both the election results (0.95) and the polls (0.96), although the polls still outperform the prediction based on tweet counts: the correlation between election results and polls was 0.98.

---

[1] http://twitter.com.

© Springer International Publishing AG 2016
E. Spiro and Y.-Y. Ahn (Eds.): SocInfo 2016, Part II, LNCS 10047, pp. 259–268, 2016.
DOI: 10.1007/978-3-319-47874-6_18

Gayo-Avello discusses a number of aspects that are usually not taken into account in the election predictions based on tweets [3], among which demographics. Barberá and Rivero show the inbalance in distribution of gender in tweets about the 2011 Spanish legislative elections and the 2012 US presidential elections [1]. Tjong Kim Sang and Bos attempted to incorporate demographics in their study [14], but they did not have demographic data of Twitter users. Wang et al. forecasted the 2012 USA presidential elections based on highly non-representative polls conducted on the Xbox using multilevel regression and post-stratification [19]. In this paper we report on a case study on two recent political elections in the Netherlands in which we take demographic distributions of Twitter users into account when predicting election results from the Twitter stream. The demographics are estimated using machine learning methods [11,12] and are therefore error-prone.

For using demographics[2] in the prediction of the election results based on tweets to be meaningful, there should be a difference in voting behavior of different demographic groups. Furthermore, there should be a difference in demographics between voters and Twitter users for the demographic data to have a correcting effect on the basic count-based estimates. These two conditions are discussed in Sect. 2. In Sect. 3 we introduce our data, and in Sect. 4 we show how we obtain demographics in the voter and tweeter populations. In Sect. 5 we explain how we adapt our predictions to the demographic distributions and show the results. We finish with conclusions in Sect. 6 and a discussion in Sect. 7.

1.2   Elections Used in This Study

Two Dutch elections have been investigated in this study: (1) The national parliamentary elections of 12 September 2012, which are the most important elections in the Netherlands; (2) The provincial elections of 18 March 2015, where the parliaments of the twelve Dutch provinces are elected. Because the Dutch national senate is elected in these elections indirectly (as the provincial electees elect the senate members), they were also important at the national level.

2   Demographic Bias

2.1   Voting Behaviour for Different Gender and Age Groups

Different voting behaviour of different age and gender groups is assumed as a boundary condition for this study. There is ample evidence in the literature though that different voting behaviour across demographic groups is real. For instance, Inglehart and Norris show in their study in 19 countries that men and women have different political orientations [6]. Also, men show more interest

---

[2] In this study we only studied gender and age because these are the two most basic demographic data and because these are the only two that are automatically retrievable to a certain extent.

in politics than women [18]. Webster et al. show that people tend to vote for people close to their age [20] while Goerres argues that people tend to vote more when they are older [5]. Furthermore people in Great Britain tend to vote more conservative when they age [15].

2.2 Demographic Distribution on Twitter

Users on social media in general and Twitter specifically do not mirror the people in society in all demographic aspects [9]. More specifically, people are younger and there are slightly more men on Twitter than women [10]. Duggan and Brenner did surveys for different social media and found out that Twitter is especially appealing for adults from 18 to 29 [2].

3 Political Tweets, Polls and Election Results

In this study we compare counts of political parties mentioned in tweets with election results and polls. In this section we explain how we acquired the data.

The tweets that were retrieved for this study were taken from TwiNL, a large collection of about 40 % of all Dutch tweets [16]. TwiNL filters Dutch tweets on typical Dutch words that rarely appear in other languages, and follows a dynamically updated group of Dutch Twitter users that post often. Political tweets were selected from TwiNL by looking for party names in the tweets using regular expressions of the 11 political parties in parliament (see [13] for details). A tweet that matches one of the political party regular expressions is called a "political tweet" throughout this paper. The person who posted the tweet is called a "political tweeter". For both elections political tweets were gathered from ten days preceding and including election day. In total we collected 159,826 political tweets for 2012 and 183,602 for 2015.

The polling data was retrieved from allepeilingen.com, which is a website that keeps track of all polls from all well known polling institutes in the Netherlands since 2000. The election results were retrieved from kiesraad.nl[3], the official website that presents the election results in detail.

4 Demographics in Elections and in Tweets

4.1 Demographics

If the demographics of the political tweeters would be the same as that of the electorate that turns up for voting, there would be no need for adaptation. Hence, first we need to compare whether (as the literature suggests) and how these demographics differ. In the next subsections we explain how we retrieved and validated the demographic data for the elections of 2012 and 2015 and for the political tweeters.

---

[3] http://www.verkiezingsuitslagen.nl/Na1918/Verkiezingsuitslagen.aspx.

4.2   Voter Demographics

Information about the age and gender distribution of the voters in the 2012 and 2015 elections were obtained from TNS-Nipo[4]. TNS-Nipo is a market research company well-known for election polling. They used a large user panel that was asked for their age, gender and what they voted (among many other things). TNS-Nipo divided their data in three age groups: 18–35, 35–55 and 55+. These age groups are used throughout this paper. Table 1 lists the distributions over age groups and gender for voters in 2012 and 2015.

Table 1. Demographics of elections of 2012 and 2015 according to TNS-Nipo in percentages

	2012			2015		
Age group	Men	Women	Total	Men	Women	Total
18–34	6.1	5.6	11.7	6.2	6.5	12.7
35–54	15.7	17.5	33.2	13.6	15.3	28.9
55+	24.8	30.3	55.1	26.8	31.6	58.4
Total	46.6	53.4	100	46.6	53.4	100

The table shows that the gender balance is about even –with slightly more women than men voting– and over half of the voters is older than 55 years. The demographic distribution of the voters is fairly constant among the two elections. For most of the gender–age groups the differences between the two elections is less than two percent.

4.3   Twitter User Demographics

To automatically estimate the gender and age of the political tweeters, an offline version of TweetGenie [11,12] was used. TweetGenie is a machine-learning system that uses the language in the aggregated set of Dutch tweets posted by a Twitter user to identify the age and gender of the user, based on a training set of users with known ages and gender. TweetGenie was used "as is"; it was not adapted or retrained for the purpose of these experiments.

Table 2 lists the age and gender distributions of users who posted political tweets in the two elections as produced by TweetGenie. Because the age is derived from the most recent tweets of the user in 2016, while the elections were in 2012 and 2015, respectively four and one year are deducted from the age estimated by TweetGenie to arrive at the estimated ages during the elections.

The table shows that three in four of the political tweeters is male, which is in sharp contrast with the election turnout in which male and female voters are almost at a par. Furthermore the political tweeters are on average much

[4] www.tns-nipo.com.

Table 2. Demographics of political tweeters in 2012 and 2015 according to TweetGenie's estimates based on 2016 user accounts, in percentages

	2012			2015		
Agegroup	Men	Women	Total	Men	Women	Total
18–34	51.3	17.6	68.9	28.7	8.1	36.8
35–54	21.3	8.5	29.8	45.5	15.8	61.3
55+	1.0	0.4	1.4	1.4	0.5	1.9
Total	73.6	26.5	100	74.2	25.8	100

younger than the real voters. Notable is the low number of 55+ political tweeters according to TweetGenie. It is a known systematic weakness of TweetGenie that it generates too low age estimates for older people [12], so the real number of tweeters in this age category is probably higher. More generally, the statistics in Tables 1 and 2 highlight that there is a difference between voters and political tweeters and that correcting Twitter-based counts on the basis of differences in the two demographics is in principle a good idea.

4.4 Quality Control by Human Annotators

We use the data provided by TweetGenie as the demographics for the political tweeters, but we do not know how accurate the estimation of gender and age by TweetGenie is. In the absence of a golden standard (a set of Twitter users of which we have certainty on their gender and age) we compared the TweetGenie estimates to demographics determined by human annotators. Two annotators guessed the gender and age of 3,000 political tweeters. This was done in a webtool, created for this purpose, that showed the profile from 2012 or 2015 (depending from what year the political tweets was) and a link to the current profile. Sometimes there is age information in the profile description, but most demographic clues were derived from the profile picture. Note that the profile picture in Twitter is often not up to date so that the guessed age based on it might be several years too young. The demographics of a number of tweeters could not be established by either the annotator or TweetGenie, and about 10 % of the tweeters was annotated as non-human (but as an institution, newspaper, etc.). Leaving these out, we ended up with a set of 1,815 Twitter accounts for which the demographics were compared. Table 3 shows the age and gender distributions for the 2012 and 2015 as estimated by the human annotators and TweetGenie. Figure 1 shows a scatter plot with the age as indicated by humans against that of TweetGenie.

For gender, the distributions as annotated by the human annotators and TweetGenie do not differ very much for both elections, which indicates that the gender distribution as produced by TweetGenie is reliable. For age, on the other hand, the differences are larger. TweetGenie assigns some of the youngest

Table 3. Gender and age distributions as annotated by humans and TweetGenie for 2012 and 2015 in percentages

	Human	TweetGenie
Gender		
Men	70.6	74.4
Women	29.4	25.6
Age		
18–	18.7	10.3
18–34	33.4	36.7
35–54	35.8	51.6
55+	14.1	1.3

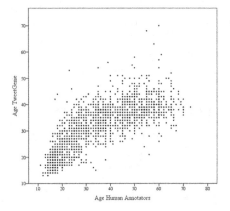

Fig. 1. Scatterplot of age annotation by human annotators and TweetGenie

group to an older group, and assigns some of the oldest group to a younger group. This same pattern is visible in the scatter plot in Fig. 1.

To verify that human annotators can guess the age and gender of a tweeter better than TweetGenie, we computed intra-annotator agreement between the two human annotators and TweetGenie. A set of 100 accounts were annotated by both annotators and TweetGenie. After removal of unannotated accounts we ended up with a comparison set of 72 accounts for which gender was annotated and 61 for which age was annotated by the two annotators and TweetGenie. Table 4 lists the Cohen's kappa values for the agreement between the annotators and TweetGenie for age and gender.

Cohen's kappa between the two human annotators is near-perfect for gender and substantial for age, which makes the annotation of the humans reliable for comparison. The kappa between TweetGenie and the humans is moderate for gender, and only slight for age. From this we conclude that the age annotations of the human annotators is more reliable than those of TweetGenie.

Table 4. Cohen's Kappa between two human annotators and TweetGenie for gender and age

	Gender		Age	
	Annotator2	TweetGenie	Annotator2	TweetGenie
Annotator1	0.97	0.52	0.64	0.09
Annotator2	-	0.56	-	0.14

5 Prediction of Election Results

5.1 Without Adaptation

Following Sanders and Van den Bosch [13], the outcome of the elections are predicted by counting how often a political party is mentioned in a tweet in the ten days before and including election day. These counts are compared with the polls of one day before the elections and the outcome of the elections.

5.2 Adaptation of Tweet Counts Based on Demographics

Using the distributions in Tables 1 and 2 we can use post-stratification [4] to compute weights with which we adapt our tweet counts for the prediction of election results, so they reflect the correct demographic distribution. Tables 5 and 6 present the results with and without adaptation for the 2012 and 2015 elections, respectively. The quality of the prediction is represented in Pearson's

Table 5. Prediction with Twitter, polls and outcome of elections of 2012 with and without demographic adaptation in percentages. At the bottom the correlation (with 95 % confidence interval) and mean absolute error with the election results.

Political party	Election result	Polls	Twitter	Twitter adapted
VVD	26.8	23.4	22.7	23.1
PVDA	25.1	22.7	20.1	21.9
SP	9.8	14.3	11.2	11.8
PVV	10.2	11.5	10.2	7.7
CDA	8.6	8.1	8.1	8.7
D66	8.1	7.9	9.0	7.5
GL	2.4	2.7	8.1	8.1
CU	3.2	4.1	3.6	4.6
SGP	2.1	1.8	2.6	1.9
PVDD	2.0	1.6	3.0	3.2
50PLUS	1.9	2.1	1.4	1.6
Correlation elections		0.98 (0.94–1.0)	0.97 (0.79–1.0)	0.97 (0.70–1.0)
MAE elections		1.3	1.8	1.9

Table 6. Prediction with Twitter, polls and outcome of elections of 2015 with and without demographic adaptation in percentages. At the bottom the correlation (with 95 % confidence interval) and mean absolute error with the election results.

Political party	Election result	Polls	Twitter	Twitter adapted
VVD	16.7	16.7	26.1	25.3
PVDA	10.6	10.1	14.1	13.9
SP	12.2	13.4	7.0	8.8
PVV	12.3	15.5	11.2	8.8
CDA	15.4	12.4	9.9	11.6
D66	13.0	15.5	15.6	13.3
GL	5.6	4.1	5.7	6.1
CU	4.2	4.0	3.4	4.3
SGP	2.9	2.6	2.2	2.1
PVDD	3.6	2.9	3.3	3.6
50PUS	3.5	2.9	1.2	2.3
Correlation elections		0.96 (0.89–0.99)	0.84 (0.68–0.99)	0.86 (0.75–0.98)
MAE elections		1.3	2.9	2.3

correlation over all political parties and the Mean Absolute Error (MAE); the mean absolute difference in percentages.

Table 5 shows that the prediction of the election outcome is very good for 2012. The correlation of 0.97 is only slightly worse than that between the polls and the election outcome, 0.98. The MAE shows a clear difference, though. Importantly, correcting the demographics does not improve the prediction; the MAE ends up marginally worse. For 2015, the prediction based on tweets is clearly worse than the polls. Yet, here the adaptation for gender and age does improve the predictive power of the tweets. The correlation is slightly improved from 0.84 to 0.86, and the MAE is decreased by 21 % from 2.9 to 2.3.

6   Conclusions

In this study we attempted to improve the prediction of election results on the basis of counts of mentions of political parties in tweets by using demographic (age and gender) information. We did this for two Dutch national elections. For the parliamentary elections of 2012 the prediction without demographic adaptation was precise to begin with. Adding demographic information did not result in an improvement. For the provincial/senate elections of 2015 the prediction was not as good as for the 2012 election. Here, adding demographic information did improve the prediction. The Pearson's correlation with the election results improved, and the MAE was reduced.

7   Discussion

In a case study in which we compare two national Dutch elections we find a small improvement of adding demographic information in the predictive power of Twitter for election results when the prediction without the demographic information is not very high to begin with. Obviously, correcting for demographics is only one of the possible and perhaps necessary adaptations that could be done to improve the predictive power of tweets, as suggested by Gayo-Avello. In this respect, the results we found are encouraging. Further research is needed to find out whether adapting for demographics really helps in predicting election results based on tweets.

The first avenue for improvement appears to lie in achieving a better estimate of the age distribution of the tweeters. TweetGenie appeared to guess the gender of political tweeters well, but for age the agreement with human annotators was somewhat low, especially for the oldest age category. TweetGenie may be retrained on more data, or its predictions could be calibrated itself to correct its biases.

In this study we only took age and gender of Twitter users into account, while neglecting education, social status, ethnicity etc. Including these demographic factors might increase the predictions, but they will likely be harder to automatically retrieve with sufficient accuracy.

For correcting the Twitter count we used the actual voter turnout demographic distribution of the election we tried to predict the results from. These numbers are not known in advance. However, the distributions for the 2012 and 2015 were fairly similar, so we presume that using the distribution of a recent election will be a good estimate for a new election.

Acknowledgments. The authors would like to thank Dong Nguyen who provided the TweetGenie data, TNS-Nipo who provided the demographic data of the election turnout, Ruut Brandsma of allepeilingen.com for the polling information and Eline Pilaet, who did part of the demographic annotations of the political tweeters.

References

1. Barberá, P., Rivero, G.: Understanding the political representativeness of twitter users. Soc. Sci. Comput. Rev. 0894439314558836 (2014)
2. Duggan, M., Brenner, J.: The demographics of social media users, vol. 14. Pew Research Center's Internet & American Life Project, Washington, DC (2013)
3. Gayo-Avello, D.: A meta-analysis of state-of-the-art electoral prediction from Twitter data. Soc. Sci. Comput. Rev. **31**(6), 649–679 (2013)
4. Gelman, A.: Struggles with survey weighting and regression modeling. Stat. Sci. **22**, 153–164 (2007)
5. Goerres, A.: Why are older people more likely to vote? The impact of ageing on electoral turnout in Europe. Br. J. Politics Int. Relat. **9**(1), 90–121 (2007)
6. Inglehart, R., Norris, P.: The developmental theory of the gender gap: womens and mens voting behavior in global perspective. Int. Political Sci. Rev. **21**(4), 441–463 (2000)

7. Jungherr, A.: Predictor of electoral success and public opinion at large. In: Jungherr, A. (ed.) Analyzing Political Communication with Digital Trace Data: The Role of Twitter Messages in Social Science Research. Contributions to Political Science, pp. 189–210. Springer, Heidelberg (2015)
8. Jungherr, A., Jürgens, P., Schoen, H.: Why the pirate party won the German election of 2009 or the trouble with predictions: a response to tumasjan, a., sprenger, to, sander, pg, & welpe, im predicting elections with twitter: what 140 characters reveal about political sentiment. Soc. Sci. Comput. Rev. **30**(2), 229–234 (2012)
9. Mellon, J., Prosser, C.: Twitter and Facebook are not representative of the general population: political attitudes and demographics of social media users. Available at SSRN (2016)
10. Mislove, A., Lehmann, S., Ahn, Y.Y., Onnela, J.P., Rosenquist, J.N.: Understanding the demographics of Twitter users. In: ICWSM 2011, p. 5 (2011)
11. Nguyen, D., Trieschnigg, D., Meder, T.: Tweetgenie: development, evaluation, and lessons learned. In: Proceedings of the 25th International Conference on Computational Linguistics (COLING 2014), pp. 62–66. Association for Computational Linguistics, August 2014. http://doc.utwente.nl/94056/
12. Nguyen, D.P., Gravel, R., Trieschnigg, R., Meder, T.: How old do you think i am? A study of language and age in Twitter. In: Proceedings of the Seventh AAAI Conference on Weblogs and Social Media. AAAI Press (2013)
13. Sanders, E., Van den Bosch, A.: Relating political party mentions on Twitter with polls and election results. In: Proceedings of DIR-2013, pp. 68–71 (2013). http://ceur-ws.org/Vol-986/paper_9.pdf
14. Tjong Kim Sang, E., Bos, J.: Predicting the 2011 Dutch senate election results with Twitter. In: Proceedings of the Workshop on Semantic Analysis in Social Media, pp. 53–60. Association for Computational Linguistics (2012)
15. Tilley, J., Evans, G.: Ageing and generational effects on vote choice: combining cross-sectional and panel data to estimate APC effects. Electoral Stud. **33**, 19–27 (2014)
16. Tjong Kim Sang, E., Van den Bosch, A.: Dealing with big data: the case of Twitter. Comput. Linguist. Neth. J. **3**, 121–134 (2013)
17. Tumasjan, A., Sprenger, T.O., Sandner, P.G., Welpe, I.M.: Predicting elections with Twitter: what 140 characters reveal about political sentiment. In: ICWSM 2010, pp. 178–185 (2010)
18. Verge Mestre, T., Tormos Marín, R.: The persistence of gender differences in political interest. Revista Española de Investigaciones Sociológicas 138 (2012)
19. Wang, W., Rothschild, D., Goel, S., Gelman, A.: Forecasting elections with non-representative polls. Int. J. Forecast. **31**(3), 980–991 (2015)
20. Webster, S.W., Pierce, A.W.: Older, younger, or more similar? The use of age as a voting heuristic. Tech. rep., working paper (2015)

On the Influence of Social Bots in Online Protests

Preliminary Findings of a Mexican Case Study

Pablo Suárez-Serrato[1]([✉]), Margaret E. Roberts[2], Clayton Davis[3],
and Filippo Menczer[3,4]

[1] Instituto de Matemáticas, Universidad Nacional Autónoma de México,
Mexico City, Mexico
pablo@im.unam.mx
[2] Department of Political Science, University of California, San Diego,
La Jolla, USA
[3] School of Informatics and Computing, Center for Complex Networks
and Systems Research, Indiana University, Bloomington, USA
[4] Indiana University Network Science Institute, Bloomington, USA

Abstract. Social bots can affect online communication among humans. We study this phenomenon by focusing on #YaMeCanse, the most active protest hashtag in the history of Twitter in Mexico. Accounts using the hashtag are classified using the *BotOrNot* bot detection tool. Our preliminary analysis suggests that bots played a critical role in disrupting online communication about the protest movement.

Keywords: Social bots · Twitter · Protests

1 Introduction

On November 7th, 2014, the Mexican Federal District Attorney, José Murillo Karam, cued one of his aides towards the end of a press conference by saying *"Ya me cansé"* (I am tired). The press conference was about the status of the investigation into the disappearance of 43 teachers in training from the rural normal school in Ayotzinapa, Guerrero on September 26th, 2014. This gesture of fatigue catalyzed the largest use of a protest hashtag on Twitter in Mexico to date.

1.1 #YaMeCanse

In 2014 Twitter had just over 7 million users in Mexico. According to Crimson Hexagon, the #YaMeCanse hashtag was used in over 2 million tweets during the month following November 7th, 2014, and a total of about 4.4 million times to this date. Its use peaked on 21st November 2014 with 500 thousand posts. Figure 1 shows the recorded volume and period of activity for #YaMeCanse

© Springer International Publishing AG 2016
E. Spiro and Y.-Y. Ahn (Eds.): SocInfo 2016, Part II, LNCS 10047, pp. 269–278, 2016.
DOI: 10.1007/978-3-319-47874-6_19

and some related hashtags. It was claimed during this period that Twitter was being gamed, flooded by accounts that mainly tweeted this hashtag repeatedly, in an attempt to make it more difficult for human users to communicate and find each other using it. As an innovative response, the human users of the hashtag switched to using the #YaMeCanse2 hashtag. As the alleged spamming accounts moved too, the human users subsequently iterated this evasion strategy using #YaMeCanse3, #YaMeCanse4, and so on. This continued with strong use through #YaMeCanse25.

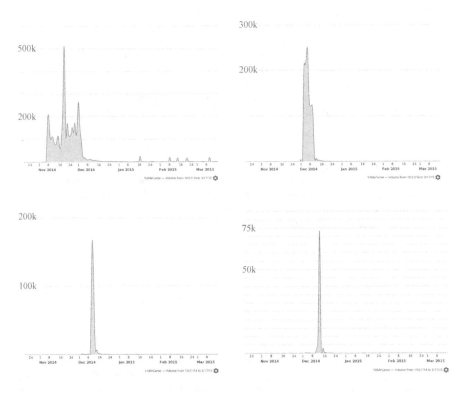

Fig. 1. Recorded volume and period of activity for four hashtags: #YaMeCanse (upper left) had strongest use between November 7th and December 14th, 2014; #YaMeCanse2 (upper right) had strongest use between December 1st and December 16th; #YaMeCanse3 (lower left) had strongest use between December 8th and December 16th; #YaMeCanse4 (lower right) had strongest use between December 9th and December 16th. These diagrams were obtained through Crimson Hexagon.

1.2   Social Automation

As companies and institutions strive to automate service processes, there has been a rise in the use of automated social media accounts (for example, chat bots). While some of these accounts can be benign and helpful, some have been designed and deployed with intentions that are not benevolent. Reports of political protests in Russia being swamped by spam on Twitter [1] and of mass pro-government Twitter campaigns in Turkey [8] have already appeared.

A bot account can display various levels of automation and content sophistication. They range from spammers that just repeat a hashtag and include irrelevant characters, like punctuation, to bots that repeat a crafted message over and over using multiple—hundreds or even thousands—accounts. *Sybil* accounts attempt to disguise themselves as humans. *Cyborg* accounts mix automation and human intervention. For example they can be programmed to post at certain intervals. Twitter accounts of news organizations often fall under this category. A review of the pervasiveness of social bots and the state of the art in detecting them can be found elsewhere [9].

BotOrNot is a general supervised learning method for detecting social bot accounts on Twitter [6]. This system exploits over 1,000 features that include user meta-data, social contacts, diffusion networks, content, sentiment, and temporal patterns. Based on evaluation on a large set of labeled accounts, *BotOrNot* is reported to have high accuracy in discriminating between human and bot accounts, as measured by the Area Under the ROC Curve (AUC 94 %).

It is now understood that the influence of social bots in political discourse, such as a protest, can impede free speech and fracture activist groups [18]. Instances of these outcomes were reported to have also taken place in Mexico in 2015 [15]. Recently, Freedom House added pro-government commentators and bots to their analysis of government censorship on the web because these methods can alter the nature and accessibility of the conversation [11].

The goal of this research is to establish whether bots interfered with communication between real Twitter users in the context of the #YaMeCanse protest. We hope to empirically test whether increases in bot activity cause users to lose track of the conversation. If this is the case, the bots would be functioning as a form of censorship, distracting users from the conversation, similar to online propaganda in other countries like China [13]. The ultimate goal of our project is to quantify bot influence on suppression of communication.

In the preliminary analysis presented here, we use bot identification techniques to verify the involvement of bot accounts in the #YaMeCanse protest. To this end we present data visualization techniques, using multivariate kernel decomposition estimates and hexagonal bins, to identify regions of potential bots accounts in the phase space obtained by considering pairs of classification probabilities based on different subsets of *BotOrNot* features. These techniques allow us to flag potential bot accounts by focusing on the different outputs produced by *BotOrNot*. This way we can focus on language-independent classifiers, which is important in the present case study because the tweet corpus is in Spanish whereas *BotOrNot* is trained on English content.

1.3   Related Work

Twitter bots have been alleged to influence the political discourse surrounding *Brexit* in the United Kingdom [12]. To evaluate the role of bots, opinions were clustered based on hashtag use. It was found that a very small fraction of most active accounts was responsible for a large fraction of pro-Brexit content: fewer than 2,000 accounts in a collection of 300,000 users (less than 1%) generated up to 32% of Twitter traffic about Brexit.

Network decomposition techniques were used to conclude that the success of social protests depends in part on activating a critical periphery [2]. These peripheral participants may be as essential to the communication of the protest message as the most connected and active members. The influence of bots, hindering communication and blocking potential adhesion of new members to a community, could lead to a halt in the movement's growth.

In relation to Mexico, the use of Twitter as a vital communication channel was investigated in the context of urban warfare related to the ongoing war on drugs [14]. Users would tweet the location of conflicts as they erupted, so that their followers could then avoid these violent zones. The authors emphasized that warfare is also a conflict over the control of information, and provided a longitudinal survey of the adoption of Twitter to create safe networks of information.

The appearance of fake Twitter accounts among the followers of political figures is common, with 20–29% fake followers in the cases of some prominent people. Accounting for social bots is crucial, for example, in electoral polls and in the identification of influential users on topics of interest. Post sentiment has been studied to separate human from non-human users in Twitter [7]. A machine learning process was developed, trained on a retrospective analysis of 867 suspended accounts. Political science methods have been used to study Twitter in relation to political issues, with a focus on electoral periods [10].

The SentiBot tool employed a combination of graph-theoretic, syntactic, and semantic features to discern between humans, cyborgs, and bots [3]. 19 out of the top 25 variables that determine if an account is a bot were found to be related to sentiment. Another classifier exploits natural language processing for social bot detection [4]. Such techniques were recently used to investigate the presence and effect of social bots promoting vaporizers and e-cigarettes [5]. These are examples of tools that would have to be retrained in order to establish the same results in languages other than English. Porting them to Spanish would be necessary for application to a corpus of tweets such as the one analyzed in this paper. One of our present contributions is the observation that in considering the different classification scores produced by *BotOrNot*, we can use the language-independent features to flag potential bot accounts in Spanish. This technique could potentially be used in other languages as well.

2   Bot Analysis

2.1   Data and Methods

We had access to datasets of streamed tweets collected by activists using Twitter's streaming API during November and December 2014. The tweets were collected as each hashtag was rising in popularity and only for limited periods, hence

Table 1. Total unique users in each dataset

Dataset	Unique users
#YaMeCanse	14756
#YaMeCanse2	1605
#YaMeCanse3	8831
#YaMeCanse4	2530
#YaMeCanse5	437

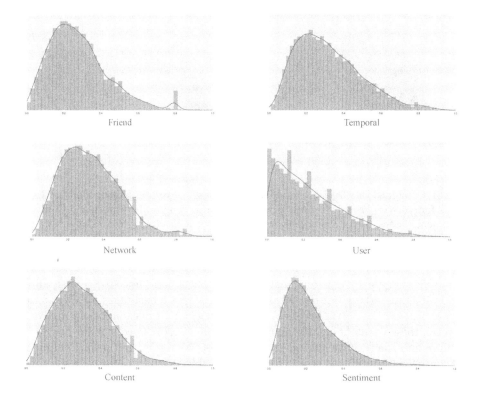

Fig. 2. *BotOrNot* probability densities for its classifiers using Friend, Temporal, Network, User, Content, and Sentiment features. The distributions are based on 14,756 unique accounts using the #YaMeCanse hashtag between November 26 and 30, 2014.

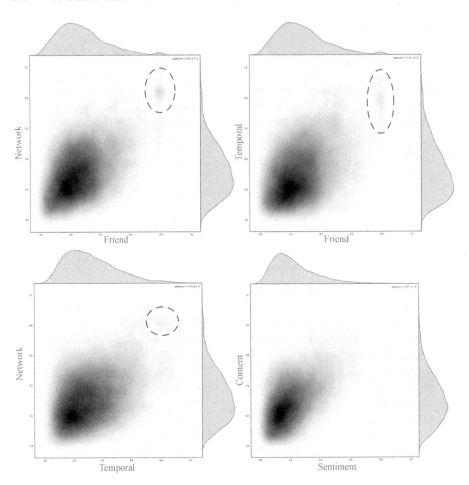

Fig. 3. Combined *BotOrNot* probability densities: bi-variate Kernel Density Estimates for pairwise Friend-Network, Friend-Temporal, Temporal-Network and Content-Sentiment classes, for 14,756 unique accounts using the hashtag #YaMeCanse between November 26 and 30, 2014.

the preliminary nature of our analysis. The tweets were not stored with all of the Twitter metadata, but the datasets about the first 5 hashtags (#YaMeCanse, #YaMeCanse2, …, #YaMeCanse5) have sufficient information for our analysis. In total, these datasets include information from 152,757 tweets. They provide a glimpse into the activity around these hashtags. We report on the 28,159 unique accounts that mentioned these five hashtags (Table 1).

The user accounts were fed to the *BotOrNot* API [6], taking care to comply with the Twitter API limits. The *BotOrNot* API includes scores for classifiers trained on subsets of features related to Friends, Network, Time, Content, and Sentiment, as well as an overall score obtained from a classifier trained on all

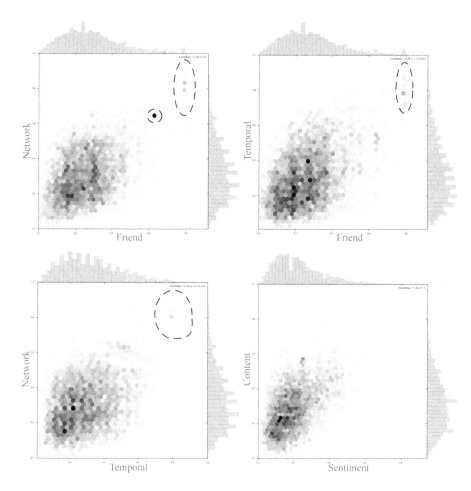

Fig. 4. Combined *BotOrNot* probability densities. Bi-variate Hexagonal bins for pair-wise Friend-Network, Friend-Temporal, Temporal-Network and Content-Sentiment classes, for 8,831 unique accounts using the hashtag #YaMeCanse3 between December 11 and 13, 2014.

features. Each score can be interpreted as the likelihood that the given account is a bot, according to the features in each of these categories [9].

2.2 Preliminary Results

Since most if not all of the tweets are in Spanish, the distributions of Content and Sentiment scores do not register any significant signal pointing to the existence of bots, as illustrated in Fig. 2 for the first hashtag, #YaMeCanse. This is not surprising, as the Content and Sentiment classifiers have been trained for English language tweets. The distribution of Friend scores, however, displays a clear signal: a significant number of accounts have score above 0.8, strongly suggesting the presence of bots. The Friend classifier considers four types of friends (contacts): users retweeting, mentioning, being retweeted, and being mentioned. For each group separately, *BotOrNot* extracts features about number of languages used, local time, account age distribution, popularity distributions, and other user metadata.

The *BotOrNot* distributions for the Friend, Network, and Temporal classifiers display discernible modes for high bot scores. Therefore we plot these pairwise using bivariate kernel density estimates (Fig. 3). The same is done for hexagonal bin plots (Fig. 4). This analysis is carried out with the Seaborn library [17]. The bot regions lie in the upper-right quadrants above 0.65×0.65 scores. Table 2 summarizes the numbers of bots in these regions. In both figures, we observe clusters of potential bots, grouped away from the human users (marked by dashed lines). In the Content-Sentiment case there is no clear bot cluster. In the Friend-Network hexagonal bin plot, we also notice a second, sharp cluster, suggesting that there may be two distinct types of bots.

Table 2. Numbers of likely bot accounts with *BotOrNot* score above 0.65, according to each classifier and in each dataset. For the Temporal classifier we also show the corresponding percentages out of all accounts in each dataset.

Dataset	Temporal	Friend	Network	Content	Sentiment	User
#YaMeCanse	610 (4 %)	340	467	313	162	365
#YaMeCanse2	79 (4 %)	40	58	44	27	42
#YaMeCanse3	394 (4 %)	259	312	200	110	240
#YaMeCanse4	130 (5 %)	56	83	67	31	63
#YaMeCanse5	24 (5 %)	13	19	9	7	12

Table 3. Deleted accounts found in the datasets.

Dataset	Deleted accounts
#YaMeCanse	2084 (14 %)
#YaMeCanse2	235 (14 %)
#YaMeCanse3	1203 (13 %)
#YaMeCanse4	259 (10 %)
#YaMeCanse5	51 (11 %)

3   Conclusions

It is clear that there was a substantial bot presence affecting the online discussion about the #YaMeCanse protest. In fact, the above estimates are to be considered as lower bounds. While computing bot scores, we ran into many accounts that had been deleted by Twitter (Table 3). Many of these were probably bots.

A further analysis that incorporates a more representative sample will be carried out in future work, in order to quantify the influence of these bots on the discourse and on the fragmentation among human users of these hashtags. We would also like to look for causal links between volume of tweets produced by bots and shifts between hashtags. To this end we plan to apply an information theoretic approach similar to one previously used for measuring influence in social media [16].

Acknowledgments. We thank IPAM at UCLA and the organizers of the Cultural Analytics program, where this work was first conceived, for a wonderful working environment and for bringing diverse fields together. PSS is thankful to the Primrose Foundation for coding help. We thank Alberto Escorcia for collecting the tweet data and giving us access to it, and also Twitter for allowing access to data through their APIs.

References

1. BBC News, Technology Section, Russian Twitter political protests 'swamped by spam', March 2012. http://www.bbc.com/news/technology-16108876
2. Barberá, P., Wang, N., Bonneau, R., Jost, J.T., Nagler, J., Tucker, J., González-Bailón, S.: The critical periphery in the growth of social protests. PLoS ONE **10**(11), e0143611 (2015)
3. Chu, Z., Gianvecchio, S., Wang, H., Jajodia, S.: Detecting automation of Twitter accounts: are you a human, bot, or cyborg? IEEE Trans. Dependable Secure Comput. **9**(6), 811–824 (2012)
4. Clark, E.M., Williams, J.R., Jones, C.A., Galbraith, R.A., Danforth, C.M., Dodds, P.S.: Sifting robotic from organic text: a natural language approach for detecting automation on Twitter. J. Comput. Sci. **16**, 1–7 (2016)
5. Clark, E.M., Jones, C.A., Williams, J.R., Kurti, A.N., Norotsky, M.C., Danforth, C.M., Dodds, P.S.: Vaporous marketing: uncovering pervasive electronic cigarette advertisements on Twitter. PLoS ONE **11**, e0157304 (2016)
6. Davis, C., Varol, O., Ferrara, E., Flammini, A., Menczer, F.: BotOrNot: a system to evaluate social bots. In: Developers Day Workshop at WWW Montreal (2016)
7. Dickerson, J.P., Kagan, V.: Using sentiment to detect bots on Twitter: are humans more opinionated than bots? In: Advances in Social Networks Analysis and Mining (2014)
8. Entwickelr.de.: Turkeys Twitter-Bot army and the politics of social media, Montag, WebMagazin, 30 Juni 2014. https://entwickler.de/online/webmagazin/turkeys-twitter-bot-army-and-the-politics-of-social-media-1153.html
9. Ferrara, E., Varol, O., Davis, C., Menczer, F., Flammini, A.: The rise of social bots. Commun. ACM **59**(7), 96–104 (2016)
10. Gayo-Avello, D.: A meta-analysis of state-of-the-art electoral prediction from Twitter data. Soc. Sci. Comput. Rev. **31**(6), 649–679 (2013)
11. Freedom House. Freedom on the Net 2015. Privatizing Censorship, Eroding Privacy (2015)
12. Howard, P.N., Kollanyi, B.: Bots, #StrongerIn, and #Brexit: computational propaganda during the UK-EU Referendum. Available at SSRN 2798311 (2016)

13. King, G., Pan, J., Roberts, M.E.: How the Chinese government fabricates social media posts for strategic distraction, not engaged argument. http://j.mp/1Txxiz1
14. Monroy-Hernandez, A.: The new war correspondents: the rise of civic media curation in urban warfare, pp. 1–10, December 2012
15. Porup, J.M.: How Mexican Twitter Bots Shut Down Dissent, Motherboard, 24 August 2015. http://motherboard.vice.com/read/how-mexican-twitter-bots-shut-down-dissent. Accessed 26 Jul 16
16. Ver Steeg, G., Galstyan, A.: Information transfer in social media. In: WWW 2012 (2012). http://arxiv.org/abs/1110.2724
17. Waskom, M.: Seaborn. https://github.com/mwaskom/seaborn
18. Wooley, S.: Automating power: Social bot interference in global politics. First Monday **21**(4), 4 April 2016. http://firstmonday.org/ojs/index.php/fm/article/view/6161/5300. Accessed 4 Sep 2016

What am I not Seeing? An Interactive Approach to Social Content Discovery in Microblogs

Byungkyu Kang[1]([✉]), Nava Tintarev[2], Tobias Höllerer[1], and John O'Donovan[1]

[1] Department of Computer Science, University of California, Santa Barbara, USA
{bkang,holl,jod}@cs.ucsb.edu
[2] Department of Informatics and Computing, Bournemouth University, Poole, UK
ntintarev@bournemouth.ac.uk

Abstract. In this paper, we focus on the informational and user experience benefits of user-driven topic exploration in microblog communities, such as Twitter, in an inspectable, controllable and personalized manner. To this end, we introduce "HopTopics" – a novel interactive tool for exploring content that is popular just beyond a user's typical information horizon in a microblog, as defined by the network of individuals that they are connected to. We present results of a user study (N=122) to evaluate HopTopics with varying complexity against a typical microblog feed in both personalized and non-personalized conditions. Results show that the HopTopics system, leveraging content from both the direct and extended network of a user, succeeds in giving users a better sense of control and transparency. Moreover, participants had a poor mental model for the degree of novel content discovered when presented with non-personalized data in the Inspectable interface.

Keywords: Communities · Content discovery · Explanations · Interfaces · Microblogs · Visualization

1 Introduction

Twitter is a microblogging service where users post messages (tweet) about any topic within the 140-character limit and follow others to receive their tweets. As of Feb. 2016, Twitter has around 320M active users, and 500M tweets are sent every day. This noisy, user-generated content contains valuable information. The majority (over 85 %) of trending topics are headline news or persistent news [15], and Twitter is frequently used as a news beat for journalists [5].

With large amounts of noisy, user generated content, we have no choice but to rely on automated filters to compute relevant and personalized information that are small enough to avoid cognitive overload. However, once an automated information filtering mechanism of any type is applied, there is a real risk that useful, or critical information will never reach the end user. This problem is not new: there is a sweet-spot between similarity and diversity in personalization. Smyth [22] and Herlocker [11] refer to it as a general black-box problem with

E. Spiro and Y.-Y. Ahn (Eds.): SocInfo 2016, Part II, LNCS 10047, pp. 279–294, 2016.
DOI: 10.1007/978-3-319-47874-6_20

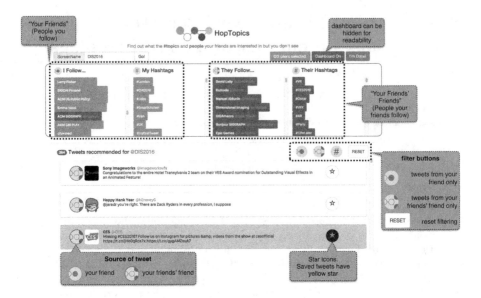

Fig. 1. Screenshot of the system (condition D), with explanatory labels. Top: Network Dashboard controlling the source of the tweets using community structure (1-hop and 2-hop followers) and topics (hashtags). Bottom: Content Viewer with resulting tweets, that can be filtered and starred by users.

recommender systems, and more recently, Pariser [19] describes it as a *filter bubble* problem, wherein personalized filtering algorithms narrow a user's window of information.

In social networks such as Twitter, a user's information feed is populated with content from the other users that they follow directly. Here, filtering can be seen as a two step process. First, the user elects to follow another user, and second, that (second) user acts as an information curator by either authoring or propagating messages. Both steps in this process are subject to failures (*c.f.*, [9,23]). Allowing people to see how their social network influences the information they receive may help alleviate these issues. A high profile example of the filter bubble effect occurred during the 2016 "Brexit" vote in the United Kingdom. Through automatic personalization of content via algorithms, or via selected friends serving as news curators, many people incorrectly assumed that the 'vote remain' campaign would win by a large majority, and expressed shock on Twitter after the result was announced. We believe that explanation of filtering processes, and user-interaction with information filtering algorithms can help to mitigate filter-bubble effects.

To this end, the main contribution of this paper is the introduction and evaluation of a novel interface for Twitter, *HopTopics*, that addresses the filter bubble problem. HopTopics enables users to leverage their network to source novel and potentially relevant topics from both the local and extended social network. The approach can be viewed as a hybrid of strong and weak ties (c.f., [10]) for personalized information seeking. The main scientific contribution is the use of community structure to support content discovery, while improving control and inspectability of navigation. In our study, we use an example from journalism, but we note that the system can be generalized to most information-seeking tasks, such as understanding public opinion about political events, as in the above example.

The remainder of this paper is organized as follows: First, we present a discussion of related work. Next, we introduce the HopTopics system, including the design choices for the interactive user interface (Fig. 1). This is followed by a user experiment (N=122) in which we evaluate the system on real data and users from Twitter in a 4 x 2 mixed design experiment. We conclude by discussing key results and ideas for next steps.

2   Background

To frame this research in the context of related work, we look at three key areas. First, we discuss related work on *inspectability and control in intelligent systems*. Second, we focus on inspectability and control of data in *microblogs* – a topic which is central to our research and has also received much attention in recent years. Finally, we present a discussion of related work in the area of *community based content discovery*.

2.1   Inspectability in Intelligent Systems

Mechanisms for improving inspectability and control have been introduced to different classes of intelligent systems from open learner models [6,8], to autonomous systems [7], decision support [3,14], and recommender systems [13,25]. These studies have found that inspectability and control can have a positive effect on user experience as well as improved mental models.

There has been a shift toward supporting more open searches and users' understanding of novel domains evolving through use [1]. This has also meant an evolution from static explanations to more dynamic forms of explanation such as interactive visualization. For example, [26] has looked at how interaction visualization can be used to improve the effectiveness and probability of item selection when users are able to explore and interrelate multiple entities – *i.e.* items bookmarked by users, recommendations and tags.

2.2   Inspectability in Microblogs

In order to better deal with the vast amounts of user-generated content in microblogs, a number of recommender systems researchers have studied user experiences through systems that provide inspectability of and control over recommendation algorithms. Due to the brevity of microblog messages, many systems provide summary of events or trending topics with detailed explanations [16]. This unique aspect of microblogs makes both inspectability and control of recommender algorithms particularly important, since they help users to more efficiently and effectively deal with fine-grained data. Schaffer *et al.* found that in addition to receiving transparent and accurate item recommendations in a microblog, users gained information about their peers, and about the underlying algorithm through interaction with a network visualization [20]. The Eddi system [4], a Twitter dashboard that supported topic-based browsing on Twitter, and was found to be more efficient and enjoyable way to browse an update feed than the standard chronological interface.

2.3   Community-Based Content Discovery

Serendipity is defined as the act of unexpectedly encountering something fortunate. In the domain of recommender systems, one definition has been the extent to which recommended items are both useful and surprising to a user [12]. This paper investigates how exploration can be supported in a way that improves serendipity, and maintains a sense of inspectability and control. The intuitions guiding the studies in this paper are based on findings in the area of social recommendations, *i.e.*, based on people's relationships in online social networks (*e.g.*, [17]) in addition to more classical recommendation algorithms.

The *first intuition* is that weak rather than strong ties are important for content discovery. This intuition is informed by the findings of the cohesive power of weak ties in social networks, and that some information producers are more influential than others in terms of bridging communities and content [10]. Results in the area of social-based explanations also suggest that mentioning which friend(s) influence a recommendation can be beneficial (*e.g.*, [21,27]). In this case, we support exploring immediate connections or friends, as well as friends-of-friends.

The *second intuition* is that the intersection of groups may be particularly fortuitous for the discovery of new content. This is informed by exploitation of cross-domain exploration as a means for serendipitous recommendations [2]. It is in these two intuitions that the novelty of the work in this paper lies: using community structure in microblogs to aid content discovery, by supporting inspectability and control.

3   HopTopics System

In the following sections, we first describe the UI design behind HopTopics, from the initial design used in a formative study, to the final design, shown in condition D of the main study, and in Fig. 1. After that, we describe the interaction design.

The system architecture was designed to support real-time network-based, topic-specific data exploration, including caching algorithms in order to prevent exceeding the given rate limit[1], and a cluster of back-end servers to increase the number of possible concurrent requests.

---

[1] https://dev.twitter.com/rest/public/rate-limiting, retrieved July 2016.

3.1 Formative User Study

We first conducted a formative user study (N=12) to evaluate the interface and interaction design [24]. We used a layered evaluation approach [18], focusing on the decision of an adaptation and how it was applied (in contrast to which data was collected or how it was analyzed). So, to isolate some aspects of user interface and interaction design, the HopTopics interface was evaluated using obfuscated (*Lorem Ipsum*) data. We briefly summarize the previous results and their impact on the interface and interaction design below.

Participants in two countries interacted with the interface in semi-structured interviews. In two iterations of the same study (n=4, n=8), we found that the interface gave users a sense of control. Users asked for an active selection of communities, and a functionality for saving individual 'favorite' users. Users found the community-based exploration feature to be particularly useful (feature retained), but re-ranking of tweets less so (feature omitted).

Based on this formative study, a number of improvements have been implemented in the system presented in this paper. The current system uses a scrolling mechanism to allow users to see more data, and clearer icons to annotate whether a tweet comes from someone the user follows, or whether it is from someone two hops away. This addresses issues with previous annotations about group membership being unclear. The current interface also considers participants' requests to better integrate with the existing twitter website: it now includes a facility to favorite tweets, and includes multimedia and URLs. As a consequence, the system contains both a *Content Viewer* (for Tweets) and a *Network Dashboard* pane (for Inspectability and Control) (*c.f.*, Sect. 3.2).

3.2 User Interface Design

Figure 1 shows a screen shot of the training screen for the system and indicates the various components. The dark grey speech boxes illustrate the basic components of the system. The system has two core components: (1) A *Network Dashboard* shows the active user's one and two hop network along with the topics/hashtags that are prominent in them; (2) A *Content Viewer* panel shows a collection of the messages that are derived from the current set of selections made in the dashboard view.

The *Content Viewer* shows an iconized combination of tweets from the different network groups: one-hop, two-hop, and global tweets (outside the user's network). Messages from each group are shown with a source provenance icon, shown on the left side of Fig. 1. Within this viewer, participants can filter messages based on group (one-hop, two-hop, global).

Due to limited screen space for most web users, an important feature of the system is the ability to retract the Network Dashboard – which is the inspectability and control mechanism – and focus only on the Content Viewer – which contains the tweets: the information they are typically interested in. In addition to the icons, a color coding scheme is applied to the Network Dashboard to indicate links between hashtags and the groups (one-hop, two-hop, global) they originate from, as shown in dark blue and cyan in the Network Dashboard of Fig. 1.

Since the number of nodes increase exponentially as one traverses the Twitter network, query complexity and data relevance were primary design considerations[2]. Node selections were limited to three selections for each column. The selection limit is shown dynamically on the top right of the view window.

3.3 Interaction Design

A user begins by typing a query into the system. This is typically their own Twitter ID, but could also be another user whose network they are interested in exploring. The API is queried for network and content information, which are then displayed in the Dashboard and Content panes respectively. A user can mouse-over any of the user profiles listed in the columns to find out more (*e.g.*, view the profile description and picture). Users interact with the Dashboard by first selecting up to 3 people in the left column, which consists of all the people they follow (one-hop group). As a user clicks on one of the people in the first column, the third column is populated with people who they follow (two-hop group, or friends-of-friends). When the user selects someone from the two-hop group, further hashtags get shown in the "their hashtags" column furthest to the right. This also adds more tweets in the Content Viewer.

As in typical Twitter feed interfaces, users can "star" or favorite tweets in the Content Viewer. A user can also select the 'reset' button at any point in a data exploration session to return the system to its default view of the network.

4 Experiment

In this section, we describe an experiment to evaluate versions of the interface, and the effect of personalization, using real world data. The experimental toolkit was deployed as a web service and the link was made available on Amazon Mechanical Turk (AMT).

4.1 Experiment Design

The experiment used a mixed design, as shown in Table 1. The interface variant was assigned between participants and was one of: A) Baseline - standard Twitter feed only, B) Data - augmented feed including topics mentioned by friends of friends, C) Inspectable - dashboard visible but not interactive, D) Controllable - interaction with dashboard, this is the full system introduced in the previous section. These conditions were compared between rather than within participants, in order to avoid learning and ordering effects for a specific Twitter account.

---

[2] In an ideal scenario, the HopTopics system would be connected via a firehose, https://dev.twitter.com/streaming/firehose (retrieved July 2016), connection where complex queries would not pose quite as much of a constraint. However, given limited bandwidth for our real-time experimental setup, node selections had to be restricted.

Table 1. Overview of conditions. Degree of personalization is within subjects, system type is between subjects

	Baseline	Augmented Data	Inspectable	Controllable
Non-personalized	A1	B1	C1	D1
Personalized	A2	B2	C2	D2

In addition we compared, within participants, the effect of personalization of the data, comparing (1) a non-personalized id (always the same id: @ACMIUI) with (2) a personalized (their own Twitter ID). The condition using the data for the non-personalized ID was always shown first. While this data was retrieved live, it was not currently in progress. This is also a relatively small community on Twitter, so the dataset is relatively static, and contains many topics that may be unfamiliar to the average Twitter user.

The motivation for using such a dataset is (1) to create familiarity with the interface through training, and (2) to have a condition where we expect participants to have a consistent level of familiarity (low) with the content, as it is not personalized. This design means, for example, that participants assigned to the Augmented Data condition would always see first B1 (non-personalized) and then B2 (personalized).

In addition to the responses we collected and computed the following indirect measures: #people saved, #tweets starred, #hashtags saved, correlation between perceived novelty and #hashtags identified as novel.

4.2   Hypotheses

We hypothesized that the system will help users discover more unexpected and useful content, and lead to better subjective perceptions when the system is a) Inspectable and Controllable (compared to two benchmarks: Baseline and Data), and when the content is personalized (compared to non-personalized content). The itemize hypotheses are described in Table 2.

Participants. Participants were recruited from the US only and were required to have a Mechanical Turk acceptance rate of greater than 90 % (at least 90 % of their HITs are considered of good quality by other requesters). They were required to correctly answer some filler questions, and to have a minimal degree of interaction with the system (2 min and 1 interaction).

155 participants completed the full study, however 33 participants were excluded from analysis as they had technical issues (most likely they interacted with the system beyond Twitter's rate limits[3]. Out of the remaining 122, the distribution across the 4 versions of the system was (A=32, B=32, C=27, D=31). The lower number of participants in conditions C and D is due to a lower completion rate in these conditions. User comments suggest that this is due to the system being slow.

---

[3] https://dev.twitter.com/rest/public/rate-limiting, retrieved July 2016.

Table 2. Overview of hypotheses (H1-H7)

H1: Perceived serendipity. "Compared to your regular twitter feed, how much does this interface help you find relevant and surprising items that you did not know about yet?" H1a: Perceived serendipity will be higher in the Inspectable and Controllable conditions. H1b: Perceived serendipity will be (slightly) higher in the personalized compared to the non-personalized conditions, across types of system.

H2: Perceived familiarity. "Compared to your regular twitter feed, how helpful is this interface for finding information that is both relevant and familiar?" H2a: Perceived familiarity will be higher in the Baseline and Data conditions. (compared to Inspectable and Controllable). H2b: Perceived familiarity will be higher in the personalized compared to the non-personalized conditions, across types of system.

H3: Perceived transparency. "The interface helped me understand where these tweets came from." Perceived transparency will be higher in the Inspectable and Controllable conditions. We do not anticipate a difference in perceived transparency between the personalized and non-personalized conditions.

H4: Perceived control. "The interface helped me change the tweets that are recommended to me." Perceived control will be higher in the Inspectable and Controllable conditions We do not anticipate a difference in perceived control between the personalized and non-personalized conditions.

H5: Content discovery. H5a: Degree of content discovery (sum of people + hashtags + tweets saved) will be greater in the Controllable condition (compared to the Interactive condition). H5b: Degree of content discovery will be greater in the personalized than the non-personalized conditions.

H6: Correctness of mental model. H6a: The correlation between perceived serendipity (subjective) and content discovery (objective) will be higher for the Controllable condition (compared to the Inspectable condition). H6b: The correlation between perceived serendipity (subjective) and content discovery (objective) will be higher for the personalized compared to the non-personalized condition.

H7: Perceived diversity. Perceived diversity will be higher in the Inspectable and Controllable conditions (compared to the Baseline and Data). We do not anticipate a difference in perceived diversity between the personalized and non-personalized conditions.

The majority of participants (50 %) were 25–35, 24 % were 18–25, 22 % were 35–50, and only 4 % were 50–65. Participants were balanced across genders (49 % male vs 51 % female). 91 % of the participants reported that they used Twitter ("Sometimes" (27 %), "Often" (39 %), or "All of the Time" (25 %)).

4.3 Materials

Tweets were retrieved live at the time that the experiment was run (Oct 2015), and used to populate the interface described in Sect. 3. Tweets were either retrieved for the @ACMIUI account (in the non-personalized condition), or the user's own Twitter id (personalized condition). The default ordering of tweets and people given by the API was used (chronological order).

4.4   Procedure

The procedure contained the following steps, described in detail below: Pre-survey $\Rightarrow$ Instructions $\Rightarrow$ HopTopics Non-personalized $\Rightarrow$ Post-survey 1 $\Rightarrow$ Instructions $\Rightarrow$ HopTopics personalized $\Rightarrow$ Post-survey 2. Participants started the experiment with a pre-survey[4], including demographics. They were then taken to an instruction screen (see Fig. 1), where they were given an open-ended task (Fig. 2):

Imagine that you have just taken on a new role as a freelance journalist. You need to write a few pieces for a client. You can write them on any topics you find interesting and surprising. However, you need to send your boss a short summary on these topics by tomorrow! Your job is to find people and topics that help you with your task:

- Save people and hashtags by clicking on them.
- Star any tweets that you would use as the basis of the articles you are going to write.

Once they closed the introduction screen, the main interface became visible with the non-personalized content. Participants could not move on to the next screen if they had interacted with the system for less than 2 min. In A and B conditions only interactions with tweets could be logged, while in conditions C and D, interactions with people and hashtags could also be logged.

Participants moved forward to a post-survey[5] after they selected the "I'm Done!" button. Here they were asked about their perceptions of the system and its contents. Next, participants were taken to the personalized variant in their condition where they were asked to enter their own Twitter ID. They performed the same task a second time, and were taken to an identical post-survey for this second interaction.

4.5   Results

The distribution of responses were not normally distributed for any of the variables, and non-parametric tests (Kruskal-Wallis or Wilcoxon) are used to consistently compare between conditions.

H1: Perceived serendipity. H1a. There was no significant difference between versions of the interface with regard to the degree of perceived serendipity (Kruskal-Wallis chi-squared = 5.8, df = 3, p-value = 0.12). H1b. There was also no significant difference between the perceived serendipity for the non-personalized and personalized conditions (Wilcoxon rank, W = 5876.5, p-value = 0.71). Tables 3 and 4 summarize the means. While we did not find an effect of condition on perceptions of serendipity, these differences may be revealed in a more longitudinal study.

---

[4] Pre-survey, https://ucsbltsc.qualtrics.com/SE/?SID=SV_819BEUBNzqmvzed.
[5] Post-survey 1, https://ucsbltsc.qualtrics.com/SE/?SID=SV_8kvPK1O6qGiW2Z7.

(A) Baseline: This is the typical message list view.

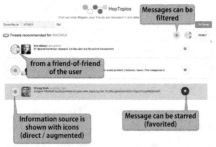

(B) Augmented Data: Augmented with new messages.

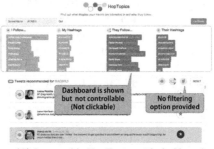

(C) Inspectable HopTopics with dashboard.

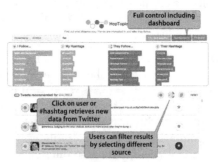

(D) Controllable HopTopics with interaction.

Fig. 2. Screenshots of the four between subjects system versions (A-D).

Table 3. Mean (SD) perceived serendipity by system level (A-D), (0=low, 100=high).

A	B	C	D	All
59.81 (30.58)	50.97 (32.54)	62.59 (28.95)	60.38 (26.33)	57.95 (30.34)

Table 4. Mean (SD) of perceived serendipity for personalized and non-pers. conditions.

Non-personalized	Personalized	All
57.69 (29.38)	58.22 (31.39)	57.95 (30.34)

H2: Perceived familiarity. Table 5 summarizes the mean familiarity of hash-tags. H2a. In a comparison between the Inspectable and Controllable interface conditions, we did not find a significant effect on the perception of being able to find familiar tweets (Wilcoxon rank, $W = 1944.5$, p-value $= 0.07$).

Table 5. Mean (SD) of content identified as relevant and familiar. (1=low, 100=high)

	Non-Pers.	Personalized	Both
C	59.96 (28.03)	55.37 (27.69)	57.67 (27.69)
D	38.67 (34.40)	52.07 (34.18)	45.37 (34.66)
Combined	48.75 (33.05)	53.63 (31.04)	51.19 (32.02)

H2b. There was no significant difference w.r.t. perceived familiarity compar-ing the personalized compared to the non-personalized data (*c.f.* Table 5), across types of system (Wilcoxon, $W = 1497.5$, p-value $= 0.47$). A deeper understanding of the sets of selected hashtags, and associated user interests may be necessary to explore differences in familiarity perception.

H3: Perceived transparency. There was a significant difference between the versions of the interface (A-D) w.r.t. the degree of perceived transparency (Kruskal-Wallis chi-squared $= 8.5456$, df $= 3$, p-value < 0.05). The means in Table 6 show higher means for the Controllable and Inspectable conditions (pair-wise comparisons were not significant after correction was applied). We did not anticipate a difference in perceived transparency between the personalized and non-personalized conditions.

H4: Perceived control. There was a strong and significant difference between the versions of the interface (A-D) w.r.t. the degree of perceived control (Kruskal-Wallis chi-squared $= 13.562$, df $= 3$, p-value < 0.01). Table 7 summarizes the means per condition, demonstrating a greater sense of control in the Inspectable (C) and Controllable (D) conditions. *Post-hoc* pairwise comparisons show a sig-nificant effect of conditions only between the Baseline and both the Inspectable (C) and Controllable (D) conditions (Wilcoxon, p < 0.01, Bonferroni corrected).

Table 6. Mean (SD) perceived transparency by system level (A-D), (1=low, 7=high).

A	B	C	D	All
4.45 (2.05)	5.12 (1.82)	5.41 (1.52)	5.45 (1.37)	5.09 (1.76)

We did not anticipate a difference in perceived control between the personalized and non-personalized conditions.

Table 7. Mean (SD) perceived control by system level (A-D), (1=low, 7=high).

A	B	C	D	All
3.97 (1.67)	4.38 (1.88)	4.88 (1.63)	5.03 (1.29)	4.49 (1.71)

H5: Content discovery. Participants could not select topics or people in conditions A and B, so we compared conditions C and D only. H5a. There was a trend toward greater content discovery in the Inspectable (C) condition compared to the Controllable (D) condition (Wilcoxon rank, $W = 1168.5$, p-value = 0.07), but the Inspectable condition also had a much larger standard deviation. H5b. There was also no significant difference for the degree of content discovery between the non-personalized and personalized conditions (Wilcoxon rank, $W = 6131.5$, p-value = 0.84), means are shown in Table 9.

Table 8. Mean (SD) of content discovered by system level (C, D). (1=low, 7=high)

C	D	All
6.37 (5.17)	5.00 (3.49)	5.65 (4.40)

H6: Correctness of mental model. Participants could not select hashtags or people in conditions A and B, so we compared conditions C and D only. H6b. There were significant correlations between perceived serendipity and degree of content discovery in the personalized condition (Table 10). This suggests that participants were better at estimating the amount of novel hashtags and tweets they were marking as favorite in this condition (Table 8).

H6a (revised) *Post-hoc* comparisons show that for the Inspectable interface and the non-personalized condition this correlation was negative and significant (Spearman, $p<0.05$, rho=-0.42, Bonferroni corrected). This suggests that participants who could inspect their data, but had a non-personalized profile, were poor at estimating the amount of novel content they marked as favorites. Participants in the non-personalized condition discovered as much novel content

Table 9. Mean (SD) content discovered for pers. and non-pers. conditions. (1=low, 7=high)

Non-personalized	Personalized	All
5.65 (3.81)	5.65 (4.95)	5.65 (4.40)

Table 10. Spearman correlations b/w perceived serendipity and content discovery.

Comparison	p	rho
Condition C	0.16	-0.19
Condition D	0.57	0.10
Personalized	0.02	0.15
Non-Pers	0.16	0.08

(Table 9) as for the personalized condition, but underestimated the perceived novelty (*c.f.* Table 4).

H7: Perceived diversity. There was no significant difference between the versions of the system (A-D) w.r.t. the degree of perceived diversity (Kruskal-Wallis chi-squared = 3.8267, df = 3, p-value = 0.28). We did not anticipate a difference in perceived diversity between the personalized and non-personalized conditions.

5   Limitations

The focus of our study was on the impact of inspectability and control of a recommender system on content discovery and user experience. The recommendation process itself was treated as a black box, enabling us to establish that the *interface and interaction* (decoupled from the algorithm) are effective in terms of user perceptions.

We did not apply our own recommendation algorithm to filter people to follow, the hashtags, or the tweets. Rather, the content was based on social network structure, and the chronological order normally used by Twitter. While Twitter is reputed to modify its ranking algorithms, details of the current tweet selection algorithm are available in their online documentation[6]. With similarity-based ranking applied to the user lists, the HopTopics system would behave similarly to a standard user-based collaborative filtering algorithm. While outwith the scope of this paper, the next natural steps will be to implement this system with different algorithms, such as user and item-based collaborative filtering, and compare user perceptions for these. Note that in our experiment, the *same* black box algorithm was used across all conditions, creating a fair comparison in a between-subjects design for the different levels of inspectability.

Our definitions of serendipity and familiarity follow the definition of *e.g.*, [12], and put an emphasis on "relevance", *e.g.*, serendipity is defined as both

---

[6] https://support.twitter.com/articles/164083, retrieved July 2016.

Table 11. Significance levels, - if not significant. NA=Not Analysed.

Variable	Interface	Personalization
Serendipity	-	-
Familiarity	-	-
Transparency	0.05	NA
Control	0.01	NA
Content	-	-
Mental Model	-	0.05

relevant and surprising. Given the relatively weaker personalization, through the regular Twitter content selection, and selection by social network, it is likely that there were fewer *personally relevant* items than might appear in a system that applied a better personalization algorithm. In this paper, the main point was however to compare between conditions. The results demonstrate that there were enough relevant tweets to compare (fairly) between conditions, and we were able to evaluate familiarity and serendipity. An alternative could have been to ask only whether a tweet is surprising (/familiar). However, this would not capture the impact of an inspectable interface on deciding whether a tweet is relevant, instead it would answer the question if they have seen it before (or not) regardless of its informational value.

6   Conclusion and Future Work

In this study, we designed a novel interactive tool, called HopTopics, for social content discovery on Twitter. The system is designed to combat the filter-bubble effect by sourcing relevant information from beyond a user's typical information horizon in microblogs. Specifically, we leverage n-hop social connections and hashtags to create augmentations to the "traditional" Twitter feed that include opinions on relevant topics from both local (1 hop) and extended (>1 hop) networks. We conducted a 4 by 2 mixed design user experiment to evaluate the impact of feed augmentation, inspectability and control on a variety of UX metrics compared to a baseline, while varying personalization. The study was conducted on a crowd-sourced platform, and a fictional information gathering scenario was used to promote user engagement with the system. Our findings suggest that our interface and interaction model are significantly more effective for improving user experience in terms of both perceived transparency and perceived control, compared to baseline interfaces.

Inspectability can sometimes also be harmful. Participants in the Inspectable condition underestimated how much novel content they discovered when the twitter data was not personalized to them. While the results for the Inspectable and Controllable interface conditions were largely identical throughout the experiment, the error in mental models found for the Inspectable condition did not appear for the Controllable condition. This suggests that the Controllable interface might help users form better mental models, and that the measure of serendipitous content objectively discovered, versus the perception of serendipity, could be a good proxy for evaluating when transparency is helpful. Table 11 summarizes the key findings of the experiment. In our next experiments, we will investigate the effect of specific algorithms (*e.g.* kNN and content-based filtering) on the selection of top users and hashtags on interaction with the interface. We also plan to investigate the effect of the global/trending use of hashtags mentioned in the network on novel and relevant content discovery.

Acknowledgements. This work was partially supported by the U.S. Army Research Laboratory under Cooperative Agreement No. W911NF-09-2-0053; The views and conclusions contained in this document are those of the authors and should not be interpreted as representing the official policies, either expressed or implied, of ARL, NSF, or the U.S. Government. The U.S. Government is authorized to reproduce and distribute reprints for Government purposes notwithstanding any copyright notation here on.

References

1. Amar, R.A., Stasko, J.T.: Knowledge precepts for design and evaluation of information visualization. IEEE Trans. Vis. Comput. Graph. **11**, 432–442 (2005)
2. André, P., Schraefel, M.C., Teevan, J., Dumais, S.T.: Discovery is never by chance: Designing for (un)serendipity. In: Creativity and Cognition (2009)
3. Bennett, S.W., Scott, A.C.: The rule-based expert systems: The MYCIN experiments of the stanford heuristic programming project, chapter 19 - specialized explanations for dosage selection, pp. 363–370. Addison-Wesley Publishing Company (1985)
4. Bernstein, M.S., Suh, B., Hong, L., Chen, J., Kairam, S., Chi, E.H.: Eddi: Interactive topic-based browsing of social status streams. In: User Interface Software and Technology, UIST 2010, pp. 303–312 (2010)
5. Broersma, M., Graham, T.: Twitter as a news source. J. Pract. **7**(4), 446–464 (2013)
6. Brusilovsky, P., Schwarz, E., Weber, G.: ELM-ART: an intelligent tutoring system on world wide web. In: Frasson, C., Gauthier, G., Lesgold, A. (eds.) ITS 1996. LNCS, vol. 1086, pp. 261–269. Springer, Heidelberg (1996). doi:10.1007/3-540-61327-7_123
7. Cerutti, F., Tintarev, N., Oren, N.: Formal arguments, preferences, and natural language interfaces to humans: an empirical evaluation. In: ECAI, pp. 207–212 (2014)
8. Dimitrova, V.: Style-olm: Interactive open learner modelling. Int. J. Artif. Intell. Educ. **17**(2), 35–78 (2003)
9. Garcia Esparza, S., O'Mahony, M.P., Smyth, B.: Catstream: Categorising tweets for user profiling and stream filtering. In: International Conference on Intelligent User Interfaces, IUI 2013, pp. 25–36 (2013)
10. Granovetter, M.S.: The strength of weak ties. Am. J. Sociol. **78**(6), 1360–1380 (1973)
11. Herlocker, J.L., Konstan, J.A., Riedl, J.: Explaining collaborative filtering recommendations. In: ACM Conference on Computer Supported Cooperative Work, pp. 241–250 (2000)

12. Herlocker, J.L., Konstan, J.A., Terveen, L., Riedl, J.T.: Evaluating collaborative filtering recommender systems. ACM Trans. Inf. Syst. **22**(1), 5–53 (2004)
13. Knijnenburg, B.P., Willemsen, M.C., Gantner, Z., Soncu, H., Newell, C.: Explaining the user experience of recommender systems. User Model. User Adap. Inter. **22**(4–5), 441–504 (2012)
14. Kulesza, T., Burnett, M., Wong, W.-K., Stumpf, S.: Principles of explanatory debugging to personalize interactive machine learning. In: IUI (2015)
15. Kwak, H., Lee, C., Park, H., Moon, S.: What is twitter, a social network or a news media? In: International Conference on World Wide Web, WWW 2010, pp. 591–600 (2010)
16. Marcus, A., Bernstein, M.S., Badar, O., Karger, D.R., Madden, S., Miller, R.C.: Twitinfo: Aggregating and visualizing microblogs for event exploration. In: Conference on Human Factors in Computing Systems, CHI 2011, pp. 227–236 (2011)
17. Nagulendra, S., Vassileva, J.: Providing awareness, understanding and control of personalized stream filtering in a P2P social network. In: Antunes, P., Gerosa, M.A., Sylvester, A., Vassileva, J., Vreede, G.-J. (eds.) CRIWG 2013. LNCS, vol. 8224, pp. 61–76. Springer, Heidelberg (2013). doi:10.1007/978-3-642-41347-6_6
18. Paramythis, A., Weibelzahl, S., Masthoff, J.: Layered evaluation of interactive adaptive systems: Framework and formative methods. User Model. User Adap. Interact. **20**, 2–12 (2010)
19. Pariser, E.: The filter bubble: What the Internet is hiding from you. Penguin Books, New York (2011)
20. Schaffer, J., Giridhar, P., Jones, D., Höllerer, T., Abdelzaher, T., O'Donovan, J.: Getting the message?: A study of explanation interfaces for microblog data analysis. In: Intelligent User Interfaces, IUI 2015, pp. 345–356 (2015)
21. Sharma, A., Cosley, D.: Do social explanations work? studying and modeling the effects of social explanations in recommender systems. In: World Wide Web (WWW) (2013)
22. Smyth, B., McClave, P.: Similarity vs. diversity. In: Aha, D.W., Watson, I. (eds.) ICCBR 2001. LNCS (LNAI), vol. 2080, pp. 347–361. Springer, Heidelberg (2001). doi:10.1007/3-540-44593-5_25
23. Teevan, J., Morris, M.R., Azenkot, S.: Supporting interpersonal interaction during collaborative mobile search. Computer **47**(3), 54–57 (2014)
24. Tintarev, N., Kang, B., ODonovan, J.: Inspection mechanisms for community-based content discovery in microblogs. In: IntRS15 Joint Workshop on Interfaces and Human Decision Making for Recommender Systems at ACM Recommender Systems (2015)
25. Tintarev, N., Masthoff, J.: Recommender Systems Handbook (second ed.), chapter Explaining Recommendations: Design and Evaluation (2015)
26. Verbert, K., Parra, D., Brusilovsky, P., Duval, E.: Visualizing recommendations to support exploration, transparency and controllability. In: International Conference on Intelligent User Interfaces, IUI 2013, pp. 351–362 (2013)
27. Wang, B., Ester, M., Bu, J., Cai, D.: Who also likes it? generating the most persuasive social explanations in recommender systems. In: AAAI Conference on Artificial Intelligence (2014)

Poster Papers: Markets, Crowds, and Consumers

Targeted Ads Experiment on Instagram

Heechul Kim[1], Meeyoung Cha[1(✉)], and Wonjoon Kim[1,2]

[1] Graduate School of Culture Technology, KAIST, Daejeon, South Korea
{heechul90,meeyoungcha,wonjoon.kim}@kaist.ac.kr
[2] School of Business and Technology Management,
KAIST, Daejeon, South Korea

Abstract. Ensuring media is brought appropriately and directed toward the "right people" is an important challenge. Marketers have traditionally employed demographic-based strategies such as age and gender to find target ad viewers. This research explores an alternative method by utilizing the embedding of brand relationships drawn from rich social media data. We presume that co-mentioned brands reflect the interest relationships of people and seek to exploit such information for targeted advertisements. Our 3-week experiment demonstrates the efficacy of the relationship-based ad campaign in yielding high click-through-rates. We also discuss the implications of our finding in designing social media-based marketing strategies.

Keywords: Social media · Targeted advertisement · Ad experiment

1 Introduction

The marketing landscape has changed significantly in the recent years. Traditional media such as radio, TV, and newspapers have less influence than before and instead a countless number of social media platforms have become more active [10,12]. The daily time spent on social media has surpassed that of TV for youth as well as other growing populations. Likewise it is expected that social media will play an increasing role in influencing the purchasing decisions of individuals as they become the primary go-to place to find information about products and services [3,5].

Following this communication paradigm, marketers have shown great interest in understanding the power and influence of user-generated discourse on social media [7]. Companies utilize social media for digital advertising, as well as for wholehearted communication with their consumers such as handling complaints and collecting feedback [11]. Interactive and targeted marketing capabilities are considered effective for social media marketing [4]. According to eMarketer, the total expenditure for social media advertising in 2017 is expected to reach 35.98 billion USD globally, comprising 16 % of the total digital advertising expenditures. Yet, social marketing strategies so far have relied largely on basic demographic features, like age and gender.

© Springer International Publishing AG 2016
E. Spiro and Y.-Y. Ahn (Eds.): SocInfo 2016, Part II, LNCS 10047, pp. 297–306, 2016.
DOI: 10.1007/978-3-319-47874-6_21

Recent studies suggest advantages of new behavioral targeting and personalization in online advertising. For example, researchers have suggested a cost-effective methodology to predict demographic profiles of website visitors for targeted ads [2]. Their methodology transforms the website visitors' clickstream patterns to a set of features and trains classifiers to generate predictions for gender, age, level of education, and occupation. From the industry side, already a number of patents exist regarding targeted ads on the web and social media. These patents use users' search history, IP addresses, location, status, real-time keywords to customize advertising and target users.

A few studies have proposed targeted advertisements specific to social media. For example, one survey study investigated whether it is possible to predict user demographics from social media data and whether such predictions are effective for targeted advertising [1]. The authors found that people with high openness and low neuroticism responded more favorably to targeted advertisements. Another recent study provided empirical evidence regarding the role of various advertising methods on social networks (i.e. earned vs. paid) [9]. The authors concentrated on the act of friend tagging and found it a powerful user-initiated solution for matching products with potential buyers.

Moving forward, data embedded in social media platforms like Facebook and Instagram can identify consumers of specific brands on its platform (e.g., via analyzing posts) and enable targeting based on demographic information, web traffic, and various in-app activities. For example, recognizing a demographic-level affinity to cultural products can be effective. Even when demographic information is not complete, services can make use of previous log history such as the kinds of posts or brands individuals (or their connected network) have "liked" on the site, which form the basics of social marketing.

In this study, we make an attempt to utilize such social context and measure the effect of several social media advertising strategies. We investigate whether and to what extent co-mentioned brand relationships in user-generated content can identify individuals who are potentially interested in particular products. This idea was tested via a real-world ads experiment on Instagram, thanks to support from Lucky Chouette, a notable fashion brand in South Korea. Lucky Chouette is popular among young women both nationally and internationally and is owned by KOLON Industries, Inc.

The brand name #luckychouette was mentioned over 18,000 times as of August 2016. As target ad keywords, 17 other brands that were frequently co-mentioned with Lucky Chouette were identified. When setting target keywords as brands, online users who have interacted with those brands get exposed to targeted ads. The advertising content utilized for our experiment was a 119-piece ad campaign collection featuring a popular Korean celebrity figure that has not been released prior to this study. The ad content was produced by professionals hired by the brand's marketing team. The ad campaign cost nearly 4,300 USD and a total of 1,016,974 Instagram users were exposed with the ads over the course of 3 weeks.

Our experiment shows that utilizing brands that are highly co-mentioned with the target fashion brand (Lucky Chouette) yield better click-through-rates in ad campaigns than those brands that are less frequently co-mentioned. We also find that non-fashion brands to be also a useful candidate for promoting fashion ads in that 'Starbucks' was one of the prominent keywords that yielded high click-through-rates. These findings suggest that social data-driven keywords may be a useful method to find new target keywords, compared to using only general demographic keywords like age and gender. We discuss challenges and difficulties in conducting an online ad experiment.

2   Data Methodology

An ad experiment was conducted in collaboration with Lucky Chouette. The brand's mother company, KOLON Industries, Inc. is one of the largest manufacturing companies in South Korea. To understand the chatter around this brand, we collected Instagram posts that mention the brand name via the hashtag #luckychouette (both Korean and English translation) using the service's API. We focused on Instagram because the platform is well known for its fashion-loving community [6] and the marketing team of the brand was particularly interested in exploiting a targeted ad campaign on this platform.

The gathered data comprise 23,765 photo posts and 238,599 hashtags shared from July 2011 to May 2015. The number of users who posted #luckychouette at least once was 9,155. Text descriptions of these posts were used to identify co-mentioned brands. Prior to analysis, we excluded all posts from the official account of Lucky Chouette as well as 10 avid brand advocates, who were voluntarily participating as the brand's social media promoters.

From the gathered text data, we then extracted hashtags that co-appeared. Hashtags are a popular practice on Instagram. Users employ hashtags to add rich context information and description to their photos. Hashtags allow other users to easily find interesting content [8]. In a way, adding and searching hashtags on Instagram is considered equivalent to joining a virtual community of users discussing the same topic [13].

We listed the top co-occurring hashtags and then manually identified other brands from this list. Three major kinds of brands were co-mentioned frequently with #luckychouette: apparel brands, shoe brands, and coffee places. This relationship is shown in Fig. 1, where the word clouds represent which other brands are co-mentioned with Lucky Chouette on Instagram.

Among the apparel brands, global brands (e.g., Zara, Uniqlo) were mentioned more frequently than regional competitors identified by the brand's marketing team. Interestingly, certain brands the marketing team did not consider competitors were also co-mentioned frequently (e.g., Chanel). Another prominent category was shoe and sports brands. Vans was the most frequently mentioned shoe brand, which may be because Lucky Chouette launched a collaborative collection with Vans in 2015. The third category that was mentioned frequently was cafes, where Starbucks had the most prominent share.

(a) Apparel brands   (b) Shoes/sports brands   (c) Cafe brands

Fig. 1. Word clouds of the co-occurrence frequency for apparel, sports, and cafe brand hashtags, where text size represents co-mention frequency

Overall Instagram users associated #luckychouette the most frenetically with global SPA brands in the following brand order: American Apparel, Zara, and Uniqlo. Their frequencies are higher than luxury brands like Chanel or premium casual brands likes Saint James and Lacoste. Among the shoes which can be worn with Lucky Chouette, sneakers brands such as Vans, Bensimon, Superga, Converse had high frequency. Starbucks was the highest in frequency among cafe brand hashtags.

Examining the co-mention relationships helped the brand marketing team understand two new trends. First, the rank of brands confirmed existing assumptions of marketers in that young casual apparels such as Zara were mentioned frequently. Yet, brand marketers identified some brands that were not in their competitor landscape and found the co-mention relationship useful. Second, the reach of brand names beyond apparels indicated that there might be a new, untapped audience if one were to consider exploiting cafe and shoe brands for advertisement. This cross-genre ad opportunity is an area that has not been deeply explored in marketing.

3   Ads Experiment

Lucky Chouette began promoting its products in November 2015. At the time, the firm relied on limiting its advertising exposure to a general target demographic of women in their 20 s and 30s, who were believed to be highly perceptive to changes in fashion and likely to make purchases. Target advertisements were geared towards potential online shoppers rather than generic online consumers. Several key brands that are often mentioned alongside Lucky Chouette were identified, and consumers who have shown interest in those particular products or services would be targeted for subsequent advert exposure. This may be considered an attempt to discover a Personalized Ads Strategy for the Lucky Chouette brand.

For our targeted ads experiment we utilized the top 17 brands that were frequently brought up in Lucky Chouette affiliated chatter. The ads were run on the Instagram platform, yet were delivered to a subset of its users who have connected their Instagram and Facebook accounts. This is because Instagram

Fig. 2. Examples of Lucky Chouette ads used in the study

does not require users to give demographic information upon registration, yet Facebook profiles contain a rich set of details about the gender, age, and region of users. Furthermore, this allows the ad platform to target users who have "liked" or "followed" any of the select brands on Facebook.

The ads experiment was launched over a 21 day period (Nov 19th, 2015 - Jan 8th, 2016). A comprehensive set of 119 photo images were newly prepared featuring a popular actress in South Korea, Jeong-ahn Chae. For this launch (1) none of the ad images we employed in the experiment were released in public, and (2) the ad quality was controlled so as to be consistent throughout the experiment. These two conditions are important because pre-exposed photographs or inconsistent ad images likely lead to different levels of engagement.

The 119 photo shoot campaign was used for a diverse set of ad combinations with a fixed set of demographic settings (i.e., age, gender) and varying target keywords (i.e., 17 brands). Each ad campaign had a budget of 50,000 KRW (nearly $50) and ran for 3 consecutive days, meaning that ads were delivered until this budget was exhausted. The budget was not used entirely when the target keyword we chose only yielded a small audience (i.e., fewer online users have previously liked or interested with the brand set as target keyword). Each keyword was employed in multiple different ad sets. Altogether, the ads experiment consumed a total of 4,300,000 KRW (nearly $4,300).

3.1   Considered Target Keywords

Determining brands to be employed as target keywords was a joint effort with the brand's marketing team. We first included all brands that had prominent levels of co-mentions with #luckychouette, including American Apparel (apparel), Tomboy (apparel), Zara (apparel), Chanel (apparel), Uniqlo (apparel), Vans (shoe), Converse (shoe), Adidas (sports), Nike (sports), and Starbucks (cafe). Then we added brands from the lower ranks to include more brands that are of interest to the brand marketers for comparison purposes. They included Bean Pole (apparel), Giordano (apparel), Ralph Lauren (apparel), New Balance (sports), Reebok (sports), Angel-in-Us (cafe), and Caffe Bene (cafe). These

17 selected brands were ranked and categorized according to their frequency of joint mention with Lucky Chouette before the experiment began.

To compare advertising effects, we employed three criteria to rank brands and divide them into two groups. The first method is examining the raw total count of brand names jointly mentioned with #luckychouette (co-mention). The second method measured the total frequency of individual brand-related hashtags (frequency). The last method normalized the first methods total count by taking the second methods frequency into account, quantifying how often these hashtags were mentioned alongside #luckychouette while considering their inherent frequency of mentions (normalized).

1. Co-mention: the count of brand hashtags mentioned jointly with #luckychouette
2. Frequency: the total count of brand mentions as of Feb 17th, 2016
3. Normalized: the total co-mention frequency divided by the total frequency

Each of the three methods was used to group the 17 select brands into two groups: high and low. The top 8 brands above the median value of the co-mention measure were included in the high group, and the remaining 9 groups were included in the low group. The grouping for the respective methods can be found in the chart below. This resulted in 3 sets of high and low groupings, based on three different methods for counting co-frequency. Depending on the method, a given brand could appear in the high or low group. For instance, Starbucks was in the high group for the co-mention and frequency measures, yet it was included in the low group in the normalized measure.

3.2   Considered Ad Metrics

To assist advertisers in gauging the effectiveness of their marketing campaigns, Instagram offers a number of users metrics as listed below.

- *Impressions*: the number of exposures for each ad
- *Click counts*: the number of clicks ads received
- *Like counts*: the number of likes ads received
- *Comment counts*: the number of comments ads received
- *Cost Per Click (CPC)*: the average cost of ads
- *Click Through Rate (CTR)*: the rate at which people click on viewed ads

Quantifiable metrics such as click counts and like counts increase with every successful content delivery to a consumer (i.e., impression) and are directly influenced by budget constraints. To quantify ad effectiveness, normalized metrics may provide more comprehensive statistics. Of the metrics mentioned above, CPC and CTR belong to this category. CPC measures the advertising cost paid per click, and CTR shows the percentage of users that actually clicked on the advert after being exposed to it. Hence, an ad is considered effective when it achieves a low CPC but a high CTR.

4   Results

Having explained the ad experiment setting, we will now compare how the 17 target keywords performed. Figure 3 shows the result of CTR across brands, which are divided into two groups (high and low) based on the co-mention method. The vertical axis represents the average CTR of ad experiments of the particular keyword. We compare ad effectiveness at the group level using the Wilcoxon's Rank Sum Test, because the ad sample size is limited. Note that the ad experiment was highly controlled (e.g., visuals, content quality, demographic setting) and that the only variation was target keywords.

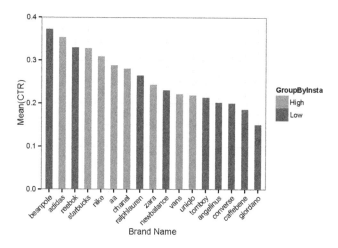

Fig. 3. Effectiveness of the 17 target keywords based on the co-mention strategy

When grouped according to the co-mention method, the high group produced better advertising effects than the low group as indicated by the rank of red bars. The difference in CTR between the two groups were statistically significant (P<.1). This indicates that brands that are jointly mentioned frequently–Adidas, Starbucks, Nike, American Apparel (which we refer to as 'aa' in figures), Chanel–are also effective in terms of generating more clicks per ad than other brands that appeared less frequently with #luckychouette.

Grouping target keywords by frequency resulted in no statistical difference between the high and low groups (p-value=0.176). This is because the method does not consider any co-mention rates. Finally, grouping keywords in a normalized fashion as in the normalized method also did not incur any significant difference between the high and low group (p-value=0.172). This result indicates that the raw co-mention frequency was most predictive of how target keywords performed on advertisement. We found similar results when it came to CPC - the co-mention method yielded the highest gain in ad efficiency. This result is promising given that marketers often employ basic demographic keywords in targeting and may not know what other keywords might be useful to reach new audiences.

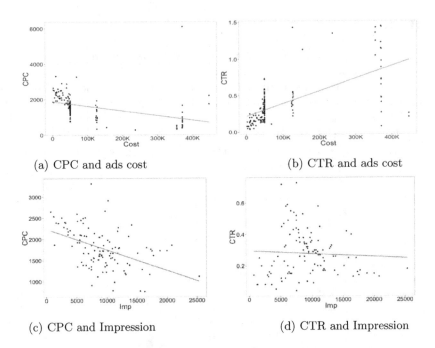

(a) CPC and ads cost   (b) CTR and ads cost

(c) CPC and Impression   (d) CTR and Impression

Fig. 4. Ads efficiency in terms of the cost-per-click (CPC) and click-through-rate (CTR) as a function of the total ads cost and impression (the number of times ads reached any audience)

Our experiment confirms that it is possible to identify potential customers for fashion brands and deliver online adverts more effectively from studying hashtag co-mentions. However, our particular findings about why the co-mention method yielded the best performance needs to be carefully taken into account. We may interpret the result from the inherent tendency of large targeting ad campaigns to perform better than smaller ones. Targeted ad campaigns are known as complex systems that take into account numerous factors in real-time, including factors such as how many ads compete for the same target keyword concurrently, and how much of the target audience is online, etc.

Note that ad keyword matching is determined by proprietary auction algorithms which tend to favor ad campaigns with large budgets and large audiences. Hence, given an identical target keyword, an ad campaign with a larger budget will perform better in terms of CPC and CTR. Conversely, given the same budget, an ad campaign with a larger audience will perform better. To understand this system better, we compared the overall ad effectiveness in terms of CPC and CTR as a function of ad impression and the total ad cost.[1] Figure 4 clearly demonstrates that with increasing ad cost, campaigns are able to achieve better performance with smaller CPC and higher CTR. When it comes to increasing

---

[1] We utilized our experiment results as well as previous ad records from the studied brand.

impressions (e.g., how many times ad reached audience), the larger audience meant smaller CPC, yet there was no obvious trend for CTR.

This inherent bias toward favoring larger ad campaigns may explain why in particular the co-mention method was far more effective than the normalized method, as the large raw frequency in this method could represent the popularity of the target keyword in reaching audiences. Our findings also indicate that to compare the value of target keywords themselves, one needs to carefully control the budget as well as the actual cost involved in ad experiments. Comparing the CTR and CPC of ad keywords that consumed similar amounts of budget may help better assess the reach of target keywords.

5 Conclusion

The most noticeable observation we made is that the brands that were identified to have higher frequency of being co-mentioned with our brand of interest, Lucky Chouette, on Instagram had comparatively higher advertising effects than other brands. This may be interpreted as the users who have shown interest in brands with greater frequency of joint mentions with Lucky Chouette responded more favorably to these advertisements than those who did not.

Through these results we were able to confirm it is possible to identify potential customers for fashion brands and deliver online adverts more effectively with the Instagram's hashtag data. The experiment not only discovered a statistically significant difference between the two groups but also managed to identify an affiliated area for further research. In an era where offline sale growths are dwindling while online sales annually grow, these findings may be said to pose a profound implication for a fashion industry trying to increase its online sales in ways such as social marketing reinforcement.

It is interesting to note that Bean Pole, categorized in the low group, actually showed the greatest advertising effects. A further analysis into similarities between Instagram users who indicate an interest in Bean Pole's content/page with Lucky Chouette customers may assist in discovering potential consumers. Subsequent research into outliers such as these who severely deviate from the rest of the results is recommended.

Although this research used only the frequency of joint mentions within Instagram to select similar brands, a quality study of the nature of these similar brands used to supplement the experiment may yield even more effective results. Suggestions for potential follow-up research include attempts to identify the nature of brands based on content and users who post them on Instagram. For example, the results from this experiment show that advertisements that targeted Bean Pole, a brand that offers products in the same price range as Lucky Chouette, were the most effective. As brand images are formed by public perception it is a difficult task to identify which brands are "similar". Yet because individuals from a wide range of gender and age demographics share content on social media, it is possible to define a brands image by analyzing who uploads the content and how a brand is discussed on Instagram.

Acknowledgement. This work was partly supported by the Institute for Information & communications Technology Promotion (IITP) grant funded by the Korea government (MSIP) (R0115-15-100).

References

1. Chen, J., Haber, E., Kang, R., Hsieh, G., Mahmud, J.: Making use of derived personality: The case of social media ad targeting. In: International AAAI Conference on Web and Social Media (2015)
2. De Bock, K., Van den Poel, D.: Predicting website audience demographics forweb advertising targeting using multi-website clickstream data. Fund. Inf. **98**(1), 49–70 (2010)
3. Foux, G.: Consumer-generated media: Get your customers involved. Brand Strategy **8**, 38–39 (2006)
4. Hanna, R., Rohm, A., Crittenden, V.L.: Were all connected: The power of the social media ecosystem. Bus. Horiz. **54**(3), 265–273 (2011)
5. Lempert, P.: Caught in the web. Progressive Grocer **85**(12), 18 (2006)
6. Mahmud, J., Fei, G., Xu, A., Pal, A., Zhou, M.: Predicting attitude and actions of twitter users. In: International Conference on Intelligent User Interfaces, ACM (2016)
7. Mangold, W.G., Faulds, D.J.: Social media: The new hybrid element of the promotion mix. Bus. Horiz. **52**(4), 357–365 (2009)
8. Morstatter, F., Pfeffer, J., Liu, H., Carley, K.M.: Is the sample good enough? comparing data from Twitter's streaming API with Twitter's firehose. In: International AAAI Conference on Web and Social Media (2013)
9. Park, J.Y., Sohn, Y., Moon, S.: Power of earned advertising on social network services: a case study of friend tagging on facebook. In: International AAAI Conference on Web and Social Media (2016)
10. Rashtchy, F., Kessler, A.M., Bieber, P.J., Shindler, N.H., Tzeng, J.C.: The user revolution: The new advertising ecosystem and the rise of the Internet as a mass medium. Piper Jaffray Investment Research, Minneapolis (2007)
11. Solis, B.: Engage: The Complete Guide for Brands and Businesses to Build, Cultivate, and Measure Success in the New Web. Wiley, Hoboken (2010)
12. Vollmer, C., Precourt, G.: Always on: Advertising, Marketing, and Media in an Era of Consumer Control. McGraw Hill Professional, New York (2008)
13. Yang, L., Sun, T., Zhang, M., Mei, Q.: We know what@ you# tag: does the dual role affect hashtag adoption?. In: International conference on World Wide Web (2012)

Exploratory Analysis of Marketing and Non-marketing E-cigarette Themes on Twitter

Sifei Han[2] and Ramakanth Kavuluru[1,2(✉)]

[1] Division of Biomedical Informatics, Department of Internal Medicine,
University of Kentucky, Lexington, KY, USA
[2] Department of Computer Science, University of Kentucky, Lexington, KY, USA
{sehan2,ramakanth.kavuluru}@uky.edu

Abstract. Electronic cigarettes (e-cigs) have been gaining popularity and have emerged as a controversial tobacco product since their introduction in 2007 in the U.S. The smoke-free aspect of e-cigs renders them less harmful than conventional cigarettes and is one of the main reasons for their use by people who plan to quit smoking. The US food and drug administration (FDA) has introduced new regulations early May 2016 that went into effect on August 8, 2016. Given this important context, in this paper, we report results of a project to identify current *themes* in e-cig tweets in terms of semantic interpretations of topics generated with topic modeling. Given marketing/advertising tweets constitute almost half of all e-cig tweets, we first build a classifier that identifies marketing and non-marketing tweets based on a hand-built dataset of 1000 tweets. After applying the classifier to a dataset of over a million tweets (collected during 4/2015 – 6/2016), we conduct a preliminary content analysis and run topic models on the two sets of tweets separately after identifying the appropriate numbers of topics using *topic coherence*. We interpret the results of the topic modeling process by relating topics generated to specific e-cig themes. We also report on themes identified from e-cig tweets generated at particular places (such as schools and churches) for geo-tagged tweets found in our dataset using the GeoNames API. To our knowledge, this is the first effort that employs topic modeling to identify e-cig themes in general and in the context of geo-tagged tweets tied to specific places of interest.

1 Introduction

Electronic cigarettes (e-cigs) are an emerging smoke-free tobacco product introduced in the US in 2007. An e-cig essentially consists of a battery that heats up liquid nicotine available in a cartridge into a vapor that is inhaled by the user [12], an activity often referred to as *vaping*. The broad topic of e-cig use has become a major fault line among clinical, behavioral, and policy researchers who work on tobacco products. There are arguments on either side given their reduced harm aspect ([28] claims they are 95 % less harmful than combustible cigarettes) may

© Springer International Publishing AG 2016
E. Spiro and Y.-Y. Ahn (Eds.): SocInfo 2016, Part II, LNCS 10047, pp. 307–322, 2016.
DOI: 10.1007/978-3-319-47874-6_22

help addicted smokers quit smoking [24] while the long term effects of e-cigs are not yet thoroughly understood. However, there is recent evidence that vaping is linked to suppression of genes associated with regulating immune responses [27]. Furthermore, based on recent news releases from the Centers for Disease Control (CDC) [37], there is an alarming 900 % increase in e-cig use from 2011 to 2015 by middle and high school students who might be acquiring nicotine dependence albeit through the new e-cig product. There is also recent evidence that never smoking high school students are at increased risk of moving from vaping to smoking [2]. In light of these findings, the FDA has recently introduced a final deeming rule [13] that went into effect on 8/8/2016 when regulations were extended to many electronic nicotine delivery systems including e-cigs. In this context, surveillance of online messages on e-cigs is important both to monitor the spread of false/incomplete information [22] about them and to gauge prevalence of any adverse events related to their use [6, 36] as disclosed online.

For an emerging product like e-cigs, the follower-friend connections and "hashtag" functionality offer a convenient way for Twitter users (or "tweeters") to propagate information and facilitate discussion. An official quote we obtained earlier this year from Twitter Inc. indicates there are over 30 million public tweets on e-cigs since 2010. In our prior effort [19], we found that there is a 25 fold increase in e-cig tweets from 2011 to 2015 indicating the popularity of e-cig messages on Twitter. A major amount of chatter on e-cigs on Twitter surrounds their marketing by vendors making it generally difficult to analyze regular e-cig tweets that are not dominated by marketing noise. As such, building and using a classifier that separates marketing tweets is an important pre-processing step in several efforts. We are aware of at least four such efforts [10, 14, 18, 20] on building automatic classifiers for e-cig marketing tweets for various end-goals. Other researchers who studied e-cig tweets focused on sentiment analysis [14, 30] and diffusion of messages from e-cig brands on Twitter [8]. In our current effort

1. We manually estimated the proportion of marketing and non-marketing tweets to be 48.6 % (45.5–51.7 %) : 51.4 % (48.3–54.5 %) from a sample of 1000 randomly selected tweets selected from over one million e-cig tweets collected through Twitter streaming API between 4/2015 and 6/2016 (Sect. 2). The ranges in parentheses show 95 % confidence intervals of the proportions calculated using Wilson score [38].
2. We built a classifier that achieves an accuracy of 88 % (Sect. 3) in identifying marketing and non-marketing tweets using a variety of approaches ranging from traditional linear text classifiers to recent advances in classification with convolutional neural networks based on word embeddings [21]. Prior efforts [18, 20] that seem to report similar or slightly superior ($< 2.5\%$) results estimate the proportion of marketing tweets in the dataset to be 80 %–90 %[1], which we find unrealistic in the current situation (based on our own assessment mentioned earlier) as public awareness and their participation in the conversation have increased.

---

[1] Although achieving high F-scores for the minority class is generally difficult in heavily skewed datasets, they typically lend themselves to building classifiers with high overall accuracy across all classes or high F-score for the majority class.

3. After applying the binary classifier to over a million e-cig tweets, we conducted a rudimentary analysis of differences in content and user traits in both subsets (Sect. 4). We then ran topic modeling algorithms tailored for short texts [7] on the two separate subsets by determining the ideal numbers of topics using average topic coherence scores [32]. We manually examined the topics generated to identify themes in general and also based on subsets of geotagged tweets at popular places of interest as identified through the GeoNames geographical database (http://www.geonames.org). Although prior efforts identified broad themes through manual analyses [9], we believe our current effort is the first to employ topic modeling to discover more specific e-cig themes (Sect. 5). Thus, rather than having investigators predetermine which themes to look for in the dataset, our approach lets the dataset determine the prominent themes.

2   Dataset and Annotation

We used the Twitter streaming API to collect e-cig related tweets based on following key terms: `electronic-cigarette`, `e-cig`, `e-cigarette`, `e-juice`, `e-liquid`, `vape` and `vaping`. Variants of these terms with spaces instead of hyphens or those without the hyphens (for matching hashtags) were also used. A total of 1,166,494 tweets were obtained through the API calls from 4/2015 to 6/2016. From this dataset we randomly chose 1000 tweets to manually annotate them as marketing or non-marketing. For our purposes, marketing tweets are those that

- promote e-cig sales (coupons, free trials, offers),
- advertise new e-cig products (liquid nicotine or vaping devices), or
- review different flavors or vaping devices aiming to sell.

We (both authors) independently annotated the 1000 tweets. The labels matched for 87.3 % of the tweets with an inter-annotator agreement of $\kappa = 0.726$, indicating substantial agreement [23]. Conflicts for the 127 tweets where we chose different labels were resolved based on a subsequent face to face discussion resulting in a consolidated labeled dataset of 1000 tweets. Disagreements occurred when the marketing/advertising intent is not explicit or clear. For example, a simple message that encourages the followers to also follow the tweeter's Instagram account is not explicitly promoting e-cigs in and of itself but is nevertheless aimed toward marketing. Conflicts also occurred with reviews/recommendations when it was not clear whether a user is genuinely recommending a particular flavor that he/she has tried or whether it is the message from a manufacturer simply drawing followers' attention to their product line. While the former is not a marketing tweet, the latter would definitely fit our notion of such a message. Our final consolidated dataset has 486 marketing and 514 non-marketing tweets.

3   Marketing Tweet Classifier

The measure of performance used in this effort is accuracy, which is essentially the proportion of correctly classified tweets. We did not use the popular F-measure given we wanted to give equal importance to both classes given our aim is to study themes in both subsets of tweets. We first used linear classifiers such as support vector machines (SVM) and logistic regression (LR) classifiers as made available in the scikit-learn [33] machine learning framework. Tweet text was first preprocessed to replace all hyperlinks with the token URL and user mentions with the token TARGET. This is to minimize sparsity of very specific tokens having to do with links and user mentions and is in line with other efforts [1]. Besides uni/bi-grams we also used as features, counts of emoticons, hashtags, URLs, user mentions, sentiment words (positive/negative), and different parts of speech in the tweet. These additional features were useful in our prior efforts in tweet sentiment analysis [15] and spotting e-cig proponents [19] on Twitter. However, in this effort, considering average accuracy over hundred distinct 80 %–20 % train-test splits of the dataset, we did not observe any improvements with these additional features. So our final mean and 95 % confidence intervals for accuracies are 88.10 ± 0.40 with LR and 87.14 ± 0.45 with SVM.

Recent advances in deep learning approaches specifically convolutional neural networks (CNNs) have shown promise for text classification [21]. Given our own positive experiences in replicating those approaches for biomedical text classification [35], we also applied CNNs with word embeddings to generate feature maps for marketing tweet classification. The main notion in CNNs is of so called *convolution filters* (CFs) that are traditionally used in signal processing. The general idea is to learn several CFs which are able to extract useful features from a document for the specific classification task based on the training dataset. In the training phase, the inputs to the CNN are projections of constituent word vectors (which are typically randomly initialized) from a fixed size sliding window over the document. Model parameters to learn include the word vectors, the convolution filters (which are typically modeled as matrices), and the connection weights from the convolved intermediate output to the two nodes (for binary classification) in the output layer. Due to the nature of this particular paper, we refer the readers to our recent paper [35, Section 3] for a detailed description of CNN models including specifics of parameter initialization and drop-out regularization (to prevent overfitting). Averaging the [0, 1] probability estimates of the corresponding classes from several (typically ten) CNNs seems to help in getting a more robust model. We ran ten such models (each with ten CNNs, so a total of 100 CNNs) on ten different 80 %–20 % train-test splits of the dataset. The corresponding accuracies were: 89, 88.5, 85.5, 86, 87, 90.5, 87.2, 88.5, 90.5, and 89 with an average of 88.17 %, which is only slightly better than the mean accuracy obtained using logistic regression.

4   Characteristics of Marketing/Non-marketing Tweets

As discussed earlier, although the ability to separate marketing tweets from those that do not have that agenda is of interest in and of itself, in this effort, we wanted to study themes evolving from both subsets of the dataset. We applied all three classifiers (SVM, LR, and CNN) built in Sect. 3 using all hand-labeled tweets to all 1,166,494 tweets in our full dataset. We considered those tweets for which all three classifiers predicted the same label, which turned out to be for 1,021,561 (87.56 % of the full dataset) of which 456,290 (44.66 %) were predicted to be marketing and 565,271 (55.34 %) belonged to the other class. To get a basic idea of the tweet content, we simply counted and sorted the words in each subset in descending count values. The top 20 words in both subsets are

- *Marketing*: win, vaporizer, free, mod, get, enter, giveaway, new, premium, code, shipping, bottles, USA, use, box, promo, kit, available, follow, DNA
- *Non-marketing*: smoking, new, use, rips, like, cigarettes, via, man, get, tobacco, health, video, study, FDA, ban, one, smoke, people, news, explodes

Even with this simple exercise, we notice that the marketing tweets are dominated by e-cig promotions and sales terms or devices for vaping (mod, vaporizer, kit). On the other hand, terms in the non-marketing tweets are about tobacco smoking, health studies, and FDA regulations.

Table 1. Content and user characteristics of the datasets

	Marketing	Non-marketing
E-cig flavors	25472	4612
Harm reduction	19	2256
Smokefree aspect	553	3201
Smoking cessation	6363	22421
Contain "FDA"	204	18297
Number of unique users	66,957	231,982
User handles containing e-cig terms	4777 (7.1 %)	3859 (1.7 %)
Avg. # tweets per user	6.81 ($\sigma = 197$)	2.44 ($\sigma = 84$)

Next, we look at specific content and user characteristics of both subsets. In our prior work [19], we analyzed the tweets generated by e-cig proponents tweeters along four well known broad themes. We developed regular expressions (please see [19, Section 5.3]) in consultation with a tobacco researcher to capture tweets belonging to these themes. As part of the preliminary analysis, in this effort, we applied those regexes to the two subsets of tweets and obtained the corresponding numbers of thematic tweets shown in the first four rows of Table 1. Except for e-cig flavors, which are a well known major selling point, the non-marketing datasets contain more tweets in the three other themes (even after accounting for the slight variation in dataset sizes). It is still disconcerting to see the 6363 (1.4 %) marketing tweets discussing smoking cessation when long term consequences of e-cig use are still being investigated. We also looked at how many tweets mention FDA and as expected the majority belong to the non-marketing class.

The last three rows of Table 1 deal with user characteristics of both datasets. We notice that there are 3.5 times as many unique tweeters in the non-marketing set as in the marketing class (row 6). We clarify that some users can belong to both the marketing and non-marketing class if they generate tweets in both datasets. In fact, the top non-marketing tweeter @ecigitesztek has 37,949 such tweets but is also ranked 2nd among tweeters in the marketing group with 27,019 tweets. A cursory examination of this public profile indicates that it belongs to a Hungarian vaping aficionado who almost exclusively tweets about e-cigs and at the time of this writing (re)tweeted over 153,000 times. However, with 11,186 tweeters common to both datasets corresponding to counts from row 6, the Jaccard similarity coefficient is only 0.03. Given marketers tend to use appealing user handles that indicate their purpose, we counted the number of user handles that contain e-cig popular terms such as ecig, vapor, vapour, vape, vaping, eliquid, ejuice, and smoke as substrings of the user handle. 15 out of the top 20 tweeters in both datasets contain one of these terms as a substring. From row 7, we see that more than 7 % of the marketing profiles satisfy this compared with only 1.7 % from the other class.

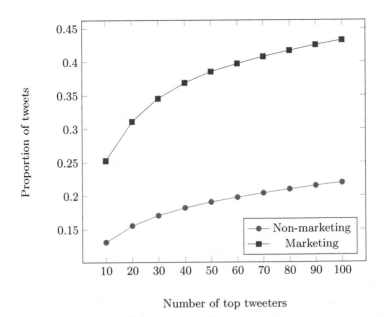

Number of top tweeters

Fig. 1. Proportion of tweets via top 10, . . . , 100 tweeters

The final row indicates the average number of tweets per user with standard deviations in parentheses; the difference in the averages is not surprising but the standard deviation magnitude in the marketing set being more than twice that in the other class is revealing in that few users are responsible for many marketing tweets. To further examine this phenomenon, we plotted the cumulative proportion of tweets in the corresponding datasets contributed by the top $10, \ldots, 100$ tweeters in Fig. 1. It is straightforward to see that the top tweeters in the marketing dataset generate twice the proportion of tweets as generated by the corresponding top users in the non-marketing dataset. Although the Jaccard coefficient between tweeter sets from both datasets in only 0.03, when considering only top 100 tweeters from both datasets, 84 of the top 100 marketing tweeters have generated non-marketing tweets; 88 of the top 100 non-marketing tweeters also authored marketing tweets.

5   Themes in Marketing/Non-marketing Tweets

To dig more into these two subsets of tweets, we applied the Biterm Topic Modeling (BTM) [7] approach, which is specifically designed for short text messages like tweets, to these marketing and non-marketing tweets subsets separately. Given recent results that demonstrate that aggregating short text messages such as tweets can lead to better modeling [17], we partitioned the datasets into groups of ten tweets each where each such group is treated as a short document before applying BTM. Besides using the same tweet pre-processing techniques used for classification, we additionally removed commonly occurring terms from the tweets such as stop words and frequent terms such as the key words used to search for e-cig tweets (e.g., e-cig, vape, vaping, vapor, eliquid) given we already know the tweets are on the general topic of e-cigs.

5.1   Topic Modeling Configuration

Most topic modeling approaches have the inherent requirement that the user suggest the number of topics k to fit to the corpus. It is often tricky to pick a specific k, which is generally chosen by trial and error based on human examination of topics generated with different settings of k. We circumvented this potentially tedious and subjective exercise by using a recently introduced measure of topic coherence by O'Callaghan et al. [32] based on neural word embeddings. Topic coherence is a direct measure of intrinsic quality of a topic. For each topic T generated, let $w_1^T, \ldots, w_N^T$ be the set of top N words according to the $P(w|T)$ distribution resulting from the topic modeling process. Then the coherence of T parameterized by N is

$$C_N^T = \frac{1}{\binom{N}{2}} \sum_{i=2}^{N} \sum_{j=1}^{i-1} \cos(\boldsymbol{w_i^T}, \boldsymbol{w_j^T}),$$

where $\boldsymbol{w}_i^T \in \mathbb{R}^d$ is the dense vectorial representation for the corresponding words learned through the continuous bag-of-words (CBOW) word embedding approach, which is part of the popular word2vec package [29]. We picked dimensionality $d = 300$ and word window size of five for the CBOW configuration in word2vec and ran it on the full corpus of e-cig tweets. Given this definition of average coherence, the idea is to pick $k \in \{10, 20, 30, \ldots, L\}$ that maximizes the weighted average coherence (WAC) across k topics

$$\sum_{i=1}^{k} P(T_i) \cdot \mathcal{C}_N^{T_i}, \tag{1}$$

where $P(T_i)$ is the probability estimate of the prominence of topic T_i in the corpus (from BTM output), N is the number of top few terms chosen per topic (typically 10 or 20, the latter is used in this paper), and L is chosen to be 50. Note that cosine similarity measure we use here scores term pairs that are semantically similar higher than pairs of words related in a different fashion. This does not, however, affect the validity of our topic coherence approach given topics that contain highly similar words are generally more coherent and simpler to interpret than those that contain words that are related in a more associative manner.

5.2   Prominent E-Cig Themes

We recall that topic models output several parameters [3] including a distribution of topics per document (topic proportions: $P(T|d)$) in the corpus and also the distribution of words per topic (per-topic term probabilities: $P(w|T)$), where T is a topic, d is a document, and w is a word. In general, a topic is visualized by displaying the top N (the variable in Eq. 1) words w according to $P(w|T)$. However, a human agent still needs to look at the top N terms of the topic and identify/interpret a semantic *theme*. This is the distinction we use in this effort too – a topic is a group of N words/terms sorted in descending order according to $P(w|T)$ and a theme is a semantic interpretation of what the topic represents based on our manual review. Even though topic modeling research has come a long way, interpretation of resulting topics for exploratory purposes involves significant manual effort, albeit guided by output distributions mentioned earlier. The rest of this paper involves such exploration to grasp the underlying themes.

Based on our experiments, we found that $k = 10$ maximizes the WAC in Eq. 1 for the marketing tweets in the corpus. The corresponding value for non-marketing tweets is $k = 50$. This is not surprising given marketing tweets are expected to contain fairly predictable themes that are favorable to e-cigs in general encouraging tweeters to buy/try them or sign-up for more offers. However, the non-marketing subset is more diverse given it is essentially a catch-all for all other topics about e-cigs. Next, we discuss some topics from both subsets.

Marketing Themes: Upon manual examination of the ten topics from the marketing tweets (MT) corpus, we notice a few that are clear and reflect expected themes from this subcorpus. Here we show three of those topics enumerating some of the top 20 words in the topic. The words are rearranged slightly to better reflect the theme on hand. (However, all words are still from the list of top 20 terms for the topic; otherwise, our analysis would be self-deceiving.)

MT1: free, shipping, code, promo, win, purchase, prizes, enter, giveaway
MT2: vaporizer, pen, mod, kit, battery, portable, starter, electronic, atomizer
MT3: premium, line, lab, certified, AEMSA, cleanliness, consistency, wholesale

The first topic represents the theme of promotional activities involved in marketing e-cigs. The second theme involves vape pens or devices that actually vaporize the liquid nicotine to be inhaled by vapers. The third topic surfaces an unexpected theme of marketing activities that also highlight the quality of the e-liquid products through independent lab certifications offered by the registered nonprofit organization American E-liquid Manufacturing Standards Association (AEMSA), which was established in 2012 for the purpose of promoting safety and standardization in manufacturing liquid nicotine products.

Non-marketing Themes: The following is the list of major topics in the non-marketing tweets (NT) corpus.

NT1: lungs, cells, flavors, toxic, effects, exposure, study, damage, aerosols
NT2: FDA, poisonings, calls, surge, skyrocket, nicotine, poison, children
NT3: explodes, coma, teen, mouth, burns, injured, suffers, neck, hole, hospital
NT4: FDA, tobacco, industry, market, regulation, product, ban, deeming, rule
NT5: tobacco, laws, CASAA, smoke, healthier, alternative, FDA, grandfather
NT6: quit, smoking, help, current, smokers, cigarette, users, NHS, review
NT7: teen, smoking, CDC, study, middle, school, students, tripled, fell
NT8: ban, Wales, government, public, enclosed, spaces, pushes, ahead
NT9: gateway, drug, doing, cocaine, bathroom, lines, puffin, Wendy, heroin

Note that we only report nine topics here because we found these to be most interesting and also given several others seemed very similar to these nine. There are also a few that do not seem to indicate a specific non-trivial theme and hence were excluded. The first theme NT1 is about toxic effects of e-cigs. An examination of biomedical articles with the search terms `e-cigarettes` AND `toxic` AND `lungs` returned several articles discussing experiments that demonstrated how flavoring agents of e-cigs, and not the liquid nicotine itself, are responsible for toxic effects of inhaling e-cig vapors. NT2's theme relates to a news piece that diffused through Twitter about FDA receiving many calls involving poisoning complaints by e-cig users. NT3 and several related topics (not displayed here) discussed explosions of the vaping devices while in use resulting in burns and hospitalizations [36]. NT4 represents a general theme involving FDA regulatory activities and the new deeming rule [13], which was thought to be impending throughout the past few years.

In NT5, we see a very specific theme that involves the non-profit organization Consumer Advocates for Smokefree Alternatives Association (CASAA) and the

general harm-reduction perspective of e-cigs as an healthier alternative to cigarettes for people who want to quit smoking. The last term 'grandfather' in NT5
refers to new regulations extending to any product introduced/modified on or
after the so called grandfather date set to 2/15/2007 by the FDA [13]. This date
is critical to many e-cig businesses as all those products (already in market) will
now be subject to the new FDA regulations and hence need to be approved by
it. NT6 represents the theme of using e-cigs as an aid to smoking cessation. The
term NHS refers to UK's National Health Service, which has taken a favorable
stance to e-cig use for treating addicted smokers [28]. NT7 is about research
reports by the CDC indicating tripling of current e-cig use by middle and high
school students from 2013 to 2014 [4]. NT8 highlights another news piece on
Wales (of UK) government passing a law to ban vaping in enclosed spaces.

Fig. 2. Tweet leading to topic NT9 on e-cigs as a gateway drug

The final topic NT9 is unusual and seems to indicate e-cigs as a gateway
drug to use other more harmful products such as cigarettes, cocaine, and heroin.
Although there is some evidence [2] to support this idea, this particular topic
appeared atypical with words like bathroom, lines, and Wendy. A deeper examination revealed that most of the words in this topic are mostly coming from
one tweet shown in Fig. 2. As can be seen, this tweet was retweeted more than
1000 times. Given retweets are essentially a reasonable and natural mechanism
to add more weight to a particular topic, we decided to not to delete them in our
analysis. However, this particular topic led us to dig deeper into manifestations
of topics of this nature. There were two other non-marketing topics like this
based on frequent retweets or many tweets involving some minor modifications
of a very specific tweet: one involved a picture of film actor Ben Affleck vaping
after getting a traffic rule violation ticket (the topic had words Ben, Affleck,
and ticket) and another involved the URL of an online petition offering support
to the then UK prime minister David Cameron and other politicians trying to
block certain e-cig regulations in the UK.

Effect of Excluding Retweets: Given this observation involving NT9, we
wanted to study the effect of retweets on topic modeling. We found that 36 %

of marketing tweets and 43 % of non-marketing tweets were due to retweets. Thus we see that retweets constitute a significant proportion of the full datasets. We generated new topic models with these subsets excluding all retweets to see if there is a noticeable difference in the themes. Although the themes did not change significantly, the words used to represent the topics have changed slightly in most cases. For example, the theme in this new set of topics corresponding to NT9 had the following top words: gateway, drug, smoking, heroin, cocaine. None of the specific words (bathroom, doing, puffin, lines, Wendy) from the highly retweeted message in Fig. 2 showed up in the new topic. There were no other topics indicating a gateway theme. There was no topic involving Ben Affleck's traffic ticket but the petition related topic involving former UK prime minister David Cameron was apparent with slightly different words. All other themes NT1–NT8 were evident in the new set of topics. There was only one new theme that wasn't already in the topic set from the full dataset. This was mostly about vaporizer/e-liquid brand names with top terms including: sigelei, hexohm, flawless, ipvmini, districtf, tugboatrda, appletop, longislandbrewed. There was no major change in the themes for the marketing tweet subset.

Finally, we wanted to see who is tweeting on various themes identified through our approach. To this end, we picked two different non-marketing themes, NT6 (e-cigs for smoking cessation) and NT7 (CDC reporting on increasing teen vaping). For each of the corresponding topics T, we ranked all tweets s according to $P(T|s)$. Based on the authorship of the top 10,000 tweets according this ranking, we sorted tweeters in descending order based on the counts of top 10,000 tweets they authored. We manually examined the top few ranked tweeters in this list. For theme NT7, 11 out of 20 top tweeters are regular people tweeting about e-cigs but only 2 out of 20 top tweeters for NT6 are regular tweeters; the other tweeters being institutions or companies that have a clear positive stance for and commercial interest in e-cigs. This indicates that regular tweeters (even if they are in favor of e-cigs) are more inclined to tweet about news involving e-cigs, even when it is not favorable. Commercial tweeters tend to exclusively focus on propagating favorable news pieces besides promoting their products.

Overall, our effort offers a complementary approach by surfacing specific themes in comparison to manual coding [9, 19] where only broad topics such as smoking cessation, flavors, and safety are typically used. This is our main contribution – demonstrating the feasibility of topic modeling based thematic analysis of e-cig chatter on Twitter. Some of our extracted themes may already be common knowledge for tobacco researchers who regularly follow e-cig related news. But we believe the topic modeling approach can help surface a more comprehensive set of themes with less manual exploration burden. It also gives a better sense of the strength of a theme (as observed by the the corresponding topic's ranking) and main tweeters authoring the corresponding thematic tweets.

5.3   Themes in Geotagged Tweets

Geotagged tweets with the associated latitude and longitude information offer a different lens to understand e-cig messages. There have been very few studies examining the locations where e-cigs are used. There is only particular study [20] that we are aware of where prepositional phrase patterns were used over tweet text to identify e-cig use in a class, school, room/bed/house, or bathroom. In our effort, we are not necessarily concerned about e-cig use, but are generally interested in knowing themes from tweets generated near different types of places of interest. Our dataset has a total of 3208 geotagged tweets which is less than 1 % of the full dataset. Using the GeoNames API (http://www.geonames.org/export/web-services.html), we identified the nearest *toponym* for each of the corresponding geocodes using the `findNearby` method. In our dataset, the average distance between the geo-code and nearest toponym was 300 meters. Toponyms can be names of larger geographical areas such as cities or rivers, but can also refer to small locations such as a school, hospital, or a park. Each toponym (e.g., University of Kentucky) is associated with a corresponding feature code (e.g., UNIV).

We aggregated tweets based on feature codes (http://www.geonames.org/export/codes.html) of the toponyms returned and obtained the following distribution (top ten codes) where counts are shown in parentheses:

hotel (596), populated-place (411), church (314), school (311), building (286), mall (158), park (109), lake (91), library (80), and post office (74).

In addition to these we also considered, travel end-points (81) as a single class (airports, bus stations, and railway stations), restaurants (39), hospitals (45), museums (13), and universities (11). A simple string search revealed that in very few cases the geotagged tweet content actually made explicit connection to the corresponding feature code. We were able to find 2–3 tweets at hotels, schools, and airports indicating the location type as part of the tweet (e.g., "vaping in class" and "flight is full"). Except for schools, parks, restaurants, hospitals, and airports, all locations had more marketing tweets than regular tweets. Overall, 52 % of geotagged tweets belonged to the marketing class, a 7.5 % increase compared with the corresponding proportion in the full dataset as discussed in the beginning of Sect. 4.

For each of these different location types, we identified top topics by fitting topic models to the corresponding sets of tweets. Given marketing tweets have a clear agenda, we only look at non-marketing top topics. For clarity, we simply outline the theme without listing all the keywords

- Church: Ban on e-cigs for minors in Texas
- Hotel: E-cig use rising among young people
- Park: Pros and cons of E-cig regulations
- School: Smoking rates fall as e-cig use increases among teens

Other locations either did not have a significant number of tweets or had tweets without any dominant theme. We realize that our analysis in this section may not be precise in the sense that tweets originating from different types of places may not be from people who are visiting those places for relevant purposes; tweeters might simply be around those places when they tweet. However, we believe with a large exhaustive dataset spanning multiple years, given we only look at top themes, we can arrive at themes that are representative of people visiting those places.

6   Conclusion

E-cigs continue to survive as a controversial tobacco product and are currently subject to new FDA regulations since 8/8/2016 with a grandfathering date set to 2/15/2007. The FDA, biomedical researchers, physicians, tobacco industry, and most important the nation's public are all key players whose activities will be affected with these products for the foreseeable future. Public health and tobacco researchers are split in their opinions regarding e-cig use by smokers who would otherwise continue with regular cigarettes. Computational social science and informatics approaches can offer a more objective lens through which the social media landscape of e-cigs can be gleaned for online surveillance of both product marketing practices and adverse events.

Although prior efforts exist in content analysis based on pre-determined broad themes, we do not see results on automatic extraction of themes from social media posts on e-cigs. We believe computational approaches provide an important avenue that can complement traditional survey based research efforts considering the cost and time factors involved in the latter case. Twitter in particular has been well studied in the context of public health informatics efforts and provides a major platform for e-cig chatter on the Web.

In this paper, we conduct thematic analysis experiments involving over a million e-cig tweets collected during a 15 month period (4/2015 – 6/2016). To deal with the major presence of marketing chatter, we first built a classifier that achieved an accuracy of over 88 % in identifying marketing and non-marketing tweets based on a manually labeled dataset. We conducted preliminary content and user analysis of marketing and non-marketing tweets as classified by our model. Subsequently, we fit topic models to the two subsets of tweets and interpreted them to identify specific themes that were not apparent in manual efforts. This is not surprising given the fast changing discourse on e-cigs creates a corresponding rapidly evolving social media landscape. This, however, points to an important weakness of our approach – it is not *online*, where new e-cig tweets continuously collected through the Twitter streaming API are used to generate new topics as enough evidence accumulates. As part of future work, we plan to employ online topic models [16] and facilitate their exploration using well known topic browsing approaches [5,26]. Nevertheless, here we provide what we believe is a first strong proof of concept for employing topic models to comprehend evolving e-cig themes on Twitter. Given gender, age group, race and ethnicity can be predicted with reasonable accuracy [11,25,31], an important future research direction is to use these methods to classify e-cig tweeters into these demographic categories and identify e-cig themes in tweets authored by specific subpopulations. For example, given african american teenagers are an active group on Twitter [34], identifying popular e-cig themes authored by them (including retweets and favorites) may yield insights specific to that demographic segment. Similar analysis can also be conducted with tweets originating from rural areas given the typical firehose is dominated by urban tweeters.

Acknowledgements. We thank anonymous reviewers for constructive criticism that helped improve the presentation of this paper. This research was supported by the National Center for Research Resources and the National Center for Advancing Translational Sciences, US National Institutes of Health (NIH), through Grant UL1TR000117 and the Kentucky Lung Cancer Research Program through Grant PO2-415-1400004000-1. The content of this paper is solely the responsibility of the authors and does not necessarily represent the official views of the NIH.

References

1. Agarwal, A., Xie, B., Vovsha, I., Rambow, O., Passonneau, R.: Sentiment analysis of twitter data. In: Proceedings of the Workshop on Languages in Social Media, pp. 30–38. Association for Computational Linguistics (2011)
2. Barrington-Trimis, J.L., Urman, R., Berhane, K., Unger, J.B., Cruz, T.B., Pentz, M.A., Samet, J.M., Leventhal, A.M., McConnell, R.: E-cigarettes and future cigarette use. Pediatrics **138**, e20160379 (2016)
3. Blei, D.M., Lafferty, J.D.: Topic models. In: Srivastava, A., Sahami, M. (eds.) Text Mining:Classification, Clustering, and Applications, chapter 4, pp. 71–93. CRC Press, Chapman and Hall (2009)
4. Centers for Disease Control. E-cigarette use triples among middle and high school students in just one year. http://www.cdc.gov/media/releases/2015/p0416-e-cigarette-use.html
5. Chaney, A.J.-B., Blei, D.M.: Visualizing topic models. In: International Conference of Weblogs and Social Media, ICWSM 2012 (2012)
6. Chen, I.-L., et al.: FDA summary of adverse events on electronic cigarettes. Nicotine Tob. Res. **15**(2), 615–616 (2013)
7. Cheng, X., Yan, X., Lan, Y., Guo, J.: BTM: Topic modeling over short texts. Knowl. Data Eng. IEEE Trans. **26**(12), 2928–2941 (2014)
8. Chu, K.-H., Unger, J.B., Allem, J.-P., Pattarroyo, M., Soto, D., Cruz, T.B., Yang, H., Jiang, L., Yang, C.C.: Diffusion of messages from an electronic cigarette brand to potential users through twitter. PloS One **10**(12), e0145387 (2015)
9. Cole-Lewis, H., Pugatch, J., Sanders, A., Varghese, A., Posada, S., Yun, C., Schwarz, M., Augustson, E.: Social listening: A content analysis of e-cigarette discussions on twitter. J. Medi. Int. Res. **17**(10), e243 (2015)
10. Cole-Lewis, H., Varghese, A., Sanders, A., Schwarz, M., Pugatch, J., Augustson, E.: Assessing electronic cigarette-related tweets for sentiment and content using supervised machine learning. J. Med. Int. Res. **17**(8), e208 (2015)
11. Culotta, A., Kumar, N.R., Cutler, J.: Predicting the demographics of twitter users from website traffic data. In: Twenty-Ninth AAAI Conference on Artificial Intelligence, pp. 72–78 (2015)
12. Etter, J.-F., Bullen, C., Flouris, A.D., Laugesen, M., Eissenberg, T.: Electronic nicotine delivery systems: a research agenda. Tob. Control **20**(3), 243–248 (2011)

13. Food and Drug Administration, HHS et al.: Deeming tobacco products to be subject to the federal food, drug, and cosmetic act, as amended by the family smoking prevention and tobacco control act; restrictions on the sale and distribution of tobacco products and required warning statements for tobacco products. final rule. Federal Reg. **81**(90), 28973 (2016)

14. Godea, A.K., Caragea, C., Bulgarov, F.A., Ramisetty-Mikler, S.: An analysis of twitter data on e-cigarette sentiments and promotion. In: Holmes, J.H., Bellazzi, R., Sacchi, L., Peek, N. (eds.) AIME 2015. LNCS (LNAI), vol. 9105, pp. 205–215. Springer, Heidelberg (2015). doi:10.1007/978-3-319-19551-3_27

15. Han, S., Kavuluru, R.: On assessing the sentiment of *general* tweets. In: Barbosa, D., Milios, E. (eds.) CANADIAN AI 2015. LNCS (LNAI), vol. 9091, pp. 181–195. Springer, Heidelberg (2015). doi:10.1007/978-3-319-18356-5_16

16. Hoffman, M., Bach, F.R., Blei, D.M.: Online learning for latent Dirichlet allocation. Adv. Neural Inf. Proc. Syst. **21**, 856–864 (2010)

17. Hong, L., Davison, B.D.: Empirical study of topic modeling in twitter. In: Proceedings of the 1st Workshop on Social Media Analytics, pp. 80–88. ACM (2010)

18. Huang, J., Kornfield, R., Szczypka, G., Emery, S.L.: A cross-sectional examination of marketing of electronic cigarettes on twitter. Tob. Control **23**, 26–30 (2014). (suppl 3)

19. Kavuluru, R., Sabbir, A.: Toward automated e-cigarette surveillance: Spotting e-cigarette proponents on Twitter. J. Biomed. Inf. **61**, 19–26 (2016)

20. Kim, A.E., Hopper, T., Simpson, S., Nonnemaker, J., Lieberman, A.J., Hansen, H., Guillory, J., Porter, L.: Using twitter data to gain insights into e-cigarette marketing and locations of use: An infoveillance study. J. Med. Int. Res. **17**(11), e251 (2015)

21. Kim, Y.: Convolutional neural networks for sentence classification. In: Proceedings of the 2014 Conference on Empirical Methods in Natural Language Processing (EMNLP), pp. 1746–1751, October 2014

22. Klein, E.G., Berman, M., Hemmerich, N., Carlson, C., Htut, S., Slater, M.: Online e-cigarette marketing claims: A systematic content and legal analysis. Tob. Regul. Sci. **2**(3), 252–262 (2016)

23. Landis, J., Koch, G.: The measurement of observer agreement for categorical data. Biometrics **33**(1), 159–174 (1977)

24. Levy, D.T., Cummings, K.M., Villanti, A.C., Niaura, R., Abrams, D.B., Fong, G.T., Borland, R.: A framework for evaluating the public health impact of e-cigarettes and other vaporized nicotine products. Addiction (2016)

25. Liu, W., Ruths, D.: What's in a name? using first names as features for gender inferencein twitter. In: Proceedings of the AAAI Spring Symposium: Analyzing Microtext, pp. 10–16 (2013)

26. Malik, S., Smith, A., Hawes, T., Papadatos, P., Li, J., Dunne, C., Shneiderman, B.: Topicflow: visualizing topic alignment of twitter data over time. In: Proceedings of the 2013 IEEE/ACM International Conference Onadvances in Social Networks Analysis and Mining, pp. 720–726. ACM (2013)

27. Martin, E., Clapp, P.W., Rebuli, M.E., Pawlak, E.A., Glista-Baker, E.E., Benowitz, N.L., Fry, R.C., Jaspers, I.: E-cigarette use results in suppression of immune and inflammatory-response genes in nasal epithelial cells similar to cigarette smoke. Am. J. Physiol. Lung Cell. Mol. Physiol. **311**, L135–L144 (2016)

28. McNeill, A., Brose, L., Calder, R., Hitchman, S., Hajek, P., McRobbie, H.: E-cigarettes: an evidence update. Report from Public Health England (2015)

29. Mikolov, T., Sutskever, I., Chen, K., Corrado, G.S., Dean, J.: Distributed representations of words and phrases and their compositionality. Adv. Neural Inf. Process. Syst. **21**, 3111–3119 (2013)
30. Myslín, M., Zhu, S.-H., Chapman, W., Conway, M.: Using twitter to examine smoking behavior and perceptions of emerging tobacco products. J. Med. Int. Res. **15**(8), e174 (2013)
31. Nguyen, D., Gravel, R., Trieschnigg, D., Meder, T.: how old do you think i am? a study of language and age in twitter. In: Proceedings of the Seventh International AAAI Conference on Weblogs and Social Media (ICWSM), pp. 439–448 (2013)
32. OCallaghan, D., Greene, D., Carthy, J., Cunningham, P.: An analysis of the coherence of descriptors in topic modeling. Expert Syst. Appl. **42**(13), 5645–5657 (2015)
33. Pedregosa, F., Varoquaux, G., Gramfort, A., Michel, V., Thirion, B., Grisel, O., Blondel, M., Prettenhofer, P., Weiss, R., Dubourg, V., Vanderplas, J., Passos, A., Cournapeau, D., Brucher, M., Perrot, M., Duchesnay, E.: Scikit-learn: Machine learning in Python. J. Mach. Learn. Res. **12**, 2825–2830 (2011)
34. Pew Research Internet Project. Part 1: Teens and social media use. http://www.pewinternet.org/2013/05/21/part-1-teens-and-social-media-use/
35. Rios, A., Kavuluru, R.: Convolutional neural networks for biomedical text classification:application in indexing biomedical articles. In: Proceedings of the 6th ACM Conference on Bioinformatics, Computational Biology and Health Informatics, pp. 258–267. ACM (2015)
36. Rudy, S., Durmowicz, E.: Electronic nicotine delivery systems: overheating, fires andexplosions. Tob. Control (2016) (in press)
37. Singh, T., Arrazola, R., Corey, C., Husten, C., Neff, L., Homa, D., King, B.: Tobacco use among middle and high school students - United States, 2011–2015. MMWR Morb. Mortal. Wkly. Rep. **65**(14), 361–367 (2016)
38. Wilson, E.B.: Probable inference, the law of succession, and statistical inference. J. Am. Statist. Assoc. **22**(158), 209–212 (1927)

Obtaining Rephrased Microtask Questions from Crowds

Ryota Hayashi[1], Nobuyuki Shimizu[2], and Atsuyuki Morishima[1(✉)]

[1] University of Tsukuba, Tsukuba, Ibaraki, Japan
`morishima-office@ml.cc.tsukuba.ac.jp`
[2] Yahoo Japan Corporation, Tokyo, Japan

Abstract. We present a novel method for obtaining and ranking rephrased questions from crowds, to be used as a part of instructions in microtask-based crowdsourcing. Using our method, we are able to obtain questions that differ in expression yet have the same semantics with respect to the crowdsourcing task. This is done by generating tasks that give a hint and elicit instructions from workers. We conduct experiments with data used for a real set of gold standard questions submitted to a commercial crowdsourcing platform and compared the results with those of a direct-rewrite method. The results show that extracted questions are semantically ranked at high precision and we identify cases where each method is effective.

1 Introduction

Microtask-based crowdsourcing is a promising approach to solve problems in many applications. A typical procedure for using a *crowdsourcing platform* is as follows: First, *requesters* design *microtasks* that have *questions* to be asked to *workers*. Then, they put the microtasks into the crowdsourcing platform, such as Amazon Mechanical Turk and Yahoo! Crowdsourcing. Finally, workers perform the microtasks and the platform receives the results.

One of the essential problems in microtask-based crowdsourcing is how to design questions for microtasks. This paper proposes a novel method to obtain a variety of questions that are different but have the same semantics. Here, the semantics of a question is defined by the *gold standard data*, which is a set of expected answers to the question with associated data items. For example, the semantics of "Is this green tea?" is defined by the gold standard data that states that the answer is yes if and only if the photo associated with the question is green tea. "Is this green tea?" and "Is this Japanese tea?" are expected to have the same semantics.

This paper discusses how to effectively use the crowd power to obtain a variety of rephrased questions. In the discussion, we devise a method to use the gold standard data to generate a variety of questions that have the intended semantics. The idea behind our proposed method is that we use the gold standard data to generate *question-eliciting tasks (QE-Tasks)* to elicit questions from workers

© Springer International Publishing AG 2016
E. Spiro and Y.-Y. Ahn (Eds.): SocInfo 2016, Part II, LNCS 10047, pp. 323–336, 2016.
DOI: 10.1007/978-3-319-47874-6_23

by showing them a hint and asking them to guess the questions. Then, we rank the elicited questions so that top-ranked questions have the intended semantics. Theoretically, this type of human computation is known as *input agreement* [5], where the workers are shown hints to suggest the original data item and asked to guess what it is. We compare the method with a direct-rewrite method in which workers directly rewrite the original question and identified the cases where the methods are effective.

Obtaining different questions with the same semantics has a wide range of interesting applications, especially in microtask-based crowdsourcing. First, when planning crowdsourcing tasks, quality control measures must be considered. In a conventional survey or questionnaire, several questions are often related to each other in a way that an answer to one question implies an answer to another [6]. For example, questionnaires may include a mix of positive and negatively worded statements, asking for the same thing in different ways. Such a construction is intended to reduce acquiescence bias; a type of response bias in which users tend to agree with a statement when in doubt, thereby possibly answering "yes" to all the questions [10]. In designing crowdsourcing tasks too, inserting differently worded questions expressing the same concept would help requesters to remove workers who do not produce thoughtful or adequate responses, forming a basis for ensuring quality of the results.

Second, the proposed method is effective in the case where the *gold standard question* is not necessarily correct and clear for workers. Gold standard questions are questions whose answers (gold standard data) are known in advance and are matched with the task results in order to choose desirable workers to perform other tasks. An internal survey of a commercial crowdsourcing service shows that it is not rare for gold standard questions to not work well, often in the initial attempts, because of their inappropriate wordings. For example, workers often (mis)interpret the question "Is [Olive Garden] a restaurant?" as "Does [Olive Garden] sound like a restaurant name?" However, the requester's intention might be "Does the restaurant named [Olive Garden] really exist?" A set of rephrased questions defined by the gold standard data will give hints to requesters in revising the original questions. For example, Fig. 1 shows some of the results of our proposed method compared to the original questions. The point is that workers for the proposed method give question sentences solely based on the gold standard data, while the original question is not necessarily thought by the crowd to have the intended semantics.

The contributions of this paper are as follows.

(1) Methods for exploiting inference capabilities of the crowd. Generating QE-Tasks is not only a novel approach to obtain appropriate questions, but also interesting in that it is required to appropriately exploit inference capabilities of the crowd. An extreme approach is to provide one big task that shows a large number of data items and asks people to come up with a question compatible with the data. However, this is not applicable to the microtask-based crowdsourcing context where each task is expected to be small enough. Therefore, our focus is how to combine the results of smaller tasks to obtain good

Original	Rephrased questions	Why the original question is not good
Is this a seasoning?	Is this something added to foods?	Positive items include Jam and similar things which are not usually called seasonings
Is this an instant food?	Can you begin to eat it in ten minutes after you start cooking?	Some positive items require other than microwave ovens and hot water for cooking. Some workers believe that instant foods require them for cooking.

Fig. 1. Examples of rephrased questions that give hints on revising the original ones

questions. Obviously, another extreme approach does not work: as shown in our experiment, QE tasks that show only one positive example chosen from the gold data did not result in semantically equivalent questions. A moderate solution is to show several positive and negative examples to generate QE-Tasks. We found that we can obtain surprisingly good results using a small number of examples.

(2) Experimental results with real data. We conducted experiments with two real data sets. One contains restaurant data and was used in a real set of gold standard questions for a commercial crowdsourcing service. The other contains photos of goods sold by a large e-commerce company in Japan. In our experiments, the proposed method performed well with the data sets. In addition, we compared the proposed method with the direct-rewrite method, which asks workers just to rephrase the shown question sentence. Our results clarify that each of them has its own strength and we identify cases where each is effective.

We used Yahoo! Crowdsourcing, a major commercial crowdsourcing service in Japan. It has 240 thousands registered workers as of Feb. 2015, and we estimate a half of microtasks contain some type of simple questions such as Yes/No questions and multiple-choice questions. Note that although we used a crowdsourcing platform whose workers are Japanese, the problem discussed in the paper is universal.

2   Related Work

To our knowledge, there is no prior work on rephrasing questions in microtask crowdsourcing. In the previous literature on natural language processing and information retrieval, there are data-driven approaches that automatically induce rephrased questions and concept hierarchy collections. As our work aims to rephrase questions, the most relevant prior work is found in the domains of information retrieval and Q&A repositories. In these domains, an important task is to identify semantically similar queries and questions. For instance, Wen et al. [11] rely on user click through logs, based on the idea that queries that result in identical clicks are bound to be similar. Zhao et al. [12] uses user click logs from the Encarta web site for rephrasing questions. Instead of click logs, Jeon et al. [4]

identify questions that have similar answers in the Naver Question and Answer database. Lytinen and Tomuro [7] suggest determining a question's category. In order to match questions in their method, questions must belong to the same category, providing yet another contextual information we may use for rephrasing question. The extracted query are often used for suggesting related queries to users [9]. Other comprehensive surveys on paraphrase and near-paraphrase extraction can be found in [1] and [8].

Compared to prior studies in natural language processing and information retrieval, our work is unique in that we have data items, that is, task-specific contexts to be crowdsourced along with the question-style instruction. We will show this contextual information to the crowd to obtain a more variety of results. Our result is compatible with the result by Chen et al. [2], which shows that people tend to give a more variety of tags when they are given by contextual information. Note, however, that our work is unique in that rephrasing microtask questions requires input-agreement style human computation.

In our approach, the original question is used by the requester for a comparison with the new ones *after* the proposed method outputs them. An alternative approach is to show the original question to the crowd and ask them improve it. In a different problem setting, [3] takes this approach to improve workflows. The assumption behind this approach is that the original one is a good approximation enough to reach the ideal result. However, this assumption is not guaranteed to hold in general. In addition, as shown in the experimental results, our approach has an advantage in that it can output a variety of results that have different expressions but semantically equivalent. Therefore, the two approach are complementary to each other.

3 Formalization and the Overview

Microtasks. In this paper, a microtask $t_{q,i}$ is represented as a pair (q, d_i), where q is the question common to a particular set of tasks, and $d_i \in D_q$ (the data set for q) is a data item unique to each $t_{q,i}$. For example, a task ('Is this picture inappropriate?', d_i) has a question sentence "Is this .." and a picture data item d_i unique to the task. We omit q and use t_i if q is obvious from the context.

Gold standard data. Some of the tasks, named *gold standard questions*, are associated with *gold standard data*, which are answers that are given in advance by requesters. The gold standard questions are used to filter out inappropriate workers such as spams and to find workers that are qualified for the tasks. In this paper, we model gold standard data as a function $C_q : D_q \to \{yes, no\}$. We can easily extend our proposed method to deal with more general cases with N choices. The extension is easy and omitted.

Overview of the method. The proposed method takes as inputs a set D_q of data items and gold standard data C_q and outputs a ranked list of rephrased questions $q'_1 \ldots q'_n$. The method consists of two phases. Phase 1 elicits questions by showing the gold standard data to the crowds. Phase 2 ranks the elicited questions and presents them to the task requester.

4   Phase 1: Query-Eliciting Tasks

A QE-task (Fig. 2) is a microtask that shows workers several data items in D_q and their answers defined by C_q, and asks them to guess a question that, together with each d_i of the shown data items provided, produces the same answer as $C_q(d_i)$. A possible result elicited from a worker with the example task could be "Is the object in the picture is man-made?" The task is denoted by $(q_{\text{query elicit}}, d_{\text{examples}})$, where each component is as follows:

- $q_{\text{query elicit}}$ = "Upon being asked a certain question, some person answered *yes* to the left group of data items and *no* to the right group of data items. Please guess a question and submit it in the form below."
- d_{examples} contains data items in D_q grouped by $C_q(d_i)$ (i.e., yes and no).

In general, given C_q, and D_q, we generate a set of QE-Tasks with Y positive and N negative examples taken from the gold standard data. We use $QE\text{-}Tasks_q(Y, N)$ to denote the set of QE-Tasks. To construct $QE\text{-}Tasks_q(Y, N)$, we need to choose the examples in some way (e.g., random sampling). As shown later, our experiments used QE-Tasks with a very small number of samples (such as one positive sample and one negative sample). We may omit (Y, N) where the number of shown examples does not matter.

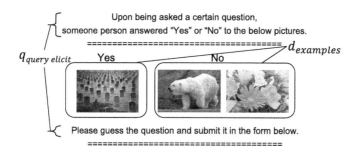

Fig. 2. QE-Task

Phase 1 crowdsources every question-eliciting task t_i in $QE\text{-}Tasks_q(Y, N)$. We ask more than one worker to perform each t_i.

5   Phase 2: Ranking Elicited Questions

Phase 2 ranks the elicited questions. Ideally, we would like to rank at the top the elicited question to which the answers coincide with the given gold standard data for every data item. If the way in which an elicited question answers every data items matches that of another one, these two questions are effectively same in terms of semantics.

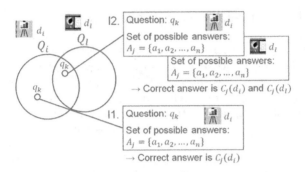

Fig. 3. Intuition I1 and I2

However, if elicited questions are elicited on one or a handful of data items, each of them tends to be more specific to the data items than the original gold standard question. In order to obtain the semantic similarity, we will try to find elicited questions that cover many gold standard data. To do so, we will treat an elicited question as a bag of words and merge all elicited question by taking the intersection of bags of words. As shown later, our experimental results show that the scheme works well in a real setting.

5.1   Intuition for Discovering the Central Concept that Encompasses Many Gold Standard Data

Let us assume that every worker we asked to perform t_i returns the set of *all* questions that can be obtained by t_i. Then, we use Q_i to denote the result set of questions.

Given $t_i, t_l \in QE\text{-}Tasks_q$, we have the following intuitions on $q_k \in Q_i \cup Q_l$.

I1. If question $q_k \in Q_i$ is elicited using data item d_i, then q_k supplies d_i with the correct answer $C_q(d_i)$.

I2. If question q_k can be elicited using data item d_i and d_l, then q_k supply correct answers $C_q(d_i)$ and $C_q(d_l)$ respectively to the data items d_i and d_l.

Note that in I2, a question with the intended semantics can take place of q_k, as it is supposed to give correct answers to all data items. That is, if one question can be elicited from every data items, it has the semantics defined by the gold standard data. We call a set of all such questions the *central concept* for the gold standard data. This observation gives us some cues on choosing elicited questions that have the intended semantics (Fig. 3).

5.2   Bag-of-Words Scheme for Ranking

In the previous section, our assumption was that every worker returns the set of all questions that can be obtained by each t_i. However, enumerating all possible questions for the given data and answers is not realistic, and it is often the case that the intersection is empty. In addition, since we cannot expect that we obtain all expressions that are apparently different but have the same meaning, we have to determine whether two questions in different expressions have the same semantics. However, this is not a trivial matter. To cope with this issue, we adopt a bag-of-words scheme to deal with elicited questions.

Let $w(q)$ be a bag of words contained in question q. Then, we compute set operations for question sets (denoted by B_i and B_j) in the following way.

- $B_i \equiv \bigcup_{q_k \in Q_i} w(q_k)$
- $B_i \cap B_j$ and $B_i \cup B_j$ are ordinary intersection and union operations.

To return a ranked list of elicited questions, we use the intuition obtained from I2. Since the set of questions that is elicited from all data items is usually empty due to the wide variations of natural language expressions, we take the intersection of bags of words to obtain rough estimate of the central concept.

Given $QE\text{-}Tasks_q(where|QE\text{-}Tasks_q| = m)$, let $B = B_1 \cap B_2 \cap \cdots \cap B_m$. This B is the estimate of the central concept. In order to rank the elicited questions according to I2, we use similarity of each question to this estimate B. The results are sorted in the descending order on $sim(w(q_k), B)$.

6   Experiments

We conducted experiments to evaluate our proposed scheme in terms of the precision of the top ranked questions. The experimental results suggest that the proposed scheme is promising and show that the quality of outputs is good.

We also compared the results with those of the direct-rewrite method, which shows workers the original gold standard questions and asks them to directly rephrase the shown questions. While the proposed method output a more diverse set of rephrased questions, the direct-rewrite method output rephrased questions that faithfully preserve the semantics of the original one. Interestingly, the proposed method often output more appropriate questions than the original ones.

In the experiment, some parameters are fixed to the constants that we thought would be appropriate. Evaluating effects of varying such parameters is out of the scope of the paper and will be discussed in the forthcoming papers.

6.1   Settings

Data sets. We used two sets of data.

(1) Restaurant data (text): One set of data items (written in text), denoted by D, contains words and phrases categorized into restaurant names, food, and drink. The data set was taken from the data used for gold standard questions in Yahoo! crowdsourcing.

In this dataset, we are given five original gold standard questions, denoted by $r_j(1 \leq j \leq 5)$ (Fig. 4). Each r_j asks workers if the shown data item is in a specified category. For example, r_3 shows terms such as "Sushi, Steak" and

ID	Original question
r_1	Is the following word or phrase related to styles of cooking?
r_2	Is the following word or phrase a type of liquor or drink?
r_3	Is the following word or phrase a type of food?
r_4	Is the following word or phrase related to a menu item or a food na me of a restaurant?
r_5	Is the following word or phrase a name of a restaurant?

Fig. 4. Original questions for the restaurant data

Name	#Data items	#Positive data items in the gold standard data
Restaurant Data for r_j	$|D| = 25$	5 (20%) for every C_j
Goods Data for g_j	$|D'| = 34$	8.5% to 28.5% depending on C'_j

Fig. 5. Statistics on experimental data. Every r_j (or g_j) shares the same set D (or D') of data items, but each question has a unique set of gold standard data C_j (or C'_j). The ratios of positive answers in the gold standard data for r_js are computed by $|\{d_i|C_j(d_i) = \text{"Yes"}, d_i \in D_j\}| = 5$ (i.e., 20 % of 25) for every $r_j(1 \leq j \leq 5)$. Those for g_js were computed in the same way.

asks workers whether it is classified as a cuisine. The statistics on the data set is given in Fig. 5.

(2) Goods data (pictures): The other set of data items (pictures), denoted by D', contains photos of goods sold by Askul, a large e-commerce company in Japan. There are 34 data items grouped into seven categories[1]. We used the seven categories to define seven gold standard questions g_j $(1 \leq j \leq 7)$ each of which has the intended semantics. An example task shows a picture of a bottle of tea and asks "Is this green tea?". Figure 5 also shows the statistics of the goods data.

In the following, we use r (g) instead of r_j (g_j) to simplify notation when there is no confusion.

QE-Tasks. For each original question, we constructed variations of QE-Tasks with different numbers of shown examples. As explained, let $QE\text{-}Tasks_q(Y, N)$ be a set of QE tasks that shows Y positive examples and N negative examples

---

[1] The items are actually associated with an hierarchical directory. We selected subcategories of "Food and Drink." Each chosen category satisfies the following conditions: (1) it is not a top-level category, (2) it has at least three data items, and (3) the category name is not a composition of two ore more different category names (such as "Beer and Wine").

Task set	#QE-tasks
$QE\text{-}Tasks_r(1,1)$	$_5C_1 \times _{20}C_1 = 100$ for each r
$QE\text{-}Tasks_r(2,0)$	$_5C_2 \times _{20}C_0 = 10$ for each r
$QE\text{-}Tasks_r(3,3)$	$_5C_3 \times 1 = 10$ for each r
$QE\text{-}Tasks_g(2,0)$	2,040 in total

Fig. 6. QE-tasks used in the experiment. The number of positive data items for the restaulant data is five for every r.

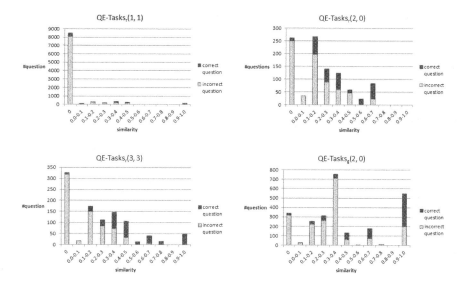

Fig. 7. Frequency distribution of Jaccard similarity to the central concept

for question q (r or g in our experiments). Each positive example is taken from data items in a category (e.g., restaurant names) specified by q.

For each r on the restaurant data, we constructed three sets of tasks: $QE\text{-}Tasks_r(1,1)$, $QE\text{-}Tasks_r(2,0)$ and $QE\text{-}Tasks_r(3,3)$ (Fig. 6). We chose the three patterns to examine several typical patterns with a small number of examples shown to workers. The QE-tasks were constructed in the following way. Give r, we constructed a set of L combinations of Y positive examples and a set of M combinations of N negative examples and computed the Cartesian product to generate its $L \times M$ QE-Tasks in $QE\text{-}Tasks_r(Y, N)$.

The number of $QE\text{-}Tasks_r(3,3)$ would be large ($_5C_1 \times _{20}C_3 = 5700$) if we generated tasks for all of the combinations. Therefore, we generated $QE\text{-}Tasks_r(3,3)$ in accordance with our observation that the results are more affected by positive examples than by negative ones: we combined every possible combination of positive data items to compute the set of combinations of three positive examples, while we just picked out three examples randomly from a category different from the correct category so that the set of combinations of three negative examples includes only one combination (i.e., $M = 1$). Therefore, the number of $QE\text{-}Tasks_r(3,3)$ for each r is $_5C_3 \times 1 = 10$. Note that this reduction does not give unfair advantages to our approach, since the approach is expected to work better if we generally have a larger number of QE-Tasks.

For the goods data, we constructed $QE\text{-}Tasks_g(2,0)$ only, because (1) we need to include pictures in the microtask, and (2) as shown later, the experiment with the restaurant data showed that $QE\text{-}Tasks_r(2,0)$ generated reasonably good results.

Worker Recruitment. We submitted our QE-tasks to Yahoo! crowdsourcing. We generated 20 duplicates of each QE-Task. Therefore, the number of questions we obtained per each QE-Task is 20. For example, since the number of $QE\text{-}Tasks_r(3,3)$ for each r is 10, we obtained $10 \times 20 = 200$ questions as the results of QE-Tasks for each r.

For both data sets, each worker was given a set of two QE-Tasks at a time and paid 5 yen (about 5 cents) for doing it. Each worker was allowed to perform two sets at most. For the restaurant data, the number of workers participated in the task is 255. For the goods data, the number is 521. Nationality of subject pool is all assumed as Japanese.

Computing similarities of questions. We used Jaccard coefficient with the threshold $\theta = 0.5$ to compute $sim(w(q_k), Q)$.

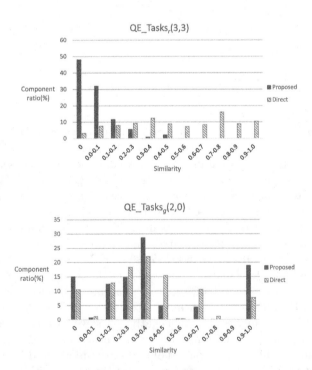

Fig. 8. Frequency distribution of Jaccard similarity to the original questions

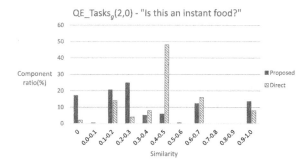

Fig. 9. Jaccard similarity distribution for an original question

6.2   Results

As explained below, the results showed that (1) the Jaccard similarity worked well, (2) the number of samples shown in each task matters, but (3) overall the proposed method worked well even if used with a small number of positive and negative samples.

Restaurant data. Figure 7 shows the frequency distribution of $sim(w(q_k), Q)$ for each q_k obtained for all r_js with $QE\text{-}Tasks_r(1,1)$, $QE\text{-}Tasks_r(2,0)$ and $QE\text{-}Tasks_r(3,3)$. The result suggests two things. First, Jaccard similarity works well; in all settings, the results show that a question having a higher similarity score is more likely to be a correct query. Second, the number of shown example matters; clearly, we can obtain more questions with higher scores when we show more examples in QE-Tasks.

Our method with each set of QE-tasks generated five ranked-lists of questions in total (i.e., a ranked-list for every r_j), except that the method with $QE\text{-}Tasks_r(1,1)$ generated ranked lists for only two out of five r_js.

The precision of top-ranked questions compared with the original questions is 0.83 (for $QE\text{-}Tasks_r(1,1)$). However, we obtained the question list for only two r_js), 0.63 (for $QE\text{-}Tasks_r(2,0)$) and 0.81 (for $QE\text{-}Tasks_r(3,3)$), respectively.

The ratios of the number of r_js that succeeded in obtaining at least one correct question with the intended semantics in the top three questions[2] out of five (the total number of r_js) are 2/5 (for $QE\text{-}Tasks_r(1,1)$), 3/5 (for $QE\text{-}Tasks_r(2,0)$), and 4/5 (for $QE\text{-}Tasks_r(3,3)$). Since we found that $QE\text{-}Tasks_r(3,3)$ is superior to the others, we examined the ten questions with the highest similarity values for every r_j with $QE\text{-}Tasks_r(3,3)$. The numbers of correct questions having the intended semantics in the top ten questions are 3, 0, 4, 3, and 7 for r_1 to r_5, respectively. They are different questions with the same semantics. For example, we obtained "Is this a name of food type?" and "Is this a kind of food?" for r_3.

---

[2] If there are more than three questions with the same support score, we included all the questions.

For r_2, there is no question with a high similarity value having the intended semantics. The original question is "Does this name a type of liquor or drink?" and the top question was "Is this a name of a drink?". The reason is that we do not distinguish between the name and the type of drink (*e.g.*, beer or wine as opposed to Bud Light).

Figure 8 (left) compares the proposed and direct-rewrite methods in the distribution of similarity of the rephrased questions to the original ones. The figure shows that the proposed method output a large number of questions that are not similar to the original ones, even though all of them do not necessarily have the intended semantics. In contrast, with the direct-rewrite method, workers typically replaced one term with an equivalent but different term or a dictionary-definition-like phrase, resulting in higher similarity to the original questions. Note that the results with the restaurant data r ($QE\text{-}Tasks_r(3,3)$ in Figs. 7 and 8) show that there are rephrased questions that are highly ranked (i.e., close to the central concept) and semantically correct but not similar to the original question.

Goods data. The precision of top-ranked questions compared with the intended semantics is 0.62. The ratio of the number of g_js that succeeded in obtaining at least one correct question in the top three against the number of all g_js is 6/7. The numbers of questions having the intended semantics in the top ten are 5, 3, 4, 4, 3, 0, and 3 for g_1 to g_7, respectively. For example, we obtained "Is this a seasoning?" and "Is this something to season food with?" for g_2.

The intended question g_6 was "Is this green tea?" but we did not obtain correct questions. The reason is that the Japanese word for tea usually refers to green tea and all workers were Japanese. Taking the background of workers into consideration is an interesting potential future work.

In contrast to the result with the restaurant data, the results with the goods data (Fig. 8 (right)) did not show a prominent difference between the proposed and the direct-rewrite method in the similarity distributions to the original questions. However, the similarity distribution varies depending on characteristics of the original questions. We found that our method produces a more variety of rephrased questions when the original question refers to not-so-clear concepts for workers even if dictionaries give their definitions. (See the similarity distributions in Fig. 9). We also found that the proposed method often output more appropriate questions than the original gold standard questions (Fig. 1).

Given these results, we argue that the proposed method is effective in the case where the original gold standard question is not necessarily correct and clear for workers, while the direct-rewrite method is effective when the original question is guaranteed to be reasonably good. An interesting future work is to develop a hybrid approach that have both of the advantages of the two methods, based on the results from the experiments. For example, a possible approach is to show workers the original question with those examples with which the answers given in the past were split and ask them to revise the question to deal with such a case.

7   Conclusion

Focusing on question-style instructions often used in microtask crowdsourcing, this paper proposed a method for eliciting rephrased instructions from the crowd. Given a real set of gold standard questions submitted to a commercial crowdsourcing platform, our experimental results showed that extracted questions are semantically ranked at high precision. In addition, we compared the proposed method with a direct-rewrite method, and the results clarified that each of them has its own strength and we identified cases where each is effective.

Future work includes optimization schemes for the proposed method that reduce the number of microtasks, and exploration of the design space of microtasks to rephrase questions, including various combinations of the example-based and direct-rewrite methods.

Acknowledgments. We are grateful to the project members of Yahoo! Crowdsourcing, including Masashi Nakagawa and Manabu Yamamoto. This work was supported by JSPS KAKENHI Grant Number 25240012 in part.

References

1. Androutsopoulos, I., Malakasiotis, P.: A survey of paraphrasing and textual entailment methods. J. Artif. Intell. Res. **38**, 135–187 (2010)
2. Chen, D., Dolan, W.B.: Collecting highly parallel data for paraphrase evaluation. In: The 49th Annual Meeting of the Association for Computational Linguistics: Human Language Technologies, Proceedings of the Conference, 19–24, Portland, Oregon, USA, pp. 190–200 (2011). http://www.aclweb.org/anthology/P11-1020
3. Dai, P., Lin, C.H., Mausam, Weld, D.S.: Pomdp-based control of workflows for crowdsourcing: Artif. Intell. **202**, 52–85 (2013)
4. Jeon, J., Croft, W.B., Lee, J.H.: Finding similar questions in large question and answer archives. In: Proceedings of the 14th ACM International Conference on Information and Knowledge Management 2005, pp. 84–90 (2005)
5. Law, E., von Ahn, L.: Input-agreement: a new mechanism for collecting data using human computation games. In: Proceedings of the 27th International Conference on Human Factors in Computing Systems, CHI 2009, Boston, MA, USA, 4–9 April 2009, pp. 1197–1206 (2009). http://doi.acm.org/10.1145/1518701.1518881
6. Lewis, J.R., Sauro, J.: The factor structure of the system usability scale. In: Kurosu, M. (ed.) HCD 2009. LNCS, vol. 5619, pp. 94–103. Springer, Heidelberg (2009). doi:10.1007/978-3-642-02806-9_12
7. Lytinen, S.L., Tomuro, N.: The use of question types to match questions in faqfinder. In: Proceedings of the IEEE 2002, p. 1 (2002)
8. Madnani, N., Dorr, B.J.: Generating phrasal and sentential paraphrases: a survey of data-driven methods. Comput. Linguist. **36**, 341–387 (2010)
9. Sahami, M., Heilman, T.D.: A web-based kernel function for measuring the similarity of short text snippets. In: Proceedings of the 15th International Conference on World Wide Web 2006, pp. 377–386 (2006)
10. Sauro, J., Lewis, J.R.: When designing usability questionnaires, does it hurt to be positive? In: Proceedings of the SIGCHI Conference on Human Factors in Computing Systems, pp. 2215–2224. CHI 2011, NY, USA (2011). http://doi.acm.org/10.1145/1978942.1979266

336 R. Hayashi et al.

11. Wen, J.R., Nie, J.Y., Zhang, H.J.: Query clustering using user logs. In: ACM Transactions on Information Systems (TOIS) 2002, pp. 59–81 (2002)
12. Zhao, S., Zhou, M., Liu, T.: Learning question paraphrases for qa from encarta logs. In: IJCAI 2007 Proceedings of the 20th International Joint Conference on Artifical Intelligence 2007, pp. 1795–1800 (2007)

To Buy or Not to Buy? Understanding the Role of Personality Traits in Predicting Consumer Behaviors

Zhe Liu[1]([⊠]), Yi Wang[1], Jalal Mahmud[1], Rama Akkiraju[1], Jerald Schoudt[1], Anbang Xu[1], and Bryan Donovan[2]

[1] IBM Almaden Research Center, 650 Harry Road, San Jose, CA 95120, USA
{liuzh,wangyi,jumahmud,akkiraju,jschoudt,anbangxu}@us.ibm.com
[2] Acxiom Corporation, 601 East 3rd Street, Little Rock, AR 72201, USA
bryan.donovan@acxiom.com

Abstract. In this paper, we explore the role of derived personality traits from Twitter in determining consumer behaviors. We conduct comprehensive analysis on a large industry dataset containing 188,654 individual level purchasing records across over 100 product categories. For each category, we build classifiers to distinguish buyers from non-buyers based on their derived personality traits. We use the models with demographic features as baseline to evaluate the predictive power of personality traits. Our results prove the decisive power of derived personality on a variety of consumer behaviors.

Keywords: Personality · Consumer behavior · Social media · Twitter

1 Introduction

Consumer behavior analysis is the study to explain how and why consumers act in particular ways under certain circumstances [1]. It is not an easy task, considering the large number of factors that would affect an individual's purchasing attitudes and decisions [2]. Among those factors, demographics, such as age, gender and income, have been extensively studied [3,4]. However, up until now, it has been difficult to go beyond the correlations between individual's demographic characteristics and consumption preferences. To enrich the existing studies in consumer behavior analysis, this paper explores the role that personality traits play in determine consumer behaviors. More specifically, in this study we are interested in how individual's intrinsic characteristics would impact their purchasing behaviors.

Compared with the demographic attributes, personality traits are even harder to collect. For instance, in order to derive an individual's personality, one has to ask the subject to finish a long survey containing usually at least 50 items [8], which could be very time and effort intensive. To avoid such costly process, a number of recent studies attempted to identify individual's personality traits from their social media fingerprints in an automatically way [9–12].

© Springer International Publishing AG 2016
E. Spiro and Y.-Y. Ahn (Eds.): SocInfo 2016, Part II, LNCS 10047, pp. 337–346, 2016.
DOI: 10.1007/978-3-319-47874-6_24

Leveraging the newly developed personality prediction models, in this work, we look specifically at classifying buyers and non-buyers based only on their derived personality traits using social media analytics.

To achieve our research goals, we perform extensive analysis on a large industry dataset, which contains 188,654 individual level purchasing records. We train and test a large amount of prediction models on consumer behaviors across over 100 product categories using the big 5 personality traits as well as their corresponding facets. To better evaluate the predictive power of personality traits, we further compare the proposed classifiers with the ones built with gender and age features. Our results demonstrate that on average using personality traits alone can generate a prediction accuracy of almost 60 % in differentiating potential buyers and non-buyers. To the best of our knowledge, we are one of the very first studies to examine the predictive role of derived personality traits in consumer behaviors on large real-world dataset.

2   Literature Review

2.1   Derived Personality Traits from Social Media

The traditional personality questionnaires are expensive and time consuming and may not be applicable to large-scale investigations. To avoid such costly process, a number of recent studies have been proposed to automatically predict personality traits from social network usage. A majority of these existing studies used textual features as predictive variables. Yarkoni [21] reported in his work the results of a large-scale analysis on the correlation between personality and word use in blogs. Golbeck, Robles, and Turner [9], and Qiu, Lin, Ramsay, and Yang [22] used Linguistic Inquiry and Word Count (LIWC) dictionary [23] categories to build classifiers that identify individual's personality from Facebook and Twitter. By extracting words, phrases and topics from Facebook messages, Schwartz et al. [11] built a model with higher prediction accuracy on individual's personality traits.

In addition to the works relying on lexical features, there are other studies predicting human personality traits using behavioral and social patterns. With individual's follow relationship on Twitter, Quercia et al. [10] accurately predict a user's personality. Bachrach et al. [24] correlated the user's personality with a number of behavioral features, such as the number of uploaded photos, number of events attended, number of group memberships, and number of times user has been tagged in photos. Kosinski, Stillwell, and Graepel [25] in their recent work showed that social behaviors such as Facebook likes can be successfully used to predict human personality traits, especially individual's agreeableness. Ortigosa, Carro, and Quiroga [26] developed classifiers that are able to predict user personality using parameters related to social interactions, such as number of posts the user has in his wall, number of different friends that have written in the user's wall, etc.

2.2 Personality Traits and Consumer Behaviors

While limited in number, evidence from qualitative studies confirmed the relationship of personality and consumer behaviors. Verplanken and Herabadi [27] investigated how impulse buying tendency is influenced by personality. They found that the impulse buying behavior is correlated with extraversion, negatively with conscientiousness. Matzler, Bidmon and Grabner-Kräuter [28] specifically studied two personality traits openness and extraversion. They empirically confirm the link between these two traits and perceived hedonic values of products (which is related to brand effect). Barkhi and Wallace [5] studied online purchase intent using personality traits. They found that neuroticism, openness and agreeableness have significant influences on the willingness to buy online.

The relationship of personality and purchasing behavior has been studied across a variety of products and services. While testing individual's preferences concerning organic foods, Chen [6] indicated that an individual's personality traits plays an important role in establishing personal food choice criteria. Cohen and Avieli [29] suggested food-related personality traits play an important role when predicting the likelihood of future food consumption. Schlegelmilch, Bohlen, and Diamantopoulos [30] explored the relationship between personality variables and pro-environmental purchasing behavior. He showed that consumers' overall environmental consciousness has a positive impact on green purchasing decision. For products with strongly symbolic meanings, such as wine, consumer's purchasing decision will be strongly affected by specifically perceptions of personality [31]. Among all the survey-based studies, Yang et al. [32] proposed one of the very few studies relying on derived personality to predict brand preferences. They showed that personality traits have played very important role in their task of brand preference prediction. Inspired by Yang et al.'s work [32], in this study, we differentiate buyers from non-buyers based only on their derived personality traits using social media analytics.

3 Data Preparation

3.1 Consumer Behavior Data Preparation

The data that we use in this study is obtained from Acxiom Corporation, an industry-leading supplier of demographic data for client's marketing needs. By collecting, parsing and analyzing customer and business information, the company helps its clients to accurately identify relevant audience and to conduct targeted advertising campaigns. Their main product, InfoBase, is the nation's largest repository of customer intelligence, contains hundreds of millions of records which covers information including: individual demographics (age, gender, education), household characteristics (household size, number of children), financial situations (income ranges, net worth), interests (sports, leisure activities, pets), and a number of purchasing activities (products bought, method of payment), etc. Our dataset contains a subset from Acxiom's InfoBase, which covers over 250 million household-level records on 1,048 dimensions over a 24 months' period.

Among the 1,048 dimensions, 184 are binary categories indicating whether or not a household has made any purchase across a variety of product categories, ranging from apparel, books, housewares, to charity donations. From these 184 dimensions, we randomly sample 100 product categories to test the predictive power of personality traits in consumer behaviors.

3.2 Personality Traits Data Preparation

In order to derive the personality traits of each input individual from Acxiom's dataset, we adopt the Personality Insights service[1] from IBM Watson. The Personality Insights API can automatically infer, from personal writings, portraits of individuals that reflect their personal traits, including big 5 personalities, their corresponding facets, needs, and values. Ever since its release, the Personality Insights service has received high intention and has been adopted in a number of research studies [33,34]. However, the major restrictions of directly applying Personality Insights service to the consumer data is that, first, we have no writing samples from the input individuals within the consumption dataset; second, most of the consumption behaviors in the consumer dataset are at the household level, instead of the individual level.

To solve the first restriction, we rely on the help of IBM InfoSphere Big Match[2], an industry-leading Probabilistic Matching Engine, that helps organizations to match customer identities across unstructured and structured data in a Hadoop environment. By correlating the individual's characteristics, such as first name, last name, email, and address, from the consumption data to the unstructured personal twitter data, Big Match returns us the corresponding Twitter accounts of the input individuals from the consumption dataset. After collecting all individual tweets using Twitter API, we then use these tweets as input to the Personality Insights service and calculated the personality score for each input individual. We delete those individuals that Big Match cannot identify or match with very low confidence (matching probability <80%), and this left us with 784,971 records. Regarding the second restriction, the mismatch between household level consumption attributes and individual level personality traits, we came up with the solution of focusing on those households with one person only. This type of households only has one single individual, so it is reasonable to assume that such individual performs all the consumption behaviors. We also checked the distribution of personality traits of the single household individuals versus that of the original population to make sure that we didn't introduce any bias when making the above-mentioned assumption. In our consumption dataset, 188,654 out of 784,971 records are on the single person household level. By matching these 188,654 individual's purchasing records from the consumption

---

[1] http://www.ibm.com/smarterplanet/us/en/ibmwatson/developercloud/personality-insights.html.

[2] http://www-03.ibm.com/software/products/en/infosphere-big-match-for-hadoop.

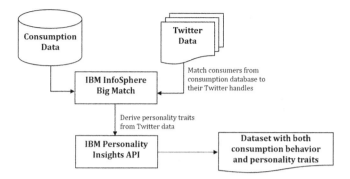

Fig. 1. Data preparation process

data with their personality traits returned from the Personality Insights service, we derived the dataset used in this study. Figure 1 depicts the matching process in more details.

4 Classification Experiments

We design a number of binary classifiers to automatically label an individual as a potential buyer or non-buyer for each of the 100 randomly sampled product categories (e.g. wine, jewelry, low fat cooking materials). We conduct experiments using different sets of features, including: personality traits, demographics, and personality plus demographics. We implement the classifiers using a number of supervised learning algorithms provided in Weka [14]: AdaBoost, Decision Tree (C4.5), Logistic Regression, Naive Bayes, Random Forest and SVM. All learning parameters use the default values in Weka and all experiments use 10-fold cross-validation, with 90 % training and 10 % testing data. Since our dataset is highly unbalanced in terms of the proportion of buyers versus non-buyers, to avoid classification bias, we apply random undersampling to balance the data before supplying them to the classifiers. For better illustration purpose, in the following sections, we only show the results from the classifiers with the best performance.

4.1 Personality Traits as Features

Personality is defined as individual's distinctive and enduring dispositions that cause characteristic patterns of interaction with environment [15]. Among the differnt personality taxonomies proposed in the past [16–18], the big 5 factor constructs [19] is the most established and used model of normal personality traits. The big 5 personality model consists of five personality factors, namely, *openness, conscientiousness, extraversion, agreeableness, and neuroticism*. All factors of the big 5 personality model are thought to contain several correlated but distinct lower level dimensions, called facets. A full list of big 5 personality facets can be found in Goldberg's work [8]. In this study, we adopt the big 5 personality as well as all 30 corresponding facets as features to predict consumer behaviors.

From all six classification techniques, Logistic Regression achieved the best performance on 72 out of 100 categories. As shown in Fig. 2, with individual's personality traits as features, on average our models achieve a classification accuracy of 58.7 % across 100 product categories, with a standard deviation of 1.90 %, which is much higher than the 50 % level that would result from a random classification choice. While looking at each individual classifier, we list in Table 1 the top 10 consumption categories on which our models achieve the highest accuracies. In addition, to quantitatively understand the usefulness of each individual personality trait in identifying potential buyers, we also perform feature analysis by calculating the information gain criterion [20] of each feature. For reasons of space, we only present in Table 1 the topmost personality feature with the highest contribution.

Table 1. Top 10 consumption categories with the highest classification accuracies based on personality traits, and their top contributed personality feature.

Consumption categories	Accuracy	Top feature
Computing software	0.622	Openness
Nutraceuticals	0.621	Self consciousness
Gardening	0.619	Immoderation
Camping & hiking	0.614	Openness
Travel	0.613	Adventurousness
Science & space	0.612	Imagination
Home furnishings	0.612	Immoderation
Sports apparel	0.612	Openness
Donation	0.611	Intellect
Music	0.610	Openness

By associating the top contributed features with the coefficient from the logistic regression models, we observe that individuals with higher level of openness tend to have higher probabilities of purchasing camping and hiking, sports, and music related products, but lower probability in buying computing software. Besides, adventurous people are more willing to take challenges, and thus have higher probabilities of purchasing travel related services. Last but not least, we notice that self-consciousness is positively related to the purchasing behaviors of nutraceuticals and vitamins.

4.2 Demographics as Features

To better understand the predicative power of personality traits, we next build classification models using two demographic features: gender and age. We chose these two specific variables because they are relatively easier to get or infer, compared to other demographic features, such as income and education.

Gender as Feature. As can be seen from Fig. 2, we find that on average the classification models built with individual's gender as predictor achieve an accuracy of 56.86 %, which is significantly lower than the accuracy derived using personality traits $(t = -4.0663, p < 0.05)$. While looking into the performance of each individual model, we notice that personality traits achieve better prediction results on most of the gender-neutral categories, for example, magazines, green living, and DVD videos. However, for consumption categories, such as jewelry, beauty and cosmetics, and fashion, etc., gender features outperformed personality features in most cases.

Age as Feature. In addition to individual's gender, we also compared the predictive power of the derived personality traits with age. It turns out that on 43 out of 100 product categories, models using the personality features achieve better prediction accuracy than using the age feature. On average, the age-based models achieve a classification accuracy of 60.81 % (as shown in Fig. 2), which is significantly higher than the performance of using personality traits as features $(t = 2.916, p = 0.01)$.

To better understand the limitations of the personality traits in predicting individual's consumption behaviors, we further inspect the classification results. We notice that compared with age, the derived personality traits have more impact on individuals while purchasing goods satisfying their deficiency needs compared to growth needs. Here by saying goods with deficiency needs, we mean purchases that can satisfy individual's physical needs, safety needs, and belonging needs, such as apparel, housewares, linens, etc. In contrast, growth needs

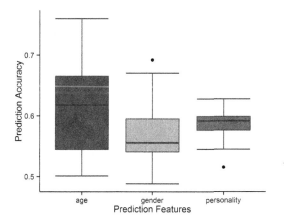

Fig. 2. Prediction accuracy distribution for consumer behaviors using personality, gender, and age as features

consist of self-esteem needs and self-actualization needs. Examples of goods satisfying individual's growth needs, include: religious and inspirational goods, career improvement, and travel.

5   Conclusion

In this study, we explored the role of derived personality from Twitter in predicting consumer behaviors. Multiple classifiers were implemented across 100 consumption categories to determine the predictive power of personality and demographic features. Through carefully analysis of our classification results, we concluded that, first, personality traits do have profound effects on predicting various consumer behaviors. Second, even though demographic variables, especially age, display more predictive power than personality traits when predicting purchases with growth needs, due to its sensitivity levels, they are usually hard to collect or even infer. Derived personality traits using social media analytics, in such a case, could be a good alternative for consumption prediction.

Like many other studies, the present work is not without its limitations. We acknowledge that our current study is done based on two major assumptions: inferred user profiles and inferred personality. Any bias that might be introduced in either inferred values can affect the performance of our models. So, for future work it would be interesting to replicate the present study with actual ground truth data to contrast results.

Despite the limitations discussed above, we believe that this research makes several major contributions to the social marketing and advertising fields. To our knowledge, this is one of the very few papers that addresses the link between online traces and offline purchasing behaviors. One major reason for the paucity of studies that attack this problem is the fact that it is hard to match individuals on social media and their purchasing actions in the real world. This issue is surpassed here by using IBM InfoSphere Big Match and Personality Insights. We believe that using online activity to predict actions in the real world is a major issue, important both for the challenge it poses and the potential applications. In addition, our work proved the possibility of predicting individual purchasing behaviors using their derived personality traits. This could be beneficial to the marketers to target potential consumers with more personalized marketing messages or campaigns.

References

1. Proctor, T., Stone, M.A.: Marketing Research. Macdonald and Evans Ltd., Phymouth (1984)
2. Furaiji, F., Małgorzata, L., Agata, W.: An empirical study of the factors influencing consumer behaviour in the electric appliances market. Contemp. Econ. **6**(3), 76–86 (2012)
3. Duesenberry, J.S.: Income, Saving, and the Theory of Consumer Behavior. Harvard University Press, Cambridge (1949)

4. Glanz, K., Basil, M., Maibach, E., Goldberg, J., Snyder, D.A.N.: Why americans eat what they do: taste, nutrition, cost, convenience, and weight control concerns as influences on food consumption. J. Am. Diet. Assoc. **98**(10), 1118–1126 (1998)
5. Barkhi, R., Wallace, L.: The impact of personality type on purchasing decisions in virtual stores. Inf. Technol. Manage. **8**(4), 313–330 (2007)
6. Chen, M.F.: Consumer attitudes and purchase intentions in relation to organic foods in taiwan: moderating effects of food-related personality traits. Food Qual. Prefer. **18**(7), 1008–1021 (2007)
7. Schlegelmilch, B.B., Bohlen, G.M., Diamantopoulos, A.: The link between green purchasing decisions and measures of environmental consciousness. Eur. J. Mark. **30**(5), 35–55 (1996)
8. Goldberg, L.R.: A broad-bandwidth, public domain, personality inventory measuring the lower-level facets of several five-factor models. Pers. Psychol. Europe **7**(1), 7–28 (1999)
9. Golbeck, J., Robles, C., Turner, K.: Predicting personality with social media. In: CHI 2011 Extended Abstracts on Human Factors in Computing Systems, pp. 253–262. ACM, Vancouver (2011)
10. Quercia, D., Kosinski, M., Stillwell, D., Crowcroft, J.: Our twitter profiles, our selves: predicting personality with twitter. In: 3rd IEEE International Conference on Social Computing, pp. 180–185. IEEE, Karlsruhe (2011)
11. Schwartz, H.A., Eichstaedt, J.C., Kern, M.L., Dziurzynski, L., Ramones, S.M., Agrawal, M., Shah, A., Kosinski, M., Stillwell, D., Seligman, M.E., Ungar, L.H.: Personality, gender, and age in the language of social media: the open-vocabulary approach. PloS One **8**(9), e73791 (2013)
12. Gou, L., Zhou, M.X., Yang, H.: KnowMe and ShareMe: Understanding automatically discovered personality traits from social media and user sharing preferences. In: 32nd Annual ACM Conference on Human Factors in Computing Systems, pp. 955–964. ACM, Toronto (2014)
13. Yang, C., Pan, S., Mahmud, J., Yang, H., Srinivasan, P.: Using personal traits for brand preference prediction. In: 2015 Conference on Empirical Methods in Natural Language Processing, pp. 1422–1432 (2015)
14. Hall, M., Frank, E., Holmes, G., Pfahringer, B., Reutemann, P., Witten, I.H.: The WEKA data mining software: an update. ACM SIGKDD Explor. Newsl. **11**(1), 10–18 (2009)
15. Goldberg, L.R.: The structure of phenotypic personality traits. Am. Psychol. **48**(1), 26 (1993)
16. Eysenck, H.J.: The scientific study of personality. British J. Stat. Psychol. **6**(1), 44–52 (1953)
17. Myers, I.B.: The Myers-briggs Type Indicator. Consulting Psychologists Press, Palo Alto (1962)
18. Keirsey, D., Bates, M.M.: Please Understand Me. Prometheas Nemesis (1984)
19. Costa, P.T., MacCrae, R.R.: Revised NEO Personality Inventory (NEO PI-R) and NEO Five-factor Inventory (NEO FFI): Professional Manual. Psychological Assessment Resources (1992)
20. Yang, Y., Pedersen, J.O.: A comparative study on feature selection in text categorization. In: 14th International Conference on Machine Learning, pp. 412–420 (1997)
21. Yarkoni, T.: Personality in 100,000 words: a large-scale analysis of personality and word use among bloggers. J. Res. Pers. **44**(3), 363–373 (2010)
22. Qiu, L., Lin, H., Ramsay, J., Yang, F.: You are what you tweet: personality expression and perception on twitter. J. Res. Pers. **46**(6), 710–718 (2012)

23. Pennebaker, J.W., Francis, M.E., Booth, R.J.: Linguistic Inquiry and Word Count: LIWC 2001. Mahway: Lawrence Erlbaum Associates, 71 (2001)
24. Bachrach, Y., Kosinski, M., Graepel, T., Kohli, P., Stillwell, D.: Personality and patterns of facebook usage. In: 4th Annual ACM Web Science Conference, pp. 24–32. ACM, Evanston (2012)
25. Kosinski, M., Stillwell, D., Graepel, T.: Private traits and attributes are predictable from digital records of human behavior. Proc. Natl. Acad. Sci. **110**(15), 5802–5805 (2013)
26. Ortigosa, A., Carro, R.M., Quiroga, J.I.: Predicting user personality by mining social interactions in facebook. J. Comput. Syst. Sci. **80**(1), 57–71 (2014)
27. Verplanken, B., Herabadi, A.: Individual differences in impulse buying tendency: feeling and no thinking. Eur. J. Pers. **15**(S1), S71–S83 (2001)
28. Matzler, K., Bidmon, S., Grabner-Kräuter, S.: Individual determinants of brand affect: the role of the personality traits of extraversion and openness to experience. J. Prod. Brand Manage. **15**(7), 427–434 (2006)
29. Cohen, E., Avieli, N.: Food in Tourism: attraction and impediment. Ann. Tourism Res. **31**(4), 755–778 (2004)
30. Schlegelmilch, B.B., Bohlen, G.M., Diamantopoulos, A.: The link between green purchasing decisions and measures of environmental consciousness. Eur. J. Mark. **30**(5), 35–55 (1996)
31. Boudreaux, C.A., Palmer, S.E.: A charming little cabernet: effects of wine label design on purchase intent and brand personality. Int. J. Wine Bus. Res. **19**(3), 170–186 (2007)
32. Yang, C., Pan, S., Mahmud, J., Yang, H., Srinivasan, P.: Using personal traits for brand preference prediction. In: 2015 Conference on Empirical Methods in Natural Language Processing, Lisbon, pp. 86–96 (2015)
33. Hu, T., Xiao, H., Luo, J., Nguyen, T.V.T.: What the language you tweet says about your occupation. In: 10th International AAAI Conference on Web and Social Media, pp. 181–190, Cologne (2016)
34. Appel, A.P., Candello, H., de Souza, B.S., Andrade, B.D.: Destiny: a cognitive mobile guide for the olympics. In: 25th International Conference Companion on World Wide Web, pp. 155–158, Montreal (2016)

What Motivates People to Use Bitcoin?

Masooda Bashir[1(✉)], Beth Strickland[1], and Jeremiah Bohr[2]

[1] School of Information Sciences, University of Illinois,
Urbana-Champaign, IL, USA
{mnb,mestric2}@illinois.edu
[2] Department of Sociology, University of Wisconsin,
Oshkosh, WI, USA
bohrj@uwosh.edu

Abstract. Bitcoin, a virtual currency that employs a novel technology that engenders trust and value in a decentralized peer-to-peer network, has exploded in use since it was first introduced in 2009. Yet, despite this explosive growth, researchers have barely started to investigate who uses Bitcoin, how and why they are used, and what the growth of virtual currency means for society. This paper examines Bitcoin as a currency, software platform, and community. It discusses the background of the virtual currency and presents the results of one of the first exploratory surveys to ask Bitcoin users and nonusers alike their opinions about Bitcoin and how these opinions relate to personal and cultural values. We examine how these views about Bitcoin are informed by attitudes towards money, technology, government, and social structure.

Keywords: Bitcoin · Cryptocurrency · Virtual currency · Libertarian · Attitudes

1 Introduction

Bitcoin, a virtual currency that employs a novel technology that engenders trust and value in a decentralized peer-to-peer network, has exploded in use since it was first introduced in 2009. As of February 2015, there were more than 13.88 million Bitcoins in circulation, valued at around $3.50 billion [1]. Thousands of retailers accept Bitcoins for payment, including Overstock.com and Expedia.com, and a number of start-up companies have developed services to allow Bitcoin users to shop at the retailers that have not yet begun to accept the virtual currency [2, 3]. While a number of other virtual currencies have failed altogether or enjoy no more than marginal use, Bitcoin has sustained for seven years and has garnered the attention of the mainstream financial world [4, 5]. The future of Bitcoin as a currency is still undecided, but the technology that makes Bitcoin possible, cryptographic hash functions and the blockchain, may have a number of other potential applications that could prove sustainable. Enthusiasts

---

By comparison, according to the World Bank, the U.S. Gross Domestic Product for 2013 was $16.8 trillion. See http://data.worldbank.org/indicator/NY.GDP.MKTP.CD.

© Springer International Publishing AG 2016
E. Spiro and Y.-Y. Ahn (Eds.): SocInfo 2016, Part II, LNCS 10047, pp. 347–367, 2016.
DOI: 10.1007/978-3-319-47874-6_25

of the technology posit that Bitcoin's technology could be applied to a number of services or activities that currently require a trusted third party to administer.

Yet, despite this explosive growth and potential, researchers have barely begun to investigate who does and does not use Bitcoin, how and why Bitcoins are used, and what the growth of this type of technology means for society. There are no simple and easy ways to begin attempts at answering these questions, and with a lack of theory about this technology and its related social phenomena, there is a need to start somewhere in the process of examining the relationship between users of this technology and the technology itself. Researchers need to not only understand the technical feats which have to be overcome to make the use of such a tool plausible, but they also need to examine the potential of what a mass shift in the use of electronic currency means for society as a whole. What would a use of this currency tell us about the users of this technology? What values and beliefs and attitudes do they demonstrate when choosing to engage or not engage with this form of currency? So little is known about this community that starting with an empirical study gives researchers a concrete place to start. What this study does is provide the first empirical building blocks to begin telling the story of Bitcoin and its community.

Although no initial exploratory study can answer all of these questions in a single large-scale survey, what it can do is give us the first pieces of empirical evidence we need so we can know how to best approach futures studies about how and why people use this tool. For example, from this study we learned that Bitcoin is more than just a currency and means different things to different users. It is a platform for new applications and it represents a unique community. The novelty and multiplicity of meanings of "Bitcoin" is reflected in the lack of standardization of the capitalization of the word. Many will use "Bitcoin" to refer to the Bitcoin protocol and network, and use "bitcoin" to refer to the coins themselves, but others do just the opposite. Some even use "BitCoin".[1] We use "Bitcoin" throughout this paper to refer to both the protocol and the currency. This seems to be the most common usage. Many Bitcoin-related topics are ripe for research, and this paper will attempt to begin to explore this broad subject area by examining how one group, university students, have responded to the advent of Bitcoin. This group represents an age group which is on the cutting edge of technological developments and may help give us our first clues into the attitudes, beliefs, and behaviors of potential and current cryptocurrency users. First, we briefly explain the basics of the Bitcoin technology and the social and cultural context in which it was invented; including some of its primary criticisms. Then, we present the results of one of the first known academic surveys to ask Bitcoin users and nonusers alike their opinions about Bitcoin and to examine how these opinions relate to personal and cultural values. We examine how these attitudes towards Bitcoin are informed by attitudes towards money, technology, government, and social structure. We conclude by discussing the implications that these first glimmers into the minds and actions of cryptocurrency users (and non-users) tells us about these phenomena and we present avenues for further research.

---

[1] Paul Krugman (see references) uses both "Bitcoin" and "BitCoin" in just one New York Times op-ed.

2 The Bitcoin Technology

2.1 Development

In late 2008 an Internet forum user going by the name "Satoshi Nakamoto" posted a white paper that claimed to describe a secure online payment system and currency that would provide an alternative to government-backed "fiat" systems [6]. Nakamoto's proposal was one of many suggested virtual currencies since the advent of the Internet [7, 8]. Those working to develop virtual currencies have differed in their motivations, but the effort is often associated with a technological libertarian ideology that advocates the use of electronic communications systems to (1) secure communications from government surveillance, (2) avoid state-controlled monetary, financial, and taxation systems, and (3) support free markets, which they posit are necessary for other freedoms [9]. Technological libertarianism is not the only political viewpoint associated with virtual currencies [10], but much of the rhetoric around the technology echoes this point of view. A distrust of centralized governments and financial institutions has provided some the motivation to develop open source peer-to-peer systems that replace government-backed "fiat" systems, as they are sometimes called, such as the U.S. Federal Reserve system. The goal of such a system would be to locate the trust necessary to provide value to a currency in an information network and its protocols rather than in a government. Generally speaking, the U.S. dollar has value because users trust the U.S. government and its ability to extend its power globally; the idea is that currency in an online peer-to-peer network would have value because users trust the network. The U.S. dollar is highly valued in the world because of the stability of the U.S. government and the trust that the government, particularly the Federal Reserve, will make appropriate decisions regarding the issuance and regulation of currency. Irresponsible behavior, such as issuing money without regard for the impact on value, creating extreme hyperinflation and devaluing a currency, undermines trust in a government and its currency. Advocates of virtual currencies argue that protocols such as Bitcoin would handle the issuance and regulation of currency better than (at least some) governments [11].

Many people, in fact, do not trust that governments are making appropriate decisions, and instead believe that the "fiat" system benefits governments and the financial industry at the expense of the rest of society [7]. The economic recession that developed in 2007-08 was in part based on a disconnect between the amount of debt assumed by individuals and businesses and real-world values [12]. When it became apparent that many debts, particularly millions of sub-prime mortgages, would never be paid, a financial structure built on debt failed. Many blamed a lack of regulatory oversight that allowed financial institutions to profit despite this disconnect from reality. As one credit bubble after another burst, the entire world economy entered a recession [13]. Whether intentional or not, Bitcoin, introduced mere weeks after the economic collapse of October 2008, was seemingly designed with these perceived failures specifically in mind. In a forum post, Nakamoto expressed his reasons for designing a currency system that avoided using banks:

The root problem with conventional currency is all the trust that's required to make it work. The central bank must be trusted not to debase the currency, but the history of fiat currencies is full of breaches of that trust. Banks must be trusted to hold our money and transfer it electronically, but they lend it out in waves of credit bubbles with barely a fraction in reserve. We have to trust them with our privacy, trust them not to let identity thieves drain our accounts. Their massive overhead costs make micropayments impossible [14].

Nakamoto's reasons echoed beliefs from a number of groups, including libertarians, open source and peer-to-peer online communities, and cryptoanarchists [15]. Nakamoto referred to his system as "trustless" because there is no need to trust a third party to verify transactions. While there are no centralized third-party institutions involved, users must still place trust in the system – including both the protocol and other actors in the Bitcoin ecosystem – to fulfill their goals. The object of trust has shifted from an intermediary person or institution to other parts of the system, but trust remains the foundation of value for the system. The issue of trust in Bitcoin is a topic ripe for further research.

Even those who do not hold strong libertarian opinions about government and central banks see flaws in the current financial system that may be solved through more efficient and trustworthy systems. Trusted third parties such as banks and credit card companies spend a great deal of their time and energy mediating disputes and dealing with fraud, both of which add significant transaction costs to the system [6]. Credit card companies charge merchants in the area of 2 % of every transaction to process payments, while people who want to transfer money (perhaps to family in less developed countries) can pay significant fees [16]. Seventy-five percent of the world's poor do not even have access to a bank account [17] but do have access to cell phones and mobile services. A currency system such as Bitcoin, argue proponents, could reduce transaction costs of transfers significantly and allow those currently excluded by the financial system to participate in the global economy [18].

2.2   How Bitcoin Works

With these motivations in mind, there have been numerous attempts over the last three decades to design a virtual currency system. Such a system is dependent upon encryption for security, and so these network-based currencies are called "cryptocurrencies." Many attempts to design a cryptocurrency system have failed for a variety of reasons. A major roadblock was developing a system that would ensure that the cryptocurrency, which is a digital object, could not simply be copied and used multiple times – the problem of "double spending" [19]. In the mainstream financial system, trusted third parties such as banks fulfill this verification function and resolve disputes if someone attempts to transfer the same money to more than one party. In pre-Bitcoin cryptocurrency systems a centralized clearinghouse was still required to verify transactions and prevent double spending.

In 2008, however, Nakamoto proposed to solve this problem by introducing the concept of a "blockchain." A blockchain is a sort of public ledger that records the proof-of-work necessary to demonstrate that a digital object, such as a currency, is unique and has not been copied or double spent [20]. Nakamoto proposed a peer-to-peer system in which the network took responsibility for publicly doing the work, using particular algorithms, to verify that each transaction was unique and proper. Thus, cryptographic proof, rather than a trusted third-party financial institution, verifies payments made between buyers and sellers. Third parties become superfluous to the network, and transaction costs are minimized. Users of the system are motivated to provide the work necessary to verify transactions by being compensated for their efforts with newly generated currency and transaction fees. Nakamoto called this system, and its currency, Bitcoin.

The Bitcoin system consists of a network of nodes, all of which run the Bitcoin software. When someone wants to conduct a transaction using Bitcoin, they announce the transaction (using their digital signature and the public key of the intended recipient) to every node in the system, which collects all these transactions in a pool and begins working to verify them. Each node is competing to be the first to verify the last ten-minutes' or so worth of transactions – one "block" of transactions – and have that block be accepted into the chain [21]. Nodes employ a hash function that integrates the current block's transactions with all previous transactions in the chain to verify that the user who announced the transaction owns the rights to the Bitcoins at issue and that they have not previously been transferred to another party. Bitcoins exist only within the blockchain itself. Owners possess the cryptographic key to access those Bitcoins and change their ownership; they do not possess the coins themselves (which do not exist in any tangible form). Users employ "wallets" to manage their cryptographic keys. If an owner loses access to his keys, whether because they were stolen, lost, or deleted, he loses his rights to his Bitcoins. This is similar to losing cash. An entire industry has developed to provide secure wallet services, and issues of trust are particularly relevant at this layer of the ecosystem.

Only the first transaction in a block for a particular Bitcoin will be accepted and worked on by any particular node, so users who attempt to double spend their Bitcoins within one block will have all but the first transaction rejected. All nodes are trying to integrate the new transactions into the chain, but only the first node to do so in a way that meets an arbitrarily determined standard will "win" that block and have their particular version of the block added to the chain. When a node finds a hash that satisfies the requirement, it announces it to the entire network and the other nodes work to verify it. While it is very difficult to calculate the hash that will satisfy the threshold requirement, once it is found, it can easily be verified [22] This is a type of game where the likelihood of winning is based on how much CPU power is being dedicated to the task, and theoretically the system cannot be manipulated unless one party controls more than 50 % of all computing power in the network [20]. The randomness of winning also provides incentives for the losing nodes to accept the winner and move onto the next transaction rather than trying to game the system. When a majority of the other nodes confirm that a node has found the appropriate hash, that node's proof-of-work is added to the chain and each node moves to work on the next block of transactions.

The block added to the chain includes the proof-of-work needed to verify the transactions and sets up the inputs for the next round of verification. Because the nodes working to verify the transactions are rewarded with newly minted Bitcoins, they are often called "miners" [23]. Miners are also rewarded with transaction fees when their block is added to the chain. Within a short amount of time anyone who wants to conduct a Bitcoin transaction will have that transaction verified and accepted into the official register of transactions. While one block takes about ten minutes to verify, with the amount of time being built into the protocol and where increased difficulty is met with increased computer power within the network, the possibility that different double spend transactions will be recognized by different nodes, and that two nodes could win the game at the same time, means that verification can actually take up to an hour [21]. When there is a dispute between two versions of one block, the version that is ultimately incorporated by the next winning block becomes the official version [20]. The system works this out itself without any dispute mediation by a third party and everyone agrees to accept the resolution. It's the Bitcoin version of Orson Scott Card's "The enemy's gate is down" [24]. What everyone (or every node) accepts as true becomes true.

Unlike traditional transactions with a credit card company or bank, there can be no chargebacks within the Bitcoin system. The transaction that is included in the blockchain is official no matter what occurred in the real world. The protocol allows for including conditions precedent to the transfer of Bitcoins (for example, delaying the transaction until certain goods are shipped), though this feature is not yet being fully exploited [25]. The transaction is included in the pending pool of transactions when first made and satisfaction is determined by the Bitcoin system itself and not by a real-world third party. The end result of a Bitcoin transaction, as argued by its proponents, is a secure, relatively private, transfer of money done completely by the system and without the need for a centralized clearinghouse to process transfers.

2.3   Criticisms of Bitcoin

Since Nakamoto's blockchain breakthrough, other cryptocurrencies and systems have developed using similar technology, but Bitcoin remains the most popular and most known [26]. The development of cryptocurrencies such as Bitcoin is still too immature to know whether they are a passing fad or part of an "electronic cash" revolution, but they are starting to catch the attention of national governments and major financial corporations alike. The New York State Department of Financial Services conducted hearings in January 2014 to assess appropriate regulatory oversight and in July 2014 proposed the first state regulations governing Bitcoin [27, 28]. In late October 2014, the U.S. Department of the Treasury's Financial Crimes Enforcement Network issued a ruling that determined that a virtual currency-trading platform is considered to be a money transmitter subject to attendant regulations [29]. China has prohibited financial institutions from conducting transactions in Bitcoin [30], while in July 2014 the United States auctioned 30,000 Bitcoins confiscated in the takedown of an illegal online market [31].

This unsettled treatment of Bitcoin by governments reflects not only the nascent nature of cryptocurrency but also the challenges of regulating a peer-to-peer virtual technology such as Bitcoin. The first issue is what kind of "thing" governments should consider Bitcoin to be. Is it really a currency, or should it be regarded as a commodity or a security? In 2014, the Internal Revenue Service (IRS) declared that it would treat

Bitcoin as property, rather than currency, for the purposes of U.S. federal taxes [32]. To recognize Bitcoin as a legitimate currency could be viewed as ceding an aspect of sovereignty that has in part defined the modern nation-state. In the United States, it could present a significant constitutional issue (see U.S. Const., Art. I., Sec. 8). But Bitcoin is a decentralized peer-to-peer network that exists everywhere and nowhere. Like other peer-to-peer networks, such as The Pirate Bay, the network functions as a hydra – cut off one head and two more will grow back. Even if a government were able to shut down nodes operating within its territory, the blockchain is distributed across millions of other nodes around the world. "It's not clear if [B]itcoin is legal, but there is no company to control and no one to arrest." [33] This resiliency may provide the opportunity to use blockchain technology for many applications other than virtual currency, but it presents a challenge for regulators [34]. One method used to confront the hydra-like quality of Bitcoin is to prohibit companies from accepting Bitcoin as payment or banks from conducting any transactions in virtual currencies [35]. Even if governments cannot regulate the Bitcoin network itself, they can regulate the environment in which Bitcoin functions to make it so inhospitable as to render Bitcoin impractical to use.

Bitcoin has also gathered negative attention for its association with illegal activity. Because Bitcoin offers users a great deal of anonymity relative to other types of virtual payments, it has become a popular medium of exchange for illegal goods online. Most infamously, The Silk Road, an online drug market, evolved into a multi-million dollar business offering a variety of illegal drugs through an interface reminiscent of eBay or Amazon [36, 37]. Although shut down by law enforcement in 2013, the website re-launched in little over a month, displaying the difficult nature of regulating international flows of narcotics sold through websites afforded by the existence of Bitcoin (or other peer-to-peer virtual currencies). The benefits of virtual currencies – more anonymity, easier transactions, and less regulatory oversight – are attractive to those motivated to avoid legal oversight. In general, there is a sort of "Wild West" atmosphere around Bitcoin, as it is used both as a tool for real-world criminals for payments and money laundering and a platform for other types of fraud conducted within the Bitcoin ecosystem itself.

There have also been a number of technical challenges identified in the current Bitcoin system. Some think that the ten-minute verification time is too long, and other cryptocurrencies that employ the same blockchain architecture but use shorter blocks to obtain faster verification have been developed. Because the number of Bitcoins that it is possible to mine is capped at 21 million (which will be reached around 2140), there are concerns that there will be significant hoarding or other economic dangers due to the fixed amount of coins [38]. The value of Bitcoins has fluctuated wildly over short periods of time, making it difficult to use Bitcoins in commerce and raising concerns about speculation [15]. This inability to hold steady value has led some commentators to dismiss Bitcoin altogether [39].

Of great concern, however, is evidence that the anonymity touted by Nakamoto is not as anonymous as he thought [40, 41]. Only cryptographic keys, not personal real-world identities, identify Bitcoin users and so the system does provide more anonymity than mainstream financial services. However, this anonymity is not absolute. In fact, the founder of The Silk Road was convicted of criminal charges in part because prosecutors were able to trace some of his Bitcoin activity [42]. There are several possible avenues for tracing identities in the Bitcoin ecosystem. For example, because users must undertake transactions using denominations of Bitcoins obtained in prior transactions (or mining) and receive change, it is possible to use one-to-many mapping to create a map of a particular user's activities. For example, if a user received 60 Bitcoins in three transactions of 20 Bitcoins each and wanted to pay another party 50 Bitcoins, that user would have to conduct three transactions of 20 Bitcoins and receive 10 Bitcoins in change. Mapping this process may reveal user identity. Bitcoin has built-in protocols to try to prevent this, but researchers have demonstrated work-arounds to these protocols [41]. Once an individual user's map has been identified, further research can be conducted to identify the real-world identity of the user [40]. Other implementations of the blockchain protocol, such as Darkcoin, have tried to make it even more difficult to trace users.

There is also a very real concern about the vast amount of computer power that must be dedicated to performing the hash function and which consumes precious resources like electricity. The proof-of-work required by the system is important to regulate the Bitcoin economy, but it requires that a significant amount of computing power be dedicated to what is essentially useless computations. The hash function is essentially a scratch-off game; success is based on luck and dedication of CPU time and effort. This monopolizes a lot of computing power and consumes electricity and time, among other resources. Beyond the sheer cost of all of this, the environmental impact of electricity use is significant – researchers have calculated that the electricity used for a proof-of-work system increases CO_2 emissions to the same degree as global air traffic [43]. One proposed option for addressing the concern of the vast waste of resources proof-of-work computations involve is to change the work to something useful, such as distributed data preservation [44]. In general, when computer scientists have identified technical flaws in the system, suggestions for adjustments to the protocol have been forthcoming and sometimes adopted by the community [45], but the anonymity and resource problems seem to be built into the system. While Bitcoin may fail because of economic or regulatory problems, these other problems could also undermine the system.

2.4   Other Uses for the Blockchain Protocol

While the initial and primary use of the blockchain protocol is Bitcoin, technologists – including computer scientists, entrepreneurs, tech-savvy lawyers, and political reformers – have envisioned myriad applications for the Bitcoin platform and blockchain technology. Some of these applications are merely pipedreams at this point, and many technical challenges remain to be solved. But a few applications beyond just currency transactions are beginning to be developed. The most enthusiastic about blockchain have even called this technology the most important development since the invention of the Internet itself because it provides a whole new platform for managing digital assets in a decentralized manner [46].[2] Those most dismissive of Bitcoin, like Paul Krugman, have generally only looked at the technology through a financial lens and have not addressed other potential applications.

However, despite the criticisms, perhaps Bitcoin can be analogized to Napster, where the poster child application of the technology (in Napster's case, peer-to-peer networks) was ultimately a failure, but the technology itself proved sustainable. Bitcoin could be "Napster for finance" – a protocol upon which an entirely decentralized system can be built [47]. If the blockchain technology is developed for uses beyond currency, it could dramatically change asset management. The key to this is understanding that is that the digital item in a blockchain ledger does not need to be a currency – it could be property deeds, assurance contracts, intellectual property rights, or any number of other objects, rights, or obligations that can be expressed as digital objects [7, 48, 49]. Computer scientists and entrepreneurs are beginning to develop protocols for these types of applications [50, 51]. Many of these applications have been long envisioned, but their practicability was often stymied by the need to administer a large central database. Blockchain could provide the technological breakthrough needed to overcome this obstacle. Each application, though, also has other usability challenges that blockchain may not fix.

3 Related Work

Despite the growing attention to Bitcoin and the enthusiasm over the technology's possibilities discussed above, research into the sociotechnical aspects of the cryptocurrency and its ecosystem has developed slowly. Almost no empirical research regarding how members of the general public view cryptocurrencies such as Bitcoin, or what types of individuals actually use cryptocurrencies, has been conducted. In early 2015, *Wall Street Journal* reporters Paul Vigna and Michael J. Casey published a book-length treatment of the subject, focusing on the history, development, and some implications for the technology. Some research has been conducted regarding Bitcoin owners [52–56] and a number of commercial research projects have examined issues such as consumer payment trends [57].

Although almost no empirical analyses have been conducted exploring the psychological and sociological aspects of cryptocurrencies like Bitcoin, other research with currency as a focus can be used to inform our analysis. For example, some parallels can be made with the failures of other types of new currency in the United States, such as the dollar coin. For example, in analyzing the attempt to institute the Susan B. Anthony dollar coin in the late 1970 s and early 1980 s, Caskey and St. Laurent argue that its

---

[2] Venture capitalist Marc Andreessen has been one of the most enthusiastic advocates for Bitcoin and the promise of its technology, but, as many note, his firm has invested a significant amount of money in Bitcoin-related ventures and has an interest in seeing the system succeed. See, for example, the comments of computer scientist and virtual currency researcher Robert McGrath at http://robertmcgrath.wordpress.com/2014/10/19/andreessen-changes-tune-agrees-with-me-about-bitcoin/. See also Alec Liu. Jan. 13, 2014. Why Silicon Valley (and Google) Loves Bitcoin. *Motherboard.* Available at http://motherboard.vice.com/blog/why-silicon-valley-and-google-loves-bitcoin.

failure rested upon a misconception of how money operates within networks [58]. Their argument—which easily extends to the more recent unpopularity of the Sacagawea dollar coin—refutes the decontextualized rationality that focuses on new money as better meeting transaction needs; instead they argue that the wide adoption of money requires network externalities where individual benefit is dependent upon how many others are using the same currency.

Another point of comparison could come from research on the meaning of money. For example, psychologists have analyzed the emotional associations with money in terms of security, power, love, and freedom [59]. Under this schema, individuals may differentially view money as a security blanket, leading them toward compulsive saving (security), a means of self-efficacy (power), a substitute for affection (love), or the means of escape to pursue personal interests (freedom). Sociologists have analyzed money beyond its ability to commensurate objects into numerical values in terms of how people separate their money and designate specific uses for it, or the social relationships involved in monetary compensations (accountability), entitlements (autonomy), and gifts (subordination) [60].

The finite design of Bitcoin has drawn comparisons to currencies based on gold standards, making it popular among individuals subscribing to libertarian ideology (which generally prescribes curbing the fiscal powers of the state) [9, 61, 62]. The use of this type of currency may also support the view of some that such a system ensures for users a type of "freedom" from the state given its decentralized and autonomous design [63]. The supply of Bitcoin is finite by design ostensibly to protect against inflation [6], though some popular criticism argues that Bitcoin will be susceptible to deflationary pressures. The ideological preference for particular types of money speaks to a longstanding concern, going back at least to the classical theory of Georg Simmel, on the relationship between monetary and social systems that allow for varying concentrations of political power [64, 65]. At least for some users, Bitcoin is politically attractive because it offers to the promise to facilitate a social order primarily organized around individuals entering voluntary associations and relying less upon state institutions that may coerce individuals to take certain actions in pursuit of collective goods.

The adoption of online payment systems such as PayPal can also inform research into Bitcoin adoption and attitudes. Economists have applied diffusion of innovations theory to PayPal adoption to explore how relative advantages, the complexity of use, and compatibility with other payment methods affects adoption of new mobile payment systems [66]. The slow adoption rates of mobile payment services, compared to other mobile services, has been explained as resulting from perceptions that mobile payments will be incompatible with current payment methods and hesitance to adopt payment services that have not yet become the norm [67].

Our study is one of the first to ask Bitcoin owners and non-owners alike their opinions regarding Bitcoin and to assess particular psychological and sociological factors in relation to these opinions and attitudes. As an initial exploration into these issues, this study was structured around two primary motivations. First, we explore factors that predict whether university students have positive or negative attitudes towards using Bitcoin. Next, we perform an analysis of the students who actually own Bitcoin. We explore both of these research questions in light of personality type, individualistic orientation, ideological identification, financial behavior preferences, past experiences with Bitcoin, and expressed motivations to use Bitcoin.

4   Methods

During April 2014, 7,500 students at a large Midwestern University were randomly invited to participate in a web-based survey about virtual currencies. University students were chosen as an exploratory population because they tend to have a greater familiarity with new online services when they first appear within the marketplace. Since studies about Bitcoin users have yet to be completed before this one, there was a need to remain open to exploring what type of population would be best to approach for this type of initial exploratory study. Ultimately, a total of 520 people responded to the survey, while 8 emails failed to send and this resulted in a final response rate of 7 %. The question of what constitutes a successful response rate varies greatly by type of survey administered and the familiarity (i.e. previous contact) researchers have or not have with the studied population [68]. Therefore, the authors consider this response rate to be successful for this type of large-scale survey with a previously unknown group of participants. Since Bitcoin is not uniformly understood by everyone, we provided respondents wtih the following description prior to being asked about their experience with and opinion of Bitcoin:

> Bitcoin is an open-source, denationalized, peer-to-peer payment system and currency that can be spent in any country where vendors accept Bitcoin. Anyone can set up a Bitcoin account, and accounts do not necessarily require personal identification. Bitcoin transactions are processed almost instantly over the Internet and carry almost no processing fees. Some users of Bitcoin equate it to "digital cash."

Three questions were asked regarding the respondent's general orientation toward Bitcoin to make up the scale used as the dependent variable for linear regression. Respondents were asked, (1) "How willing would you be to use Bitcoins or some other virtual currency?" (2) "In five years, how many of the businesses that you regularly shop at do you think will accept Bitcoins or some other virtual currency?" and (3) "Would you be willing to be paid in Bitcoins at your job if the value of Bitcoins were worth more than what you were offered in dollars?" These questions were anchored with 5-point Likert scales ranging from strongly disagree to strongly agree. Goodness of fit statistics from a confirmatory factor analysis showed a RMSEA score below 0.05 and a CFI score greater than 0.95, with Cronbach's alpha = 0.65. A final regression-weighted Bitcoin Orientation Scale was constructed from these items and used for analysis. A dummy variable measuring whether respondents own Bitcoin was used as an independent variable in the analysis of Bitcoin orientation and constitutes the dependent variable in a follow-up logistic regression analysis.

Several measures of individual dispositions were gathered to assess their relationship with Bitcoin attitudes. Without previous studies to guide this exploration, oft-used scales from psychology and sociology were adopted to determine if cultural orientation or personality made a difference in the use of and views about Bitcoin. Individualistic-Collectivistic orientation was measured using Triandis and Gelfand's

16-item measurement scale for horizontal- and vertical- individualism and collectivism, regularly used to determine cultural orientation [69]. Information on ideological orientation was gathered by asking respondents to identify which came closest to their political views among the following labels: conservative or traditionalist, libertarian, liberal or progressive, centrist or moderate, green, socialist, communist, anarchist, or other. Using these measures were key in determining the more likely type of person to use a high-secretively cryptocurreny like Facebook. Two dummy variables were created to represent libertarian and conservative ideological orientations, with centrist and left of center identities as the reference category. A 10-item short version of the Big Five Inventory was used to gather data on personality traits, providing measures for conscientiousness, openness, agreeableness, neuroticism, and extraversion [70].

In order to assess whether Bitcoin orientation was contingent upon technical skills, respondents were asked to self-assess their strengths along a 5-point scale for programming, spreadsheets, web design, and hardware operation. Although the validity of self-assessments may vary [71], for an initial exploration, this scale provides some context for the technical proficiency of respondents. These measures constitute a technical skills scale (RMSEA < .08, CFI > .95, Cronbach's alpha = .70). Although on the low side, these reliability scores are acceptable for an exploratory scale.

Two regression-weighted scales were constructed measuring behavioral preference for different kinds of money. Respondents were given the option to express whether cash, credit or debit cards (including services like PayPal), or Bitcoin was their preferred method of payment for the following activities: small purchases under $10, going out to bars or clubs, gambling, nonprofit or charity, giving money to friends or family, groceries, utilities, rent or mortgage, and health or medical expenses. A "discretionary cash" scale was created out of the first five items (RMSEA < .05, CFI > .95, Cronbach's alpha = .72) and a "necessities credit card" scale was created out of the last four items (RMSEA < .01, CFI > .99, Cronbach's alpha = .77).

Several other Bitcoin-specific measures were collected. In addition to Bitcoin ownership, dummy variables were created for "Bitcoin friend" (1 = respondent has at least one friend who has owned Bitcoins) and "devaluation" (1 = respondent believes Bitcoin will be worthless or less than its current value in five years). Several dummy

Table 1. Description of variable used in analysis

Variable	Description	M	SD	Range
Bitcoin orientation scale	Willingness to use Bitcoin and predictions re: Bitcoin future success	2.54	0.84	0.97–4.87
Female	Gender	0.51	0.50	0–1
Conscientious	Tendency to show self-discipline, act dutifully, and aim for achievement against measures or outside expectations	7.09	1.53	2–10
Vertical individualism	Tendency to value competition and hierarchies	12.27	2.72	4–20

(*Continued*)

Table 1. (*Continued*)

Variable	Description	M	SD	Range
Libertarian	Self-identification of libertarian political views	0.13	0.33	0–1
Conservative	Self-identification of conservative political views	0.12	0.33	0–1
Technical skills	Self-assessment of technical skills	2.79	0.95	1.10–5.49
Discretionary cash	Frequency of cash purchases and amount of cash carried and kept at home	0.75	0.40	0–1.19
Necessities credit card	Likeliness to use credit cards for necessities (e.g. rent and groceries)	0.97	0.28	0–1.10
Owns Bitcoin	Ownership of Bitcoin	0.09	0.29	0–1
Bitcoin friend	Bitcoin known to be owned by a friend	0.37	0.48	0–1
Devaluation	Opinion re: likeliness that Bitcoin will devalue	0.31	0.46	0–1
Anonymity	Likeliness to use Bitcoin because of anonymity of transactions	0.30	0.46	0–1
Borderless finance	Likeliness to use Bitcoin because it allow borderless financial transactions	0.35	0.48	0–1
Virtual money	Likeliness to use Bitcoin because it is virtual	0.30	0.46	0–1
Novelty	Likeliness to use Bitcoin because it is novel	0.24	0.43	0–1

variables were also included that reflect expressed motivations for wanting to use Bitcoin, including anonymity, borderless finance, virtual money (defined as the lack of physical money), and novelty. Table 1 provides a summary of relevant variable used in the analysis.

5   Results

Data was initially explored through an analysis of zero-order correlations. The measure of Bitcoin orientation had statistically significant and positive bivariate correlations with vertical individualism, libertarian ideology, technical skills, Bitcoin ownership, Bitcoin friends, anonymity, borderless finance, virtual money, and novelty. On the other hand, Bitcoin orientation had statistically significant and negative bivariate correlations with conscientiousness, gender (female), the discretionary cash scale, the necessities credit card scale, and the devaluation measure. The measure of anonymity had the strongest correlation, with r = .36. All correlations in the matrix were .36 or lower, indicating that multicollinearity is not a serious concern for the regression models.

Results from ordinary least squares regression are presented in Table 2. Missing data was handled using listwise deletion. Variables were chosen for Model 1 based upon a bivariate correlation with the dependent variable, which revealed that only the conscientiousness among personality types and vertical individualism among the individualist-collectivist scales were statistically significant. Gender was included as a control, along with ideological orientations. Conscientiousness and vertical individualism were positively significantly but weakly correlated with Bitcoin orientation

Table 2. Unstandardized coefficients and standard errors (in parentheses) from linear regression predicting bitcoin orientation.

MODEL	(1)	(2)	(3)	(4)	(5)
Female	−0.12 (.08)	−0.05 (.08)	−0.01 (.08)	−0.05 (.08)	0.04 (.07)
Conscientious	−0.05 (.02)*	−0.05 (.02)*	−0.04 (.02)	−0.03 (.02)	−0.02 (.02)
Vertical individualism	0.04 (.01)**	0.03 (.01)*	0.02 (.01)	0.02 (.01)	0.02 (.01)
Libertarian	0.52 (.11)**	0.51 (.11)**	0.47 (.11)**	0.36 (.11)**	0.31 (.10)**
Conservative	0.15 (.12)	0.17 (.12)	0.18 (.12)	0.16 (.11)	0.11 (.10)
Technical skills		0.11 (.04)**	0.08 (.04)*	0.07 (.04)	0.02 (.04)
Discretionary cash			−0.28 (.10)**	−0.21 (.09)*	−0.19 (.08)*
Necessities credit card			−0.58 (.13)**	−0.47 (.12)**	−0.33 (.12)**
Owns Bitcoin				0.33 (.13)*	0.36 (.12)**
Bitcoin friend				0.13 (.07)	0.08 (.07)
Devaluation				−0.57 (.08)**	−0.46 (.07)**
Anonymity					0.29 (.07)**
Borderless finance					0.33 (.07)**
Virtual money					0.33 (.07)**
Novelty					0.13 (.08)
Intercept	2.44 (.25)**	2.16 (.27)**	3.03 (.31)**	3.00 (.29)**	2.53 (.27)**
N	474	473	456	455	455
R^2	0.09	0.10	0.15	0.27	0.40

*$p < .05$ **$p < .01$

(b = -0.05 and 0.04, respectively), while libertarian ideology strongly predicted the outcome (b = 0.52). Altogether, this model explained about 9 % of the total variance. Added to Model 2, technical skills positively predicted Bitcoin orientation, but only explained about 1 % additional variance.

Two measures of financial behavioral preference are introduced in Model 3, including the discretionary cash scale and the necessities credit card scale. Both of these measures are statistically significant and negatively predict Bitcoin attitudes. This should not be surprising given that these measures represent different types of financial transactions for which respondents had the opportunity to express a preference for using Bitcoin rather than cash or credit cards. Of note is the degree to which respondents who strongly prefer to use credit cards for their necessities (rent or groceries) are less disposed toward using Bitcoin. Adding the discretionary cash and necessities credit card preference scales led to the conscientious personality and vertical individualism to fall out of statistical significance. Together, adding these measures accounts for an additional 5 % of the variance seen in Bitcoin orientation.

Experience and perceptions of the Bitcoin network were added to Model 4. Respondents with actual experience using Bitcoin have statistically significant positive attitudes toward Bitcoin, reflecting a generally positive experience with the currency. By contrast, and unsurprisingly, pessimistic attitudes about the future value of Bitcoin strongly and negatively predicted the dependent variable. Yet having at least one friend with Bitcoin experience had no statistically significant effect on an individual's Bitcoin orientation. These variables explained an additional 12 % of the variance in the outcome, reflecting their importance in understanding Bitcoin attitudes.

In the final model, variables indicating what may motivate a respondent to use Bitcoin were added. These included preferences for anonymity, engaging in borderless finance, owning virtual money with no physical basis, and the novelty provided by Bitcoin. Anonymity, borderless finance, and virtual money were all statistically significant measures positively predicting Bitcoin attitudes. This model explained 13 % of the variance in addition to the previous model, explaining 40 % of the total variance in the outcome.

Results from logistic regression predicting whether or not members of the university population own Bitcoin are presented in Table 3. Several divergences between factors predicting Bitcoin attitudes versus Bitcoin ownership are observed in this model. Women were only about 15 % as likely as men to have owned Bitcoin, while conscientious personality types were also less likely to be Bitcoin owners to a statistically significant degree. Of particular note is that libertarians were more than two and a half times as likely to own Bitcoins compared with those identifying as moderates or left-of-center ideologies. While having a friend who owns Bitcoin was not significant in predicting Bitcoin attitudes, this factor plays a big role in predicting whether or not an individual will own Bitcoin. Among the motivational factors, only novelty emerged as a statistically significant predictor of ownership.

Table 3. Unstandardized coefficients and standard errors (in parentheses) from logistic regression predicting bitcoin ownership.

Female	−1.73 (.59)**
Conscientious	−0.28 (.13)*
Vertical individualism	0.04 (.08)
Libertarian	1.02 (.49)*
Conservative	0.93 (.59)
Technical skills	0.72 (.23)**
Discretionary cash	−0.46 (.48)
Necessities credit card	−0.58 (.60)
Bitcoin friend	1.25 (.43)**
Devaluation	−0.88 (.49)
Anonymity	0.25 (.42)
Borderless finance	−0.64 (.45)
Virtual money	−0.30 (.45)
Novelty	1.14 (.41)**
Intercept	−2.83 (1.54)
N	468
McFadden's R^2	0.34

*$p < .05$ **$p < .01$

6 Discussion

This study provides the first empirical exploration of attitudes toward Bitcoin and likelihood of ownership. Of note in the results is a clear divergence between the factors predicting positive attitudes toward the use of Bitcoin and actual experience with using Bitcoin. For example, while gender and friendship networks were not important predictors of Bitcoin attitudes, they were strong predictors of whether or not an individual was a Bitcoin owner.

Clearly, there is a politically normative element involved in shaping attitudes toward Bitcoin as well as its adoption to date. Libertarian ideology was the only consistent factor for both attitudes and experience with the virtual currency. In the American political context, libertarians may associate the decentralized nature of Bitcoin with facilitating individualism as well as providing the means to defy the existing balance of power among financial organizations and institutions. This is in accordance with the stated viewpoints of Nakamoto and other Bitcoin developers. Of concern for developing new applications for blockchain technology is whether Bitcoin can expand its reach beyond the libertarian community.

Anonymity and novelty provide two interesting points of analysis for Bitcoin. Anonymity captures both the threatening nature of Bitcoin in the eyes of financial regulators and law enforcement as well as freedom for individuals to escape most forms of financial surveillance. While attraction to the anonymity afforded by Bitcoin positively predicted attitudes toward the currency, it was an insignificant predictor of whether or not someone had actually owned Bitcoin. Yet while novelty was not a great predictor of attitudes, its significance in predicting Bitcoin ownership probably reflects the still immature stage of development for the virtual currency. At least with this sample, Bitcoin owners appear to be more curious about the currency rather than attracted to clandestine activity.

The negative association between cash and credit card preferences with Bitcoin attitudes may shed light on how individuals view the virtual currency, particularly its association with security. While we might expect preferences for cash to indicate a desire for anonymous transactions (thus positively predicting Bitcoin orientation), the opposite trend was present in this study. Cash preferences in this sample could reflect a desire for tangible money and broader distrust of virtual transactions or could reflect a preference for local or face-to-face transactions. The negative association between preferences for using credit cards to pay for necessities and Bitcoin attitudes may reflect a hesitant view toward Bitcoin's (lack of) institutional embeddedness. Individuals wanting third parties to mediate transactions involving goods and services necessary for personal and social stability appear more distrustful of Bitcoin. This poses yet another challenge for Bitcoin—how to enculturate trust in the reliability of Bitcoin as a medium of exchange for essential services, and not simply a novel payment method used to purchase discretionary items.

7 Limitations and Future Research

The opportunity to expand the field of research about Bitcoin as a technology and community is wide open. There is an important need to examine this newly innovative way exchanging currency across the world; however, in the same way the potential for this area of study is exciting in its novelty, it is also challenging to determine the best way to start moving forward. In this study, we embarked on an exploratory mission to gather some of the first empirical data on this important topic. However, we would be remiss if we did not reflect on an important limitation of this study before suggesting ideas for future research.

One of the limitations of this study is that the sample group contains only university students. This group has been criticized by scholars as problematic given its lack of generalizability to the larger human population. This is not a position that we disagree with, however, the motivation of this study at this time was not to gather data needed to make that level of generalization. In the spirit of exploration of a novel topic, we approached a group of people who we believed would be familiar with Bitcoin as a tool whether or not they actually used it. Additionally, the intent of this study was not to investigate people with a high familiarity of Bitcoin, but rather to gain a baseline measure of what a group of individuals highly familiar with online tools think about Bitcoin and whether or not it is or would be something they would use. Perhaps a future study would benefit from the use of the information systems theoretical model of technology acceptance as a way to explore deeper the reasons as to why or why not some individuals choose to use Bitcoin.

The results of this study do provide some helpful empirical evidence when strategizing ways to move forward with related research. What we were able to find out about users of Bitcoin is that gender and friendship networks correlate to Bitcoin ownership but not necessarily attitudes. So this begs the question, what happens for an individual who ultimately ends up deciding to own Bitcoins? Also, why is it that men are so much more likely to make this decision? Do female Libertarians own more Bitcoins than do females of other political dispositions? Now that some empirical research has been completed which highlight these correlations, there is a need for future studies to go further into why these differences exist. For example, it's possible that since men are more likely than women to have computer programing skills that they are more comfortable engaging in the technical world of Bitcoin [72]. There is also a likely cultural influence on these gendered behaviors which should be considered. For example, in a recent study about gender and privacy protection behaviors online it was found that women were significantly more likely to display different information behaviors than men when navigating online resources [73]. The authors of this study used the theory of planned behavior and a discussion of social norms to explore these gendered behavioral differences online. Future studies about Bitcoin and gender should consider using this approach in their analysis to help explain the gender differences found in our study.

In addition to delving deeper into the why questions related to the findings of this study, there is a definite need to study additional groups of individuals who have different relationships to this technology. For example, a survey targeting the users of

HackerNews (a social news website focused on computer science and entrepreneur-ship), or Reddit (a social media/news website and discussion board), or Somethin-gawful (a comedic social media site and community board) might increase researcher access to actual Bitcoin users and/or access to different attitudes and beliefs about this revolutionary approach to currency exchange.[3] What about general attitudes about Bitcoin among the mainstream populous? Since students are in no way representative sample of mainstream culture, there is a need to broaden the scope of who is included in some futures studies about Bitcoin. What this study does is start a very important conversation about the Bitcoin technology, community and culture, and opens the space for additional investigation.

References

1. Blockchain (2015). Bitcoin Charts. https://blockchain.info/charts. Accessed 27 Feb 2015
2. Coindesk: What Can You Buy with Bitcoins? Coindesk (2014). http://www.coindesk.com/information/what-can-you-buy-with-Bitcoins/. Accessed 4 Mar 2015
3. Clare O'Connor: How to Use Bitcoin to Shop at Amazon, Home Depot, CVS, and More, 2 Feb. 2014. http://www.forbes.com/sites/clareoconnor/2014/02/17/how-to-use-Bitcoin-to-shop-at-amazon-home-depot-cvs-and-more/. Accessed 1 Mar 2015
4. Andolfatto, D.: Bitcoin and Beyond: the possibilities and pitfalls of virtual currencies. Dialogue with the Fed, 31 March 2014. http://www.stlouisfed.org/dialogue-with-the-fed/assets/Bitcoin-3-31-14.pdf. Accessed on 1 March 2015
5. Velde, F.: Bitcoin: A primer. The Federal Reserve Bank of Chicago Essay on Issues #317 (2013). www.chicagofed.org/digital_assets/publications/chicago_fed_letter/2013/cfldecember2013_317.pdf. Accessed 1 Mar 2015
6. Nakamoto, S.: Bitcoin: A Peer-to-Peer Electronic Cash System (2008). https://Bitcoin.org/Bitcoin.pdf. Accessed 1 Mar 2015
7. Vigna, P., Casey, M.J.: The Age of Cryptocurrency: How Bitcoin and Digital Money Are Challenging the Global Economic Order. St. Martin's Press, New York (2015)
8. Chaum, D.: Blind signatures for untraceable payments. In: Chaum, D., Rivest, R.L., Sherman, A.T.: Advances in Cryptology: Proceedings of Crypto 1982, pp. 199–203. Springer, New York (1983)
9. Karlstrøm, H.: Do libertarians dream of electric coins? The material embeddedness of Bitcoin. Distinktion 15(1), 23–36 (2014). doi:10.1080/1600910X.2013.870083
10. Smyth, L.: The Politics of Bitcoin, 7 March 2014b. Simulacrum.cc. http://simulacrum.cc/2014/03/07/the-politics-of-bitcoin/. Accessed 4 Mar 4 2015
11. Grinberg, R.: Bitcoin: An innovative alternative digital currency. Hastings Sci. Technol. Law J. 4(1), 160–208 (2011)
12. Kostakis, V., Giotitsas, C.: The (A)political economy of bitcoin. TripleC: communication, capitalism & critique. Open Access J. Global Sustainable Inf. Soc. 12(2), 431–440 (2014). http://triplec.at/index.php/tripleC/article/view/606

---

[3] The authors would like to thank the reviewers for their insight and for suggesting these groups as possible future groups to study.

13. Bernanke, B.S.: Causes of the Recent Financial and Economic Crisis. Testimony Before the Financial Crisis Inquiry Commission, 2 September 2010. http://www.federalreserve.gov/newsevents/testimony/bernanke20100902a.htm. Accessed 1 Mar 2015
14. Nakamoto, S.: Bitcoin open source implementation of P2P currency, 11 February 2009. (Online forum comment). http://p2pfoundation.ning.com/forum/topics/bitcoin-open-source. Accessed 1 Mar 2015
15. Wallace, B.: The Rise and Fall of Bitcoin. Wired, 23 November 2011. http://www.wired.com/2011/11/mf_bitcoin/all/. Accessed 1 Mar 2015
16. Freund, C., Spatafora, N.: Remittances, transactions costs, and informality. J. Dev. Econ. 86(2), 356–366 (2008). doi:10.1016/j.jdeveco.2007.09.002
17. Renzenbrink, A.: World Bank: 75 % of poor don't have bank accounts, 20 April 2012. CNN.com. http://edition.cnn.com/2012/04/19/business/poor-bank-accounts/. Accessed 1 Mar 2015
18. Van Alstyne, M.: Why bitcoin has value. Commun. ACM 57(5), 30–32 (2014). doi:10.1145/2594288
19. Dwyer, G.: The Economics of Bitcoin and Similar Private Digital Currencies. J. Financial Stability 17, 81–91 (2015)
20. Tapscott, D.: Blockchain Revolution: How The Technology Behind Bitcoin is Changing Money. Penguin, New York (2016)
21. Kondor, D., Posfai, M., Csabai, I., Vattay, G.: PLoS ONE 9(2), 1–10 (2014)
22. Bradbury, D.: Blocks we trust. Eng. Technol. 10(2), 68–71 (2015)
23. Extance, A.: The future of cryptocurrencies: bitcoin and beyond. Nature 526(7571), 21–23 (2015)
24. Orson Scott Card: Enders Game. Doherty Associates, New York (1985)
25. Harwick, C.: Cryptocurrency and the problem of intermediation. Independent Rev. 20(4), 569–588 (2016)
26. Champaign, P.: The Book of Satoshi: The Collected Works of Bitcoin Creator Satoshi Nakamoto. E53 Publishing, LLC (2014)
27. Ember, S.: New York Proposes First State Regulations for Bitcoin. New York Times, 17 July 2014. http://dealbook.nytimes.com/2014/07/17/lawsky-proposes-first-state-regulations-for-Bitcoin/. Accessed 1 Mar 2015
28. New York State Department of Financial Services (NYSDFS). Virtual Currency Hearing, New York City, 28–29 January 2014. http://www.dfs.ny.gov/about/hearings/vc_01282014_indx.htm. Accessed 27 Feb. 2015
29. FinCEN, FIN-2014-R011, 27 October 2014. http://www.fincen.gov/news_room/rp/rulings/html/FIN-2014-R011.html. Accessed 1 Mar 2015
30. Bloomberg news: China Bans Financial Companies from Bitcoin Transactions, 5 December 2013. Bloomberg. http://www.bloomberg.com/news/2013-12-05/china-s-pboc-bans-financial-companies-from-Bitcoin-transactions.html. Accessed 1 Mar 2015
31. Ember, S.: Single Winner of All Bitcoin in U.S. Auction. New York Times, 14 July 2014. http://dealbook.nytimes.com/2014/07/01/single-winner-of-all-Bitcoins-in-u-s-auction/. Accessed 1 Mar 2015
32. IRS 2014. Notice (2014). http://www.irs.gov/uac/Newsroom/IRS-Virtual-Currency-Guidance. Accessed 7 May 2014
33. Davis, J.: The Crypto-Currency: Bitcoin and its mysterious inventor. The New Yorker, 10 October 2011. http://www.newyorker.com/magazine/2011/10/10/the-crypto-currency. Accessed 1 Mar 2015
34. Bohme, R., Christin, N., Edelman, B., Moore, T.: Bitcoin: economics, technology, and governance. J. Econ. Perspect. 29(2), 213–238 (2016)

35. Kaplanov, N.: Nerdy Money: bitcoin, the private digital currency, and the case against its regulation. Loyola Consum. Law Rev. **25**(1), 111–174 (2012)
36. Christin, N.: Traveling the Silk Road: a measurement analysis of a large anonymous online marketplace. In: Proceedings of the 22nd International World Wide Web Conference, pp. 213–224 (2013)
37. Martin, J.: Lost on the Silk Road: online drug distribution and the 'cryptomarket'. Criminol. Crim. Justice **14**(3), 35–367 (2014). doi:10.1177/1748895813505234
38. Barber, S., Boyen, X., Shi, E., Uzun, E.: Bitter to better - how to make bitcoin a better currency. In: Keromytis, A.D. (ed.) FC 2012. LNCS, vol. 7397, pp. 399–414. Springer, Heidelberg (2012)
39. Krugman, P.: Bitcoin Is Evil. New York Times, 28 December 2013. http://krugman.blogs. nytimes.com/2013/12/28/bitcoin-is-evil/. Accessed 1 Mar 2015
40. Androulaki, E., Karame, G.O., Roeschlin, M., Scherer, T., Capkun, S.: Evaluating User Privacy in Bitcoin. In: Sadeghi, A.-R. (ed.) FC 2013. LNCS, vol. 7859, pp. 34–51. Springer, Heidelberg (2013)
41. Reid, F., Harrigan, M.: An analysis of anonymity in the bitcoin system. In: Altshuler, Y., Elovici, Y., Cremers, A.B., Aharony, N., Pentland, A.: Security and Privacy in Social Networks, pp. 197–223. Springer, New York (2013). doi:10.1007/978-1-4614-4139-7_10
42. Pagliery, J.: Bitcoin fallacy led to Silk Road founder's conviction, 5 February 2015. Money. cnn.com. http://money.cnn.com/2015/02/05/technology/security/bitcoin-silk-road/. Accessed 1 Mar 2015
43. Becker, J., Breuker, D., Heide, T., Holler, J., Rauer, H.P., Böhme, R.: Can we afford integrity by proof-of-work? Scenarios inspired by the Bitcoin currency. In: Böhme, R. (ed.) The Economics of Information Security and Privacy, pp. 135–156. Springer, Heidelberg (2013). doi:10.1007/978-3-642-39498-0_7
44. Miller, A., Juels, A., Shi, E., Parmo, B., Katz, J.: Permacoin: repurposing bitcoin work for long-term data preservation. In: Proceeding of the 2014 IEEE Symposium on Security and Privacy, pp. 475–490. IEEE (2014). doi:10.1109/SP.2014.37
45. Castillo, A.: Bitcoin's untapped potential. The Umlaut, 21 May 2013. http://theumlaut.com/ 2013/05/21/bitcoins-untapped-possibilities/. Accessed 1 Mar 2015
46. Andreessen, M.: Why Bitcoin Matters. New York Times, 21 January 2014. http://dealbook. nytimes.com/2014/01/21/why-Bitcoin-matters/. Accessed 1 Mar 2015
47. Morris, D.: Bitcoin is not just digital currency. It's Napster for Finance, 21 January 2014. Fortune.com. http://fortune.com/2014/01/21/bitcoin-is-not-just-digital-currency-its-napster-for-finance/. Accessed 1 Mar 2015
48. Fairfield, J.: Smart contracts, bitcoin bots, and consumer protection. Washington Lee Law Rev. Online **71**(2), 35–50 (2014). http://scholarlycommons.law.wlu.edu/wlulr-online/vol71/ iss2/3
49. Smith, S.V.: Bitcoin, what is it good for? NPR, 14 April 2014. http://www.marketplace.org/ topics/tech/Bitcoin-what-it-good. Accessed 1 Mar 2015
50. Aron, J.: What's wrong with Bitcoin? New Sci. **221**(2955), 19–20 (2014). doi:10.1016/ S0262-4079(14)60271-2
51. Alisie, M.: Crypto renaissance (2014). Ethereum Blog. https://blog.ethereum.org/2014/09/ 02/crypto-renaissance/. Accessed 1 Mar 2015
52. Bohr, J., Bashir, M.: Who Uses Bitcoin? An exploration of the Bitcoin community. In: Proceedings of the 2014 Twelfth Annual International Conference on Privacy, Security and Trust (PST), pp. 94–101. IEEE (2014). doi:10.1109/PST.2014.6890928
53. Hernandez, I., Bashir, M., Jeon, G., Bohr, J.: Are Bitcoin Users Less Sociable? An analysis of users' language and social connections on twitter. In: Stephanidis, C. (ed.) HCI 2014. CCIS, vol. 435, pp. 26–31. Springer, Heidelberg (2014). doi:10.1007/978-3-319-07854-0_5

54. Wilson, M.G., Yelowitz, A.: Characteristics of Bitcoin Users: an analysis of Google search data (2014). SSRN 2518603. Accessed 27 Feb 2015
55. Smyth, L.: Trust, Organisation, and Community within Bitcoin, 2 April 2014. Simulacrum. cc. http://simulacrum.cc/2014/04/02/bitcoin-trust/. Accessed 1 Mar 2015
56. Smyth, L.: The Politics of Bitcoin, 7 March 2014. Simulacrum.cc. http://simulacrum.cc/2014/03/07/the-politics-of-bitcoin/. Accessed 4 Mar 2015
57. Accenture. My Pay. My Way.: how consumer choice will shape the future of payments. In: 2014 North America Consumer Payments Survey (2014). http://www.accenture.com/SiteCollectionDocuments/accenture-2014-north-america-consumer-payments-survey.pdf. Accessed 1 Mar 2015
58. Caskey, J.P., St. Laurent, S.: The Susan B. Anthony Dollar and the theory of coin/note substitutions. J. Money Credit Banking 26(3), 495–510 (1994)
59. Furnham, A., Wilson, E., Telford, K.: The meaning of money: The validation of a short money-types measure. Personal. Individ. Differ. 52(6), 707–711 (2012). doi:10.1016/j.paid.2011.12.020
60. Zelizer, V.: The social meaning of money: "Special Monies". Am. J. Sociol. 95(2), 342–377 (1989)
61. Maurer, B., Nelms, T.C., Swartz, L.: "When perhaps the real problem is money itself!": The practical materiality of Bitcoin. Soc. Semiotics 23(2), 261–277 (2013). doi:10.1080/10350330.2013.777594
62. Yelowitz: Characteristics of bitcoin users (2015)
63. Garrod: Real world decentralized autonomous society (2016)
64. Dodd, N.: Simmel's Perfect Money: fiction, socialism and Utopia in the philosophy of money. Theory Cult. Soc. 29(7–8), 146–176 (2012). doi:10.1177/0263276411435570
65. Simmel, G.: The Philosophy of Money. Routledge, London (2011)
66. Mallat, N.: Exploring consumer adoption of mobile payments–A qualitative study. J. Strateg. Inf. Syst. 16(4), 413–432 (2007). doi:10.1016/j.jsis.2007.08.001
67. Schierz, P.G., Schilke, O., Wirtz, B.W.: Understanding consumer acceptance of mobile payment services: an empirical analysis. Electron. Commer. Res. Appl. 9(3), 209–216 (2010). doi:10.1016/j.elerap.2009.07.005
68. McMinn, M.R., Bearse, J., Heyne, L.K., Smithberger, A., Erb, A.L.: Technology and independent practice: survey findings and implications. Prof. Psychol. Res. Pract. 42(2), 176–184 (2011)
69. Triandis, H., Gelfand, M.: Converging measurement of horizontal and vertical individualism and collectivism. J. Pers. Soc. Psychol. 74(1), 118–128 (1998). doi:10.1037/0022-3514.74.1.118
70. Rammstedt, B., John, O.P.: Measuring personality in one minute or less: a 10-item short version of the Big Five Inventory in English and German. J. Res. Pers. 41(1), 203–212 (2007). doi:10.1016/j.jrp.2006.02.001
71. Mabe, P.A., West, S.G.: Validity of self-evaluation of ability: a review and meta-analysis. J. Appl. Psychol. 67(3), 280–296 (1982). doi:10.1037/0021-9010.67.3.280
72. Kane, C.L.: The computer boys take over: computers, programmers, and the politics of technical expertise. Contemporary Sociol. 40(5), 579–580 (2011)
73. Burns, S., Roberts, L.: Applying the theory of planned behaviour to predicting online safety behaviour. Crime Prev. Community Safety 15(1), 48–64 (2013)

Spiteful, One-Off, and Kind: Predicting Customer Feedback Behavior on Twitter

Agus Sulistya[1,2(✉)], Abhishek Sharma[2], and David Lo[2]

[1] Human Capital Center, PT Telekomunikasi Indonesia, Bandung, Indonesia
[2] School of Information Systems, Singapore Management University,
Singapore, Singapore
{aguss.2014,abhisheksh.2014,davidlo}@smu.edu.sg

Abstract. Social media provides a convenient way for customers to express their feedback to companies. Identifying different types of customers based on their feedback behavior can help companies to maintain their customers. In this paper, we use a machine learning approach to predict a customer's feedback behavior based on her first feedback tweet. First, we identify a few categories of customers based on their feedback frequency and the sentiment of the feedback. We identify three main categories: spiteful, one-off, and kind. Next, we build a model to predict the category of a customer given her first feedback. We use profile and content features extracted from Twitter. We experiment with different algorithms to create a prediction model. Our study shows that the model is able to predict different types of customers and perform better than a baseline approach in terms of precision, recall, and F-measure.

Keywords: Social media · Customer relationship management · Machine learning

1 Introduction

The use of social media in the customer relationship context has gained popularity nowadays. A report by VB Insight [1] reveals that modern consumers complain about brands 879 million times a year on Facebook, Twitter, and other social media portals. About 10 % of those consumers make a complaint on social media every day. With this extensive use of social media by customers, opportunities arise for companies to engage with their customers and be aware of the issues that they face. For example, a customer can complain on social media after experiencing a failure of service; this complaint notifies the company and prompts it to take necessary actions to prevent further damage to the company's reputation and customer base. Therefore, it is important for the company to continuously monitor the voices of their customers, which refer to an activity called *social listening and monitoring*.

It would be interesting to be able to predict different types of customer feedback behavior. Such prediction can help a company to formulate a suitable

© Springer International Publishing AG 2016
E. Spiro and Y.-Y. Ahn (Eds.): SocInfo 2016, Part II, LNCS 10047, pp. 368–381, 2016.
DOI: 10.1007/978-3-319-47874-6_26

strategy to manage and improve customer satisfaction and retention. For example, some users may complain many times to a company's Twitter account if the users are not given sufficient attention in a short period of time, others may complain only once, and yet others may express their thanks after a good service has been rendered by a company. From the company's side, multiple complaints are considered as something that should be avoided, since it can affect the company's reputation. Having an ability to predict this type of customer would allow the company to take preventive action before the user spreads negative opinion about the company in social media. A company may also want to provide good reasons for the third category of customers to publicize good service and improve the company's reputation.

In this study, we try to address the aforementioned prediction problem by employing a two-stage machine learning algorithms. In the first stage, our approach clusters social media users into several categories based on their feedback frequency and sentiment polarity. We identify three categories of users: spiteful (i.e., the user complains many times in social media), one-off (i.e., the user only provides negative feedback once), and kind (i.e., the user provides positive feedback). In the second stage, our approach builds a prediction model that can assign a user into one of the three categories based on his/her first feedback. We experiment with different supervised machine learning algorithms (i.e., Naive Bayes, Logistic Regression, and Random Forest), to build an automated prediction model.

As a case study, we use an internal data from a state-owned telecommunication company in Indonesia to evaluate the effectiveness of our proposed approach. The company named Telkom extensively uses social media such as Facebook and Twitter, to interact with its customers. To facilitate social listening, the company has set up a dedicated unit to actively monitor customer feedback. Our work extends the current social listening platform that the company has by adding some predictive capabilities. Under 10-fold cross validation, our experiments show that our proposed approach can predict customer feedback behavior categories with a weighted precision, recall, and F-measure of up to 0.797 (Random Forest), 0.881 (Naive Bayes), and 0.800 (Random Forest) respectively. Our approach outperforms a baseline that randomly assign categories to customers based on the distribution of customer feedback behavior categories in a training data.

Extracting knowledge from microblogs has been one of active research areas. We believe that this study would be important towards the development of techniques that make use of social media data to improve product and service quality. Specifically, our contributions are as follows:

1. We propose a new problem of predicting different types of customer feedback behavior on Twitter.
2. We use a clustering algorithm to identify different types of customer feedback behavior.

3. We propose a set of features, i.e. content features and profile features, that can be used to predict customer feedback behavior by leveraging a supervised machine learning algorithm to build a prediction model.
4. We have evaluated our proposed approaches on a dataset containing 11,809 tweets. Our proposed approaches can achieve reasonable precision, recall and F-measure which are higher than those of a baseline approach.

The structure of the remainder of this paper is as follows. In Sect. 2, we describe social listening activities in a company used as our case study and data analysis techniques that we leverage for this work. We describe how we cluster customers to create several categories in Sect. 3. In Sect. 4, we explain our approach which extract features from customer Twitter accounts and their corresponding tweets and use them to build a prediction model to predict customer categories based on their first feedback tweet. We describe our experiments which evaluate the prediction accuracy of our approach in Sect. 5. Related work is presented in Sect. 6. We finally conclude and mention future work in Sect. 7.

2   Preliminaries

2.1   Social Listening at Telkom

In this paper, we experiment with a dataset collected and annotated by a state-owned telecommunication service provider in Indonesia, namely Telkom[1]. The company serves tens of millions of customers throughout Indonesia, offering a wide range of products including broadband internet connections, cable TV, and land line telephone connections.

Telkom has set up a system that actively monitors what customers say on social media, and handles each issues raised by forwarding the problem to a back-room unit. To monitor customer voices, the company uses tools provided by Brand24[2] and BrandFibres[3]. The first tool is used to crawl any contents containing keywords related to the company's product from different platforms, including Facebook, Twitter, blog posts, and news media. These crawled records are then filtered by removing irrelevant posts. The filtering process requires manual work performed by several social media analysts. The analysts use a second tool called BrandFibres dashboard. Using this tool, they evaluate each post, and then assign a sentiment score to each post. They give scores ranging from "+5" (very positive feedback) to "−5" (very negative feedback). The analysts also assign a post into one of the 8 different categories shown in Table 1. Note that a tweet can be assigned to more than one category, and an analyst will assign a sentiment score for every category that applies to a tweet.

Figure 1 shows an example of a customer complaint on Twitter. In the figure, the tweet mention a company's account (@telkomcare). The tweet also mentions

---

[1] http://www.telkom.co.id/en/tentang-telkom.
[2] http://www.brand24.com/.
[3] http://www.brandfibres.org/.

Fig. 1. A tweet posted by a customer to a company's customer care channel

Table 1. Customer feedback categories

Category	Description
Quality Evaluation	Tweets related to general quality of a product or service (for example, slow or unstable internet connection).
Offer Evaluation	Tweets providing feedback to a product offering (such as an ongoing promotion of a certain product)
Activity Disturbance	Tweets reporting specific disturbance in a user's activity while using a product (for example, trouble when browsing or downloading).
Invoice Related	Tweets reporting issues related to product or service invoicing (such as reports of incorrect billing).
Customer Service Quality	Tweets reporting issues related to quality of customer service (such as quality of customer service agents, and how the company handles current problems experienced by the customer)
Actions	Tweets related to action taken by the customer (such as comparing product provided by the company with other competitors).
Social Media	Tweets related to social media interactions between company and customers.
Others	Tweets about other issues related to the company and subsidiaries

other users (@detikcom and @telkompromo). The first one is an online news media account and the latter is the company's other account that focuses on disseminating the company's promotional events and deals.

This study analyzes data consisting of tweets collected and annotated by Telkom for a 3 month period from June-August 2015. In total, there are 12,634 posts. We consider only the posts that have been collected from Twitter, which results in about 11,809 posts (or tweets) constituting about 93.4 % of the total posts. These tweets are those that mention the official company's customer care account on Twitter, namely @telkomcare. For the tweets in our dataset, we extract distinct twitter users who posted them, resulting in 6,031 distinct users. We use this set of users as the input to our clustering and prediction tasks described in the next two sections. The data provided by Telkom did not include the profile of these 6,031 Twitter users. To get these profiles, we call the standard Twitter API using Tweepy[4] Python module.

2.2 Handling Imbalanced Data

Imbalanced data problem typically refers to a classification problem where the classes are not represented equally. For customer feedback, typically we would see more negative feedback rather than positive ones. One way to deal with imbalanced data is by using sampling methods, which modifies the distribution of the original training samples to obtain a relatively balanced data. There are two types of sampling methods: oversampling and undersampling [11]. Oversampling is conducted by adding more samples to the minority class, while undersampling is done by creating a subset of the majority class. One popular oversampling algorithm to handle imbalanced data is SMOTE (Synthetic Minority Over-sampling Technique) [6]. This oversampling algorithm creates synthetic samples from the minority class instead of creating copies. SMOTE works by finding the k nearest neighbors of each sample in the minority class. Next, artificial samples are then generated along the line of some or all of the k nearest neighbors, depending on the amount of oversampling required.

3   Clustering Customers

In the first stage of our work, we cluster customers in our dataset (i.e., the 6,031 users described in Sect. 2.1) into several categories based on their feedback frequency and the sentiment polarity of these feedback. Figure 2 shows our overall approach to cluster customers.

Fig. 2. Our approach for clustering customers

We represent each customer as a set of metrics: NumOfFeedback, NumOf-PosFeedback and NumOfNegFeedback. These metrics are listed and defined in Table 2. Next, based on this representation, we cluster the users together. To cluster the users, we use Expectation-Maximization (E-M) algorithm. E-M algorithm assigns a probability distribution to each instance which indicates the probability of it belonging to each of the clusters. A previous study conducted by Meilă and Heckerman [13] has found that the E-M algorithm often performs

---

Table 2. Metrics used for clustering users

Features	Description
NumOfFeedback	Number of feedback tweets generated by a user
NumOfPosFeedback	Number of feedback tweets that are of positive sentiment polarity
NumOfNegFeedback	Number of feedback tweets that are of negative sentiment polarity

better than other clustering methods such as k-means and model-based hierarchical agglomerative clustering.

We use the implementations of E-M Algorithm in Weka [10]. We do not initiate number of cluster and let the E-M algorithm decides the best number of clusters. All parameters are set into Weka default setting.

Table 3 shows the results of the E-M clustering algorithm. We verify the result by manually investigating the properties of each cluster. Based on this manual investigation, we conclude general properties for each group as shown in the fourth column of the table.

Table 3. Clusters of users based on their tweets mentioning the company

Cluster	Count	Percentage	Observed Properties
0	1235	20.48 %	Post one or two times, with at least one positive feedback
1	152	2.52 %	Post more than 2 tweets, with more than two possitive feedback
2	481	7.98 %	Post at least 4 tweets, with majority of negative feedback
3	82	1.36 %	Post at least 9 tweets, with majority of negative feedback
4	2837	47.04 %	Post only one tweet with negative feedback
5	1244	20.63 %	Post 2 or 3 times with majority of negative feedback

Note that there are similarities among these clusters. Cluster 4 represents the majority of customers who only provide one negative feedback, without posting further tweets. Clusters 2, 3 and 5 correspond to customers who post more than one tweet with negative sentiment. These customers are typically the group of customers that may damage a company's reputation if they are not managed well. The other two groups (clusters 0 and 1) are groups of customers that post at least one positive feedback such as thanking the company for its good service. These customers can improve the company's reputation. Based on this observation, we decide to group the clusters further into three groups based on

Table 4. Three main categories of customers

Class	Cluster	Percentage
Kind	0,1	23.0 %
One-Off	4	47.0 %
Spiteful	2,3,5	30.0 %

how the customers complain or behave. This new groups are shown in Table 4. We will use these three groups as class labels for the second stage of our approach that predicts customer feedback behavior.

4   Predicting Customer Categories

In the second stage, our approach builds a prediction model that can assign a customer into one of the three categories based on their first feedback tweet. With our prediction model, a company would be able to know the category of a customer early and take necessary actions. Our approach first extracts a number of features that characterize a customer and his/her first feedback tweet. Features of customers belonging to the three categories are then used to train a prediction model that can differentiate each category. The following subsections explain features used and our approach to build the prediction model.

4.1   Feature Engineering

We use two types of features: profile features (i.e., features that we extract from a customer's Twitter profile) and content features (i.e., features that we extract from a customer's first feedback tweet).

Profile Features. Twitter provides several information about its user which include the user's number of followers, number of followee, etc. We consider five profile features to infer customer categories. These five features are described in detail below.

- **TweetCount:** This feature is the number of tweets or re-tweets generated by a user. This metric represents a user's level of activity on Twitter.
- **FollowerCount:** This feature is the number of followers that a user has. If A follows B on Twitter, all B's tweets would be propagated to A. This feature is a basic measure of a user's popularity on Twitter.
- **FolloweeCount:** This feature is the number of people a user follows. It represents the user's level of interest on others and correlates to the number of tweets that the user would receive daily.

- **FavCount:** Twitter users may express their liking of a tweet by marking the tweet as a favorite. This feature is the number of the tweets that a particular user has favorited. A higher value of this metric indicates that this user often gives positive feedback to others and may indicate his/her level of agreeableness.
- **ListCount:** A Twitter user can create lists of other Twitter users whom he/she follow. Each of this list typically contains related Twitter users who belong to a particular topic or interest (e.g., a list of friends, co-workers, celebrities, athletes, etc.). This feature is the number of lists that a user creates. We use this feature to capture another aspect of a user's level of activity on Twitter.

Content Features. Content features characterize a Twitter user's first feedback tweet. The tweets in the dataset that we use has been annotated by Telkom's social media analysts (see Sect. 2.1). We leverage the annotations and use them as content features. We use a total of eight content features; each of them corresponds to one of the eight possible categories of tweets listed in Table 1. The value of each of these eight features is the sentiment polarity score that is assigned manually by Telkom's social media analyst.

4.2   Methodology

In general, our methodology contains of two phases: a model building phase and a prediction phase, as shown in Fig. 3. In the model building phase, our goal is to build a prediction model based on a training set of customers along with their profiles, first feedback tweets and category labels. In the prediction phase, this model is used to predict the category of a new customer based on his/her profile and first feedback tweet.

In the model building phase, we first extract profile and content features from customers in the training data. Next, for the profile features (i.e., *TweetCount, FollowerCount, FolloweeCount, FavCount, ListCount*), since the variation of the feature values is high, we normalize them to have values between 0 and 1. However, we do not normalize the content features, since we want to preserve the actual sentiment polarity scores and the variation of these scores is not high. After the features are extracted, we apply SMOTE (described in Sect. 2.2) to handle imbalanced data. Finally, we use a classification algorithm to build a prediction model.

We explore three classification algorithms, namely Logistic Regression, Naive Bayes and Random Forest. These algorithms are widely used in data mining research such as in [3,5,20].

In the prediction phase, we extract values of profile and content features for a new customer whose category is to be inferred. These feature values are extracted from the new customer's Twitter profile and his/her first feedback tweet. Next, we apply the prediction model that we have learned in the model building phase on the new customer's feature values. This model will output a prediction, which is one of the three categories listed in Table 4.

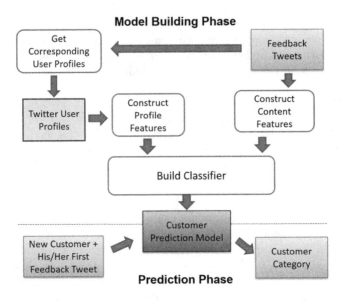

Fig. 3. Our approach for predicting customer categories

5   Experiments and Results

5.1   Dataset and Experiment Setting

There are 6,031 distinct Twitter users in our Telkom dataset. However, we could not collect profile features of a number of them. This is the case since not all Twitter accounts are public. Among the 6,031 users, we are able to get 5,813 user profiles. This represents 96.39 % of distinct users in our dataset. For each of these users, we identify his/her tweet that will be used as input to the prediction task. We consider the earliest feedback tweet that is posted during the observation period (i.e., June–August 2015) as such tweet.

We use the implementations of Logistic Regression, Naive Bayes and Random Forest in Weka [10]. We apply SMOTE filter for all of the three variants. All parameters are set into Weka default settings. We also perform 10-fold cross validation to investigate the effectiveness of our approach.

As a baseline, we use an approach which we refer to as *WeightedRandom-Picker*. This baseline picks one of the three categories randomly based on the percentage of customers of each category in our dataset (see Table 4). For example, given a new customer, *WeightedRandomPicker* predicts that the customer belongs to Class 1 (Kind) with a probability of 0.10, Class 2 (One-Off) with a probability of 0.47 or Class 3 (Spiteful) with a probability of 0.30.

5.2   Evaluation Metrics

As yardsticks to measure the effectiveness of our approach and the baseline, we use precision, recall, and F-measure. These metrics are common metrics that have been widely used in many past studies such as [8,12,17,18,21].

These metrics are calculated based on four possible outcomes of a Twitter user in an evaluation set: True Positive (TP), True Negative (TN), False Positive (FP) and False Negative (FN). For example, in case of predicting spiteful customer, TP is when a spiteful customer is correctly predicted as such; FP is when a non spiteful customer is wrongly predicted as a spiteful customer; FN is when a spiteful customer is wrongly predicted as a non spiteful customer; TN is when a non-spiteful customer is correctly predicted as non-spiteful customer.

Since we deal with multi-class classification, we also calculate weighted precision, weighted recall and weighted F-measure. We use the following formula to calculate weighted F-measure (similarly for weighted precision and recall):

$$WeightedFM = \frac{\sum_c FM(x) \times x))}{n} \tag{1}$$

In the above equation, c is total number of classes (in our case: 3), $FM(x)$ is the F-measure score for class x, x is total number of data instances that belong to a particular class, and n is the total number of instances in the dataset.

5.3   Research Questions and Results

RQ1: How well does our approach perform in predicting different categories of customers?

Approach: In this research question, we investigate three variants of our approach which uses three supervised classification algorithms (Logistic Regression, Naive Bayes and Random-forest), and compare its performance (measured in terms of precision, recall, and F-measure) with that of the *WeightedRandomPicker* baseline.

Results: Table 5 shows the results of our experiments. From the table, we can see that the three variants of our approach consistently outperform the *WeightedRandomPicker* baseline. Among the three classes, determining spiteful customers (C3) is the hardest problem. In this case, Logistic Regression performs the worst when compared to the other two supervised algorithms. But still, it outperforms the baseline by 13 % in terms of weighted F-measure. Meanwhile, determining kind customers (C1) is the easiest task, and all three variants of our approach outperform the baseline by more than 53 % in terms of F-measure.

RQ2: How effective is the oversampling strategy to improve classification accuracy?

Approach: We apply SMOTE to handle imbalanced class problem. In this research question, we compare results obtained by our approach when SMOTE is used and when it is not used.

Table 5. Effectiveness of various variants of our approach which uses different underlying classification algorithms to predict customer categories. C1: Kind customers, C2: One-Off customers, C3: Spiteful customers.

Algorithm	Metrics	C1(Kind)	C2(One-Off)	C3(Spiteful)	Weighted
Logistic Regression	precision	0.969	0.732	0.574	0.738
Logistic Regression	recall	0.769	0.971	0.372	0.744
Logistic Regression	F-measure	0.858	0.835	0.451	0.724
Random Forest	precision	0.973	0.754	0.697	0.787
Random Forest	recall	0.819	0.926	0.667	0.824
Random Forest	F-measure	0.890	0.832	0.682	0.800
Naive Bayes	precision	0.881	0.733	0.525	0.704
Naive Bayes	recall	0.767	0.925	0.897	0.881
Naive Bayes	F-measure	0.820	0.818	0.663	0.772
Baseline	precision	0.452	0.474	0.296	0.415
Baseline	recall	0.212	0.474	0.366	0.382
Baseline	F-measure	0.288	0.472	0.322	0.385

Table 6. Weighted F-Measure of our approach with and without SMOTE

Algorithm	No SMOTE	With SMOTE	Improvement
Logistic Regression	0.542	0.724	33.58 %
Random Forest	0.591	0.800	35.33 %
Naive Bayes	0.534	0.772	44.49 %

Results: Table 6 shows results that our approach achieves when SMOTE is turned off and on. We can note that by applying SMOTE the effectiveness of our approach (measured in terms of weighted F-measure) can be improved by 33–44%. This result gives evidence that handling imbalanced data by using minority-class oversampling improves the accuracy of the constructed prediction model.

5.4 Discussion

Our experiments show that among the three variants of supervised classification algorithm, Random Forest performs the best with F-Measure of 0.890, 0.832 and 0.682 for predicting kind, one-off, and spiteful customers respectively. This finding is consistent with previous study by Caruana et al. [5] which observed that random forests tended to perform well across different settings. Even for the variant that uses the most basic machine learning approach among the three (i.e., Naive Bayes), the prediction performance is 30 % better than that of *WeightedRandomPicker*. These results highlight a promising potential of applying machine learning techniques to identify different categories of customer based on their first feedback tweets.

Our prediction model relies on sentiment polarity of customer feedback. To ensure the correctness of sentiment polarity of user's tweet, we decided to use labeled/annotated data that Telkom provides and none of the authors are involved in the labeling/annotation process.

A limitation of our study is the sample used in the case study. We have evaluated the effectiveness of our approach to infer customer categories from tweets that mention one company in Indonesia. In the future, we plan to address this limitation by considering a larger set of tweets collected over a longer period of time. We also plan to experiment with other companies situated in different countries.

6 Related Work

Inferring User's Attributes and Behavior. Pennacchiotti et al. presented an approach to infer the values of a Twitter user's hidden attributes such as political orientation or ethnicity by analyzing observable information such as the user behavior, network structure and the linguistic content of the user's Twitter feed [16]. Another work by On et al. studies interactions in email network [15]. They use internal company dataset, and build a model to predict email reply order. However, our work differs with previously mentioned works since we use different sets of features taken from user profile and feedback contents. We also focus on a different problem, namely the prediction of customer category based on feedback tweets.

Handling Imbalanced Data. Van et al. [19] and Huang et al. [11] have investigated the effectiveness of oversampling strategies to handle issues with imbalanced datasets. Our findings in this work further demonstrate the value of using an oversampling method to deal with imbalanced dataset. In RQ2 (see Sect. 5.3), we show that oversampling substantially improves the prediction accuracy of our customer category prediction approach.

Social Listening Framework. Bhatia et al. develop a system that automatically monitors social network platforms, analyzes data from the platforms, and triggers events that lead to corrective actions [4]. Ajmera et al. analyze posts and messages in social network platforms and identify posts relevant to an enterprise [2]. Einwiller et al. examined the complaining behavior and complaint management on Social Media, focusing primarily on how companies manage the complaints [9]. Millard et al. found that customers engage with brands not only to complain but also to complement [14]. Chen et al. introduce a brand-specific intelligent filters on Twitter which is called CrowdE using a common crowd-enabled process [7]. Our work highlights another framework that has been implemented and currently used by a large telecommunication company using customized commercial tools. Our work extends the company's social listening framework with a capability to predict customer categories.

7   Conclusion and Future Work

In this study, we propose a method to predict customer categories (i.e., kind, one-off, and spiteful) given a customer's profile and first feedback tweet. Our approach extracts a set of profile features and content features and use these to build a prediction model using a classification algorithm. To demonstrate the accuracy of our approach, we evaluate our approach using a real dataset of labeled tweets mentioning an official account of a large telecommunication company in Indonesia. We evaluate our approach by using common evaluation metrics in data mining research (i.e., precision, recall, and F-measure), and compare its performance with that of a weighted random picker baseline. Our experiment results show that three variants of our approach that uses different underlying classification algorithms can substantially outperform the baseline. Our approach can benefit companies to improve their customer service strategies to deal with different categories of customers. For the company in our case study, our approach extends the current capability of their social media listening system by adding a prediction functionality.

In the future, we plan to evaluate our proposed approach on more datasets. To improve the accuracy of our approach further, we plan to extract more features to better characterize different categories of users. We also plan to investigate more advanced classification algorithms.

Acknowledgments. This research is supported by the National Research Foundation, Prime Minister's Office, Singapore under its International Research Centres in Singapore Funding Initiative, and PT Telekomunikasi Indonesia (Telkom).

References

1. VentureBeat Report social media (2014). http://venturebeat.com/2014/12/12/social-media-we-complain-879-million
2. Ajmera, J., Ahn, H.i., Nagarajan, M., Verma, A., Contractor, D., Dill, S., Denesuk, M.: A CRM system for social media: challenges and experiences. In: Proceedings of the 22nd international conference on World Wide Web, pp. 49–58. International World Wide Web Conferences Steering Committee (2013)
3. Arapakis, I., Cambazoglu, B.B., Lalmas, M.: On the feasibility of predicting news popularity at cold start. In: Aiello, L.M., McFarland, D. (eds.) SocInfo 2014. LNCS, vol. 8851, pp. 290–299. Springer, Heidelberg (2014). doi:10.1007/978-3-319-13734-6_21
4. Bhatia, S., Li, J., Peng, W., Sun, T.: Monitoring and analyzing customer feedback through social media platforms for identifying and remedying customer problems. In: Proceedings of the 2013 IEEE/ACM International Conference on Advances in Social Networks Analysis and Mining, pp. 1147–1154. ACM (2013)
5. Caruana, R., Niculescu-Mizil, A.: An empirical comparison of supervised learning algorithms. In: Proceedings of the 23rd International Conference on Machine Learning, pp. 161–168. ACM (2006)
6. Chawla, N.V., Bowyer, K.W., Hall, L.O., Kegelmeyer, W.P.: SMOTE: synthetic minority over-sampling technique. J. Artif. Intell. Res. **16**, 321–357 (2002)

7. Chen, J., Cypher, A., Drews, C., Nichols, J.: CrowdE: filtering tweets for direct customer engagements. In: ICWSM, Citeseer (2013)
8. Dev, H., Ali, M.E., Mahmud, J., Sen, T., Basak, M., Paul, R.: A real-time crowd-powered testbed for content assessment of potential social media posts. In: Liu, T.-Y., Scollon, C.N., Zhu, W. (eds.) SocInfo 2015. LNCS, vol. 9471, pp. 136–152. Springer, Heidelberg (2015). doi:10.1007/978-3-319-27433-1_10
9. Einwiller, S.A., Steilen, S.: Handling complaints on social network sites-an analysis of complaints and complaint responses on facebook and twitter pages of large us companies. Public Relations Rev. **41**(2), 195–204 (2015)
10. Hall, M., Frank, E., Holmes, G., Pfahringer, B., Reutemann, P., Witten, I.H.: The weka data mining software: an update. ACM SIGKDD Explor. Newslett. **11**(1), 10–18 (2009)
11. Huang, P.J.: Classication of imbalanced data using synthetic over-sampling techniques (2015)
12. Jhamtani, H., Chhaya, N., Karwa, S., Varshney, D., Kedia, D., Gupta, V.: Identifying suggestions for improvement of product features from online product reviews. In: Liu, T.-Y., Scollon, C.N., Zhu, W. (eds.) SocInfo 2015. LNCS, vol. 9471, pp. 112–119. Springer, Heidelberg (2015). doi:10.1007/978-3-319-27433-1_8
13. Meilă, M., Heckerman, D.: An experimental comparison of several clustering and initialization methods. In: Proceedings of the Fourteenth Conference on Uncertainty in Artificial Intelligence, pp. 386–395. Morgan Kaufmann Publishers Inc. (1998)
14. Millard, N.J.: Serving the social customer: how to look good on the social dance floor. In: Nah, F.F.-H., Tan, C.-H. (eds.) HCIB 2015. LNCS, vol. 9191, pp. 165–174. Springer, Heidelberg (2015). doi:10.1007/978-3-319-20895-4_16
15. On, B.W., Lim, E.P., Jiang, J., Purandare, A., Teow, L.N.: Mining interaction behaviors for email reply order prediction. In: International Conference on Advances in Social Networks Analysis and Mining (ASONAM), pp. 306–310. IEEE (2010)
16. Pennacchiotti, M., Popescu, A.M.: A machine learning approach to twitter user classification. ICWSM **11**(1), 281–288 (2011)
17. Rangnani, S., Susheela Devi, V., Narasimha Murty, M.: Autoregressive model for users' retweeting profiles. In: Liu, T.-Y., Scollon, C.N., Zhu, W. (eds.) SocInfo 2015. LNCS, vol. 9471, pp. 178–193. Springer, Heidelberg (2015). doi:10.1007/978-3-319-27433-1_13
18. Tian, Y., Lo, D., Xia, X., Sun, C.: Automated prediction of bug report priority using multi-factor analysis. Empirical Softw. Eng. **20**(5), 1354–1383 (2015)
19. Van Hulse, J., Khoshgoftaar, T.M., Napolitano, A.: Experimental perspectives on learning from imbalanced data. In: Proceedings of the 24th International Conference on Machine Learning, pp. 935–942. ACM (2007)
20. Van Vlasselaer, V., Eliassi-Rad, T., Akoglu, L., Snoeck, M., Baesens, B.: Afraid: fraud detection via active inference in time-evolving social networks. In: Proceedings of the 2015 IEEE/ACM International Conference on Advances in Social Networks Analysis and Mining 2015, pp. 659–666. ACM (2015)
21. Xia, X., Lo, D., Shihab, E., Wang, X., Yang, X.: ELBlocker: predicting blocking bugs with ensemble imbalance learning. Inf. Softw. Technol. **61**, 93–106 (2015)

Poster Papers: Privacy, Health and Well-being

Validation of a Computational Model for Mood and Social Integration

Altaf Hussain Abro$^{(\boxtimes)}$ and Michel C.A. Klein

Behavioural Informatics Group, Vrije Universiteit Amsterdam,
De Boelelaan 1081, 1081 HV Amsterdam, The Netherlands
{a.h.abro,michel.klein}@vu.nl

Abstract. The social environment of people is an important factor for the mental health. However, in many internet interventions for mental health the interaction with the environment has no explicit role. It is known that the social environment can help people to reduce the feelings of loneliness and has a positive impact on mood in particular. Participation in social activities and maintaining social interaction with friends and relatives are frequently seen as indicators of a happy and healthy life. It is also commonly accepted that being integrated in social network has a strong protective effect on health and helps to avoid feelings of loneliness. In this paper we present a computational model that can be used for analyzing and predicating the mood level of individuals by taking into account the social integration, the participation in social activities and the enjoyableness of those activities. In addition to this, we explain the method that we developed to validate the computational model. For the validation, we use real EMA data that was collected from E-COMPARED project. This model allows to make more precise predictions on the effect of social interaction on mood and might be part of future internet interventions.

Keywords: Social network · Social integration · Mood · Social interaction · Social activities

1 Introduction

The social network of an individual can help him out of feelings of loneliness and isolation, as having good friends around gives a sense of social integration within a social network and provides a reason for happiness as well. These social networks can encourage individuals to participate or engage themselves more in social activities as well as social interaction which ultimately leads to better mental as well as a better physical health. Socially isolated individuals or people with limited contacts have a higher chance of suffering from health issues [1, 2]. Socially well integrated people have more social contact, they are mentally happier and healthier than those with limited social contacts [3].

There are several influences between mental health and a person's social environment. Researchers in the field of social psychology and social science have put forwarded several theories about the association between the *loneliness* (or the comparable concept of isolation) and the *integration within in social networks* [4, 5]. The structural

© Springer International Publishing AG 2016
E. Spiro and Y.-Y. Ahn (Eds.): SocInfo 2016, Part II, LNCS 10047, pp. 385–399, 2016.
DOI: 10.1007/978-3-319-47874-6_27

characteristics of social networks such as quality and quantity of relationships have much influence on individual's lives. Relationships give a perception of social integration and have a positive influence on mental as well as physical health [6, 7]. People with higher level of quality and a larger number of relationships are considered to be well integrated within the network [8], and on the other hand people having feelings of loneliness and isolation are less integrated [9].

In addition, *social interaction* among or between friends is considered as one of the major benefits of a rich social networks. It also depends on the quality and quantity of our social interactions [10], e.g. people with less participation in social activities suffers of cognitive decline [11, 12]. So, various social environmental factors have been identified as predictors of health and wellbeing either physically or mentally.

Nowadays much attention has been paid towards the computational aspects in relation with mental and psychological well-being, [13, 14] as this can form the basis for human-aware smart applications that can assist people in their routine life and to support healthy behaviour.

In this paper we take the structural aspects of social networks into account, in relation with mood level of the individuals, and we explore these computationally. Specifically, we propose a computational model and a methodology to validate this model with real life data. The model incorporates various social aspects such as: social network strength (e.g. number of friends, strong and weak relationships), level of social integration, social interaction, involvement in social activities, enjoyableness of social activities and their effect on individual's mood level. For the validation of this model we have compared the predictions of our model with three naïve predictions. The goal is to compare all of 4 predictions and see whether our model has added value compared with the naïve approaches. Thus, our main research questions are:

- How can we computationally model the effect of social factors on mood?
- Is this model better in predicting the mood then simple?

The remainder of this paper is organized as follows: In Sect. 2, some literature about the effect of social networks on mental health in general and mood in particular is discussed. The conceptual overview of the computational model is described in Sect. 3, which includes the details of various concepts and their relationships. In Sect. 4, the data collection process is explained. In Sect. 5, simulation results are provided to show whether our model can better predict the mood level of individuals. Finally, Sect. 6 concludes the paper and identifies possibilities for future research.

2 Background

Researchers in clinical psychology are working on different interventions by considering various aspects related to the mental health. In this paper we take the social environment of the individuals into account. There are not many interventions yet that explicitly take the interaction of people with their social environment into account. In the following paragraphs, we discuss the relevant literature on the relation between social environment and mental health.

Social networks of people consist of relationships with other people, and relations are usually classified as either *strong ties* or *weak ties*. In other words, both the quality and quantity of relationships in a social network is relevant.

The level of *social integration* plays a vital role in the health and well-being. It has been observed that people who often join social groups that help them to expand their social circles by making more friends, and who involve themselves in various social activities relates to the health promotion, remain healthier either mentally or physically [2, 15].

In the literature it also has been well documented that an increase in *loneliness and social isolation* can have a serious impact on wellbeing and quality of life, with certain negative health consequences[16, 17], such as, psychological and cognitive decline. Individuals who have very less or weak relationships and do not participate in social activities are at an increased risk of cognitive decline [12], and increased risk of mortality and suicide [18]. On the other hand, people who are well integrated in social networks live longer and have a healthier life [19, 20].

Social activities can help people to increase the interaction with their social network and to maintain good relationships. In particular, meaningful activities should be enjoyable, as enjoyableness has a positive effect on the mental health in general and mood in particular. One could think of activities such as outings, having discussions with friends. etc. Enjoyable activities give a sense of happiness as well as feelings of being connected to normal life, which helps people to maintain their mood at certain level and to avoid depression. Some studies shows that people who are not much engaged in social activities or people who have dysfunctional social behaviour are more at risk of depression. [11] In contrast, people with a low mood or who have depressive symptoms are often involve more in negative social interactions[21, 22]. Numerous studies show that participation in social activities as well as enjoyment have a lasting effect on cognitive health [23]. In another study, which has been conducted on elderly people is reported that these social and productive activities are considered as an useful intervention for elderly people, as these activities require less or no physical effort [24]. Such kind of social activities have a positive influence on the psychological and mental health.

It appears that socially connected individuals engage themselves more often in positive social interactions than less connected individuals [25], so the mood level of the individuals varies on the basis of social connectedness and social interaction. Another study shows that positive social interaction on daily basis has a positive effect on daily mood, and negative social interaction has a negative effect on daily mood [26].

Also from a computational perspective attempts have been made to get an insight that how mood level varies with environmental factors. Mood level often changes with circumstances. People that are emotionally unstable may perceive situations more negatively, are unable to regulate their emotions, and are more vulnerable to low mood/symptoms of depression compared to emotionally stable people [13]. From the perspective of perceived social support: if people have the perception that people in their social circles will provide support during bad situations, it will have a positive impact on mood level [14].

3   Computational Model

In this section, an overview of the various concepts of the proposed computational model is presented, and a short description of each state is provided in Table 1. It is explained in detail how different states affect each other. The model is depicted in Fig. 1, which shows the dynamics of relationships (indicated by arrows) between the concepts. This computational model is based on the literature discussed in the sections above.

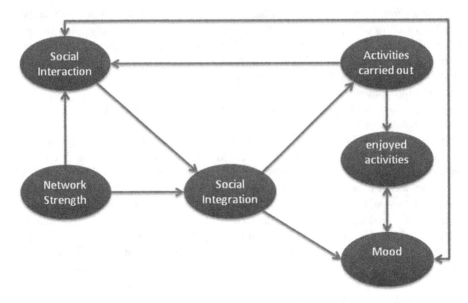

Fig. 1. Proposed computational model

Table 1. Overview of the states of the proposed model (see also Fig. 1)

Name	Description
Network strength	An abstract representation of the overall strength of social network that may include aspects such a: number of friends, strong ties, weak ties
Social interaction	The level of communication with friends, for example: how often they interact with each other
Social integration	The level of social integration of a person within his/her social network
Activities carried out	Number of social activities carried out by an individual
Enjoyed activities	The extent to which a person enjoyed his social activities
Mood	The mood level of an individual at a particular point in time

In this model, a state named *network strength* has been used. It represents an abstract concept of the overall strength of social network that include aspects such as number of friends, strong relationships and weak relationships (strong and weak ties). The network strength has influence on the perception of individuals of being integrated within social network, so it has influence on the level of *social integration*. With a high level of social integration, an individual is considered as well integrated within its social network, and on the other hand a low level of social integration is similar to isolation/loneliness. In other words, if an individual is less integrated within its social network, he/she might have more feelings of loneliness compared to a person who is well integrated. Social integration has direct influence on the *mood level* of an individual as well, as people who are well integrated within their social circles have a more positive mood and feel more happy and healthy.

It is generally accepted that if people have more friends or strong relations within their network they interact more or make often communication compared with less number of friends or having weak relationships. Therefore, the concept mood level has an influence on *social interaction*.

Participation in *social activities* is also somehow dependent on the social integration: people with more friends and who have more strong ties are encouraged to go out and carry out more social activities with friends. As a consequence, a higher participation level leads to more *social interaction* as well. As was discussed earlier, the social network strength is one of the main factors in the social environment, so the social interaction between friends also depends on the type of relationships: if a person has a larger number of friends and have more close friends, he might communicate and interact with them often. So the *activities carried out* and the *network strength* have influence on the social interaction. On the other hand, as we have seen it also depends on the mood of the person.

Enjoyment of activities is considered in this paper as the rating for the performed activities, it shows that how much a person have enjoyed these activities. This has a direct effect on mood as well. On the other hand, if the mood level is low it can reduce the level of enjoyment.

3.1 Formalization

The conceptual model of social integration and mood as depicted in Fig. 1 is formalized as follows. The following elements are part of the formalization:

- For each connection from state X to state Y a *weight* $\omega_{X,Y}$ (a number between 0 and 1), for the strength of the impact through this connection.
- For each state Y a *speed factor* η_Y (a positive value) is used that determines the speed with which the value of state changes, and
- For each state Y (a reference to) a *combination function* $c_Y(...)$ used to aggregate multiple impacts from different states on one state Y.

For a numerical representation of the model the states Y get activation values indicated by $Y(t)$: real numbers between 0 and 1 over time points t, where the time variable t ranges over the real numbers. More specifically, the conceptual representation of the model (as shown in Fig. 1 and in Table 1) can be transformed in a systematic or even automated manner into a numerical representation as follows [27]:

- At each time point t each state X connected to state Y has an *impact* on Y, which is defined as

$$\textbf{impact}_{X,Y}(t) = \omega_{X,Y}X(t)$$

where $\omega_{X,Y}$ is the weight of the connection from X to Y

- The *aggregated impact* of multiple states X_i on Y at t is determined using a *combination function* $\textbf{c}_Y(..)$:

$$\textbf{aggimpact}_Y(t) = \textbf{c}_Y\left(\textbf{impact}_{X_1,Y}(t), \ldots, \textbf{impact}_{X_k,Y}(t)\right)$$
$$= \textbf{c}_Y\left(\omega_{X_1,Y}X_1(t), \ldots, \omega_{X_k,Y}X(t)\right)$$

where X_i are the states with connections to state Y
- The effect of $\textbf{aggimpact}_Y(t)$ on Y is exerted over time gradually, depending on *speed factor* η_Y:

$$Y(t + \Delta t) = Y(t) + \eta_Y[\textbf{aggimpact}_Y(t) - Y(t)]\Delta t$$

or, in differential equation format:

$$\textbf{d}Y(t)/\textbf{d}t = \eta_Y[\textbf{aggimpact}_Y(t) - Y(t)]$$

- Thus the following *difference* and *differential equation* for Y are obtained:

$$Y(t + \Delta t) = Y(t) + \eta_Y[\textbf{c}_Y(\omega_{X_1,Y}X_1(t), \ldots, \omega_{X_k,Y}X_k(t)) - Y(t)]\Delta t$$
$$\textbf{d}Y(t)/\textbf{d}t = \eta_Y\left[\textbf{c}_Y(\omega_{X_1,Y}X_1(t), \ldots, \omega_{X_k,Y}X_k(t)) - Y(t)\right]$$

In the model considered here for all states for the standard combination function the *advanced logistic sum combination function* $\textbf{alogistic}_{\sigma,\tau}(...)$ is used [27]:

$$c_Y(V_1, \ldots, V_k) = \textbf{alogistic}_{\sigma,\tau}(V_1, \ldots, V_k) = \left(\frac{1}{1 + e^{-\sigma(V_1,\ldots,V_k - \tau)}} - \frac{1}{1 + e^{\sigma\tau}}\right)(1 + e^{\sigma\tau})$$

Here σ is a *steepness* parameter and τ a *threshold* parameter. The advanced logistic sum combination function has the property that activation levels 0 are mapped to 0 and it keeps values below 1. For example, for the mood state the model is numerically represented in difference equation form as:

$$\text{mood}(t + \Delta t) = \text{mood}(t) + \eta_{\text{mood}}[\textbf{aggimpact}_{\text{mood}}(t) - \text{mood}(t)]\Delta t$$

where:

$$\textbf{aggimpact}_{\text{mood}}(t) = \textbf{alogistic}_{\sigma,\tau}(\omega_{\text{socialInteraction, mood}}\,\text{socialInteraction}(t),$$
$$\omega_{\text{activitiesCarriedout, mood}}\,\text{activitiesCarriedout}(t),$$
$$\omega_{\text{enjoyedActivities, mood}}\,\text{enjoyedActivities}(t))$$

The *steepness* σ and *threshold* τ parameters are used to keep simulation patterns within boundaries of zero and one, and these parameters have been chosen after applying various scenarios on the model.

In this way the conceptual model presented above is transformed into a computational model in terms of differential equations. The simulations are performed by applying a computational simulation method to this numerical model representation. All states and differential equations for them have been converted in a form that can be computed in a programming environment, i.e. MATLAB™.

4   Data Collection Method

To validate the computational model we have used data that has been collected in the E-COMPARED (European Comparative Effectiveness Research on Internet-based Depression treatment) project through trials conducted by psychologists that treated depressed patients with help of an online system. In these trials a mobile phone application is used for so-called Ecological Momentary Assessment (EMA): real-time monitoring of patients' state in their natural environment.

The EMA data consists of measurements of (1) mood state, (2) the degree to which of respondents enjoyed activities, (3) the level of social interactions and (4) the degree to which respondents engaged in pleasant activities. The data has been collected during 10 weeks of psychological treatment. Table 2 shows the questions and number of time these questions were asked during 10 weeks. The total number of patients in the data set is 49. The question on the mood has been asked every day. In addition, during the first and last week of the treatment and at one random day during week 2 till 8, the mood question has been asked again (thus, once in the morning and once in the evening) and the other questions have been asked in the evening as well.

Table 2. EMA questions and number of times asked.

Questions	Total number
How is your mood right now?	92
How much have you been involved in social interactions today?	22
To what extent have you carried out enjoyable activities today?	22
How much have you enjoyed the day's activities?	22

Thus, the EMA questions assess mood, enjoyment in activities, social contacts and the level of engagement in pleasant activities. The following Figs. 2 and 3 show examples of actual data for two specific patients.

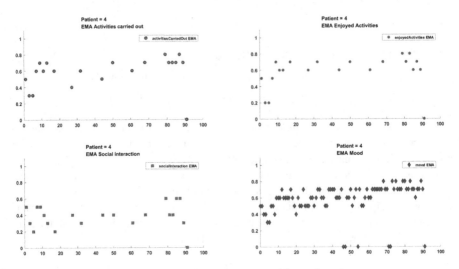

Fig. 2. Actual data points as rated by patient 4

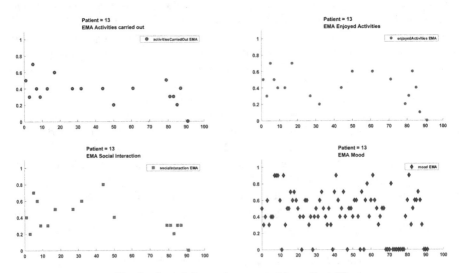

Fig. 3. Actual data points as rated by patient 13

5   Validation

This section describes the simulation results of our proposed model and a comparison to naïve models based on the EMA data.

5.1   Approach

To compare the outcomes of our model with the naïve models, we have used the Root Mean Error Square (RMSE) measure to calculate how good a model is in prediction the mood. As our model contains several parameters, it was necessary to tune these parameters first. For that purpose, we initially took the data of the first 10 patients to (manually) train our model in such a way that it resulted in the lowest RMSE for first 10 patients. Values for parameters threshold τ and steepness σ are given in Table 3. Parameter values of weights for connections between all states are provided in Table 4. After that, the model has applied to the test data set of the remaining 39 patients.

The simulation was executed for 92 time steps (assuming a step size of 1 day). The initial values for all states were set to initial actual EMA values for that specific patient.

Table 3. Values of threshhold and steepness

State	τ	σ	State	τ	σ
Network strength	0.34	4	Activities carrierout	0.34	5
Social integration	0.34	3	Enjoyed activities	0.34	4
Social interaction	0.44	3	Mood	0.50	4

Table 4. Values of parameters used: connection weights

Weight	Value	Weight	Value
$\omega_{networkStrength,socialIntegration}$	0.5	$\omega_{socialInteraction,mood}$	0.5
$\omega_{networkStrength,socialInteraction}$	0.4	$\omega_{mood,socialInteraction}$	0.5
$\omega_{socialInteraction,socialIntegration}$	0.4	$\omega_{socialIntegration,mood}$	0.3
$\omega_{socialIntegration,activitiesCarriedout}$	0.4	$\omega_{enjoyedActivities,mood}$	0.5
$\omega_{activitiesCarriedout,socialInteraction}$	0.6	$\omega_{mood,enjoyedActivities}$	0.4
$\omega_{activitiesCarriedout,enjoyedActivities}$	0.5		

The computational model presented in this paper is compared with three other simple models to predict the mood based on the basis of social aspects. The three different naïve approaches of predictions are the following:

1. **Average of EMA data:** In this first approach, the mood is predicted by taking the average of all other EMA measurements (social interaction, social activities, enjoyed activities) for the previous time point, according to the user rating of the particular question at particular time. Furthermore, in this approach it is assumed that the calculated mood is kept at same level until the next EMA measurement becomes available (this is necessary because the mood is asked more frequently than the other EMA questions).

2. **Mood average of first week:** this third approach of prediction is much simpler one, in this we have taken mood average of very first week; by assuming that mood will not change.

3. **Mood average of first and last week:** In this approach the mood average of first and last week is used as prediction of (a stable) mood level. Note that this approach, in contrast to the other methods, takes future knowledge (i.e. the average mood in the last week) into account.

For all these approaches, the RMSE is calculated based on the difference between the predicted mood and the actual mood according to the EMA questions.

5.2   Simulation Results

In addition to the concepts that are directly related to the EMA measures, our model uses two abstract concepts: the first one is *network strength* and the second one *social integration*. As we don't have real data for these concepts, we have taken a fixed value for network strength as well as for social integration, these values are 0.5 and 0.4 respectively.

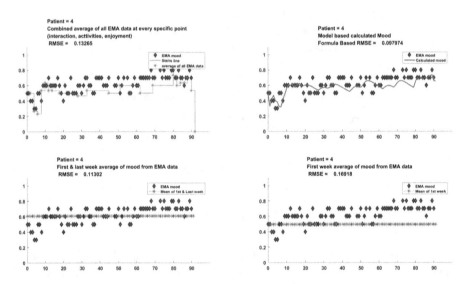

Fig. 4. Simulation results for patient 4

The graphs in Fig. 4 show a comparison between the four types of prediction for patient number 4. It shows that our computational model based approach depicted in top right corner has better predictive results then other three approached. By looking at the RMSE, we see that our model has a lower value than the others.

Apart from the scores of the model predictions, we could also look at the specific relation between the different EMA measurements. For this, we compare the model predictions in Fig. 4 with the data points as shown in Fig. 2 for patient 4. It can be seen that the mood of this patient is quite stable with a slight increase over time. Also, we see that he/she is getting involved in more social activities. This is reflected in our model predictions, which also show an gradual increase in mood.

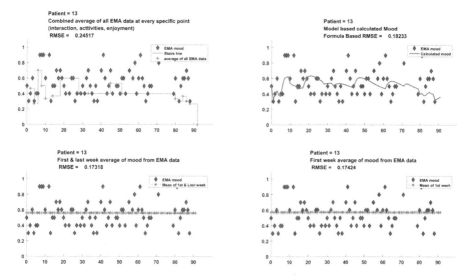

Fig. 5. Simulation results for patient 13

The results in Fig. 5 provide a comparison between the four types of prediction for patient 13. It shows that our computational model based approach depicted in top right corner still has better predictive results then the one in the top left corner, but that the predictions in the bottom row (which predict a fixed mood value) are better. It turned out that the mood of this person was quite stable, and that a prediction of a fixed value was more close to the actual mood than our dynamic prediction.

When we look at the actual EMA measurements for this patient shown in Fig. 3, it shows that patient was less active as he has less social interaction even sometimes he didn't rated the questions as well as enjoyment and activities carried out fluctuates al lot. In combination with the fluctuating mood scores, this made it difficult to predict.

In Table 5 an overview is provided of the RMSE of the different predictions for all 49 patients. In the bottom row, it can be seen that our model (second column with scores) has on average a lower error score than the "Average of all other EMA mea-surements" (first column). A standard t-test revealed that this difference is significant for the test set (patient 11 to 49) with a p-value of 0.01241. However, it can also be seen that the *fixed* predictors "average mood in the first week" and the "average mood in first and last week" are on average better predictors in this dataset, although the difference with our model prediction is not significant (p-value of respectively 0.3157 and 0.06327). When considering *dynamic* predictions, we can conclude that our model provides added value compared to a naïve approach that only uses the average EMA measurements of social interaction, activities carried out, and the enjoyment of the activities.

Table 5. RMSE values of all patients

Patient number	Average other EMA scores	Model predictions	Average mood first and last week	Average mood first week
1	0.25536	0.24906	0.24665	0.25415
2	0.31528	0.29616	0.28951	0.2902
3	0.19365	0.17494	0.16727	0.16494
4	0.13265	0.097974	0.11302	0.16918
5	0.25658	0.27311	0.22296	0.22296
6	0.20991	0.21215	0.17827	0.17075
7	0.29515	0.25034	0.22758	0.21466
8	0.17503	0.17296	0.17573	0.20845
9	0.17387	0.18691	0.16847	0.16741
10	0.15757	0.16915	0.13492	0.13341
↑training set, test set ↓				
11	0.21174	0.28907	0.27099	0.36401
12	0.25069	0.25767	0.247	0.2474
13	0.24517	0.18233	0.17318	0.17424
14	0.14286	0.14976	0.20082	0.23186
15	0.24124	0.20321	0.15393	0.17321
16	0.24592	0.20493	0.12119	0.12201
17	0.15868	0.14261	0.14153	0.14039
18	0.13453	0.12956	0.10693	0.10913
19	0.25874	0.23285	0.17609	0.17582
20	0.23651	0.22784	0.21958	0.21958
21	0.24956	0.22062	0.22649	0.25654
22	0.3107	0.21781	0.20375	0.2052
23	0.21303	0.21568	0.2226	0.23859
24	0.32293	0.31329	0.25526	0.26523
25	0.47434	0.061777	0.15	0.15
26	0.19285	0.19233	0.16037	0.17384
27	0.2103	0.17795	0.17944	0.18375
28	0.39299	0.13805	0.12472	0.12472
29	0.32924	0.27192	0.20997	0.19685
30	0.1693	0.14202	0.12134	0.15303
31	0.20721	0.21183	0.1777	0.1777
32	0.36193	0.29885	0.2643	0.2651
33	0.11777	0.10759	0.089146	0.088089
34	0.33422	0.29228	0.25897	0.25897
35	0.23552	0.19888	0.17094	0.17579
36	0.19498	0.16941	0.13622	0.16382
37	0.30148	0.25973	0.24271	0.25114
38	0.23666	0.22222	0.20887	0.21951

(Continued)

Table 5. (*Continued*)

Patient number	Average other EMA scores	Model predictions	Average mood first and last week	Average mood first week
39	0.27954	0.26553	0.23953	0.28643
40	0.23924	0.21702	0.18754	0.2212
41	0.24819	0.19186	0.19729	0.20556
42	0.25692	0.24088	0.18726	0.1868
43	0.20586	0.176	0.16342	0.16702
44	0.20894	0.23053	0.18919	0.18633
45	0.24195	0.25615	0.20407	0.19529
46	0.31625	0.28451	0.24912	0.25208
47	0.22697	0.1913	0.17938	0.17852
48	0.20646	0.20997	0.19962	0.20843
49	0.49494	0.25684	0.22509	NaN
Average RMSE total	**0.2463551**	**0.2109268**	**0.1910189**	**0.1997766**
Average RMSE test	**0.2465571**	**0.2106915**	**0.1906470**	**0.1993570**

6   Conclusion and Future Work

Our social environment plays an important in mental health: people who are going out, have good relationships and are involved in social interaction seem to have higher mood level. In this paper a computational model of the effect of social environment on the mood has been proposed. The model can be used to simulate, analyze and predict the effect of social aspects on mood level.

The model has been evaluated using a data set of EMA measurements of different social aspect and the mood of people following a depression therapy. In the results it is shown that our model is better in dynamically predicting mood with the help of other EMA measurements than a simple model that takes the average of the other measurements as prediction. However, it also turned out that predicting a fixed mood value for the whole period resulted in an even lower error score. This suggests that the dynamics in mood are not very large and that the effect of other social aspects on mood are relatively small.

In the future, an analysis of the dynamics of mood and social aspects of a larger data set is needed in order to validate the small effects. Also, an integration of the presented model with an existing model of mood dynamics based on other aspects [28] is planned. Given the fact that the presented model has only partially been validated, as yet we don't have the data regarding all concepts which have been used in the model, further data is required. Finally, it is planned to extend the model by adding some more structural social aspects e.g. (strong and weak ties/relationships).

Acknowledgement. Funding for this research work is provided by the E-COMPARED project. The E-COMPARED project is funded by the European Commission's Seventh Framework Programme, under grant number 603098.

References

1. Nicholas Jr., N.R.: Social isolation in older adults: an evolutionary concept analysis. J. Adv. Nurs. **65**(6), 1342–1352 (2008)
2. Cacioppo, J.T., James, H.F., Nicholas, A.C.: Alone in the crowd: the structure and spread of loneliness in a large social network. J. Pers. Soc. Psychol. **97**(6), 977–991 (2009)
3. Cohen, S., Willis, T.A.: Stress, social support, and the buffering hypothesis. Psychol. Bull. **98**, 310–357 (1985)
4. Zavaleta, D., Samuel, K., Mills, C.: Social isolation: a conceptual and measurement proposal. In: OPHI Working Papers, vol. 67. University of Oxford (2014)
5. Cacioppo, J.T., Patrick, B.: Loneliness: Human Nature and the Need for Social Connection. W.W. Norton & Company, New York (2008)
6. Cohen, S., Gottlieb, B., Underwood, L.: Social relationships and health. In: Cohen, S., Underwood, L., Gottlieb, B. (eds.) Measuring and Intervening in Social Support, pp. 3–25. Oxford University Press, New York (2000)
7. Uchino, B.N., Cacioppo, J.T., Kiecolt-Glaser, J.K.: The relationship between social support and physiological processes: a review with emphasis on underlying mechanisms and implications for health. Psychol. Bull. **119**, 488–531 (1996)
8. Brissette, I., Cohen, S., Seeman, T.E.: Measuring social integration and social networks. In: Cohen, S., Underwood, L., Gottlieb, B. (eds.) Measuring and Intervening in Social Support, pp. 53–85. Oxford University Press, New York (2000)
9. Cacioppo, J.T., Hawkley, L.C., Crawford, E., Ernst, J.M., Burleson, M.H., Kowalewski, R. B., et al.: Loneliness and health: potential mechanisms. Psychosom. Med. **64**, 407–417 (2002)
10. Kiecolt-Glaser, J.K., Newton, T.L.: Marriage and health: his and hers. Psychol. Bull. **127**, 472–503 (2001)
11. Steger, M.F., Todd, B.K.: Depression and everyday social activity, belonging, and well-being. J. Couns. Psychol. **56**(2), 289–300 (2009)
12. Beland, F., Zunzunegui, M.V., Alvarado, B., Otero, A., Del Ser, T.: Trajectories of cognitive decline and social relations. J. Gerontol. Ser. B Psychol. Sci. Soc. Sci. **60**, 320–330 (2005)
13. Abro, A.H., Klein, M.C.A., Manzoor, A.R., Tabatabaei, S.A., Treur, T.: Modeling the effect of regulation of negative emotions on mood. Biologically Inspired Cognit. Archit. **13**, 35–47 (2015)
14. Abro, A.H., Klein, M.C.A., Tabatabaei, S.A.: An agent-based model for the role of social support in mood regulation. In: Bajo, J., Hallenborg, K., Pawlewski, P., Botti, V., Sánchez-Pi, N., Duque Méndez, N.D., Lopes, F., Julian, V. (eds.) PAAMS 2015. CCIS, vol. 524, pp. 15–27. Springer, Heidelberg (2015). doi:10.1007/978-3-319-19033-4_2
15. Pitkala, K.H., Routasolo, P., Kautiainen, H., Tilvis, R.S.: Effects of psychosocial group rehabilitation on health, use of health care services, and mortality of older persons suffering from loneliness: a randomised, controlled trial. J. Gerontol. Med. Sci. **64A**(7), 792–800 (2009)
16. Masi, C.M., Chen, H.Y., Hawkley, L.C., Cacioppo, J.T.: A meta-analysis of interventions to reduce loneliness. Pers. Soc. Pyschol. Rev. **15**(3), 219–266 (2011)

17. Allen, N.B., Badcock, P.B.T.: The social risk hypothesis of depressed mood: evolutionary, psychosocial, and neurobiological perspectives. Psychol. Bull. **129**, 887–913 (2003)
18. Eng, P.M., Rimm, E.B., Fitzmaurice, G., Kawachi, I.: Social ties and change in social ties in relation to subsequent total and cause-specific mortality and coronary heart disease incidence in men. Am. J. Epidemiol. **155**, 700–709 (2002)
19. Fratiglioni, L., Paillard-Borg, S., Winblad, B.: An active and socially integrated lifestyle in late life might protect against dementia. Lancet Neurol. **3**, 343–353 (2004)
20. Wang, H.X., Karp, A., Winblad, B., Fratiglioni, L.: Late-life engagement in social and leisure activities is associated with a decreased risk of dementia: a longitudinal study from the Kungsholmen project. Am. J. Epidemiol. **155**, 1081–1087 (2002)
21. Nezlek, J.B., Hampton, C.P., Shean, G.D.: Clinical depression and day-to-day social interactions in a community sample. J. Abnorm. Psychol. **109**, 11–19 (2000)
22. Nezlek, J.B., Imbrie, M., Shean, G.D.: Depression and everyday social interaction. J. Pers. Soc. Psychol. **67**, 1101–1111 (1994)
23. Flatt, J.D., Tiffany, F.H.: Participation in social activities in later life: does enjoyment have important implications for cognitive health? Aging Health **9**(2), 149–158 (2013)
24. Glass, T.A., De Leon, C.M., Marottoli, R.A., Berkman, L.F.: Population based study of social and productive activities as predictors of survival among elderly Americans. BMJ **319** (7208), 478–483 (1999)
25. Rook, K.S.: Emotional health and positive versus negative social exchanges: a daily diary analysis. Appl. Dev. Sci. **5**, 86–97 (2001)
26. Hawkley, L.C., Burleson, M.H., Berntson, G.G., Cacioppo, J.T.: Loneliness in everyday life: cardiovascular activity, psychosocial context, and health behaviors. J. Pers. Soc. Psychol. **85**, 105–120 (2003)
27. Treur, J.: Dynamic modeling based on a temporal–causal network modeling approach. Biologically Inspired Cognit. Archit. **16**, 131–168 (2016)
28. Both, F., Hoogendoorn, M., Klein, M., Treur, J.: Modeling the dynamics of mood and depression. In: Proceedings of the 2008 Conference on ECAI 2008, 18th European Conference on Artificial Intelligenc, pp. 266–270. IOS Press (2008)

PPM: A Privacy Prediction Model for Online Social Networks

Cailing Dong[1]([⊠]), Hongxia Jin[2], and Bart P. Knijnenburg[3]

[1] Department of Information Systems, University of Maryland, Baltimore County, Baltimore, USA
cailing.dong@umbc.edu
[2] Samsung Research America, Mountain View, USA
[3] Human-Centered Computing, Clemson University, Clemson, USA

Abstract. Online Social Networks (OSNs) have come to play an increasingly important role in our social lives, and their inherent privacy problems have become a major concern for users. Can we assist consumers in their privacy decision-making practices, for example by predicting their preferences and giving them personalized advice? To this end, we introduce PPM: a Privacy Prediction Model, rooted in psychological principles, which can be used to give users personalized advice regarding their privacy decision-making practices. Using this model, we study psychological variables that are known to affect users' disclosure behavior: the *trustworthiness* of the requester/information audience, the *sharing tendency* of the receiver/information holder, the *sensitivity* of the requested/shared information, the *appropriateness* of the request/sharing activities, as well as several more traditional *contextual factors*.

Keywords: Privacy prediction · Decision making · Online Social Networks (OSNs)

1 Introduction

The rising popularity of Online Social Networks (OSNs) has ushered in a new era of social interaction that increasingly takes place online. Pew Research reports that 72 % of online American adults maintain a social network profile [12], which provides them with a convenient way to communicate online with family, friends, and even total strangers. To facilitate this process, people often share personal details about themselves (e.g. likes, friendships, education and work history). Many users even share their current activity and/or real-time location. However, all this public sharing of personal and sometimes private information may increase security risks (e.g., phishing, stalking [4,21]), or lead to threats to one's personal reputation [5]. It is therefore no surprise that privacy aspects of OSN use has raised considerable attention from researchers, OSN managers, as well as users themselves.

© Springer International Publishing AG 2016
E. Spiro and Y.-Y. Ahn (Eds.): SocInfo 2016, Part II, LNCS 10047, pp. 400–420, 2016.
DOI: 10.1007/978-3-319-47874-6_28

The privacy dilemma OSN users face is the choice between sharing their information (which may result in social benefits [16,18,27,38]) and keeping it private or restricted to certain users only (thereby protecting their privacy). To help users with this decision, experts recommend giving users comprehensive *control* over what data they wish to share, and providing them with more *transparency* regarding the implications of their decisions [1,10,59,60,62]. Advocates of transparency and control argue that it empowers users to regulate their privacy at the desired level: without some minimum level of transparency and control, users cannot influence the risk/benefit tradeoff. Moreover, people can only make an informed tradeoff between benefits and risks if they are given adequate information [40,54].

The privacy decisions on ONSs are so numerous and complex, that users often fail to manage their privacy effectively. Many users avoid the hassle of using the "labyrinthian" privacy controls that Facebook provides [13,53], and those who make the effort to change their settings do not even seem to grasp the implications of their own privacy settings [43,45]. There is strong evidence that transparency and control do not work well in practice, and several prominent privacy scholars have denounced their effectiveness in helping users to make better privacy decisions [8,47,57].

Are there more effective ways to assist consumers in their privacy decision-making practices? A solution that has recently been proposed, is to learn users' privacy preferences and subsequently give them *user-tailored decision support* [30]. To this end, in this paper we introduce PPM: a comprehensive *privacy prediction model* that can be applied in a multitude of privacy decision making scenarios. The model is *theoretically grounded* in psychological research that has investigated the key variables known to affect users' disclosure behavior. Consequently, PPM can be applied to a wide range of OSNs, and arguably even other privacy-sensitive systems such as e-commerce systems.

A practical application of PPM's predictions would be to provide automatic initial default settings in line with users' privacy preferences (e.g., by default, it discloses Mary's location to her best friends on weekends, but it does not disclose John's location to his boss when he is on vacation). These "smart defaults" could alleviate the burden of making numerous complex privacy decisions, while at the same time respecting users' inherent privacy preferences. Balebako et al. [6] argue that the help provided via such machine learning systems can be seen as "adaptive nudges". Indeed, Smith et al. [56] argue that "Smart defaults can become even smarter by adapting to information provided by the consumer as part of the decision-making process." (p. 167)

In this paper we will formally define PPM and then use it to comprehensively study the important psychological and contextual factors that affect privacy decision making on OSNs, with the ultimate goal of assisting users to make appropriate privacy decisions. Specifically, we validate PPM in *multiple scenarios*, testing the effect of several psychological antecedents of information disclosure behavior—the *trustworthiness* of the requester/audience, the *sharing tendency* of the user, the *sensitivity* of the information, the *appropriateness* of the request/disclosure—as well as several more traditional *contextual factors* on data collected on Twitter, Google+, and a location sharing preference study. We provide a comparative evaluation of the importance of each of these factors in determining users' privacy decisions.

2   Related Work

2.1   Privacy Decision Making

A majority of OSN users takes a pragmatic stance on information disclosure [14,55,61]. These "pragmatists" [23] have balanced privacy attitudes: they ask what benefits they get, and balance these benefits against risks to their privacy interests [14]. This decision process of trading off the anticipated benefits with the risks of disclosure has been dubbed *privacy calculus* [15,39]). In making this tradeoff, these users typically decide to share a subset of their personal information with a subset of their contacts [14,37,42,49]. The term *privacy calculus* makes it sound like users make "calculated" decisions to share or withhold their personal information. In reality though, these decisions are numerous and complex, and often involve uncertain or unknown outcomes [32]. Acquisti and Grossklags [2] identified *incomplete information, bounded rationality,* and *systematic psychological deviations from rationality* as three main challenges in privacy decision making. Consequently, people's privacy behavior is far from calculated or rational. Most OSN users share much more freely than expected based on their attitudes [1,2,9,34] (a disparity that has been labeled the "privacy paradox" [48]).

2.2   Predicting Privacy Decisions

When left to their own devices, users thus seem particularly inept at making even the simplest privacy decisions in a rational manner [34], and many users actively try to avoid the hassle of making such decisions [13]. Knijnenburg et al. have recently proposed a way to circumvent users' unwillingness or inability to make accurate privacy decisions: if one can *predict* users' privacy preferences, one can give them *user-tailored decision support* [30] in the form of recommendations [31] or adaptive defaults [29]. Regarding the first step of this proposal (i.e., predicting users' privacy decisions), scientists have had modest success using various machine learning practices. For example, Ravichandran et al. [52] applied k-means clustering to users' contextualized location sharing decisions to come up with a number of default policies. They showed that a small number of default policies learned from users' contextual location sharing decisions could accurately capture a large part of their location sharing preferences. Pallapa et al. [51] proposed a system that determines the level of privacy required in new situations based on the history of interaction between users. They demonstrated that this solution can deal with the rise of privacy concerns while at the same time efficiently supporting users in a pervasive system full of dynamic and rich interactions.

2.3   Psychological Antecedents of Privacy Decisions

While machine learning studies have had modest success predicting users' privacy decisions, their results have been scattered; each work only considers a small subset of (one or two) contextual factors, in the context of a single OSN. In this paper, we therefore make an effort to *integrate* new and existing privacy prediction factors into a single comprehensive model. To make this model generalizable across a wide range of OSNs (and, arguably, information systems in general), we theoretically ground this framework in *psychological* research that has identified factors that consistently influence the outcomes of users' privacy decisions. Existing work has found several psychological factors that influence users' decision making process. For example, Adams identified three major factors that are key to users' privacy perceptions: *information sensitivity, recipient* and *usage* [3]. These factors are in line with Nissenbaum's theory of *contextual integrity* [46], which argues that disclosure depends on *context, actors, attributes*, and *transmission principles* (cf. usage and flow constraints). Based on these theories and other existing research, we identify the following factors:

- **The user: sharing tendency.** The most widely accepted finding in privacy research is that users differ in their innate tendency to share personal information [14,55,61]. Most prominently, Westin and Harris [23] developed a privacy segmentation model which classifies people into three categories: *privacy fundamentalists, pragmatists*, and *unconcerned*. Recent work has demonstrated that this categorization might be overly simplistic [64]. In this light, Knijnenburg et al. [33] demonstrated that people's disclosure behavior is multidimensional, i.e., different people have different tendencies to disclose different types of information (see also [49]).
- **The information: sensitivity.** Several studies have found that different types of information have different levels of sensitivity, and that users are less likely to disclose more sensitive information. For instance, Consolvo et al. [14] and Lederer et al. [41] both found that users are more willing to share vague information about themselves than specific information. Users occasionally differ in what information they find most sensitive; i.e., Knijnenburg et al. [33] find that some users are less willing to publicly share their location than their Facebook posts, while this preference is reversed for others.
- **The recipient: trustworthiness.** Many studies highlight the trustworthiness of the recipient of the information as an important factor [14,25,26,48, 49,62]. Lederer et al. [41] even found that this factor overshadows more traditional contextual factors in terms of determining sharing tendency. Indeed, OSN users tend to restrict access to their profiles by sharing certain information with certain people only [28,44], and OSNs have started to accommodate this factor by introducing the facility to categorize recipients into "groups" or "circles" [28,29,63].
- **The context: appropriateness.** Nissenbaum's theory of contextual integrity [46] posits that context-relevant norms play a significant role in users' sharing decisions. Specifically, the theory suggests that disclosure depends

on whether it is deemed appropriate or inappropriate in that specific context. Several scholars argue that the appropriateness of the information request/disclosure plays an important role in determining users' sharing decisions [7,11,46,65]. Evaluations of appropriateness are based on users' perception of whether there is a straightforward reason why this recipient should have access to this piece of information in this specific context.

3   The Privacy Prediction Model

The aforementioned psychological antecedents of privacy decisions highlight the incredible complexity of the *privacy calculus*. For each privacy decision, users need to estimate the benefits and risks of disclosure by determining and then integrating all these components: their sharing tendency, the sensitivity of the information, the trustworthiness of the recipient, and the appropriateness of the disclosure in context. Arguably, this is a mentally challenging activity, and it is no surprise that our boundedly rational minds are unable to cope with such complex decisions [2]. However, the cited work suggests that these antecedents *do* hold considerable predictive value, meaning that while the privacy calculus may be *mentally* unattainable, it may very well be *computationally* feasible to make consistent predictions of users' preferred privacy decisions based on these antecedents. Indeed, machine learning algorithms are particularly suitable to provide such consistent predictions based on a multitude of anteceding factors. As mentioned, such algorithms have been used in limited cases to predict privacy behaviors [19,51,52,54]. In this section, we will formalize and operationalize this approach. Specifically, we will first provide a generic *formal definition* of the PPM, then operationalize the PPM by proposing an expandable set of *behavioral analogs* of the psychological antecedents, which allow us to unobtrusively measure these antecedents.

3.1   Theory: A Comprehensive Model Based on Psychological Antecedents

Formally, PPM models the probability of disclosure $p(D)$ by user u of item i to recipient r in context c as a function of the user's disclosure/sharing tendency $\bar{D}_u$, the sensitivity of the item S_i, the trustworthiness of the recipient T_r, and the appropriateness of the disclosure in this specific context A_c:

$$p(D_{uirc}) = f(\overline{D_u}, S_i, T_r, A_c) \tag{1}$$

Overly simplistic implementations of this model are the loglinear additive model (Eq. 2) and the full factorial loglinear model (Eq. 3):

$$ln(\frac{p(D_{uirc})}{1 - p(D_{uirc})}) = \alpha + \beta_1\overline{D_u} + \beta_2 S_i + \beta_3 T_r + \beta_4 A_c \tag{2}$$

$$ln(\frac{p(D_{uirc})}{1 - p(D_{uirc})}) = \alpha + \beta_1\overline{D_u} + \beta_2 S_i + \beta_3 T_r + \beta_4 A_c + \beta_5\overline{D_u}S_i + \beta_6\overline{D_u}T_r$$
$$+ \beta_7\overline{D_u}A_c + \beta_8 S_i T_r + \beta_9 S_i A_c + \beta_{10}T_r A_c + \beta_{11}\overline{D_u}S_i T_r$$
$$+ \beta_{12}\overline{D_u}S_i A_c + \beta_{13}\overline{D_u}T_r A_c + \beta_{14}S_i T_r A_c + \beta_{15}\overline{D_u}S_i T_r A_c$$

$$(3)$$

3.2 Practice: Large-Scale Prediction Using Behavioral Analogs

An essential step towards operationalizing PPM is to measure or estimate the parameters $\overline{D_u}$, S_i, T_r, and A_c. This is not a trivial task: these psychological antecedents are hard to quantify, especially on the large scale needed for successful machine learning. Current work defines three ways in which a system can do this: (1) The simplest solution is to directly ask the user, either by having them specify exact values for each parameter, or by allowing them to make a broad classification [28,29,63]. (2) Another option is to derive these values from static user variables. (3) A system can forego the parameters, and simply try to estimate users' disclosure behaviors based on any available contextual variables. As a finer-grained and thus ultimately more precise method, most existing work uses this approach (cf. [19,51,52,54]). The PPM approach described in this paper takes a spin on the third method by specifying specific types of data as *behavioral analogs* of the established psychological antecedents.

4 Data Collection

Friend requests are the most common and direct way to get access to a user's information in many OSNs. Accepting a friend request discloses at least a part of one's profile and online activities to the requester, so the acceptance or rejection of a friend request is an important privacy decision. In our study, we collected two real-life datasets from Twitter and Google+, targeting on the "Friend requests" activities to simulate the information disclosure behavior. Besides, location-sharing has gained popularity both in stand-alone apps (e.g. Foursquare, Glympse) and as a feature of existing OSNs (e.g. location-tagging on Facebook and Twitter), which is an information disclosure activity that is particularly strongly influenced by privacy concerns [50,66]. Unfortunately, existing location-sharing datasets often do not extend beyond check-in behaviors. We thus created three location-sharing datasets based on a study with manually-collected rich location sharing preferences that includes twenty location semantics, three groups of audiences and several contextual factors such as companion and emotion [17]. In the following, we will describe how we collect each dataset and identify the above-mentioned behavioral analogs in these datasets.

4.1 Twitter Dataset

Our Twitter dataset consists of a set of Twitter users crawled and classified as legitimate users by Lee et al. [36]. We measured the behavioral analogs of our psychological antecedents as follows:

Disclosure Behavior. In our datasets from both Twitter and Google+, we study users' responses to "Friend requests" as the target information disclosure behavior. Two general friendship mechanism exist on social networks: in *bilateral friendship requests* (e.g. Facebook), friendships are reflexive, and a friendship is only established after the user accepts the request. While this is the "cleanest" version of our scenario, the requests themselves are not accessible through the Facebook API, making it impossible to observe rejected requests. Twitter and Google+, on the other hand, use *unilateral friendship requests*: users do not need permission to "follow" or "add to circle" other users, making one-sided "friendships" possible. Users who are followed/added to a circle may respond in one of three ways: (1) they may reciprocate the request by following/adding the requester back; (2) they may delete or block the requester; (3) or they may do nothing and simply leave the friendship one-sided. Behavior 1 is observable as a separate, subsequent friendship request; behaviors 2 and 3 are indistinguishable using the Twitter and Google+ APIs. However, since users are notified of being followed, we argue that users will most commonly follow/add the requester back if they accept the request, and otherwise simply ignore the request. Our work is based on the assumption that when a user follows the requester back the request is accepted, otherwise it is rejected.

To measure this behavior, we extracted each user u's profile item settings, and following and follower lists. Each friendship request is represented by a tuple $< f(u), f(v), f(u, v), l >$, where u is the requester, v is the receiver and l is the decision label indicating if v accepts (1) or rejects (0) u's request. $f(u)$ and $f(v)$ are collections of features associated with u and v respectively, and $f(u, v)$ represents the relationship between u and v. We classify friend request decisions as follows:

Definition 1 (Disclosure behavior). *For each friend u_f on u's following list, if u_f is also in u's followers list, that is, if u_f also follows u, we say that u_f accepted u's request, otherwise u_f rejected u's request.*

Note that based on our Twitter dataset we are unable to distinguish who follows whom first; when two users u_1 and u_2 follow each other, this results in two records $< f(u_1), f(u_2), f(u_1, u_2), 1 >$ and $< f(u_2), f(u_1), f(u_2, u_1), 1 >$. Only one of these describes the actual reciprocation of a friend request, the other is spurious. In the Google+ dataset we present below, we are able to untangle the chronological order of friend requests. This dataset thus arguably provides more accurate results. In both datasets, we removed "verified" users, since most of such users are celebrities who often have many more followers than followees; our definition of "accepting/rejecting a friend request" will arguably not hold for such users.

Sharing Tendency of the Information Holder. *FollowTendency* is a behavioral analog of users' *disclosure/sharing tendency*, defined as the relative number of people they follow. Formally speaking:

Definition 2 (Sharing tendency). $followTendency = \frac{\#following}{\#follower + \#following}$

Trustworthiness of the Recipient. Our behavioral analog of the recipient's *trustworthiness* is based on the intuition that a user with relatively many followers is likely to have a higher reputation, and thus more trustworthy. Formally speaking:

Definition 3 (Trustworthiness). $trustworthiness = \frac{\#follower}{\#follower+\#following}$

Sensitivity of the Requested Information. We argue that the sensitivity S_i of a Twitter user u's profile item i depends on how common the user's value of i is in the population: the more common the value, the less sensitive the information. Formally speaking:

Definition 4 (Sensitivity). *Suppose a profile item i has m possible settings $\{i_1, i_2, \ldots, i_m\}$ ($m \geq 1$). The distribution of different settings over the whole population is $P^i = \{p_{i_1}, p_{i_2}, \ldots, p_{i_m}\}$, where $0 \leq p_{i_j} \leq 1$ and $\sum_{j=1}^{m} p_{i_j} = 1$. If user u sets his/her profile item i as i_k ($1 \leq k \leq m$), the sensitivity value of $S_i = \frac{1}{p_{i_k}}$.*

On Twitter, users have the option to set profile items *GEO*, *Protected* and *URL*. *GEO* and *Protected* are boolean values indicating whether the user has enabled the automatic geo-tagging of her tweets, and whether her profile is protected. *URL* is a field that users can use to enter a personal website. We categorize its value as either *blank*, a *personal* URL (linking to Facebook or LinkedIn), or an *other* URL. We calculate the corresponding sensitivity scores use them as our behavioral antecendents of *sensitivity*.

Appropriateness of the Request. Friendship requests are more appropriate if there is a lot of existing overlap between the two users' networks. Formally speaking:

Definition 5 (Appropriateness). *The approrpiateness of a friend request of user v to user u depends on the overlap between their networks, which can be measured with the following indicators:*

- $JaccardFollowing_{(u,v)} = \frac{|followings(u) \cap followings(v)|}{|followings(u) \cup followings(v)|}$
- $JaccardFollower_{(u,v)} = \frac{|followers(u) \cap followers(v)|}{|followers(u) \cup followers(v)|}$
- $comFollowing(u) = \frac{\#commonFollowing}{\#following(u)}$
- $comFollower(u) = \frac{\#commonFollower}{\#follower(u)}$
- $comFollowing(v) = \frac{\#commonFollowing}{\#following(v)}$
- $comFollower(v) = \frac{\#commonFollower}{\#follower(v)}$

4.2   Google+ Dataset

Gong et al. [20] crawled the whole evolution process of Google+, from its initial launch to public release. The dataset consists of 79 network snapshots, these stages can be used to uncover a rough chronological account of friendship creation (i.e. users adding each other to their circles). We focus on the first two stages to build our dataset, where "who sends the friend request to whom first" can be identified by the stage ids.

We used the same factors in our Twitter dataset to measure the behavioral analogs of our psychological antecedents. Specifically, we use the "add to circle" activity to simulate the information **disclosure behavior**. Each instance in our Google+ dataset is a tuple $< f(u), f(v), f(u, v), l >$, where l indicates v' decision to reciprocate u's request (1) or not (0). We use the same behavioral analogs as those defined in Twitter dataset for **sharing tendency**, **trustworthiness**, **sensitivity**, and the **appropriateness** of the request. For *sensitivity*, we use profile attributes available on Google+, namely *Employer*, *Major*, *School* and *Places* that are either publicly displayed (1) or not (0). We thus calculate the following four behavioral analogs for *sensitivity*: (1) *S(Employer)*, (2) *S(Major)*, (3) *S(School)* and (4) *S(Places)*.

4.3   Location Sharing Datasets

Our location sharing datasets are built based on a survey on location sharing preferences. We conducted this survey by recruiting 1,088 participants using Amazon Mechanical Turk[1]. We restricted participation to US Turk workers with a high worker reputation who had previously used a form of location sharing services.

We specifically asked the participants about their privacy concern in the location sharing survey. The distribution of their claimed *privacyLevel* is: *Very Concerned* (39 %), *Moderately* (41 %), *Slightly* (15 %), *Not Care* (5 %). Consistent with previous research, as many as 80 % of the participants claimed to be moderately or very concerned about their privacy [23, 24].

Disclosure Behavior. We constructed our dataset by requesting users' feedback to systematically manipulated location sharing scenarios. In each scenario, participants were asked to indicate whether they would share their location with three different types of audience: *Family*, *Friend* and *Colleague*.

We ran 5 different studies to collect our data:

- In study 1, each scenario consisted of one of the twenty location semantics supported by Google Places[2]: *Airport, Art Gallery, Bank, Bar, Bus Station, Casino, Cemetery, Church, Company Building, Convention Center, Hospital, Hotel, Law Firm, Library, Movie Theater, Police Station, Restaurant, Shopping Mall, Spa* and *Workplace*.
- In study 2, each scenario consisted of a location, plus a certain *time*: on a weekday during the day, on a weekday at night, on the weekend.

---

[1] https://www.mturk.com/mturk/.
[2] https://developers.google.com/places/.

- In study 3, each scenario consisted of a location, plus a *companion*: alone, with family, with friends, or with colleagues.
- In study 4, we combined location, time and companion in each scenario.
- In study 5, each scenario consisted of a location, plus an *emotion*: positive or negative.

For each targeted group of audience V (*Family*, *Friend* and *Colleague*), we collect the associated sharing records represented by tuples $< f(u), f(u, V, loc), f(loc), l(V) >$, which results in 3 location sharing datasets. $f(u)$ represents the user's features. $f(u, V, loc)$ describes the relationship between the three parties, i.e., *user, audience* and *location*. As the sharing information is the given location, we specifically include the features $f(loc)$ regarding the current location loc into each tuple. $l(V)$ is the decision label indicating if u shares her location with audience V (1) or not (0).

Sharing Tendency of the Information Holder. In our location sharing preference study, each user u only has one sharing option to each group of audiences under a specific scenario. That is, we do not have the "historical" sharing records of u with the same scenario to predict the current or future privacy decision. Therefore, we choose to use other users' sharing behavior to estimate individual u's sharing probability.

One type of estimation on sharing tendency is based on a specific feature Q over all the populations in the dataset, i.e., *overall sharing probability*, represented by $p^u(Q)$. That is, $p^u(Q)$ represents the sharing tendency of u based on feature Q. We estimate it by the sharing probability of the users in the given dataset R who have the same feature value of Q with u, regardless of other scenario information. Formally speaking,

Definition 6 (overall sharing probability). *Suppose a feature Q has m possible values $\{q_1, q_2, \ldots, q_m\}$ ($m \geq 1$). The sharing probability of the users with different feature values on Q over the whole records R is $P^u(Q) = \{p_{q_1}, p_{q_2}, \ldots, p_{q_m}\}$, where $0 \leq p_{q_i} \leq 1$ and $\sum_{i=1}^{m} p_{q_i} = 1$. That is, p_{q_i} represents the sharing probability of the users with q_i as the value of feature Q. If user u's feature value on Q is q_k ($1 \leq k \leq m$), the overall sharing probability of u based on feature Q is $p^u(Q) = p_{q_k}$.*

We consider the *disclosure tendency* of u based on demographic features *age, gender* and *marriage*, as well as the claimed *privacyLevel*. It results in four types of overall sharing probability, i.e., $p^u(age)$, $p^u(gender)$, $p^u(marriage)$ and $p^u(privacyLevel)$.

Besides the specific feature Q, the other type of estimation on sharing tendency also considers the contextual variable α. We call such estimation as α-*conditional sharing probability* and denoted it as $p_\alpha^u(Q)$. We use the sharing probability of the users in R with the same attribute value on Q who have been under the same scenario α to estimate u's sharing probability. Formally speaking,

Definition 7 (α-conditional sharing probability). *Suppose an attribute Q has m possible values $\{q_1, q_2, \ldots, q_m\}$ ($m \geq 1$). The sharing probability of the users with different attribute values on Q over the whole set of sharing records R^α under scenario α is $P_\alpha^u(Q) = \{p_{q_1}^\alpha, p_{q_2}^\alpha, \ldots, p_{q_m}^\alpha\}$, where $0 \leq p_{q_i}^\alpha \leq 1$ and $\sum_{i=1}^m p_{q_i}^\alpha = 1$. If user u's feature value on scenario Q is q_k ($1 \leq k \leq m$), the α-conditional sharing probability of u based on Q under α is $p_\alpha^u(Q) = p_{q_k}^\alpha$.*

We set α as the current location loc or the audience V. Companion, emotion, time loc and V are possible contextual variables. That is, we consider the following α-conditional sharing probabilities: $p_{loc}^u(companion)$, $p_{loc}^u(emotion)$, $p_{loc}^u(time)$, $p_V^u(companion)$, $p_V^u(emotion)$, $p_V^u(time)$, $p_{loc}^u(V)$ and $p_V^u(loc)$.

When building the PPM, we choose the corresponding *overall sharing probability* and α-*conditional sharing probability* to represent the *disclosure/sharing tendency* $\overline{D_u}$.

Trustworthiness of the Recipient. As location sharing is a voluntarily sharing activity, the information holder usually makes the privacy decisions partially based on the trustworthiness of the audience. Typically, the higher of the probability u is willing to share to the given type of audience V, the higher trustworthiness of V is believed by u. According to the above definitions on sharing tendency, we can formally define trustworthiness as follows:

Definition 8 (Trustworthiness). *The trustworthiness of u to the sharing audience V is estimated by the audience-conditional sharing probability of u under the contextual variable V without considering other features. That is, $trustworthiness(V) = p_V^u(\cdot)$.*

Sensitivity of the Shared Location. The sensitivity of the location being shared is of vital importance to the privacy decisions. Usually, the more people are willing to share a location loc, the less they think the location is sensitive. Thus, we can use the *location-conditional sharing probability* regardless of any features to estimate the sensitivity of the shared location. Formally speaking,

Definition 9 (Sensitivity). *The sensitivity of a given loc being shared by u is defined as: $sensitivity(loc) = p_{loc}^u(\cdot)$.*

4.4   Final Datasets

The final datasets used in our study is shown in Table 1, including the basic statistics on the final privacy decisions.

5   PPM: Privacy Prediction Model

Before building the privacy prediction model, we have analyzed and proved the above defined behavioral analogs work rather well on all the datasets listed in Table 1 [17].

Table 1. Statistics of datasets.

Dataset	Statistics
Twitter dataset	$D_{Twitter} = TSet_{req} \cup TSet_{rec}$ (#accepted: 4,874; #rejected: 7,914)
Google+ dataset	$D_{Google+} = GSet_{req} \cup GSet_{rec}$ (#accepted: 21,798; #rejected: 114,400)
Location dataset	D_{Family}: #shared: 8,241; #not shared: 2,554
	D_{Friend}: #shared: 8,030; #not shared: 2,845
	$D_{Colleague}$: #shared: 5,249; #not shared: 5,546

In this section we talk about how we build the PPM using the above presented behavioral analogs of the following psychological antecedents:

- (1) *Trustworthiness* of the requester/information audience
- (2) *Sharing tendency* of the receiver/information holder
- (3) *Sensitivity* of the requested/shared information
- (4) *Appropriateness* of the request
- (5) *Other contextual factors*

The PPM is aimed to help OSN users to manage their privacy by predicting their disclosure behavior and recommending privacy settings in line with this behavior. Specifically, based on these behavioral analogs of the psychological antecedents, we build a binary classification model that learns the influence of these features on privacy decisions (i.e. sharing/disclosure decisions, such as *accept* vs. *reject*, or *share* vs. *not share*).

5.1 Machine Learning Outcomes

We build a decision making model for each of the five datasets described in Table 1. One problem with the five datasets is they are imbalanced, that is, the number of accepts/shares is much larger or smaller than the number of rejects/not-shares. We employ the common machine learning practice – *under-sampling*, to balance the sets by randomly sampling items from the "large class" to match the size of the "small class". We use an adapted *10-fold cross validation* approach (detailed implementation is described in the discussion in Sect. 5.2) to split the training and testing datasets. The final results are averaged over the classification results in all the folds. The commonly used *F1* and *AUC* are employed as evaluation metrics. *F1* is the harmonic mean of precision and recall, and *AUC* is a statistic that captures the precision of the model in terms of the tradeoff between false positives and false negatives. The higher these values, the better of the performance. We use several of the binary classification algorithms provided by Weka [22] to build our models, including *J48, Naïve Bayes, Support Vector Machine (SVM)*, etc. Among them, *J48* produced the best results in terms of both *F1* and *AUC*. The corresponding results are shown in Table 2, as

Table 2. Performance of using different feature sets in decision making prediction model.

Dataset	#Tuples	F1						AUC					
		All	Removed features					All	Removed features				
			(1)	(2)	(3)	(4)	(5)		(1)	(2)	(3)	(4)	(5)
$D_{Twitter}$	9,748	0.796	0.784	**0.751**	0.796	**0.767**	-	0.850	0.835	**0.812**	0.850	**0.832**	-
$D_{Google+}$	43,596	0.898	0.889	**0.887**	0.898	0.889	-	0.899	0.892	**0.891**	0.898	**0.890**	-
D_{Family}	5,378	0.845	-	0.840	**0.833**	**0.833**		0.879	-	0.875	**0.867**	0.870	
D_{Friend}	5,822	0.810	-	**0.798**	0.802	0.800		0.844	-	0.840	0.839	**0.835**	
$D_{Colleague}$	11,054	0.737	-	0.730	**0.726**	0.727		0.752	-	0.748	**0.743**	0.745	

indicated using **All** the features. As seen, our privacy decision making prediction model has a good performance (cf. [58]).

We further verify the effectiveness of each factor by testing the privacy decision making model without the corresponding factor. Specifically, Table 2 compares the performance of the decision making models with all features (*All*) against their performance after removing the features belonging to each of the factors using *J48*: (1) *trustworthiness*; (2) *sharing tendency*; (3) *sensitivity*; (4) *appropriateness*; (5) *contextual factors*[3]. The results showed that removing some factors may reduce the prediction performance. In line with our feature ranking results in [17], this is mainly true for the *trustworthiness* of requester and the *follow tendency* of the receiver as well as the *appropriateness* of the request in the *friend requests* scenario of Twitter and Google+. These factors are more important than the *sensitivity* factors, probably because we simply use user profile to quantify it. In the *location sharing* scenario, the *sharing tendency* of the user, the *sensitivity* of the location as well as the *contextual factors* are all important predictors that reduce the *F1* and *AUC* values when excluded. For different types of recipients, the dominate factors to privacy decisions vary as well. For instance, when the recipients/audiences are colleagues and family members, the sensitivity of the shared location is more important. However, the information holder's sharing tendency has a bigger influence on the privacy decisions when the recipients are their friends.

5.2 Discussion

Class Imbalance Problem. As we mentioned earlier, class imbalance is a common but serious problem when building classification models. To deal with

---

[3] As contextual factors are not studied in the *friend requests* scenario, we have no results for removing such factors from $D_{Twitter}$ and $D_{Google+}$. Similarly, although *trustworthiness* of the audience is an important factor in the location sharing study, our privacy decision making model is built for audience separately (as these measures are repeated per scenario). Thus, *trustworthiness* is not a feature in D_{Family}, D_{Friend} and $D_{Colleague}$. Finally, the features belonging to *appropriateness* are difficult to split off from *contextual factors* in the location study, so those results are combined.

imbalanced datasets, *oversampling* and *undersampling* are the two most commonly used techniques that use a bias to select more samples from one class than from another in order to balance the dataset. Specifically, *oversampling* is to repeatedly draw samples from the smaller class until the two classes have the same size. This technique may result in biased distribution due to certain over-repeated samples from the smaller class. While, *undersampling* is to randomly select the same number of instances from the bigger class as the number of instances in the smaller class. As some instances in the bigger class are not included in the final dataset, it may lose some information and result in a biased distribution. In our approach, we choose to use *undersampling* because of another approach adopted – *adapted 10 fold cross validation*. Traditional *k-fold cross validation* splits the same balanced dataset into k parts, then each time uses 1 part of the dataset as testing dataset and the remaining k-1 parts as training dataset to evaluate the performance. This process repeats k times with different part selected as testing dataset and the final results are averaged over the k folds. In our work, instead of using the same balanced dataset and repeats 10 times, we built a different balanced dataset using *undersampling* each time and applied *10-fold cross validation*. In this way, all the balanced datasets together may cover all or the majority of the instances in the bigger class. The results are actually averaged over the 100 different training-testing datasets, which can largely reduce the potential bias caused by *undersampling*.

Missing Values in Location Datasets. There are a few differences when we built PPM for Google+/Twitter datasets and the three location datasets. In Google+ and Twitter datasets, each instance has the same number of attributes, represented by the measurements we defined for the behavioral analogs of those psychological antecedents, including *trustworthiness, sharing tendency, sensitivity* and *appropriateness*. Thus, any machine learning algorithms can be directly applied on the datasets. For the three location datasets, however, the *disclosure/sharing tendency* will be measured by the *overall sharing probability* or the α-*conditional sharing probability*, which is based on the contextual variables, e.g. *companion, time, emotion*. The five location-sharing scenarios we studied used different subsets of these contextual variables. By putting the data from these fives studies together, we thus create missing values for some of the attributes. We deal with this problem using the embedded *ReplaceMissingValues* filter in Weka. That is, Weka will deal with the missing values using different mechanisms given the different machine learning algorithms. Internally, *J48* replaces the missing values based on the weighted average value proportional to the frequencies of the non-missing values. As *J48* produces the best results not only on all the three location datasets, but also on Google+ and Twitter datasets, we may claim that the way missing values are solved by *48* is acceptable and may be a preferable approach compared with other possible approaches. Besides, the best performance of PPM achieved by *J48* does not purely rely on its mechanism of dealing with missing values.

PPM on Simulation Datasets. Table 2 shows that PPM produces better results on the $D_{Google+}$ dataset than on the $D_{Twitter}$ dataset in terms of both $F1$ score and AUC using different feature combinations. As we mentioned earlier, the Google+ dataset simulates the "friend request" scenario more accurately than the Twitter dataset, as the chronological order of friendship invitations is captured in our Google+ dataset but not in the Twitter dataset. The better performance of PPM on the more accurate dataset arguably demonstrates the effectiveness of the proposed privacy prediction model.

6   Limitations

In our studies, we demonstrated that the factors identified in the PPM are essential in *predicting* users' disclosure/sharing behavior. The ultimate goal of the PPM is to give personalized privacy decision support to overwhelmed OSN users, and our first limitation is that we did not actually implement such a decision support system. We plan to do this in our future work, and point to [35] for initial results in the context of recommender systems rather than OSNs. Aside from this, there are some other limitations with regard to our study that can be accounted for in future work.

The first limitation is our datasets: users' real sharing behavior data (rather than questionnaire data) is more accurate in identifying true factors that affect user decision making. We make use of a large dataset of Twitter and Google+ users to analyze such real behavior, but we use an imperfect proxy of users' actual "friend requests". Moreover, the location sharing data is based on imagined scenarios rather than real sharing behavior. This work takes a first step towards comprehensively and accurately study the factors affecting users' privacy decision making. In the future we hope to study these factors using real behavioral data such as true friend request acceptance and location sharing behaviors on social network sites.

The second limitation involves the identified factors. We analyzed the most important factors in privacy decision making as identified by existing work, but the study of factors determining privacy decisions has been far from comprehensive in the past. To create a more generally applicable model, we also specifically selected factors that are applicable to a wide range of privacy decision making scenarios in different social media, not restricted to online social networks. Consequently, we may miss system/domain-specific factors that could have a significant influence on privacy decision making. Finally, we tried to focus on very simple behavioral analogs in our study. More sophisticated analogs could likely be identified that further increase the prediction performance.

Our third limitation is that our privacy decision making prediction model is built as a binary classifier. In practical applications of, say, a "privacy recommender", it is not always appropriate to make "hard" recommendations to users, e.g., to pervasively recommend either "accept" or "reject". Instead, it might be more applicable to calculate a "privacy risk" score based on the identified factors and let user to make the final decision based on the score, or to only intervene when the calculated risk passes a certain (user-defined) threshold.

7   Design Implications and Conclusion

In this paper, we developed a generic Privacy Prediction Model based on known psychological antecedents of privacy decision making. We operationalized the PPM by identifying behavioral antecedents of the psychological factors, and analyzed how these factors influenced privacy decisions in several real-world and collected datasets. While our work does not move beyond prediction, it provides evidence for the feasibility of an automated tool to predict users' disclosure/sharing behavior. Such a tool can help ease the burden on the user, who on most OSNs has to make an almost unreasonable number of privacy decisions. Our investigation specifically led to the following important observations that may affect future design of OSNs:

- Consistent with previous studies in *information sharing*, *who* is the information audience is an important factor deciding if the information will be shared. Similarly, in the scenario of *information requests*, *who* is the requester determines the *trust* from the receiver, therefore determines the final privacy decision making outcome.
- As self-representation is one of the main purposes of OSNs, users' privacy decision making does not exclusively depend on their privacy concern, but more generally on the tradeoff between privacy and self-presentation. We have captured this in the definition of *sharing tendency*.
- *Sensitivity* is not an objective concept; it varies with audience. For example, in our location sharing preference study, locations such as *Bar*, *Casino* are more sensitive if shared to *Colleague* or *Family* compared to *Friend*. This effect may be captured by the *appropriateness* of the request.
- Although the assumption of "rationality" in privacy decision making has been criticized by many studies, users do consider the *appropriateness* of the request/sharing activity. Consistent with our intuition and previous studies, users tend to share information with or accept requests when this is appropriate in the current context.
- *Contextual information* is an indispensable factor in privacy decision making, primarily due to its effect on *appropriateness*. However, the effect of different contextual factors varies.

Concluding, privacy decision making is a trade-off between the potential benefit and risk; a tradeoff that is rather difficult for users to make. Our privacy decision making prediction model combines several important psychological and contextual factors that influence this tradeoff, and learns their functionality by building a binary classifier. The proposed privacy decision making prediction model produces good results based on the five identified factors, and can be used to assist users to protect their privacy in online social networks.

References

1. Acquisti, A., Gross, R.: Imagined communities: awareness, information sharing, and privacy on the facebook. In: 6th Workshop on Privacy Enhancing Technologies, pp. 36–58 (2006)

2. Acquisti, A., Grossklags, J.: Privacy and rationality in individual decision making. IEEE Secur. Priv. **2**, 24–30 (2005)
3. Adams, A.: Multimedia information changes the whole privacy ballgame. In: Proceedings of the Tenth Conference on Computers, Freedom and Privacy: Challenging the Assumptions, CFP 2000, pp. 25–32. ACM (2000)
4. Al Hasib, A.: Threats of online social networks. IJCSNS. Int. J. Comput. Sci. Netw. Secur. **9**(11), 288–293 (2009)
5. Baker, D., Buoni, N., Fee, M., Vitale, C.: Social networking and its effects on companies and their employees. Retrieved November 15 (2011)
6. Balebako, R., Leon, P.G., Mugan, J., Acquisti, A., Cranor, L.F., Sadeh, N.: Nudging users towards privacy on mobile devices. In: CHI 2011 Workshop on Persuasion, Influence, Nudge and Coercion Through Mobile Devices, Vancouver, Canada, pp. 23–26 (2011). http://www.andrew.cmu.edu/user/jmugan/Publications/chiworkshop.pdf
7. Bansal, G., Zahedi, F., Gefen, D.: The moderating influence of privacy concern on the efficacy of privacy assurance mechanisms for building trust: a multiple-context investigation. In: Proceedings of the ICIS 2008, Paris, France (2008). http://aisel.aisnet.org/icis2008/7
8. Barocas, S., Nissenbaum, H.: On notice: the trouble with notice and consent. In: Proceedings of the Engaging Data Forum: The First International Forum on the Application and Management of Personal Electronic Information (2009). http://www.nyu.edu/pages/projects/nissenbaum/papers/ED_SII_On_Notice.pdf
9. Becker, L., Pousttchi, K.: Social networks: The role of users' privacy concerns. In: Proceedings of the 14th International Conference on Information Integration and Web-based Applications & Services, IIWAS 2012, pp. 187–195. ACM (2012). http://doi.acm.org/10.1145/2428736.2428767
10. Benisch, M., Kelley, P.G., Sadeh, N., Cranor, L.F.: Capturing location-privacy preferences: quantifying accuracy and user-burden tradeoffs. Pers. Ubiquit. Comput. **15**(7), 679–694 (2011). http://dx.doi.org/10.1007/s00779-010-0346-0
11. Borcea-Pfitzmann, K., Pfitzmann, A., Berg, M.: Privacy 3.0 := data minimization + user control + contextual integrity. Inf. Technol. **53**(1), 34–40 (2011). http://www.oldenbourg-link.com/doi/abs/10.1524/itit.2011.0622
12. Brenner, J., Smith, A.: 72% of online adults are social networking site users. PewResearch Internet Project (2013)
13. Compañó, R., Lusoli, W.: The policy maker's anguish: regulating personal data behavior between paradoxes and dilemmas. In: Moore, T., Pym, D., Ioannidis, C. (eds.) Economics of Information Security and Privacy, pp. 169–185. Springer, US (2010). doi:10.1007/978-1-4419-6967-5_9
14. Consolvo, S., Smith, I.E., Matthews, T., LaMarca, A., Tabert, J., Powledge, P.: Location disclosure to social relations: Why, when, & what people want to share. In: Proceedings of the SIGCHI Conference on Human Factors in Computing Systems, CHI 2005, pp. 81–90. ACM (2005)
15. Culnan, M.J.: "how did they get my name?": An exploratory investigation of consumer attitudes toward secondary information use. MIS Q. **17**(3), 341–363 (1993). http://www.jstor.org/stable/249775
16. DiMicco, J., Millen, D.R., Geyer, W., Dugan, C., Brownholtz, B., Muller, M.: Motivations for social networking at work. In: Proceedings of the 2008 ACM Conference on Computer Supported Cooperative Work, CSCW 2008, pp. 711–720. ACM (2008). http://doi.acm.org/10.1145/1460563.1460674

17. Dong, C., Jin, H., Knijnenburg, B.P.: Predicting privacy behavior on online social networks. In: Proceedings of the Ninth International Conference on Web and Social Media, ICWSM 2015, University of Oxford, Oxford, UK, 26–29 May 2015, pp. 91–100 (2015). http://www.aaai.org/ocs/index.php/ICWSM/ICWSM15/paper/view/10554

18. Ellison, N., Steinfield, C., Lampe, C.: The benefits of facebook "friends:" social capital and college students' use of online social network sites. J. Comput. Mediated Commun. **12**, 1143–1168 (2007)

19. Fang, L., LeFevre, K.: Privacy wizards for social networking sites. In: Proceedings of the 19th International Conference on World Wide Web, WWW 2010, pp. 351–360. ACM (2010)

20. Gong, N.Z., Xu, W., Huang, L., Mittal, P., Stefanov, E., Sekar, V., Song, D.: Evolution of social-attribute networks: measurements, modeling, and implications using google+. In: Proceedings of the 2012 ACM Conference on Internet Measurement Conference, IMC 2011, pp. 131–144 (2011)

21. Gross, R., Acquisti, A.: Information revelation and privacy in online social networks. In: Proceedings of the 2005 ACM Workshop on Privacy in the Electronic Society, WPES 2005, pp. 71–80. ACM (2005). http://doi.acm.org/10.1145/1102199.1102214

22. Hall, M., Frank, E., Holmes, G., Pfahringer, B., Reutemann, P., Witten, I.H.: The weka data mining software: An update. SIGKDD Explor. **11**(1), 10–18 (2009)

23. Harris, L., Westin, A., Associates: Personalized marketing and privacy on the net: What consumers want. Privacy and American Business Newsletter (1998)

24. Ho, A., Maiga, A., Aïmeur, E.: Privacy protection issues in social networking sites. In: AICCSA, pp. 271–278. IEEE (2009)

25. Hsu, C.W.: Privacy concerns, privacy practices and web site categories: toward a situational paradigm. Online Inf. Rev. **30**(5), 569–586 (2006). http://www.emeraldinsight.com/journals.htm?articleid=1576312&show=abstract

26. Johnson, M., Egelman, S., Bellovin, S.M.: Facebook and privacy: it's complicated. In: Proceedings of the 8th Symposium on Usable Privacy and Security. ACM, Pittsburgh (2012). http://doi.acm.org/10.1145/2335356.2335369

27. Joinson, A.N.: Looking at, looking up or keeping up with people?: motives and use of facebook. In: Proceedings of the SIGCHI Conference on Human Factors in Computing Systems, CHI 2008, pp. 1027–1036. ACM (2008). http://doi.acm.org/10.1145/1357054.1357213

28. Kairam, S., Brzozowski, M., Huffaker, D., Chi, E.: Talking in circles: selective sharing in google+. In: Proceedings of the SIGCHI Conference on Human Factors in Computing Systems, pp. 1065–1074. ACM Press, Austin (2012). http://dl.acm.org/citation.cfm?id=2208552

29. Knijnenburg, B.P., Kobsa, A.: Increasing sharing tendency without reducing satisfaction: finding the best privacy-settings user interface for social networks. In: ICIS 2014 Proceedings. Auckland, New Zealand (2014)

30. Knijnenburg, B.P.: Simplifying privacy decisions: towards interactive and adaptive solutions. In: Proceedings of the Recsys 2013 Workshop on Human Decision Making in Recommender Systems (Decisions@ RecSys 2013), Hong Kong, China, pp. 40–41 (2013)

31. Knijnenburg, B.P., Jin, H.: The persuasive effect of privacy recommendations. In: Twelfth Annual Workshop on HCI Research in MIS. Milan, Italy (2013). http://aisel.aisnet.org/sighci2013/16

32. Knijnenburg, B.P., Kobsa, A.: Making decisions about privacy: information disclosure in context-aware recommender systems. ACM Trans. Interact. Intell. Syst. **3**(3), 20:1–20:23. http://doi.acm.org/10.1145/2499670

33. Knijnenburg, B.P., Kobsa, A., Jin, H.: Dimensionality of information disclosure behavior. Int. J. Hum. Comput. Stud. **71**(12), 1144–1162 (2013)

34. Knijnenburg, B.P., Kobsa, A., Jin, H.: Preference-based location sharing: are more privacy options really better? In: Proceedings of the SIGCHI Conference on Human Factors in Computing Systems, CHI 2013, pp. 2667–2676. ACM (2013)

35. Knijnenburg, B.P.: A user-tailored approach to privacy decision support. Ph.D., University of California, Irvine, United States - California (2015). http://search.proquest.com/docview/1725139739/abstract

36. Kyumin Lee, B.D.E., Caverlee, J.: Seven months with the devils: a long-term study of content polluters on twitter. In: International AAAI Conference on Weblogs and Social Media (ICWSM) (2011)

37. Lampe, C., Ellison, N.B., Steinfield, C.: Changes in use and perception of facebook. In: Proceedings of the 2008 ACM Conference on Computer Supported Cooperative Work, CSCW 2008, pp. 721–730. ACM (2008)

38. Lampe, C., Gray, R., Fiore, A.T., Ellison, N.: Help is on the way: patterns of responses to resource requests on facebook. In: Proceedings of the 17th ACM Conference on Computer Supported Cooperative Work & Social Computing, CSCW 2014, pp. 3–15. ACM (2014). http://doi.acm.org/10.1145/2531602.2531720

39. Laufer, R.S., Wolfe, M.: Privacy as a concept and a social issue: a multidimensional developmental theory. J. Soc. Issues **33**(3), 22–42 (1977)

40. Lederer, S., Hong, J.I., Dey, A.K., Landay, J.A.: Personal privacy through understanding and action: five pitfalls for designers. Pers. Ubiquit. Comput. **8**(6), 440–454 (2004). http://www.springerlink.com/index/10.1007/s00779-004-0304-9

41. Lederer, S., Mankoff, J., Dey, A.K.: Who wants to know what when? privacy preference determinants in ubiquitous computing. In: CHI '03 Extended Abstracts on Human Factors in Computing Systems, CHI EA 2003, pp. 724–725. ACM (2003). http://doi.acm.org/10.1145/765891.765952

42. Lewis, K., Kaufman, J., Christakis, N.: The taste for privacy: an analysis of college student privacy settings in an online social network. J. Comput. Mediated Commun. **14**(1), 79–100 (2008)

43. Liu, Y., Gummadi, K.P., Krishnamurthy, B., Mislove., A.: Analyzing facebook privacy settings: user expectations vs. reality. In: Proceedings of the 2011 ACM SIGCOMM Conference on Internet Measurement Conference, pp. 61–70. ACM (2011)

44. Madden, M.: Privacy management on social media sites. Technical report, Pew Internet & American Life Project, Pew Research Center, Washington, DC, February 2012. http://www.pewinternet.org/2012/02/24/privacy-management-on-social-media-sites/

45. Madejski, M., Johnson, M., Bellovin, S.: A study of privacy settings errors in an online social network. In: 2012 IEEE International Conference on Pervasive Computing and Communications Workshops (PERCOM Workshops), Lugano, Switzerland, pp. 340–345 (2012)

46. Nissenbaum, H.: Privacy in Context: Technology, Policy, and the Integrity of Social Life. Stanford University Press, Stanford (2009)

47. Nissenbaum, H.: A contextual approach to privacy online. Daedalus **140**(4), 32–48 (2011). http://dx.doi.org/10.1162/DAED_a_00113

48. Norberg, P.A., Horne, D.R., Horne, D.A.: The privacy paradox: personal information disclosure intentions versus behaviors. J. Consum. Aff. **41**(1), 100–126 (2007)

49. Olson, J.S., Grudin, J., Horvitz, E.: A study of preferences for sharing and privacy. In: CHI '05 Extended Abstracts on Human Factors in Computing Systems, CHI EA 2005, pp. 1985–1988. ACM (2005)
50. Page, X., Kobsa, A., Knijnenburg, B.P.: Don't disturb my circles! boundary preservation is at the center of location-sharing concerns. In: Proceedings of the Sixth International AAAI Conference on Weblogs and Social Media, pp. 266–273. Dublin, Ireland, May 2012. http://www.aaai.org/ocs/index.php/ICWSM/ICWSM12/paper/view/4679
51. Pallapa, G., Das, S.K., Di Francesco, M., Aura, T.: Adaptive and context-aware privacy preservation exploiting user interactions in smart environments. Pervasive Mobile Comput. **12**, 232–243 (2014). http://www.sciencedirect.com/science/article/pii/S1574119213001557
52. Ravichandran, R., Benisch, M., Kelley, P.G., Sadeh, N.M.: Capturing social networking privacy preferences: In: Goldberg, I., Atallah, M.J. (eds.) PETS 2009. LNCS, vol. 5672, pp. 1–18. Springer, Heidelberg (2009). doi:10.1007/978-3-642-03168-7_1
53. Reports, C.: Facebook & your privacy: Who sees the data you share on the biggest social network? (2012). http://www.consumerreports.org/cro/magazine/2012/06/facebook-your-privacy
54. Sadeh, N., Hong, J., Cranor, L., Fette, I., Kelley, P., Prabaker, M., Rao, J.: Understanding and capturing people's privacy policies in a mobile social networking application. Pers. Ubiquit. Comput. **13**(6), 401–412 (2009)
55. Sheehan, K.B.: Toward a typology of internet users and online privacy concerns. Inf. Soc. **18**(1), 21–32 (2002)
56. Smith, N.C., Goldstein, D.G., Johnson, E.J.: Choice without awareness: ethical and policy implications of defaults. J. Public Policy Mark. **32**(2), 159–172 (2013). http://search.ebscohost.com/login.aspx?direct=true&db=bth&AN=91886736&site=ehost-live
57. Solove, D.J.: Privacy self-management and the consent dilemma. Harvard Law Rev. **126**, 1880–1903 (2013). http://papers.ssrn.com/abstract=2171018
58. Swets, J.A.: Measuring the accuracy of diagnostic systems. Science (New York, N.Y.) **240**(4857), 1285–1293 (1988)
59. Tang, K., Hong, J., Siewiorek, D.: The implications of offering more disclosure choices for social location sharing. In: Proceedings of the SIGCHI Conference on Human Factors in Computing Systems, CHI 2012, pp. 391–394. ACM (2012)
60. Tang, K., Lin, J., Hong, J., Siewiorek, D., Sadeh, N.: Rethinking location sharing: exploring the implications of social-driven vs. purpose-driven location sharing. In: Proceedings of the UbiComp. 2010, Copenhagen, Denmark, pp. 85–04 (2010). http://portal.acm.org/citation.cfm?doid=1864349.1864363
61. Taylor, H.: Most people are "privacy pragmatists" who, while concerned about privacy, will sometimes trade it off for other benefits. Harris Poll **17**(19) (2003)
62. Toch, E., Cranshaw, J., Drielsma, P.H., Tsai, J.Y., Kelley, P.G., Springfield, J., Cranor, L., Hong, J., Sadeh, N.: Empirical models of privacy in location sharing. In: Proceedings of the 12th ACM International Conference on Ubiquitous Computing, Copenhagen, Denmark, pp. 129–138. ACM Press (2010). http://doi.acm.org/10.1145/1864349.1864364
63. Watson, J., Besmer, A., Lipford, H.R.: +your circles: sharing behavior on google+. In: Proceedings of the 8th Symposium on Usable Privacy and Security. ACM, Pittsburgh (2012). http://doi.acm.org/10.1145/2335356.2335373

64. Woodruff, A., Pihur, V., Consolvo, S., Schmidt, L., Brandimarte, L., Acquisti, A.: Would a privacy fundamentalist sell their DNA for $1000.. if nothing bad happened as a result? The Westin categories, behavioral intentions, and consequences. In: Symposium on Usable Privacy and Security (SOUPS) (2014). https://www.usenix. org/system/files/conference/soups2014/soups14-paper-woodruff.pdf
65. Xu, H., Dinev, T., Smith, H.J., Hart, P.: Examining the formation of individual's privacy concerns: toward an integrative view. In: ICIS 2008 Proceedings, Paris, France (2008)
66. Zickuhr, K.: Three-quarters of smartphone owners use location-based services. Technical report, D Pew Research Center (2012). http://pewinternet.org/~/ media//Files/Reports/2012/PIP_Location_based_services_2012_Report.pdf

Privacy Inference Analysis on Event-Based Social Networks

Cailing Dong and Bin Zhou$^{(\boxtimes)}$

Department of Information Systems, University of Maryland, Baltimore, USA
bzhou@umbc.edu

Abstract. In this paper, we provide a comprehensive study of privacy threats by bridging user's online and offline social activities. We adopt the recently emerged Event-Based Social Networks (EBSNs) such as Meetup as an example. Due to the intrinsic interrelated nature of users' online and offline social activities, our research revealed that using several simple yet effective privacy inference models, user's privacy of online group membership and offline event attendance can be inferred with high accuracy. The level of privacy threats by bridging user's online and offline social activities is remarkably severe.

Keywords: Online Soical Networks (OSNs) · Privacy attack · Privacy inference

1 Introduction

The newly emerged event-based social networks (EBSNs) [18] such as *Meetup* (www.meetup.com) and *Plancast* (www.plancast.com) have experienced rapid growth in recent years. Different from many existing online social networking sites, EBSNs can be considered as a hybrid of conventional online social networks (OSNs) such as Facebook (www.facebook.com) and Twitter (www.twitter. com) and specialized location-based social networks (LBSNs) such as Foursquare (www.foursquare.com) and Gowalla (www.gowalla.com). On one hand, EBSNs serve as an online platform for users to build their social profiles, maintain their social relationships to share interests and opinions online, and disseminate user-generated content over the network. To this extent, EBSNs function as conventional online social networks. On the other hand, EBSNs also provide some functions for users to organize face-to-face offline social events. Specifically, event organizers can utilize the services provided by EBSNs to initialize and advertise upcoming offline social events to the members. People who are interested in those local events are able to RSVP and later on check-in to attend those face-to-face local gatherings with other members. In such a case, EBSNs also behave similarly to some specialized location-based social networks which facilitate many offline social interactions between users within their local areas.

One of the distinct characteristics of EBSNs, which is largely different from those popular online social networking sites, is the mechanism to couple users'

E. Spiro and Y.-Y. Ahn (Eds.): SocInfo 2016, Part II, LNCS 10047, pp. 421–438, 2016.
DOI: 10.1007/978-3-319-47874-6_29

online and offline social interactions. In Fig. 1, we depict the different types of interactions existed between users in *Meetup*, one of the most popular EBSNs in the literature. According to the statistics in August, 2016, Meetup has attracted more than 27 million users from over 180 different countries[1]. Users in Meetup eventually formalize two different social communities. In the online virtual social community, users in Meetup can create, manage, and participate in different online groups. There are currently more than 250 K groups created in Meetup, covering a wide variety of topics all around the world. Users from the same group can interact with others by sharing comments and opinions online. In the offline physical social community, event organizers can hold various offline gatherings of Meetup users in their local areas. The interactions between users in Meetup are indeed face-to-face communications when they attend the same local events together. It is reported that in Meetup more than 560 K local events are organized each month and over 3.5 million users are attending those events monthly.

Analyzing users' social interactions has many important applications such as user behavior modeling, customer relationship management, and marketing. In recent years, due to the rich information about individuals and the power of many social network analysis methods, privacy is emerged as one of the biggest concerns. Despite a plethora of crucial privacy issues researched in the field, we have not yet fully analyzed an emerged but fundamental category of threats to people's privacy: those that inevitably bridge our online and offline social interactions. *What devastating outcomes such privacy threats by coupling user's online and offline social interactions may bring to us?* To answer this question, we conduct an empirical study on privacy issues in *Meetup* in this paper. Our contributions of the paper are summarized as follows.

First, *we provide a comprehensive study on users' privacy issues when their online and offline social activities are tightly coupled in Meetup.* We properly model user's privacy in their online and offline communities, and demonstrate how their privacy could be breached using several surprisingly simple yet effective privacy inference models.

Second, *we adopt a large-scale data collection crawled from Meetup, one of the most popular EBSNs in the literature, to illustrate the devastating outcomes of privacy breaches by coupling user's online and offline activities.* Our experimental results clearly show that the privacy threats are real.

The remainder of the paper is structured as follows. Section 2 provides the modeling of privacy issues in Meetup, and Sect. 3 discusses several simple yet effective privacy inference methods. Section 4 presents a case study of privacy analysis in Meetup by bridging user's online and offline social interactions. Some related studies are presented in Sect. 5 and conclusions and future directions are outlined in Sect. 6.

---

[1] https://www.meetup.com/about/.

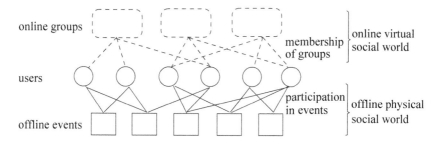

Fig. 1. The two social worlds in Meetup.

2   Modeling Privacy Threats in Meetup

To understand privacy threats in Meetup, there are several questions to be answered properly. First, what is privacy? Second, who are the victims of privacy breaches? Third, who are the privacy attackers? Last but not least, what can be utilized by the attackers to breach other's privacy? In the remaining of this section, we will provide discussions to each of those questions.

Privacy in Meetup. In the field of social network privacy research, a common categorization of privacy includes: *identity disclosure* (that is, the identity of an individual is revealed), *attribute disclosure* (that is, the value of some sensitive attributes associated with an individual is compromised), *link disclosure* (that is, the sensitive relationship between two individuals is disclosed), and *affiliation link disclosure* (that is, whether an individual is affiliated with a group is compromised).

In this paper, we focus on two types of privacy in Meetup. First, we regard the *membership of online groups* for those users in Meetup as the privacy. In general, this type of privacy can be categorized as *affiliation link disclosure*. In addition, two users within the same group often interact with each other frequently. If the membership of online group is disclosed, the potential *link disclosure* also applies. Moreover, online groups are labeled with topic tags. The topic tags of groups that one user has joined represent user's individual interest. For many users, the interest is considered to be a sensitive attribute. To this extent, the disclosure of group membership in Meetup also leads to *attribute disclosure*. Second, we treat the *attendance of local events* for those users in Meetup as the privacy as well. Intuitively, an offline social event can be considered as a local group of users that people need to physically participate. Thus, the disclosure of attendance of local events is indeed one type of *affiliation link disclosures*. This type of information disclosure can be understood as *link disclosure* as well since users attending the same events would have face-to-face communications. A subtle but hazardous consequence when user's event attendance information is compromised lies in the fact that the physical location of the user during specific time durations is compromised implicitly. That being said, the privacy of *attendance of local events* is also one type of *attribute disclosures* where *location at a time* is the sensitive attribute of the user.

Comparing with the privacy of online group membership, the privacy of offline event attendance is much more restricted by individual's physical location and time availability. In addition, attackers are more interested in breaching the privacy of event attendance for upcoming events, as user's past event history may not be beneficial for them anymore.

Victim of Privacy Breaches in Meetup. The membership of online groups and the attendance of offline events are two types of important information related to Meetup users. Therefore, all of the users in Meetup are in fact the potential victims of privacy breaching attacks. The current Meetup website also provides users with the option to hide their group membership and event's RSVP information. Users with those types of information hidden in Meetup are very susceptive to potential privacy attacks. Those users will suffer from those privacy breaches most.

Privacy Attacker in Meetup. The attackers are those people who intend to compromise victim's privacy in Meetup. The service providers of Meetup have access to all of the data shared and stored in Meetup. In this paper, we regard service providers of Meetup as legitimate and trustful. Instead, we consider the attackers as those people who have partial information about the targeted victims (i.e., name, email address, etc.). The attackers may behave similarly to those regular users in Meetup by joining some online groups and attending some offline events. However, their objective is to obtain more information about the victims and ultimately breach victims' privacy.

Background Knowledge of Attacker. The attackers are usually equipped with some background knowledge about the victim. In Meetup, such background knowledge could be the identity attributes such as name and email address, or some partial information about victim's social interactions with others either online or offline. To gather more background knowledge, the attackers may even register with Meetup and pretend to be a regular user by conducting various online and offline interactions with the victim.

3   Privacy Inference in Meetup

Privacy in people's online and offline social communities can be breached using different strategies. For example, one popular scenario is related to various privacy settings available in those OSN sites. Without proper settings of access control mechanism, user's private information shared online may become accessible to the attackers. Another popular scenario exists when rich social data are released to third parties for data analytical tasks. Data recipients can perform specialized data mining methods on the data and infer additional information which is supposed to be sensitive and protected.

In Meetup, the privacy threat may come from different scenarios due to several reasons. First, rich data in Meetup are not released to third parties. Thus, many existing privacy issues for data publishing do not apply. Second, privacy settings in

Meetup are adopted for user's online information. However, vast amount of user's offline information is not controlled by those privacy settings.

In this paper, we simulate the privacy breaching process in Meetup using inference analysis strategies. Specifically, we simulate the scenario that attackers are able to access partial data related to targeted victim, and then utilize data inference models to breach victim's privacy, including the inference of victim's online group membership in Meetup and the prediction of victim's attendance of some upcoming events.

To illustrate how devastating outcomes those privacy threats in Meetup may bring to users, we describe several simple yet effective data inference models which are able to breach user's privacy in the online and offline social communities. In Sect. 3.1, we focus on the privacy inference of user's group membership. In Sect. 3.2, we emphasize on the privacy inference of user's event attendance.

3.1   Privacy Inference of Group Membership

Content-based Privacy Inference. When a user joins an online group, the user is interested in the theme of the group. The set of groups that a user has joined captures user's preference on groups. If a new group "matches" user's preference closely, the user is very likely to join the group.

In Meetup, each online group is assigned with some tags which describes the topics of the group. We can use the tags of groups to represent user's preference. Consider a user u who is a member of some online groups $G_u \subseteq G$, where G represents all of the groups in Meetup. For each online group $g \in G$, we use T_g to represent the tags assigned to g. To quantitatively measure whether a group g matches u's preference, the popular *Jaccard similarity* can be adopted, that is,

$$P(g|u)_{group-tag} = \frac{T_g \cap (\cup_{g_u \in G_u} T_{g_u})}{T_g \cup (\cup_{g_u \in G_u} T_{g_u})}, \tag{1}$$

where $P(g|u)_{group-tag}$ $(g \in \{G - G_u\})$ is a probabilistic score indicating the likelihood that the user u will join the group g.

Equation 1 can be used to generate a list of groups as privacy inference results. Specifically, for each $g \in \{G - G_u\}$, we calculate $P(g|u)_{group-tag}$. All the groups in $\{G - G_u\}$ are then sorted using $P(g|u)_{group-tag}$ in the descending order. The k top-ranked groups are thus treated as privacy inference results, where k is a parameter which determines the number of groups to be inferred. We will refer to this method as **Group Tag Similarity (GTS)** method.

Collaborative Filtering-based Privacy Inference. The GTS method utilizes the tag contents associated with groups. It is in general a content-based method which is widely used in the recommender systems in the literature. However, the GTS method does not consider the influence of social interactions. Users joining the same online group are likely to share some common interests. Preferences of other similar users can be utilized for inference.

To estimate whether a user u will be interested in being a member of a group g, a straightforward solution is to measure how many users in g that are similar to u. If we assume that users in the same online group or involved in same offline events may behave similarly, we can qualitatively measure the likelihood that u will join the group g using:

$$P(g|u)_{collaborative} = \frac{U_g \cap (\cup_{g^* \in G(u)} U_{g^*})}{U_g \cup (\cup_{g^* \in G(u)} U_{g^*})}, \tag{2}$$

where U_g denotes the set of users that are members of the group g. The larger of $P(g|u)_{collaborative}$, the more likely that u will be interested in joining g. If we rank all the groups $g \in G - G(u)$ using $P(g|u)_{collaborative}$ in descending order, the k top-ranked groups can be returned as privacy inference results. We will refer to this method as **Group Member Similarity (GMS)** method.

3.2   Privacy Inference of Event Attendance

In order to breach victim's privacy of group membership in the online social world, we developed two difference inference models. Both the content-based and collaborative filtering-based inference models can be used to breach victim's privacy of event attendance in the offline social world as well.

Content-based Privacy Inference. Intuitively, an offline event can be considered as an offline gathering of users in Meetup. An offline event is actually a specialized "offline group" of users. Therefore, the GTS method discussed in Sect. 3.1 can also be applied on privacy inference of event attendance. However, an offline event in Meetup has several important features. On one hand, offline events are more location constrained compared with online groups. On the other hand, offline events usually have very short lifetime. Thus, privacy inference of event attendance in Meetup needs to consider the constraints of location and time duration.

In general, when a user decides to participate in an offline event, the user must be attracted by the following aspects. First, the user should be interested in the topic of the event. Second, the time-slot when the event will be held should be available to the user. Third, the location of the event should not be far away according to user's tolerance. We summarize the following features that are related to offline events in Meetup.

(1) Time duration of events. Every event in Meetup is associated with a time slot. Events in Meetup can be held periodically on a specific day of the week or a specific date of the month. To model the time of an event using "day-time" pattern, we consider the day of the week and the local starting time of the event. For example, the day-time pattern "Wed@9:00" of an event means the event is held at 9AM on Wednesday. To model the time of an event using "date-time" pattern, we consider the date of the month and the local staring time of the event. For example, the date-time pattern "1@19:00" of an event indicates the event is held at 7PM on the first day of the month.

(2) Location of events. Every event in Meetup is also associated with a specific location. We can directly extract the location entry for each event in Meetup.

(3) Topic of events. Meetup does not directly assign topic tags to events. Instead, detailed textual description is needed when an event is created. We can apply some text mining techniques to obtain topics of those events. Specifically, we use the popular *OpenCalais* Web Service API (http://www.opencalais.com) to extract several types of semantic information from event descriptions. The first type of information is **entity**, which captures the most fundamental element of the textual content. We also consider **social-tags**, which are a set of common-sense tags generalized from the whole piece of content. In addition, the *OpenCalais* API also provides **categories** which are the themes of the content using the view of document categorization.

User's historical event participation can be used for privacy inference of event attendance for upcoming events. For an upcoming event $e \in E$, if it is very "similar" to those events that a user u has participated in before, u may be interested in attending e as well. The similarity of events can be measured considering the *time duration*, *location*, and *topic* features of the events.

We use histogram to represent user's history of event participation. For the time feature, we calculate the histogram using both "day-time" pattern and "date-time" pattern. Consider a user u who has participated in $|E(u)|$ events ($E(u) \subseteq E$). The day-time histogram is $H_{E(u)}^{day-time} = [h_1^{day-time}, h_2^{day-time}, \ldots, h_{m_1}^{day-time}]$, where m_1 denotes the total number of distinct day-time patterns. $h_i^{day-time}$ ($1 \leq i \leq m_1$) is equal to $\frac{|\{e^*\}|}{|E(u)|}$, where $e^* \in E(u)$ and e^* satisfies the i-th day-time pattern. For each user u and his/her day-time histogram $H_{E(u)}^{day-time}$, we have $\sum_{1 \leq i \leq m_1} h_i^{day-time} = 1$. Similarly, we can obtain the date-time histogram $H_{E(u)}^{date-time} = [h_1^{date-time}, h_2^{date-time}, \ldots, h_{m_2}^{date-time}]$, where m_2 denotes the total number of distinct date-time patterns, and each $h_i^{date-time}$ ($1 \leq i \leq m_2$) represents the frequency of the i-th date-time pattern.

To measure the similarity between an upcoming event e and a user u's time histogram, we can simply consider whether e's time pattern is similar to any previous events. If e's day-time pattern is the same as the i-th patten ($1 \leq i \leq m_1$) in $H_{E(u)}^{day-time}$, we measure $P(e|u)_{day-time} = h_i^{day-time}$, the probability that u will attend an event corresponding to that day-time pattern. If all the day-time patterns are different from that of e, we assign $P(e|u)_{day-time} = 0$. The similar procedure can be used for date-time pattern. Thus, we measure $P(e|u)_{date-time} = h_i^{date-time}$ if e's date-time pattern is the same as the i-th pattern in the histogram, 0 otherwise. In Meetup, events are usually organized using either day-time pattern or date-time pattern. Therefore, based on the time feature of offline events, we calculate the likelihood that u will attend the event e using $P(e|u)_{time} = max\{P(e|u)_{day-time}, P(e|u)_{date-time}\}$.

Histogram can be applied on location feature as well. For a user u, we use $H_{E(u)}^{location}$ to denote the location histogram of all the events u has attended before. For an upcoming event e, if e's location is the same as i-th entry $h_i^{location}$ in the location histogram, we assign $P(e|u)_{location} = h_i^{location}$, the probability that u will attend the event based on its location. Otherwise, $P(e|u)_{location} = 0$.

Similarly, we can utilize the topic feature of events. We use $H_{E(u)}^{topic-entity}$, $H_{E(u)}^{topic-tag}$, and $H_{E(u)}^{topic-category}$ to denote the histograms of u's event history based on entities, social tags, and categories that are extracted from event descriptions. For each upcoming event e, we use the same *OpenCalais* API to obtain entities, social tags, and categories of the event. The results can be represented using the well-known *Vector Space model*. The *cosine similarity* between the vector and the topic histogram is adopted to calculate $P(e|u)_{topic-entity}$, $P(e|u)_{topic-tag}$, and $P(e|u)_{topic-category}$, the probability that the user will attend the event e by considering entity of events, social tags of events, and categories of events, respectively. Generally, these three types of semantics cover different aspects. To integrate them into a unified measure, we have $P(e|u)_{topic} = \frac{P(e|u)_{topic-entity}+P(e|u)_{topic-tag}+P(e|u)_{topic-category}}{3}$, the algorithmic mean of the three types of semantics extracted from event descriptions.

Similar to the **GTS** method described in Sect. 3.1, the content-based privacy inference of event attendance can be achieved by ranking the upcoming events based on the probabilistic scores of the events using time duration, location, and topic features. Specifically, for a user u and each upcoming event e, we calculate $P(e|u)_{time}$, $P(e|u)_{location}$, and $P(e|u)_{topic}$, respectively. Since the three features are independent and each represents a probability that u may attend the event e, we use the product of the three probability scores to estimate the overall likelihood of attending the event, that is,

$$P(e|u)_{content} = P(e|u)_{time} \cdot P(e|u)_{location} \cdot P(e|u)_{topic}. \tag{3}$$

All the upcoming events are sorted based on $P(e|u)_{content}$ in descending order. The l top-ranked events can be regarded as privacy inference results, where l is a parameter which determines the number of events to be returned. We will refer to this method as **Event Content Similarity (ECS)** method.

Collaborative Filtering-based Privacy Inference. The ECS method for privacy inference of event attendance only considers the characteristics of events. However, it does not utilize the influence of social interactions existed in user's offline social worlds. Similar to the **GMS** method for privacy inference of online group membership, users who attended local social events frequently indicate that they share some common interests. Preferences of other similar users can be utilized for inference.

To evaluate whether a user u will be interested in attending an upcoming social event e, an intuitive solution is to measure how many users who are going to attend e (i.e., RSVPed to attend e in Meetup) that are similar to u. If we assume that users attended offline social event together frequently may behave similarly, we can qualitatively measure the likelihood that u will attend the upcoming event e using:

$$P(e|u)_{collaborative} = \frac{U_e \cap (\cup_{e^* \in E(u)} U_{e^*})}{U_e \cup (\cup_{e^* \in E(u)} U_{e^*})}, \tag{4}$$

where $E(u)$ denotes u's historical event participation and U_e denotes the set of users that are going to attend the event e. The larger of $P(e|u)_{collaborative}$, the more likely that u will be interested in attending e. If we rank all the upcoming events using $P(e|u)_{collaborative}$ in descending order, the l top-ranked events can be returned as privacy inference results. We call this method **Event Attendant Similarity (EAS)** method.

4 Case Study: Privacy Analysis in Meetup

We implemented the privacy inference models described in Sect. 3 and empirically simulated the privacy breaching scenarios using a large dataset crawled from Meetup. In this section, we report a set of experimental results related to the effectiveness of privacy inference in users' online and offline social worlds.

4.1 The Meetup Dataset

We crawled a large-scale Meetup dateset from www.meetup.com. The crawling process was started with 3 randomly selected seed users in Meetup. We chose the three seed users who were members of more than 100 groups online. The purpose is to make sure the crawled dataset would be large enough to cover many users, groups and events. In addition, the three seed users were selected from different states in the United States to avoid potential geographical bias. We used the Meetup API to extract the information of users, groups and events. Specifically, in the first round, all the groups that the seed users joined were firstly extracted using the Meetup API. Then, based on the list of retrieved group IDs, the API method `members` was used to obtain all the group members. Meanwhile, user's profile information such as interest-tags, the timestamps of joining each group, etc. were also retrieved. The API method `events` was used to crawl the list of past events held by these groups in Meetup. The RSVPs were downloaded to indicate whether the user attended the event or not. The above crawling process was repeated three times to obtain a large-scale Meetup dataset. The statistics of the Meetup dataset are provided in Table 1.

Table 1. The statistics of the Meetup dataset.

#Groups	#Users	#Events
86,396	6,532,384	6,088,084
#RSVPs	#Topic-tags	#Interest-tags
45,173,583	880,176	52,654,436

To protect user's privacy, Meetup in fact provides users with the option to hide their interest tags and membership of online groups. Based on whether user's interest tag and membership of groups are set to be public or hidden, we categorize users in Meetup into four classes, as shown in Table 2. As the hidden action can only be applied on the whole set of tags or groups, the four classes of users are not overlapping to each other. More than 34 % users indeed decided to

Table 2. The statistics of different classes of users.

Class of Users	Percentage
(1) Public Interest and Public Group	56.75 %
(2) Hidden Interest and Public Group	26.79 %
(3) Public Interest and Hidden Group	8.88 %
(4) Hidden Interest and Hidden Group	7.38 %

hide their interest tags and more than 16 % users chose to hide their membership of online groups as well. The statistics verify that many users in Meetup have serious privacy concerns regarding their personal information.

4.2 Settings of Privacy Inference Simulation

In Table 2, user classes (3) and (4) hide their membership of online groups. Due to the privacy settings in Meetup, their records of event attendance are also hidden accordingly. We regard these users as potential victims of privacy inference in this paper. The simulation of privacy inference focuses on the following question: *how can we infer the hidden membership of online groups and attendance of offline events for those victims by adopting the privacy inference models described in Sect. 3?*

Selection of Privacy Inference Victims. We randomly selected 30 users from user classes (3) and (4) each and constructed two sets of testing users, denoted as D_{hidden} and D_{public}, respectively. The subscripts *hidden* and *public* represent users' settings of their interest-tags. We denote the whole set of testing users as $D = D_{hidden} \cup D_{public}$. The selected users need to meet the two requirements: (1) the user has been a member of at least 5 online groups; (2) the user has attended at least 5 events. These requirements are used to capture partial knowledge of privacy attackers in Meetup.

   To examine the accuracy of privacy inference, we extracted user's complete set online group membership and offline event attendance records as ground-truth knowledge.

Simulation of Group Membership Inference. In order to utilize the two inference models **GTS** and **GMS** for privacy inference of user's group membership, we assume that the attacker has accumulated some partial knowledge about victim's online groups. The objective of the attacker is to infer other online groups that the victim may join.

   For a specific victim u in Meetup, we use $G(u)$ to denote the ground-truth of u's membership of online groups. $|G(u)|$ represents the number of online groups u has joined. When an attacker starts to conduct privacy inference, we use $G(u)^A$ to denote the partial knowledge of the attacker regarding u's group membership. $|G(u)^A|$ represents the partial number of online groups u has joined which is known by the attacker. Apparently, $G(u)^A \subseteq G(u)$. When $G(u)^A = G(u)$, we can conclude that u's privacy of group membership is completely breached.

A brute-force strategy to conduct privacy inference of group membership is to examine every group $g \in G - G(u)^A$ and adopt either the **GTS** method or the **GMS** method to calculate the likelihood $P(g|u)$ that u may join g. All of the groups $g \in G - G(u)^A$ can thus be ranked based on $P(g|u)$. The top-k ranked groups can be considered as privacy reference results with high confidence. Among the k inferred groups, we will use $G(u)^I_+$ to denote the set of groups that are actually joined by u. In other words, $G(u)^I_+ \subseteq G(u)$. We will use $G(u)^I_-$ to represent those inferred groups that u in fact is not a member.

Simulation of Event Attendance Inference. When an attacker conducts privacy inference of event attendance, the attacker is also equipped with some background knowledge about the victim. Different from the privacy inference of online group membership, the privacy inference of offline event attendance is usually targeted on predicting whether the victim may attend an upcoming event physically.

For a specific victim u in Meetup, we use $E(u)$ to denote the ground-truth of u's attendance of offline social events. When an attacker starts to conduct privacy inference, we use $E(u)^A$ to denote the partial knowledge of the attacker regarding u's historical records of event attendance. Obviously, $E(u)^A \subseteq E(u)$. When $E(u)^A = E(u)$, the attacker knows the complete history of u's event attendance.

As the privacy inference of event attendance is usually for future prediction, we use $E(u)^F$ to denote the set of events that u will attend. When an upcoming event e is available in Meetup, the privacy inference of event attendance is to infer whether u will attend e physically during the specified time duration. Similar to the privacy inference of online group membership, a brute-force strategy to conduct privacy inference of event attendance is to examine each of the upcoming events in Meetup. This strategy is somehow mission impossible, due to the fact that more than 500 K events are held monthly in Meetup.

It is necessary to emphasize that offline events are held by specific online groups in Meetup. Users need to be the member of the group in order to RSVP and check-in later to attend the event. Beside, most of the groups in Meetup hold the events weekly. Therefore, when an attacker conducts privacy inference of event attendance, we consider all of the events held by each group $g \in G(u)^A$ in the following week as the candidate set for privacy inference. The **ECS** method and the **EAS** method can be applied to conduct privacy inference of event attendance. A top-l ranked list of events based on $P(e|u)$ can be returned as inference results. We will use $E(u)^I_+$ to denote the inferred set of upcoming events that u will actually attend, and use $E(u)^I_-$ to represent the false positive inference results (i.e., u will not attend those events in $E(u)^I_-$).

Evaluation Metrics of Privacy Inference. The privacy inference models described in Sect. 2 regard the inference of group membership and event attendance as a ranking problem. The ranking positions for those correctly inferred instances are crucial to evaluate the inference performance. In this paper, we adopt the commonly used ranking evaluation metrics such as $P@K$ (precision at rank K) and $R@K$ (recall at rank K) as evaluation metrics.

4.3   Experimental Results

In this section, we report some experimental results for privacy inference in Meetup. To capture the impact of attacker's different levels of background knowledge for privacy inference, we organize users' various online and offline activities (joining an online group, or attending an offline event) based on their timestamps. We partition the dataset and use different percentages of the data to model attacker's different levels of background knowledge. In the following, attacker's background knowledge is modeled using 20%, 40%, 60%, and 80% of the data, respectively.

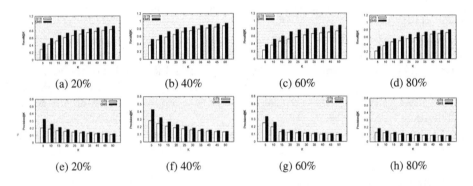

(a) 20%   (b) 40%   (c) 60%   (d) 80%

(e) 20%   (f) 40%   (g) 60%   (h) 80%

Fig. 2. Average performance of privacy inference of group membership on user set D. ((a)-(d) show $R@K$ and (e)-(h) show $P@K$, and X% represents X percentage of background knowledge).

Average Performance of Group Membership Inference. To conduct privacy inference of group membership, both the **GTS** method and the **GMS** method are considered. In Fig. 2, we show the average results of different privacy inference models of group membership over all of the users in D, evaluated by $R@K$ and $P@K$. To model different levels of attacker's background knowledge, we use different percentage of data, specifically, the attacker knows 20%, 40%, 60% and 80% of the online groups the victims have joined.

The results in Fig. 2 show that even when K is set to be very small, i.e., 5, 10, ..., 50, the metric $R@K$ can reach as low as 0.6 (when $K = 5$) but as high as 0.9 (when $K = 50$). The metric $P@K$ are not very high due to many *false positives*. However, in online social world, the attacker only requires little effort by examining the group pages so as to reduce the number of false positives. Thus, even for a low $P@K$, the privacy inference result is still severe enough to raise a red alert in people's online social world. We also notice several important findings based on the results in Fig. 2. First, the **GMS** method performs much better than the **GTS** method, which verifies the usefulness of users' online social interactions for privacy inference. Second, even with less background knowledge, the attacker is able to infer victim's privacy with very high accuracy. Third, additional background knowledge may not always be helpful. Even when the attacker is equipped with as high as 80 % knowledge, the inference result is not very high. This is partly due to the fact that in the simulation, the remaining positive instance is small.

Average Performance of Event Attendance Inference. We also conducted privacy inference of event attendance using both the **ECS** method and the **EAS** method. In Fig. 3, we present the performance of different privacy inference models, evaluated by $R@L$ and $P@L$. As the average number of events each user attends weekly is usually small (i.e., less than 6), we vary L from 1 to 5.

Based on the results in Fig. 3, we have several important findings. First, both the **ECS** method and the **EAS** method produce remarkably high *recall*, especially when L is set to be a larger value. Even when L is small (i.e., 1 or 2), the **ECS** method still can achieve high *recall* and *precision*. Recall the discussion of privacy in Sect. 2, if an attacker knows the physical location of a victim during some specific time durations, there could be some severe privacy breaching consequences such as home burglary. Second, the inference models did not show much difference in terms of the attacker's background knowledge. Even when the attacker knows as little as 20 % knowledge, the inference result is still relatively high. Still, when 80 % knowledge is acquired by the attacker, the performance of privacy inference does not increase.

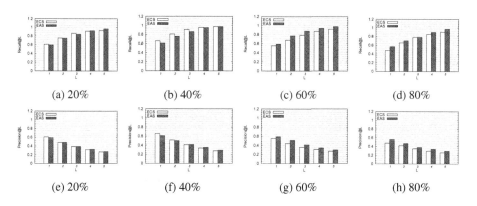

| (a) 20% | (b) 40% | (c) 60% | (d) 80% |

| (e) 20% | (f) 40% | (g) 60% | (h) 80% |

Fig. 3. Average performance of privacy inference of event attendance on user set D. ((a)-(d) show $R@L$ and (e)-(h) show $P@L$, and X% represents X percentage of background knowledge).

Case Study: Privacy Inference of Event Attendance for D_{public} and D_{hidden}. As shown in Table 2, user classes (3) and (4) have different preference regarding privacy, that is, whether user's interest tag should be hidden in Meetup. For those users who decide to make their interest tag public, intuitively, they might not worry about their privacy of group membership, as group membership largely reflects user's interest and preference. However, it is still unclear about user's opinion regarding their privacy of event attendance. In the following, we target on finding out what impact user's interest tag may have regarding their privacy of event attendance. Similar to the process reported in Sect. 3.2, we conducted privacy inference of offline event attendance on the two user sets D_{public} and D_{hidden}, respectively. Here the subscripts $hidden$ and $public$ represent users' settings of their interest-tags.

In Figs. 4 and 5, we report the performance of different privacy inference models, evaluated by $R@L$ and $P@L$. In general, the performance of privacy inference models on D_{public} is slightly higher than that on D_{hidden}, evaluated by *recall*. The difference of privacy inference performance is even more significant based on *precision*. It indicates that users with public interest tag in Meetup are indeed more vulnerable to privacy inference.

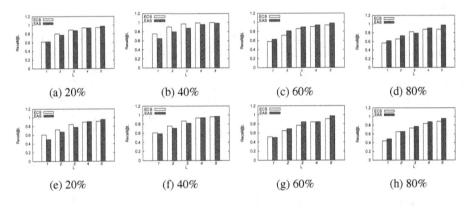

| (a) 20% | (b) 40% | (c) 60% | (d) 80% |

| (e) 20% | (f) 40% | (g) 60% | (h) 80% |

Fig. 4. Average R@L using privacy inference of event attendance ((a)-(d) on D_{public}, (e)-(h) on D_{hidden}, and X% represents X percentage of background knowledge).

5   Related Work

Attribute inference attack and link prediction attack are two types of most common attacks on OSNs. There have been many work devoted to identifying, simulating and defending such attacks [9,11,12,14,16,23]. In [21], the authors proposed to infer user's unknown profile attributes based on the observed attributes of the users in the same detected dense communities. Their experiments showed that even only knowing the attributes of as low as 20 % of the users, the unknown attributes of the remaining users within the community can be inferred with high accuracy. In [10], Bayesian network was employed to model the causal relations among users in an online weblog service provider Livejournal. The authors studied the impact of prior probability, influence strength, and society openness

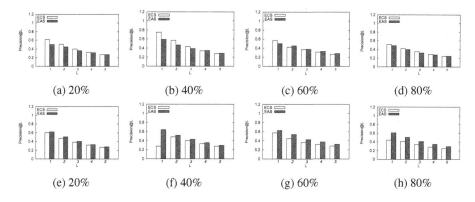

Fig. 5. Average P@L using privacy inference of event attendance ((a)-(d) on D_{public}, (e)-(h) on D_{hidden}, and X% represents X percentage of background knowledge).

to attribute inference capability, and showed that high inference accuracy can be obtained especially when users are in strong relationships. The experiments showed that incorporating both friendship links and profile details can get highest inference accuracy than using either piece of information. In addition to friendship links, group affiliations were utilized to build attribute inference models in [27]. This work simulated potential privacy attacks that an adversary can employ on OSNs with a mixture of public and private profiles. By comparing attack models under different types and amount of background knowledge, their experiments on four real-life datasets showed that online groups are another important type of information carrier besides friendship links, that could lead to potential privacy leaking. Other than utilizing links and topology of OSNs, attributes of vertices and edges have also been incorporated into link prediction techniques [1,2,6,17,22,26]. In [8], Gong et al. explicitly proposed a framework to jointly conduct link prediction and attribute inference using a Social-Attribute Network (SAN). Their experiments on a large-scale Google+ dataset showed that the accuracy of link prediction can be largely improved by first inferring missing attributes.

Users in OSNs would suffer a lot from privacy breaching [3]. The privacy issues would get even worse if users post their presence or absence location information publicly on conventional OSNs or LBSNs [4,7,13,24,25]. Targeting on Twitter, Cheng *et al.* [5] proposed a probabilistic framework to predict a Twitter user's city-level location based on the content of the user's tweets, while Mahmud *et al.* [20] successfully predicted the home location of Twitter users at different granularities via a hierarchical ensemble algorithm. In [15], Jurgens proposed to use spatial label propagation which leverages the social relationships and geographic distribution to infer user's location. The experiments on the data obtained from Twitter and Foursquare showed that nearly all of the users can be located with an estimated median error under 10KM.

How is our work different? Most of the existing privacy related studies in social networks only focused on one type, either offline physical social world or online virtual social world. EBSNs such as Meetup provide a nature way to bridge the two social worlds [19]. Privacy issues when two social worlds are bridges are facing new challenges, which thus require substantial research efforts. Our work in this paper provides a first step towards a comprehensive study of privacy issues by bridging user's online and offline social worlds.

6 Conclusion

In this paper, we focus on analyzing privacy threats when people's online and offline social interactions are bridged. The Meetup dataset was adopted for conducting the analysis of privacy threats. Based on a comprehensive simulation study of privacy inference in Meetup, our research revealed that using a series of simple yet effective privacy inference models, user's privacy of online group membership and offline event attendance can be inferred with high accuracy. The level of privacy threats by bridging user's online and offline social activities is remarkably severe. Therefore, there is a strong need of practical privacy protection mechanisms.

References

1. Al Hasan, M., Chaoji, V., Salem, S., Zaki, M.: Link prediction using supervised learning. In: SDM 2006: Workshop on Link Analysis, Counter-terrorism and Security (2006)
2. Backstrom, L., Leskovec, J.: Supervised random walks: predicting and recommending links in social networks. In: Proceedings of the Fourth ACM International Conference on Web Search and Data Mining, pp. 635–644. ACM (2011)
3. Becker, L., Pousttchi, K.: Social networks: The role of users' privacy concerns. In: Proceedings of the 14th International Conference on Information Integration and Web-based Applications and Services, pp. 187–195. IIWAS 2012, ACM (2012)
4. Carbunar, B., Rahman, M., Pissinou, N.: A survey of privacy vulnerabilities and defenses in geosocial networks. Commun. Mag. IEEE **51**(11), 114–119 (2013)
5. Cheng, Z., Caverlee, J., Lee, K.: You are where you tweet: a content-based approach to geo-locating twitter users. In: Proceedings of the 19th ACM International Conference on Information and Knowledge Management, pp. 759–768. ACM (2010)
6. Doppa, J.R., Yu, J., Tadepalli, P., Getoor, L.: Learning algorithms for link prediction based on chance constraints. In: Balcázar, J.L., Bonchi, F., Gionis, A., Sebag, M. (eds.) ECML PKDD 2010. LNCS (LNAI), vol. 6321, pp. 344–360. Springer, Heidelberg (2010). doi:10.1007/978-3-642-15880-3_28
7. Freni, D., Ruiz Vicente, C., Mascetti, S., Bettini, C., Jensen, C.S.: Preserving location and absence privacy in geo-social networks. In: Proceedings of the 19th ACM International Conference on Information and Knowledge Management, pp. 309–318. ACM (2010)
8. Gong, N.Z., Talwalkar, A., Mackey, L., Huang, L., Shin, E.C.R., Stefanov, E., Shi, E.R., Song, D.: Joint link prediction and attribute inference using a social-attribute network. ACM Trans. Intell. Syst. Technol. (TIST) **5**(2), 27 (2014)

9. He, J., Chu, W.W.: Protecting private information in online social networks. In: Chen, H., Yang, C.C. (eds.) Intelligence and Security Informatics. SCI, vol. 135. Springer, Heidelberg (2008)
10. He, J., Chu, W.W., Liu, Z.V.: Inferring privacy information from social networks. In: Mehrotra, S., Zeng, D.D., Chen, H., Thuraisingham, B., Wang, F.-Y. (eds.) ISI 2006. LNCS, vol. 3975, pp. 154–165. Springer, Heidelberg (2006). doi:10.1007/11760146_14
11. Heatherly, R., Kantarcioglu, M., Thuraisingham, B.: Preventing private information inference attacks on social networks. Knowl. Data Eng. IEEE Trans. **25**(8), 1849–1862 (2013)
12. Hoens, T.R., Blanton, M., Chawla, N.V.: A private and reliable recommendation system for social networks. In: 2010 IEEE Second International Conference on Social Computing (SocialCom), pp. 816–825. IEEE (2010)
13. Huo, Z., Meng, X., Zhang, R.: Feel free to check-in: privacy alert against hidden location inference attacks in GeoSNs. In: Meng, W., Feng, L., Bressan, S., Winiwarter, W., Song, W. (eds.) DASFAA 2013. LNCS, vol. 7825, pp. 377–391. Springer, Heidelberg (2013). doi:10.1007/978-3-642-37487-6_29
14. Jorgensen, Z., Yu, T.: A privacy-preserving framework for personalized, social recommendations. In: EDBT, pp. 571–582 (2014)
15. Jurgens, D.: That's what friends are for: Inferring location in online social media platforms based on social relationships. In: ICWSM (2013)
16. Li, D., Lv, Q., Shang, L., Gu, N.: Yana: an efficient privacy-preserving recommender system for online social communities. In: Proceedings of the 20th ACM International Conference on Information and Knowledge Management, pp. 2269–2272. ACM (2011)
17. Lichtenwalter, R.N., Lussier, J.T., Chawla, N.V.: New perspectives and methods in link prediction. In: Proceedings of the 16th ACM SIGKDD International Conference on Knowledge Discovery and Data Mining, pp. 243–252. ACM (2010)
18. Liu, B., Xiong, H.: Point-of-interest recommendation in location based social networks with topic and location awareness. In: SDM, pp. 396–404. SIAM (2013)
19. Liu, X., He, Q., Tian, Y., Lee, W.C., McPherson, J., Han, J.: Event-based social networks: Linking the online and offline social worlds. In: Proceedings of the 18th ACM SIGKDD International Conference on Knowledge Discovery and Data Mining, pp. 1032–1040. ACM, New York (2012)
20. Mahmud, J., Nichols, J., Drews, C.: Where is this tweet from? inferring home locations of twitter users. In: ICWSM (2012)
21. Mislove, A., Viswanath, B., Gummadi, K.P., Druschel, P.: You are who you know: inferring user profiles in online social networks. In: Proceedings of the Third ACM International Conference on Web Search and Data Mining, pp. 251–260. ACM (2010)
22. Scellato, S., Noulas, A., Mascolo, C.: Exploiting place features in link prediction on location-based social networks. In: Proceedings of the 17th ACM SIGKDD International Conference on Knowledge Discovery and Data Mining, pp. 1046–1054. ACM (2011)
23. Shang, S., Hui, Y., Hui, P., Cuff, P., Kulkarni, S.: Privacy preserving recommendation system based on groups. arXiv preprint (2013). arXiv:1305.0540
24. Terrovitis, M.: Privacy preservation in the dissemination of location data. ACM SIGKDD Explor. Newsl. **13**(1), 6–18 (2011)

25. Xu, D., Cui, P., Zhu, W., Yang, S.: Graph-based residence location inference for social media users. MultiMedia IEEE **21**(4), 76–83 (2014)
26. Yang, S.H., Long, B., Smola, A., Sadagopan, N., Zheng, Z., Zha, H.: Like like alike: Joint friendship and interest propagation in social networks. In: Proceedings of the 20th International Conference on World Wide Web, WWW 2011, pp. 537–546. ACM (2011)
27. Zheleva, E., Getoor, L.: To join or not to join: the illusion of privacy in social networks with mixed public and private user profiles. In: Proceedings of the 18th International Conference on World Wide Web, pp. 531–540. ACM (2009)

Empirical Analysis of Social Support Provided via Social Media

Lenin Medeiros$^{(\boxtimes)}$ and Tibor Bosse

Behavioural Informatics Group, Vrije Universiteit Amsterdam,
De Boelelaan 1081, 1081 HV Amsterdam, The Netherlands
{l.medeiros,t.bosse}@vu.nl

Abstract. Social media are an effective means for people to share everyday problems with their peers. Although this often leads to empathic responses which help alleviate the experienced stress, such peer support is not always available. As an alternative solution for such situations, this paper explores the possibilities to develop an 'artificial friend' that offers online social support through text messages. To formulate the requirements for such a system, a pilot study was performed in which 230 participants were asked (via a crowdsourcing platform) to report their experiences regarding online peer support. The results have been converted into a number of working hypotheses about online peer support. Based on these hypotheses, a conceptual framework has been developed that describes the functioning of the proposed support system.

Keywords: Social media · Everyday stress · Empirical analysis · Social support · Support agent

1 Introduction

Imagine the following situation. After a relationship of more than 5 years, your partner just informed you that (s)he will leave you for someone else. It is Friday evening, only a few hours after you heard the bad news, and you are still overwhelmed by a mixture of surprise, sadness and anger. In an attempt to cope with your emotions, you open your laptop and log in into your favourite social network. When online, you share your experience with some of your friends, hoping for some empathic reactions. Not long afterwards, you indeed receive a number of messages in response, varying from "Let it go, you were too good for that loser anyway!", to "Why don't you take some days off to do something fun?". And as if it were magic, reading these messages indeed makes you feel a bit better already.

This paper is part of a project that studies the role of peer support via online social networks in situations like the one sketched above. The project is motivated by the observation that everyday problems like broken relationships, difficult work situations, and loss of family members are important sources of stress [1]. Moreover, peer support was found to be a promising means for improving

© Springer International Publishing AG 2016
E. Spiro and Y.-Y. Ahn (Eds.): SocInfo 2016, Part II, LNCS 10047, pp. 439–453, 2016.
DOI: 10.1007/978-3-319-47874-6_30

various aspects of (mental) health and well-being [6,10,14,16]. Nowadays, one of the quickest and most frequently used approaches to provide peer support is to make use of online social networks like Facebook or Twitter [17]. After all, this type of support does not take much more effort than sending a short text message every now and then. Indeed, sharing problems and showing affection turn out to be among the most common reasons why people use social media [15].

Nevertheless, helpful peer support is not always available for all social media users, for the simple reason that some people have fewer friends than other. Moreover, even if they have many friends, users do not always want to share their problems online, especially in cases where such problems are very personal.

To improve upon this situation, our research explores the potential of artificial friends with the ability to provide social support via online social networks, thereby helping people to deal with their personal everyday problems. This vision is inspired by promising recent initiatives in developing artificial agents that support humans, varying from virtual depression therapists [5] to virtual buddies that support victims of cyberbullying [18]. We envision our system as an intelligent piece of software (possibly, but not necessarily embodied in the form of an avatar), which has the ability to analyze messages posted in social media, understand which messages potentially seek for peer support, and generate appropriate response messages with the aim to reduce the users experience of stress.

As a first step in that direction, this paper make an analysis of the requirements of such a system. Its main goals are to: (1) categorize the types of personal problems people share via social networks, (2) categorize the types of support people offer in such cases (and relate them to the categories found in the first step), and (3) provide a conceptual description of the proposed artificial friend. To acquire sufficient data to formulate an answer to the first two questions, a simple survey has been conducted via the crowdsourcing platform CrowdFlower[1].

The remainder of this paper is structured as follows. First, in Sect. 2, some background is provided about existing theories on peer support, as well as existing systems for computational (peer) support. Next, Sect. 3 describes the method used to analyze the requirements of the proposed system, and Sect. 4 presents the results. Section 5 provides a description of the system at a conceptual level, which may serve as a blueprint for implementation in follow up research. Section 6 concludes the paper with a discussion.

2   Related Work

2.1   Peer Support

Peer support, which can also be referred to as social support or peer-to-peer support, can be defined according to Kim et al. [11] as supportive information from people with strong social ties, i.e., friends, within a given network. Typically it occurs when a given person talks about their needs for support with others,

---

[1] Copyright © 2016 CrowdFlower Inc. Available at https://www.crowdflower.com/. Accessed: 2016-07-14.

which leads to a response expressing that one is loved and cared for. An example of that would be to share a stressful personal situation with friends on Facebook. Such friends may give some responses as a way to help the one who is stressed.

The relation between peer support and health is well known. In [10], several types of social network interventions are discussed, as well as examples of the use of social support for promoting health. For instance, Cohen et al. [3] confirms that there is a positive relation between social support and the way that people deal with stressful situations, and Cobb [2] presents evidence that this type of support is a good strategy to avoid health-related consequences of stress.

2.2   Emotion Regulation

Even though much is known about peer support, an open question is how we can distinguish between different types of peer support. In our work, we attempt to categorize classes of peer support by relating them to emotion regulation strategies. According to Gross [8], emotion regulation "includes all of the conscious and nonconscious strategies we use to increase, maintain, or decrease one or more components of an emotional response". Moreover, Gross identifies 5 different strategies for emotion regulation [9]: (1) *situation selection* – it means to take actions to make it more likely that potentially undesired situations will be avoided (e.g., not going to the party where you might encounter your ex), (2) *situation modification* – it means to take actions in order to try to change a given situation aiming to reduce or even to eliminate its undesired consequences (e.g., keeping some physical distance from your ex when you meet him or her), (3) *attentional deployment* – it is an "internal version of situation selection" in which the individual tries to stop thinking about a given undesired situation (e.g., trying to forget your ex), (4) *cognitive change* – it means to change one or more appraisals regarding a given situation in order to try to modify the situation's emotional significance (e.g., convincing yourself that you can find a better partner), and (5) *response modulation* – it means to influence physiological, experiential, or behavioral responses to a given undesired situation that already occurred (e.g., suppressing your tendency to start shouting at your ex).

Originally, such strategies mentioned above were meant to apply to self-regulation, but in our work we explore to what extent these strategies could be used in an inter-personal manner, hence connecting them to social support strategies in the context of people trying to cope with stress[2]. Additionally, based on our data and the concept of emotional support provided by Heaney et al. [10], we came up with the concept of *general emotional support* as another strategy for supporting friends: in this type of support the friends try to provide social

---

[2] We decided to use these strategies given by Gross [9] since it is a very well accepted work in the scientific community and we aimed to check to what extent such strategies, which are originally defined in the scope of emotional self-regulation, are also used in our context (friends trying to regulate peers' emotion). The idea to use Gross' strategies in an interpersonal manner was also adopted by other researchers, for instance in [19].

support only by providing empathy, love, trust and/or caring (e.g., by saying how much their friend means to them).

2.3 Computational Peer Support

Recently, an increasing amount of research puts forward the idea of robots and smart pieces of software being social, i.e., having social interactions with humans. Gockley et al. [7] describe the challenges regarding design decisions in developing human-like robots (or avatars) that are able to socially interact with humans. Leite et al. [12] showed that, indeed, robots acting emphatically can facilitate the perception that human users have about them in the sense that people can see such robots as friends. Accordingly, a number of initiatives actually developed such supporting agents for a variety of domains. For example, van der Zwaan et al. [18] made an effort to implement and test, with victims of cyberbullying, a domain-independent conversational model aiming to promote emotional support using an embodied conversational agent. DeVault et al. [5] created a virtual human interviewer which is able to have a conversation with a given person and can identify potential indicators of distress in order to avoid or reduce depression.

Along these lines, our intention is to develop a virtual agent which would be able to have a conversation with a given human user in order to try to help him/her to cope with 'everyday stressful situations'. In such a conversation the human user has to feel comfortable to share stressful situations as if (s)he were in a social network sharing content with friends. To the best of our knowledge, this approach is not present in the literature yet. To do that, we will implement it in a way that it can send messages based on empathetic responses, providing various types of social support (inspired by Gross' emotional regulation strategies [9]) at the appropriate moments. Our method to elicit the knowledge required for this is described in the next section.

3 Method

3.1 Data Collection and Filtering

As explained in Sect. 1 we aimed to categorize the types of potentially stressful situations people share via social networks as well as the respective responses they usually receive from their friends as support. Additionally, we were also interested in getting some insights about what type of support is typically related to a given type of situation. To obtain answers to these questions, a qualitative research design seemed to be the most suitable approach.

Inspired by the approach described in [4], we developed the following questionnaire and presented them to anonymous participants via a survey as a 'job' (i.e., a task that is accomplished via crowdsourcing) in the CrowdFlower platform. Each participant received 5 euro-cents as a reward for participating in the survey.

Question 1. Have you ever shared any difficult or stressful personal situation with one of your friends in a social network like Facebook or Twitter?

Question 2. What was the stressful situation you shared? Describe it in a short paragraph (about 5–10 lines is enough).

Question 3. What did your friend(s) exactly say to support you? Please describe it in a short paragraph (about 5–10 lines is enough).

Question 4. Via which social network did you share your situation: Facebook, Twitter or other?

Question 5. If you selected other (above), please indicate which one.

First, we performed such a survey 3 times with 10 participants each, in order to check to what extent we were getting useful responses in a reasonable time. After each of these rounds we changed some configurations regarding the criteria used to recruit participants. The CrowdFlower platform provides some functionality to manually set these criteria for the participants of a given job (e.g. average speed and quality of answers). After these three execution rounds we performed our survey one more time, this time with a group of 200 participants, giving us 230 responses in total. In this last round, the criterion for quality of responses was set to its maximum level. Also, in this last round the phrase "(e.g. the end of a relationship, a family or work problem, etc.)" was added to Question 1, because these three categories were mentioned most frequently and we wanted to reduce the chances of participants not understanding the question (with the drawback that this might have biased the remaining responses to these three categories).

After finishing the data collection, the dataset was refined in order to discard useless responses. In our context, a useless response to the questionnaire could be: (1) any response from which we could not reliably identify and classify the shared stressful situation(s) or the respective support(s) received – we discarded 17 answers based on this criteria, (2) any response written in any other language than English or Spanish – one answer was discarded because it was written in Indonesian, (3) any response from a participant who responded to the questionnaire more than one time – 3 participants answered our questionnaire twice so we discarded these 6 answers, and (4) any response from which the answer to Question 1 was not affirmative – we discarded more 147 responses because of this criteria.

Eventually, we obtained a dataset containing 59 useful responses (25.65 % useful responses in contrast to 74.35 % useless ones). These data were classified regarding the respective shared stressful situations and received support from friends, as explained in the following section.

3.2 Data Analysis

After obtaining the dataset according to the approach described above, we analyzed the data in order to classify all the shared stressful situations and the respective support given by friends, as reported by the participants.

We used three tags to (manually) classify our data: (1) *type of stressful situa-tion*, (2) *most affected individual*, and 3) *type of support*. The first two tags were used to classify the qualitative data regarding the responses to Question 2 of the survey. The third tag was used to classify the data obtained from the responses to Question 3. For each response given by the participants regarding Question 2 we allocated exactly one value for the type of stressful situation and another one for the most affected individual. For example, for the message "my aunt died suddenly", the type of stressful situation was classified as 'death' and the most affected individual as 'family member'. The classification process was performed by two analysts, and any disagreements were solved based on a discussion. The 'type of support' tag could have more than one value. More details about the tags and the values used are provided below.

After trying to cluster the data into categories of stress regarding the responses given to Question 2, we came up with 7 types of stressful situations: (1) *relationship* – any stress due to a situation involving a partner (e.g., a boyfriend, a wife, etc.), (2) *work* – any stressful situation within or about a work envi-ronment, (3) *death* – any stress caused because someone died, (4) *financial* – any stress due to financial matters, (5) *disease* – any stressful situation due to someone (i.e., either the participant or another person close to the participant) suffering from a disease, (6) *exams* – any stressful situation due to exams or a similar issue related to school, university, etc., and (7) *other* – any other type of stressful situation that does not fit into the previous categories. The types of stressful situations identified in our data are stated in Table 1, along with some respective examples (extracted from the responses given by the participants).

Table 1. Types of shared stressful situations and respective examples.

Type	Example
relationship	"the stressful situation was related with divorce"
work	"I was fired from my job"
death	"my aunt died suddenly"
financial	"I gathered a lot of receivables from customers"
disease	"I was very sick a few months ago"
exams	"stress on studies, upcoming exams, etc."
other	"we moved to a new town and I felt lonely there"

Next, it is important to note that a given stressful situation could affect the participants of our surveys themselves, as well as any individual in their environment. To be able to distinguish between these cases, the tag for the 'most affected individual' was added. Hence, the following 3 categories were used to classify the responses to Question 2 according to that tag: (1) *self* – when the most affected individual by a given stressful situation is the participant who filled out the survey, (2) *family member* – when the most affected individual

Table 2. Examples for the types of support used in our data analysis.

Type	Example
situation selection	"leave the relation"
situation modification	"study well and manage it"
attentional deployment	"forget her"
cognitive change	"the pain is not continuous and will end very soon"
general emotional support	"they said they love me"

by a given stressful situation is a participant's family member, and (3) *other* – all the cases which do not fit into the previous categories (e.g. pets and close friends).

Finally, to classify the answers to Question 3, the 'type of support' tag was introduced. For this tag, the following values were used (based on the emotion regulation strategies introduced in Sect. 2.2): (1) *situation selection* – when a friend suggests using (an instance of) situation selection as strategy to cope with a given stressful situation, (2) *situation modification* – when a friend suggests using situation modification to cope with a given stressful situation, (3) *attentional deployment* – when a friend suggests using attentional deployment to cope with a given stressful situation, (4) *cognitive change* – when a friend suggests using cognitive change to cope with a given stressful situation, and (5) *general emotional support* – when a friend only expresses empathy to try to help coping with a given stressful situation. No instances of response modulation were found, so this category is discarded in the rest of the paper. Table 2 shows examples extracted from our data for all these types of support.

4   Results

After having our dataset filtered and analyzed, the results were processed. To do that, we counted all messages allocated to each of the categories explained in Sect. 3.2, i.e. regarding the type of stressful situations, the most affected individual and the type of support received from friends. The results of this process are discussed in this section.

First, some results are presented about the survey's participants and the social networks they used to share their personal situations. It turned out that 89.09 % of the participants used Facebook to share their problems, while 3.64 % used Twitter and 7.27 % used other environments (VK, Odnoklassniki, Google+ and Skype). Concerning the nationality, we received responses from people located in 26 different countries. Apart from the fact that a surprisingly large percentage of participants (16.98 %) was from Venezuela, the distribution among the other nationalities was balanced. Figure 1 shows an overview of this. Unfortunately, CrowdFlower does not provide more details about the participants.

Fig. 1. All the nationalities that appeared in our final dataset. The closer to black, the higher the amount of participants from this country.

4.1 Types of Problems

Regarding the answers to Question 2 of our survey, we noticed that in 32.20 % of the cases people were sharing stressful situations concerning relationships. This amount is considerably higher than the other ones: 16.95 % for work, 16.95 % for death, 8.47 % for financial problems, 10.17 % for disease, 6.78 % for exams and 8.47 % for other problems. It is important to state that such types are mutually exclusive, even though a given user may share more than one stressful situation at the same time. Although we certainly need more data to draw reliable conclusions about this, this is sufficient to give us some initial insights about the most typical types of stressful situation people share via their social networks. Figure 2 shows a representation of these results in absolute numbers.

In addition to the above, we also checked the results for the most affected individual for the stressful situations. In 74.58 % of the cases, people were talking about themselves. In 18.64 % of the cases they were talking about any family member. Finally, in 6.78 % of the cases they were talking about someone else (friends or pets).

It is important to notice that we found only one occurrence of a relationship problem involving someone else than the participant himself. Based on this finding, combined with the initial results shown in Fig. 2, we came up with the following hypothesis about stressful situations shared via social networks.

Hypothesis 1 *The most typical stressful situation shared by people via their social networks concerns to their own relationships.*

We call this statement a 'hypothesis', since future work (involving a larger dataset) is required to confirm if it is statistically significant. In case this indeed

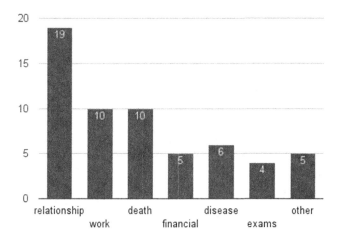

Fig. 2. Bar chart representing the occurrences (in absolute numbers) of instances of the problem categories defined in Sect. 3.2.

turns out to be the case, the finding can be taken into account in the design of a virtual support agent, e.g., by endowing such an agent with an ontology that enables it to explicitly reason about relationships problems. The next section will present a number of additional hypotheses, this time related to the support strategies used.

4.2 Support Strategies

An overview of the different types of support that were offered for each type of stressful situations (in absolute numbers) is provided in Table 3. As shown in the last row of the table, the cognitive change strategy, with 28 instances (i.e., 31.82 % of all support messages), and the use of general emotional support, with 27 instances (30.68 %), were the most common strategies people used to support friends. Situation modification, with 18 instances (20.45 %), deserves especial mention as well. Finally, attentional deployment and situation selection were used, respectively, 10 times (11.36 %) and 5 times (5.68 %).

The last column of the table presents the total number of responses to each stressful situation. Note that these number are higher than the numbers presented in Fig. 2 because some responses consisted of more than 1 support strategy; for instance, the response 'do not worry, everything will be OK' counts as an instance of attentional deployment ('do not worry') as well as cognitive change ('everything will be OK').

The numbers in the shaded cells in Table 3 provide information about possible correlations between the type of stressful situations people shared and the respective support strategies friends tend to use. Even though more data is needed to check correlations statistically, we believe that this information already provides interesting insights that can be used to state additional hypotheses.

Table 3. Frequency table for all types of support identified. The last column and row (in white) are, respectively, the amount of responses given by the participants' peers for each type of stressful situation and for each support strategy.

	general emotional support	cognitive change	attentional deployment	situation selection	situation modification	
relationship	8	6	5	4	5	28
work	3	8	3	0	4	18
death	8	4	0	0	3	15
financial	2	2	0	0	1	5
disease	1	5	2	0	1	9
exams	0	1	0	0	3	4
other	5	2	0	1	1	9
	27	28	10	5	18	

It is interesting to observe that support based on cognitive change or only empathy (general emotional support) is used much more frequently than the other ones in our dataset. Regarding these two strategies, people seem to be most likely to use general emotional support in the context of problems about relationships or death. The distribution of support actions using cognitive change is more balanced among several situations, although it seems to be most appropriate in cases of work stress and diseases. Based on these findings, we stated the following hypotheses about strategy for people helping friends to cope with stressful situations via social networks.

Hypothesis 2 *In case people share stressful situations about relationships or death, the most frequently used support strategy is general emotional support.*

Hypothesis 3 *In case people share stressful situations about work or diseases, the most frequently used support strategy is cognitive change.*

For the other stressful situations, the numbers are really too small to say anything meaningful. Nevertheless, it is interesting that attentional deployment and situation selection are most likely to be used for problems about relationships, and are never used for problems about death or financial problems. A more extensive discussion about these findings is presented in the next section.

Finally, looking at the distribution of support categories among relationship problems, the data seems to be rather balanced. This could indicate that relationship as a simple category is too generic, since these types of support could significantly differ from each other, so people could use them in completely different situations. Therefore, in future work it would be interesting to find sub-patterns for the problems fit into this category.

5   Artificial Support

Although the dataset is relatively small, the results described in the previous section provide us useful insight for the development of an artificial friend that offers peer support in social networks. This insight is mainly related to knowledge about 'which type of support works well in which situation'. For example, in case the reported stressful situation fell in the category 'death', the most frequently offered type of support was general emotional support (e.g., stating "my feelings are with you"); see Table 3. Instead, support based on attentional deployment was never offered in this situation. Intuitively, this makes sense, because when a loved one passes away, saying that one should just stop thinking about this person does not seem very appropriate. In contrast, attentional deployment does seem to be a suitable strategy in case the stress is caused by a relationship problem. Indeed, what works better than simply ignoring any thoughts about your annoying ex?

Based on these insights, in this section we present a conceptual framework describing the functioning of the proposed artificial friend. The main idea is that the software takes messages from friends in a social network as input, and generates appropriate support messages in response. To this end, a simple algorithm is put forward, which roughly uses three steps, namely *analysis* of the incoming message based on text mining and sentiment analysis [13], selection of a *support strategy* based on a computational model that matches stress categories to support strategies, and generation of *support messages* by filling in some pre-defined abstract sentences with more case-specific keywords. Below, the algorithm to process incoming messages is presented in the pseudo-code Algorithm 1.

To illustrate the algorithm in the context of a concrete example, imagine the system is confronted with the following message "Last week, my mother suddenly fell very ill". The system will then process this input in the following manner:

1. Process the message as follows (using function *ProcessIncomingMessages*):
   (a) Based on the occurrence of the keyword 'ill', the stressful situation is classified as an instance of the 'disease' category.
   (b) Since the message does not contain any information about the specific disease, the specific instance of the stressful situation cannot be defined (hence is left 'unknown').
   (c) Based on the keyword 'mother', the type of affected individual is classified as an instance of the 'family member' category.
   (d) Based on the keyword 'mother', the specific instance of the affected individual is set to 'mother'.
2. Using function *SelectStrategy*, the most appropriate support strategy is found to be 'cognitive change'.
3. Construct a response message as follows (using function *ConstructResponse*):
   (a) Select one of the predefined sentences created for the support strategy cognitive change. For example, take the following sentence: "I'm sure the situation will improve".

Algorithm 1. Processing incoming messages from a stressed user.

---
function PROCESSINGINCOMINGMESSAGES(m_1)
    situation ← CLASSIFYTHESTRESSFULSITUATION(m_1)
    ▷ Using text mining, classify the stressful situation addressed in m_1 (if any) into one of the categories from Table 1 and, if possible, try to find the specific instance of such a category.
    individual ← CLASSIFYTHEMOSTAFFECTEDINDIVIDUAL(m_1)
    ▷ Using text mining, classify the type of individual affected by the stressful situation into one of the categories *self*, *family*, and *other*. If possible, try to find the specific instance of such a category.
    strategy ← SELECTSTRATEGY(*situation*)
    m_2 ← CONSTRUCTRESPONSE(*situation*, *individual*, *strategy*)
    return m_2

function SELECTSTRATEGY(*situation*)
▷ Based on the value for *situation*, select the most appropriate support strategy using the knowledge from Table 3.
    return *strategy*

function CONSTRUCTRESPONSE(*situation*, *individual*, *strategy*)
    response ← SELECTTEMPLATE(*strategy*)
    ▷ Select a sentence template that corresponds to the support category identified previously.
    response ← SETTEMPLATEFORSITUATION(*situation*, *response*)
    ▷ Make the sentence template more specific by refining it based on the type of problem and the respective instance identified previously.
    response ← SETTEMPLATEFORINDIVIDUAL(*individual*, *response*)
    ▷ Make the sentence template more specific by refining it based on the affected type of individual and the respective instance identified previously.
    return *response*

(b) Refine the sentence by replacing the phrase "the situation will improve" by a pre-defined phrase that is more specific for the 'disease' category. This may result, for example, in the following sentence: "I'm sure the person will recover from the disease". Note that, as the specific type of disease is unknown, this is not further specified.

(c) Refine the sentence further by replacing the word 'person' by the specific individual 'mother'. This results in the following sentence: "I'm sure your mother will recover from her disease".

After the system has generated the appropriate response message m_2, it will be sent to the user via the social network. This way, for every incoming message that addresses a stressful situation, a specific response message will be generated that on the one hand is based on generic knowledge about support strategies, but on the other hand is tailored to the specific case of the individual. Note that the system assumes that the incoming messages are related to stressful situations. However, in case the system is unable to reliably classify the stressful situation into one of the pre-defined categories, it may generate a question to ask the user for additional information.

6   Discussion

This paper presented an exploration of the possibilities to have an artificial agent generate social support in situations where people share their 'everyday problems' [1] via social media. The underlying assumption is that, by generating tailored response messages, such an agent can help alleviate the stress people experience in these situations, especially in cases where users do not receive comforting responses from their human peers. As a first step to analyze the requirements for such a system, a pilot study was performed based on the CrowdFlower platform.

Based on the pilot study, we gained useful insight into the types of personal situations people typically share via social media, as well as the types of support they receive from their friends. The findings have been summarized in terms of a number of hypotheses about online peer support. In the next phase of our research, these hypotheses may function as requirements for the design of the support agent. For instance, to support people who share stressful situations in the category 'death', general emotional support seems to be the most appropriate method.

However, before the hypotheses can be considered actual requirements, more data is needed, to confirm if the correlations found so far are statistically significant. To this end, we will apply the same method as used in the pilot study, but this time with a much larger group of participants. While doing that, we will also try to minimize the number of 'useless' responses received via Crowd-Flower, for instance by providing more detailed instructions about the input that is expected from the participants.

Additionally, a conceptual framework was put forward that describes the functioning of the proposed support agent. The main idea is to process incoming messages in online social networks in three steps: analysis of the incoming message, selection of a support strategy, and generation of support messages. This approach makes it possible to generate personalized messages that are based on generic knowledge about support strategies, yet being sufficiently specific to address each individual case separately.

It should be noted that the presented approach focuses on simple interactions consisting of one incoming message and one response message. Nevertheless, it would be interesting to extend it to more complex types of interaction, eventually resulting in entire human-agent conversations. To do so, part of the approach put forward in [18] could be used. This paper distinguishes five conversation phases in so-called 'comforting conversations', namely *welcome, gather information, determine conversation objective, work out objective*, and *round off*. The current algorithm can be seen as a concrete instantiation of the work out objective phase, but in principle it can be combined with modules to address the other phases. In a similar direction, another extension would be to allow the agent to apply multiple support strategies, (e.g., to switch to the second-best option in case the best option does not work).

Finally, once the support agent is actually developed, the next step will be a systematic evaluation. Even though human peer support has already shown to be beneficial in various situations, such an evaluation is crucial to test the assumption that an agent providing such support also has an added value. This will be done both through usability studies (to investigate how users perceive the added value of the agent) and on the longer term, by actually measuring the effectiveness of the system in reducing people's experienced stress.

Acknowledgements. The authors would like to state that Lenin Medeiros' stay at Vrije Universtiteit Amsterdam was funded by the Brazilian Science without Borders program. This work was realized with the support from CNPq, National Council for Scientific and Technological Development - Brazil, through a scholarship which reference number is 235134/2014-7. We also would like to thank all the people who gave us important data by answering our questionnaire.

References

1. Burks, N., Martin, B.: Everyday problems and life change events: Ongoing versus acute sources of stress. J. Hum. Stress **11**(1), 27–35 (1985)
2. Cobb, S.: Social support as a moderator of life stress. Psychosom. Med. **38**(5), 300–314 (1976)
3. Wills, A.: Stress, social support, and the buffering hypothesis. Psychol. Bull. **98**(2), 310 (1985)
4. Dennis, M., Kindness, P., Masthoff, J., Mellish, C., Smith, K.A.: Towards effective emotional support for community first responders experiencing stress. In: 2013 Humaine Association Conference on Affective Computing and Intelligent Interaction, ACII 2013, Geneva, Switzerland, September 2–5, 2013, pp. 763–768 (2013)
5. DeVault, D., Artstein, R., Benn, G., Teresa Dey, E., Fast, A.G., Georgila, K., Gratch, J., Hartholt, A., Lhommet, M., Simsensei kiosk, et al.: A virtual human interviewer for healthcare decision support. In: Proceedings of the 2014 International Conference on Autonomous Agents and Multi-Agent Systems, pp. 1061–1068. International Foundation for Autonomous Agents and Multiagent Systems (2014)
6. Eysenbach, G., Powell, J., Englesakis, M., Rizo, C., Stern, A.: Health related virtual communities and electronic support groups: systematic review of the effects of online peer to peer interactions. BMJ **328**, 1166–1170 (2004)
7. Gockley, R., Bruce, A., Forlizzi, J., Michalowski, M., Mundell, A., Rosenthal, S., Sellner, B., Simmons, R., Snipes, K., Schultz, A.C., et al.: Designing robots for long-term social interaction. In: 2005 IEEE/RSJ International Conference on Intelligent Robots and Systems, pp. 1338–1343. IEEE (2005)
8. Gross, J.J.: Gross. Emotion regulation: Affective, cognitive, and social consequences. Psychophysiology **39**(3), 281–291 (2002)
9. Gross, J.J., Thompson, R.A.: Emotion regulation: Conceptual foundations (2007)
10. Heaney, C.A., Israel, B.A.: Social networks, social support. Health Behav. Health Educ. Theor. Res. Pract. **4**, 189–210 (2008)
11. Kim, H.S., Sherman, D.K., Taylor, S.E.: Culture and social support. Am. Psychol. **63**(6), 518 (2008)

12. Leite, I., Pereira, A., Mascarenhas, S., Martinho, C., Prada, R., Paiva, A.: The influence of empathy in human-robot relations. Int. J. Hum. Comput. Stud. **71**(3), 250–260 (2013)
13. Liu, B.: Sentiment Analysis - Mining Opinions, Sentiments, and Emotions. Cambridge University Press, New York (2015)
14. O'Dea, B., Campbell, A.: Healthy connections: Online social networks and their potential for peer support. In: Health Informatics: The Transformative Power of Innovation - Selected Papers from the 19th Australian National Health Informatics Conference, HIC 2011, 1–4, Brisbane, Australia, pp. 133–140. IOS Press, August 2011
15. Young, L.: Uses and gratifications of social media: A comparison of facebook and instant messaging. Bull. Sci. Technol. Soci. **30**(5), 350–361 (2010)
16. Takahashi, Y., Uchida, C., Miyaki, K., Sakai, M., Shimbo, T., Nakayama, T.: Potential benefits, harms of a peer support social network service on the internet for people with depressive tendencies: Qualitative content analysis and social network analysis. J. Med. Int. Res., 11(3), 2009
17. Breda, W., Treur, J., van Wissen, A.: Analysis and support of lifestyle via emotions using social media. In: Aberer, K., Flache, A., Jager, W., Liu, L., Tang, J., Guéret, C. (eds.) SocInfo 2012. LNCS, vol. 7710, pp. 275–291. Springer, Heidelberg (2012). doi:10.1007/978-3-642-35386-4_21
18. van der Zwaan, J.M., Dignum, V., Jonker, C.M.: A conversation model enabling intelligent agents to give emotional support. In: Ding, W., Jiang, H., Ali, M., Li, M. (eds.) Modern Advances in Intelligent Systems and Tools. SCI, pp. 47–52. Springer, Heidelberg (2012). doi:10.1007/978-3-642-30732-4_6
19. Williams, M.: Building genuine trust through interpersonal emotion management: A threat regulation model of trust and collaboration across boundaries. Acad. Manage. Rev. **32**(2), 595–621 (2007)

User Generated vs. Supported Contents: Which One Can Better Predict Basic Human Values?

Md. Saddam Hossain Mukta[1]([✉]), Mohammed Eunus Ali[1], and Jalal Mahmud[2]

[1] Department of Computer Science and Engineering,
Bangladesh University of Engineering and Technology, Dhaka 1000, Bangladesh
saddam944@gmail.com, eunus@cse.buet.ac.bd
[2] IBM Almaden Research Center, 650 Harry Road, San Jose, CA 95120, USA
jumahmud@us.ibm.com

Abstract. Every individual possess a set of Basic Human Values such as self-direction, power, and hedonism. These values drive an individual to commit actions in various situations in her daily lives. Values represent one's attitudes, opinions, thoughts and goals in life, and can regulate a variety of human behaviors and manners that an individual shows in the society. In this paper, we identify five higher-level values from social media interactions by analyzing two types of contents: user generated and user supported. More importantly, we identify which type of content can better predict which human values in different scenarios, which ultimately helps us to predict human values for both silent and active users. We also build a combined value prediction model by integrating different types of interaction features, which can more accurately capture the human values than that of a single feature based model. We also build separate models for silent and active users of SNS to effectively predict values for different types of SNS users. Finally, we compare the strength of different types of models to predict values from social media usage effectively.

1 Introduction

In recent times, Social Networking Sites (SNS) have become a major platform of communications for users in the web. These SNS allow a user to share ideas, thoughts, and opinions with her friends, family and acquaintances by using different types of interactions such as statuses, page-likes, link-sharing and comments. Among all the existing SNS, in the recent years, Facebook has gained tremendous popularity due to its wide variations of interaction features. The contents of these interactions in SNS provide a rich platform for the researchers to develop different applications that include identifying cognitive and psychological attributes such as values [5], personality [12,21,22], preferences [27], and emotion [2] of involved users. In this paper, we are the first to identify human values of Facebook users from different types of interaction features.

© Springer International Publishing AG 2016
E. Spiro and Y.-Y. Ahn (Eds.): SocInfo 2016, Part II, LNCS 10047, pp. 454–470, 2016.
DOI: 10.1007/978-3-319-47874-6_31

Basic human values represent a set of criteria such as conservation, hedonism, etc. that are used by individuals to take different actions in their daily lives. Values can be categorized into five higher-level dimensions: self-transcendence, self-enhancement, conservation, openness-to-change and hedonism [30]. These values provide predictive and descriptive power in the analysis of opinions and actions of people's lives. Thus identifying basic human values has many potential real-life applications that include selection of career paths, prediction of customers' buying behaviors, and detection of life style changes such as fashion and trend.

Since interactions of users in SNS resemblance their real-world charectirstics, these interactions facilitate us with the contents that can be used for identifying Basic human values. We can categorize the contents of SNS based on social networking activities into two types: user generated (UG) and user supported (US) contents[1]. When a user creates or writes a content by himself, we define the content as a *user generated content*, e.g., statuses and comments of a user. In contrast, when a user expresses her positive association with a content by a particular social network activity such as page-likes and link-sharing, we term the content as a *user supported content*.

A recent study [5] identifies five higher-level values from a user's pattern of word use collected from her statuses and comments (UG content) in Reddit. In [5], authors do not consider US contents such as *shares* and *up-votes*, while identifying values of a user. Hsieh et al. [15] show the correlation between a user's reading interest and her value scores from tweets. None of these techniques consider both UG and US contents while identifying Basic human values. This limits the applicability of existing techniques in many scenarios, particularly for the silent users who do not generate contents themselves. For example, in Facebook, many people are not interested in generating contents such as statuses and comments due to (i) privacy and safety issues, and (ii) shyness of generating contents in public [13,26]. However, we observe that most of these people interact in Facebook using other interaction features such as page-likes and shared-links. In a psychology study [32], author observes that people generally *like* in Facebook because it is quick, easy and wordless interaction feature. In another study [34], author finds that 37 % people share to let other know what they believe and 29 % people share why they support a cause, organization and belief. Thus, US content can play an important role to identify the values of such users. We consider both UG and US contents to identify values for all types of users in Facebook.

The motivation of our work comes from a key observation that different types of social media contents contribute differently in identifying different human values. Authors in [1,16] showed that some psychological attributes such as neurotic and conscientiousness personality traits, of users are difficult to predict from UG content in social network. In another study [17] authors successfully predicted highly sensitive personal attributes such as ethnic origin, political views, religion, and substances use (e.g., alcohol) from Facebook *likes* (i.e., US content). Facebook *likes* also express users positive association with online contents

---

[1] For the sake of simplicity, we would refer UG and US for "user generated" and "user supported" respectively throughout the paper.

such as product, restaurant, sports or music. In a study [13], authors defined users as *lurkers* or *silent users* who generate a little content, and prefer to consume other users' content in SNS. Understanding lurkers is important for recommender systems and targeted advertisement. In Facebook, it is observed that such users (*lurkers*) mostly interact with other by US content rather than UG content. Hence, we need to consider both UG and US contents to predict values accurately.

In this paper, we propose a technique to identify values by analyzing three types of interaction features: statuses, page-likes and shared-links, of 567 Facebook users. First, we compute correlation between users' word use in Facebook by Linguistic Inquiry and Word Count (LIWC) [25] tool and portrait value questionnaire (PVQ) [28] test result of these Facebook users. Then, we predict five higher-level values using linear regression models from different types of contents. We also apply different linear classification models to predict human values and find out the suitable classification model for each type of content. Next, we construct a weighted linear ensemble model to build a unified value score by integrating the models of both UG and US contents, which can better predict human values than that of models built from individual contents.

In this paper, we model silent users who give less status updates (i.e., 2–3 times in every two months time interval), but interact with others by liking pages and sharing links frequently (i.e., 3–5 times in a week). In particular, we build separate models of values by using different types of contents for silent and active users. Our experimental results show that we can build better value models of Facebook silent users by using US contents than that of active users.

In summary, we have the following contributions:

- We are the first to consider different types of Facebook interaction features, i.e., user generated and user supported contents to compute five higher-level values.
- We compare strength of the models that are built from different types of contents and show which features can best predict which values.
- We build an integrated model that combines different types of interaction features through a weighted linear ensemble technique, which improves the prediction accuracy significantly.
- We compare the strength of value models between silent and active users, and show which features can be effectively used for which models.

2   Preliminaries and Related Work

2.1   Basic Human Values

Our goals, actions, beliefs and behaviors depend on our values. Our values define who we are and what we do. Values signifies the importance of different aspects such as power, security, tradition, success, happiness, social status, etc., in our life. Society, culture, religion and life experiences shape up the priorities of numerous values in individuals. The priorities differ from one individual to another, which results in diversities of an individual's actions in different situations. Schwartz et al. [29] present ten root-level values by analyzing data collected from more than seventy countries with divergent cultures, languages, and customs. They also portray the discrepancies and likenesses between these values. We summarize the core ideas of these ten root-level values from the writings of several theorists and researchers as follows:

1. **Self-Direction** signifies an individual as independent, innovative, confident, deciding and controlling his own aims.
2. **Stimulation** characterizes one's preference to challenges, excitements, thrills and varieties in life.
3. **Hedonism** refers to the contentment, self-indulgence and satisfaction for oneself.
4. **Achievement** represents a person's interest to be socially recognized and the motivation, talents and accomplishments for acquiring the recognition.
5. **Power** indicates one's attraction for authority and social status, supremacy, control over other humans and resources.
6. **Security** specifies reliability, safety and peacefulness of individuals, their associations and the society.
7. **Conformity** denotes responsibility, loyalty, obedience towards others and avoid actions that can harm others and the society.
8. **Tradition** symbolizes respect and reverence for norms and customs of society, cultures or religions.
9. **Benevolence** means one's concerns for others wellbeing and happiness with whom he interacts most (e.g., family, friends, primary groups etc.).
10. **Universalism** stands for the indulgence, responsibility and accountability regarding the safety and wellbeing of all the people in world.

Schwartz et al. [29] map ten root-level Basic Human Values into five higher-level value dimensions. The five higher-level value dimensions can be obtained from ten root-level values: self-transcendence includes benevolence and universalism values, openness-to-change encompasses stimulation and self-direction values, self-enhancement encompasses achievement and power values. Conversation includes security, conformity and tradition, and hedonism remains as it appears in the root-level.

2.2 Related Works

A significant number of studies of different human attributes and cognitive aspects such as personality [12], sentiment analysis [24], and emotion [2] have been conducted from social networking sites such as Facebook and Twitter. Golbeck et al. [12] predict Big5 personality scores from the users' publicly available tweets in Twitter. The studies in [6,11,35] investigate the political alignment of a person based on his tweets.

Chen et al. [5] identify five higher-level values from user's patterns of word usage in social media. Authors analyze the statuses and comments of Reddit, an online news sharing community, by analyzing words using LIWC tool. They also conduct PVQ test among Reddit users. Later, authors compute pearson correlation analysis between LIWC categories and PVQ scores. Authors predict the value scores by computing the linear regression. To investigate the prediction potential, authors also propose classification techniques. They do not consider other interaction features such as page-likes. Boyd et al. [3] also identify values from a crowd-sourced and a Facebook dataset using natural language processing (NLP) technique. They consider user statuses while identifying values. They also do not consider other interaction features in Facebook for computing values. Hsieh et al. [15] find out relation between user's reading interest and his values by Twitter. They compute regression analysis for three different values, such as universalism, achievement, and hedonism.

However, the above studies have the following limitations: (1) They do not consider multiple interaction features in their works. To identify values from different features, we find different subset of LIWC categories. Since LIWC category subsets are different for each value, the linear regression models are also different. We cannot apply their techniques, as different features attribute to different value dimensions in different ways, (2) Silent users are significant part of the total Facebook users. Authors in previous studies do not take into account silent users during value computation, and thus suffer in lack of accuracy due to ignoring *silent* users in their studies.

3   Data Collection

We have invited 645 users to collect Facebook data through posts on Facebook, emails to relevant mailing lists, personal messages, and word of mouth communication. We send invitation through different channels for collecting users' Facebook data only. We create a Facebook application to access users' time-lines. Among 645 Facebook users, 582 users (male = 316, female = 266) authenticated the application to read their time-lines. The rest 63 users have not shown interest to share their time-lines through the application.

Among the 582 users, we identify 15 users as *churners* [13,23], who stoped using Facebook for some reasons. We observe their pattern of Facebook usage. We find that these users do not use Facebook for at least last 7 months (according to the paper [13]). Thus, we discard the churners from our dataset. Finally, we consider our dataset with a total of 567 (582-15) Facebook users. We collect a total of 62902, 51066 and 17408 English statuses (tag: "message"), page-likes (tag: "about") and shared-links (tag: "description"), respectively, as of December 20, 2015. Table 1 presents maximum, minimum and average word counts for each type of content.

We have conducted 21-item PVQ test among these 567 users as the ground truth data on the value scores. These users are the members of student and different professions (e.g., engineer, doctor, and banker), and aged between 20 and 55 years. Thus our dataset comprises different age groups, professions and gender within same ethnicity. All users are asked to fill out the survey questionnaire

Table 1. Maximum, minimum and average word counts for three different types of contents.

Content type	Maximum	Minimum	Average
Statuses	12168	7	751.33
Page-likes	1887	3	943.22
Shared-links	8949	2	347.15

via an experimental web page. In these questionnaire, a user is presented a set of statements, where each statement corresponds to a single value dimension. These statements identify the participant's value dimensions without directly asking him about a particular value [28].

Average survey scores for five higher-level values are ranged from 0.542 to 0.769 on a normalized scale of 0–1. We also calculate Cronbach's alphas of PVQ answers for five higher-level values that ranged from 0.428 to 0.553. Cronbach's alpha [7] estimates the internal consistency of reliability of test scores. Although we find that some Cronbach's alpha scores are low, these low scores are acceptable according to the study of Schwartz [5, 30].

4   Methodology

In this paper, we identify values from users' pattern of word usage of different types of interactions in Facebook. First, we extract users' UG and US contents through the Facebook application. Later we preprocess the emoticons and remove noisy special characters. For each type of content, we find out LIWC [25] category of words independently. Then, we find out best subset of LIWC categories (independent variables) and the scores of the 21-item PVQ test (dependent variable) for each type of content separately.

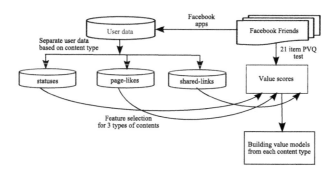

Fig. 1. Predicting value models from different types of contents.

Next, we build linear regression models that predict value dimensions using LIWC category of words from different types of contents. Figure 1 shows our methodology to build value models from different types of contents. We compare the strength of different models that are built from different contents. To improve the accuracy, we also combine all the models to compute a unified value score. Different modules of our methodology are described as follows:

- *Data Preprocessing.* We extract users' data of three interaction features through Facebook application and discard irrelevant fields such as *creation time, user id,* curly brackets *("{"),* etc. For processing *emoticons,* we use the technique of [36]. We find emoticons in the sentence level. We replace *emoticons* with a corresponding text of emotional sense (e.g., happy, sad, angry, haha, etc.).
- *Linear regression models.* In this module, we present the construction process of different linear regression models for each type of content. We also compute the strength of these models with the R^2/adjusted-R^2 measures.
- *Ensemble of the models.* In this module, we describe how to build a weighted linear ensemble model to determine a unified value score from different types of contents.

5   Building Models of Values

In this section, we build different types of value models from different interaction features. First, we identify best subset of LIWC categories for each type of content and for each human value. Later, we build regression models to identify values for each type of content. Next, we investigate prediction potential of our models using machine learning classification techniques.

5.1   Feature Selection

Users' way of expressing thoughts and ideas in social network might be different. Some may like to express their thoughts by writing statuses while others may convey their thoughts through page-likes or link-sharing. It is difficult to predict all value dimensions of a user from one type of interaction feature, as not every type of interaction feature reveals every value dimension. Thus, we need to first identify features from different types interactions that are suitable for predicting different value dimensions.

In this section, we use PVQ scores of the 397 (70 % of the total dataset) Facebook users as ground truth (dependent variable) data on value scores and best subset of LIWC categories are used as independent variables. We first select best subset of LIWC categories (predictors) using forward selection approach [8] for each type of content using *leaps* [19] R package implementation. *Leaps* package performs an exhaustive search to find out best subset of LIWC categories using an efficient *branch-and-bound* algorithm. For example, we find family, affect, human, negemo, anx, feel, etc. LIWC categories of words for computing self-transcendence values from users' statuses. Again, we find social, cogmech, insight, tentat, work, etc. LIWC categories of words to compute self-transcendence values from users' page-likes. We also find affect, posemo, bio, assent, etc. LIWC categories of words to compute self-transcendence values from users' shared-links. For each type of content and for each human value, we find a total of 15 (3×5) different subsets of LIWC categories using *leaps* R package implementation.

5.2   Regression and Classification Models

We build linear regression models to predict the score of values from each type of content. To build linear regression models, we consider 15 different subsets of LIWC features as our independent variables that are identified in Subsect. 5.1. We consider PVQ scores as our dependent variable. Figure 1 presents the technique of building linear regression models from different types of contents. For each type of content and for each type of value, we have a total of 15 different regression models. A potential problem arises when collinearity is found between values and LIWC features. When there is a perfect linear relationship exists among independent variables, the outcome for a regression model cannot be unique. To remove collinearity among independent LIWC features, we have computed lasso penalized linear regression using *glmnet* R package [5,14]. This technique reduces the coefficients to a low value or zero, thus the model does not get over fitted. Finally, for each content type, we perform a linear regression analysis with our best subsets of selected LIWC categories with a 10-fold cross-validation with 10 iterations.

Table 2 shows the results of our regression analysis. The R^2 and adjusted-R^2 of our models are reasonably moderate across all the values for each type of content. These models show that self-transcendence and openness-to-change values are predictable more accurately through UG content (i.e., status) than US contents (i.e., page-likes). In contrast, hedonism, self-enhancement and conservation values can be predictable through US content than UG content more accurately.

The above linear regression model suffers from the following limitation. Sumner et al. [33] suggested that computing mean squared error (MAE) and root mean squared error (RMSE) for error measure in regression analysis are not adequate. In particular, when the majority of the individuals are around the mean of unimodal distribution, these error measures can often mask large errors.

Since value dimensions have unimodal distributions, RMSE and MAE suffer in lack of investigating strength of a prediction model. To overcome this limitation, we apply different supervised binary machine learning algorithms on our dataset. We classify above-median level as *high* class label and below-median as *low* class label value dimension. We have experimented with several classifiers that include Logistic Regression, Naive Bayesian, Adaboost, Random Forest and RepTree classifiers. For each type of content (i.e., status, page-likes or shared-links), we have applied these classifiers to understand the prediction performance of different value dimensions.

Table 3 presents the best classifier, content type, its true positive rate (TPR), true negative rate (TNR) and area under the ROC curve (AUC) for computing each of the value dimensions [10]. TPR defines how many samples are correctly

Table 2. Strength of the linear regression models for five higher level values.

Values	Status (UG)		Page-likes (US)		Shared-links (US)	
	R^2	Adjusted-R^2	R^2	Adjusted-R^2	R^2	Adjusted-R^2
Self-Tran.	18.8%	14.1%	14.2%	10.3%	14.4%	11.2%
Openn.	22.7%	18.9%	16.8%	12.7%	17.8%	14.1%
Hedonism	13.3%	10.09%	13.6%	11.1%	12.3%	8.7%
Self-Enhan.	13.3%	11.3%	12.8%	8.3%	16.8%	12.7%
Conserv.	20.1%	16.3%	20.6%	17.71%	11.9%	8.1%

classified as *positive* among all positive samples and TNR defines how many samples incorrectly classified as *negative* among all negative samples available during the test. We conduct the performance of the classifiers using AUC values under the 10-fold cross validation. Performance of the classifiers were conducted using AUC values under the 10-fold cross validation. The curve is plotted the TPR against the TNR at different threshold.

Table 3. Best classifiers of different types of contents.

Values	Best classifying content	Best classifier	AUC	TPR	TNR
Self-Tran.	status	Adaboost	0.66	0.65	0.34
Open.	status	Naive Bayes	0.69	0.71	0.28
Hedonism	page-like	Naive Bayes	0.59	0.61	0.40
Self-Enhan.	shared-link	Naive Bayes	0.60	0.62	0.39
Conserv.	page-like	Logistic Reg	0.67	0.66	0.32

Tables 2 and 3 show the strength of the value models that are built from different types of contents. We observe that our regression models achieve moderate performance. We notice that our classifiers also achieve moderate improvement over random chances. We find the best regression models and classifiers for self-transcendence and openness-to-change values with the content type *statuses*. Again, we observe that regression models and classifiers for hedonism and conservative values show the best performance with the content type *page-likes*. We also notice that we get the best regression model and classifier for self-enhancement value from the *shared-link* content type.

6 Ensemble of Models

In this section, we combine three different linear regression models (the models that we have described in Sect. 5) to increase the accuracy of our prediction model. Among our built models, one feature may predict a better value score than the other. For example, predicting self-enhancement value dimension (according to Table 2), shared-links show the strongest (the R^2-16.8 %) and page-likes show the weakest strength (the R^2-12.8 %). Since every feature contributes to the value score based on their strength (weaker or stronger), to find out final value score, we combine all the relations obtained from the previous steps.

It is necessary to prioritize the features based on their importance, as we compute value scores from multiple interaction features in Facebook. For example, some may think that *statuses* can reveal a value of a person more accurately, while other may emphasize on *page-likes* to determine the value score correctly. Ordering among interaction features associates different weights to compute values. Weight signifies the relative importance of a particular feature/content type. To build our ensemble model, we perform the following two steps: (1) computing weights of each content type using neural networks, and (2) combining the models with a weighted linear ensemble technique. Figure 2 presents the architecture of our weighted average ensemble value model. From 30 % of our total dataset, we learn the weights for each type of content and from the rest 70 % of our total dataset, we build individual value models. Finally, we build an ensemble of value models that are derived from different types of contents using their corresponding weights.

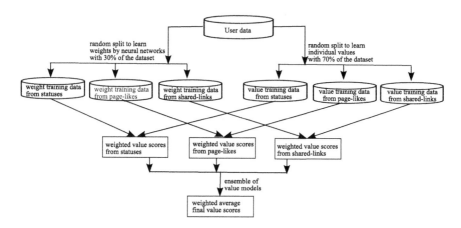

Fig. 2. Ensemble of value scores derived from models of different types of contents.

6.1   Learning Weights from Neural Networks

In this subsection, we determine the *weight* of each content type to determine a value dimension. To this end, we model a neural network using the data of 170 (30 % of the total dataset) Facebook users. The neural network builds regression model to predict value from each content type. We model our network with three types of contents and five types of values; we build in a total of 15 (3×5) neural networks using R *caret* package implementation [18]. For a single neural network, we use nine input neurons in the input layer, five neurons in the first hidden layer, three neurons in the second hidden layer and one output neuron in the output layer. For each value, we take LIWC scores as input and gives a *value* prediction score as output. We consider the *strength* (the R^2) of the regression model as the *weight* of the content type to determine a value dimension.

Table 4. Weights (the R^2) derived from neural networks.

Values	Status (UG)	Page-likes (US)	Shared-links (US)
	R^2	R^2	R^2
Self-Tran.	19.1 %	12.2 %	14.3 %
Openn.	20.7 %	15.0 %	13.8 %
Hedonism	14.7 %	15.1 %	9.07 %
Self-Enhan.	10.3 %	16.1 %	18.9 %
Conserv.	17.8 %	18.2 %	13.9 %

We first select the best subset of LIWC features for each content type and value using R *leaps* package implementation [19] by forward selection approach. Then, we normalize the LIWC scores in the interval [0,1] with max-min normalization technique to get better precision. We keep 90 % (among data of 170 Facebook users) of the data in the training set and the rest are in the test set using 10-fold cross validation with 10 iterations. For each content type and value, we compute the strength of different models. Table 4 presents the strength (the R^2) of our neural network based linear regression models that will be used as *weights* of our ensemble models in the next Subsect. 6.2.

6.2   Weighted Linear Ensemble

In this subsection, we build a weighted linear ensemble model from different types of contents of 397 (70 % of the total dataset) Facebook users [31]. We have already built different models from US and UG contents that are described in Sect. 5. Since, we train different neural networks that produce weights, we compute weighted linear ensemble score using the weights in Table 4. Finally, we build our weighted linear ensemble model using the weights generated from another dataset, thus our models do not get over-fitted.

Table 5 shows the R^2 strength of weighted linear ensemble models by using US content. We gain better result by ensembling US contents than that of a single US content. We achieve self-transcendence and self-enhancement value scores as the highest and the lowest R^2 strength, respectively. We notice that people with high self-transcendence values generally like pages for humanity such as *Save the Children* or movie pages like *The Revenant*. On the other hand, people with high self-enhancement values like pages and share links less frequently through social network. They sometimes like pages and share links regarding *technical*

Table 5. R^2 strength of ensemble model by integrating different types contents.

Values	Status (UG)	Page-likes (US)	Shared-links (US)	Ensemble of US Content	Ensemble of US and UG Contents
	R^2	R^2	R^2	R^2	R^2
Self-Tran	18.1 %	17.9 %	16.7 %	23.4 %	24.6 %
Openn	18.3 %	16.1 %	17.3 %	21.9 %	26.7 %
Hedonism	12.9 %	13.2 %	11.5 %	17.4 %	18.9 %
Self-Enhan	14.8 %	15.4 %	15.1 %	17.2 %	19.2 %
Conserv	20.2 %	20.9 %	10.2 %	22.8 %	25.2 %

tutorial, success stories, etc. We also observe the ensemble models by combining three different types (both US and UG) of contents. We find highest strength for all 5 different value dimensions by using ensemble models than that of single content type or ensemble of US contents.

Table 6 presents the performance of the respective classifiers that are built from both US and UG contents. We observe that our models obtain a substantial improvement in prediction potential over single feature based value identification models (according to the Table 2). Note that, we randomly split our dataset of 567 Facebook users into two parts, 30 % (170 Facebook users) of the dataset for learning weights and 70 % (397 Facebook users) of the dataset for building the ensembles. Splitting the dataset is somewhat similar to cross validation where we learn from one dataset and apply on another dataset. If we learn weights (i.e., contribution of different content type) from 70 % dataset and then again apply the ensemble on the same dataset, this would be like doing training/testing on the same dataset. Thus, we keep the training and testing dataset separate while building ensemble.

Table 6. Strength of ensemble model by integrating different interaction features.

Values	Best classifier	AUC	TPR	TNR
Self-trans.	RepTree	0.71	0.70	0.29
Openness.	Naive Bayes	0.78	0.77	0.25
Hedonism	Naive Bayes	0.61	0.58	0.41
Self-enhance.	RepTree	0.62	0.623	0.39
Conserv.	Naive Bayes	0.74	0.735	0.26

7   Silent vs. Active Users

In this section, we compare *active users* (who give regular status updates) with *silent users* (who give less updates through statuses) in Facebook. We show that value scores of silent users are more predictable by US contents. We observe in our dataset that silent users update statuses on average 2/3 times during every two months time interval. In contrast, they frequently like different Facebook pages such as business, brand, organization and celebrity. They like pages on average 3/5 times in a week. They also share links on average 2/3 times in every two weeks. Though these silent users generate less content, they like pages and share links regularly. Thus, page-likes and shared-links are vital interaction features to identify the values of silent users effectively in Facebook.

Table 7. R^2 strength of active and silent users for different values with different types of contents.

Values	statuses (UG)		page-likes (US)		shared-links (US)	
	active users	silent users	active users	silent users	active users	silent users
Self-Tran.	16.5 %	9.5 %	15.3 %	17.2 %	12.1 %	13.3 %
Openn.	19.3 %	13.1 %	16.1 %	18.5 %	14.5 %	15.1 %
Hedonism	11.1 %	7.1 %	12.2 %	16.9 %	13.8 %	13.7 %
Self-Enhan	12.5 %	10.5 %	14.5 %	15.1 %	17.1 %	17.7 %
Conserv.	17.3 %	13.7 %	17.1 %	19.7 %	11.1 %	12.5 %

From our dataset of 567 Facebook users, we have a total of 155 silent users who update statues irregularly. These users are regular in liking pages and sharing links. We randomly pick 155 active Facebook users from the rest of 412 (567–155) Facebook users. Table 7 presents the strength of the models that are built from the data of active and silent Facebook users. In the table, we present the best result among several random trails of active 155 Facebook users. We observe that active users achieve significant better strength of value models than that of silent users by using UG contents. In contrast, silent users gain better strengths in value prediction models than active users by using US contents.

8   Discussion

Our work is the first study to identify values from different types of contents (i.e., UG and US) in Facebook. We observe in Table 3 that US contents such as page-like and shared-link achieve better prediction potential than UG content (i.e., status) for hedonism, self-enhancement and conservation values. We find that best classifiers for hedonism and conservation value dimensions can be built from page-likes. Again, we notice that best classifier for self-enhancement value can be built from shared-links. On the other hand, we can build better prediction potential for self-transcendence and openness-to-change values from user statuses. In this paper, we have demonstrated that strength of the value prediction models could be improved (see Tables 2 and 5) by building ensemble of the interaction features. In particular, the strength of self-transcendence, openness-to-change and conservation values are increased by 32.22 %, 17.62 %

and 22.33 %, respectively than that of the highest scoring model built from single interaction features. We achieve slight improvement of prediction strength for hedonism and self-enhancement values through ensemble technique.

From Tables 2 and 3, we again observe that people with particular high value scores use particular type of interaction features (i.e., statuses, page-likes and shared-links). Some people convey their thoughts regularly by updating statuses while some may express their opinion by liking pages or sharing links in Facebook. Using a particular interaction feature depends on one's selection and satisfaction. Thus, we find better value models from those interaction features that are used frequently. For example, it is likely that the users who write long statuses, they tend to have high score in self-transcendence and openness-to-change value scores. They usually write about public awareness (i.e., well-being) and interesting insights from different observations. Again, people with high hedonism scores usually like Facebook pages such as fashion, gadget, restaurants, sports, and music that largely represent fun, enjoyment and pursuit of happiness in one's life. Similarly, people with high conservation value score like Facebook pages with heritage, religion and awareness (e.g., health tips). On the other hand, people with high self-enhancement value score have less propensity to share information in the social media through status updates frequently. Because social media foster procrastination and distract from other activities [16]. But they tend to share information (e.g., career counseling, IT tutorial) through different links in Facebook.

In a previous study, it is shown that a moderate personality prediction strength can be achieved from myPersonality [4] dataset (N = 250) with minimum and average word count of 1 and 585.004, respectively. In another well cited study [12], authors successfully predict personality from Facebook with a sample size of 279 Facebook users. Therefore, the size (N = 567) of our dataset is sufficient to predict value models from social media usage. In our experiment, we use *judgmental sampling* technique [20], because we first identify most productive Facebook friends who might response in our survey actively.

We ignore reading photo captions of users. Users' may have photo captions with two different types of photos: (1) self-tagged photos, and (2) tagged photos by others. Photo captions that are tagged by other users may not be supported (negatively associated) by the tagged user (the user who is being analyzed), since a Facebook friend can tag by default to any of his friends without permission. Thus, photo caption is such an interaction feature that may contain neither US nor UG content. We cannot collect users' comments. Users may write comments on the objects such as photos and statuses of other users who have not authorized their timeline through our Facebook application. However, we can compute more fine-grained values using other interaction features (e.g., user comments and replies if these were accessible completely).

Our approach has several limitations. We find limited prediction potential strength according to Table 2 for few values and content type (e.g., hedonism and conservation values by shared-link content). We overlook hashtags used in users' statuses, because people generally use random customized hashtags (e.g., #we_dont_care, #ramadan, etc.) in Facebook. Some of the hashtags also contain local linguistic terms. Thus, it is difficult to harmonize and normalize a diverse set of hashtags. Since we use LIWC to analyze our data, this approach usually correlates text corpus with a fixed set of words whereas a lexicon based (open vocabulary) [3] approach analyzes all the texts of user data.

9   Conclusion

In this paper, we are the first to identify five higher-level values from different types of contents among Facebook users. We have demonstrated which types of interaction contents can better predict which human values by using linear regression models and a wide varieties of classification methods. We have also built a unified prediction model by combining the values obtained from different interaction features. To produce a unified prediction model, we have integrated different linear regression models with a weighted linear ensemble technique and showed that prediction potential can be improved significantly using our ensemble technique. We have also shown that the value models of silent and active users work differently and can be derived from different content types. In future, we are interested in using *Empath* for deriving values from multiple interaction features that generates on demand new lexical categories [9]. We also plan to integrate our model with a real life application and scrutinize its usage in physical world.

Acknowledgement. This research is funded by ICT Division, Ministry of Posts, Telecommunications and Information Technology, Government of the People's Republic of Bangladesh.

References

1. Back, M.D., Stopfer, J.M., Vazire, S., Gaddis, S., Schmukle, S.C., Egloff, B., Gosling, S.D.: Facebook profiles reflect actual personality, not self-idealization. Psychol. Sci. **21**, 372 (2010)
2. Bollen, J., Mao, H., Pepe, A.: Modeling public mood and emotion: Twitter sentiment and socio-economic phenomena. In: ICWSM 11, pp. 450–453 (2011)
3. Boyd, R.L., Wilson, S.R., Pennebaker, J.W., Kosinski, M., Stillwell, D.J., Mihalcea, R.: Values in words: using language to evaluate and understand personal values. In: ICWSM (2015)
4. Celli, F., Pianesi, F., Stillwell, D., Kosinski, M.: Workshop on computational personality recognition (shared task). In: Proceedings of the Workshop on Computational Personality Recognition (2013)
5. Chen, J., Hsieh, G., Mahmud, J.U., Nichols, J.: Understanding individuals' personal values from social media word use. In: CSCW, pp. 405–414. ACM (2014)
6. Cohen, R., Ruths, D.: Classifying political orientation on twitter: it's not easy!. In: ICWSM (2013)
7. Cronbach, L.J.: Coefficient alpha and the internal structure of tests. Psychometrika **16**, 297–334 (1951)

8. Derksen, S., Keselman, H.: Backward, forward and stepwise automated subset selection algorithms: Frequency of obtaining authentic and noise variables. British J. Math. Stat. Psychol. **45**(2), 265–282 (1992)

9. Fast, E., Chen, B., Bernstein, M.: Empath: Understanding topic signals in large-scale text. arXiv preprint arXiv:1602.06979 (2016)

10. Fawcett, T.: An introduction to ROC analysis. Pattern Recogn. Lett., 861–874 (2006)

11. Golbeck, J., Hansen, D.: Computing political preference among twitter followers. In: Proceeding of CHI, pp. 1105–1108. ACM (2011)

12. Golbeck, J., Robles, C., Edmondson, M., Turner, K.: Predicting personality from twitter. In: SocialCom, pp. 149–156. IEEE (2011)

13. Gong, W., Lim, E.P., Zhu, F.: Characterizing silent users in social media communities. In: ICWSM (2015)

14. Hastie, T., Qian, J.: Glmnet vignette. Technical report, Stanford (2014)

15. Hsieh, G., Chen, J., Mahmud, J.U., Nichols, J.: You read what you value: understanding personal values and reading interests. In: CHI, pp. 983–986. ACM (2014)

16. Hughes, D.J., Rowe, M., Batey, M., Lee, A.: A tale of two sites: Twitter vs. facebook and the personality predictors of social media usage. Comput. Hum. Behav. **28**(2), 561–569 (2012)

17. Kosinski, M., Stillwell, D., Graepel, T.: Private traits and attributes are predictable from digital records of human behavior. In: Proceeding of the National Academy of Sciences, pp. 5802–5805 (2013)

18. Kuhn, M.: Caret package. J. Stat. Softw. **28**(5), 1–26 (2008)

19. Lumley, T., Miller, A.: Leaps: regression subset selection. r package version 2.9 (2009)

20. Marshall, M.N.: Sampling for qualitative research. Family Practice **13**(6), 522–526 (1996)

21. Maruf, H.A., Mahmud, J., Ali, M.E.: Can hashtags bear the testimony of personality? Predicting personality from hashtag use (2014)

22. Maruf, H.A., Meshkat, N., Ali, M.E., Mahmud, J.: Human behaviour in different social medias: a case study of twitter and disqus. In: Proceedings of the 2015 IEEE/ACM International Conference on Advances in Social Networks Analysis and Mining 2015, pp. 270–273. ACM (2015)

23. Oentaryo, R.J., Lim, E.P., Lo, D., Zhu, F., Prasetyo, P.K.: Collective churn prediction in social network. In: Proceeding of ASONAM 2012, pp. 210–214. IEEE Computer Society (2012)

24. Pak, A., Paroubek, P.: Twitter as a corpus for sentiment analysis and opinion mining. LREC **10**, 1320–1326 (2010)

25. Pennebaker, J.W., Booth, R.J., Francis, M.E.: Linguistic inquiry and word count: Liwc [computer software]. liwc.net, Austin (2007)

26. Preece, J., Nonnecke, B., Andrews, D.: The top five reasons for lurking: improving community experiences for everyone. Comput. Hum. Behav. **20**(2), 201–223 (2004)

27. Rahman, M.M., Majumder, M.T.H., Mukta, M.S.H., Ali, M.E., Mahmud, J.: Can we predict eat-out preference of a person from tweets? In: Proceedings of the 8th ACM Conference on Web Science, pp. 350–351. ACM (2016)

28. Schwartz, S.H.: A proposal for measuring value orientations across nations. In: Questionnaire Package of ESS, pp. 259–290 (2003)

29. Schwartz, S.H.: Basic human values: their content and structure across countries. In: Tamayo, A., Porto, J. (eds.) Valores e Trabalho [Values and Work], pp. 21–55. Editora Vozes, Brasilia (2005)

30. Schwartz, S.H., Melech, G., Lehmann, A., Burgess, S., Harris, M., Owens, V.: Extending the cross-cultural validity of the theory of basic human values with a different method of measurement. J. Cross Cult. Psychol. **32**(5), 519–542 (2001)
31. Sill, J., Takács, G., Mackey, L., Lin, D.: Feature-weighted linear stacking. arXiv preprint arXiv:0911.0460 (2009)
32. Smith, A.: 6 new facts about Facebook (2014). http://www.pewresearch.org/fact-tank/2014/02/03/6-new-facts-about-facebook/
33. Sumner, C., Byers, A., Boochever, R., Park, G.J.: Predicting dark triad personality traits from twitter usage and a linguistic analysis of tweets. In: ICMLA. IEEE (2012)
34. Wiltfong, J.: Global Sharers on Social Media Sites (2013). http://www.ipsos-na.com/news-polls/pressrelease.aspx?id=6239
35. Wong, F.M.F., Tan, C.W., Sen, S., Chiang, M.: Quantifying political leaning from tweets and retweets. In: ICWSM (2013)
36. Yang, C., Lin, K.H.Y., Chen, H.H.: Building emotion lexicon from weblog corpora. In: Proceedings of the 45th Annual Meeting of the ACL on Interactive Poster and Demonstration Sessions, pp. 133–136. Association for Computational Linguistics (2007)

An Application of Rule-Induction Based Method in Psychological Measurement for Application in HCI Research

Maria Rafalak[1(✉)], Piotr Bilski[2], and Adam Wierzbicki[1]

[1] Polish-Japanese Academy of Information Sciences, Warsaw, Poland
{m.rafalak,a.wierzbicki}@pjwstk.edu.pl
[2] Institute of Radioelectronics and Multimedia Technologies,
Warsaw University of Technology, Warsaw, Poland
p.bilski@ire.pw.edu.pl

Abstract. The paper presents a novel approach for creating computer adaptive version of traditional psychological tests that uses the rule induction algorithm. Currently used measures of the specific features (such as the intelligence) are based on questionnaires. Their computer versions should be short and non-repeatable. Because established methods for the computer adaptive tests show drawbacks, there is the need to propose new approaches to solve them. The proposed method uses the rule induction algorithm to generate rules for the training data, which can then be used to determine the importance of subsequent test items. They are then partitioned into groups, which allows for generating the curtailed version of the questionnaire, avoiding its repeatability. Verification results show that the proposed method significantly reduce the number of test items (by about one fifth) with relatively little loss of diagnostic accuracy.

Keywords: Computer adaptive testing (CAT) · Rules induction · Features reduction · Artificial intelligence (AI)

1 Introduction

Human-computer interaction (HCI) is a dynamically developing research area. As a multidisciplinary field it combines achievements from such disciplines as psychology, computer science and other social or technical domains. Considering the human factor in the computer system design is not a trivial task. It is often impossible to obtain explicit measures of users' psychological characteristics. Therefore HCI researchers often relay on measures applied traditionally in psychology, i.e. psychological tests and questionnaires. Academic sources abound in examples of studies with psychological questionnaires being a vital part of the experimental design. Personalization of online systems design [1–4], analyzing and predicting people's online behavior [5, 6],

---

This project has received funding from the European Union's Horizon 2020 research and innovation programme under the Marie Skłodowska-Curie grant agreement No 690962.

E. Spiro and Y.-Y. Ahn (Eds.): SocInfo 2016, Part II, LNCS 10047, pp. 471–484, 2016.
DOI: 10.1007/978-3-319-47874-6_32

improving recommender systems [7], e-learning [8–11] or creating ambient intelligence solutions [12] are examples of such studies.

Unfortunately, traditional psychological testing is time consuming. To ensure acceptable measurement reliability, the tests contain multiple questions (*test items*). During the personal contact with a psychologist it is easier to arouse and maintain adequate respondent's motivation level throughout the whole testing process. In the virtual environment, this is more difficult. When a psychological test is implemented as a part of the computer system it must be short to be usable. Therefore, traditional psychological tests applied in HCI applications should be compressed. There are many ways to do that, falling into the general category of computer adaptive testing.

The topic of the following paper is the rule induction-based approach to perform measurements of the psychological. The experimental case is shortening the psychological test and increasing the chance of completing it by as many respondents as possible. After simple modifications, our approach can also be used to analyze data from psychological questionnaires that unlike psychological tests have no right or wrong answers. Our idea is to analyze the structure of rules created during the machine learning procedure and identify the most important test items. This allows for evaluating the tested parameter (such as the intelligence quotient or the intensity of depression) in the fastest way.

The paper structure is as follows. Section 2 presents the state of the art by providing basic information on the methodology of the psychological tests construction, traditional approaches to the psychological tests curtailment and the concept of computer adaptive testing. The idea of using rule-induction based approach for shortening psychological tests is presented in Sect. 3, while Sect. 4 contains detailed information about its application in the area of intelligence testing. Description of the analyzed dataset is in Sect. 5. In Sect. 6 the obtained results are presented. The paper is concluded with Sect. 7 containing ideas for the future works.

2   State of the Art

This section presents the traditional approaches for the generation and analysis of computer adaptive testing. Two main approaches, i.e. CTT and IRT are briefly introduced and their drawbacks iterated, justifying the application of the proposed method.

2.1   Test Construction Methodological Framework

Classical Test Theory (CTT). Classical Test Theory is the basic conceptual framework for psychological test construction and test results analysis. The theory was introduced and described in [13, 14].

Score obtained by a testee (a person taking a psychological test) is defined as the sum of points assigned to answers given in the test according to the scoring key developed by the test designer. Usually every test item is scored in the same manner - regardless its difficulty or discriminating power (ability to distinguish between categories of testees).

The assumption in the CTT is that a test score X is a sum of the factual psychological trait level (T; true score), which cannot be measured explicitly, and error (E).

$$X = T + E \tag{1}$$

The error is assumed to be random and following the normal distribution. It is expected to be invariant with the true score (T).

The advantages of CTT are that it can be easily applied to the paper-pencil versions of psychological tests as calculating the score does not require any advanced computations.

Item Response Theory (IRT). Item Response Theory is an alternative theoretical framework to CTT. It is a probabilistic-based approach to the analysis of test results. Points assigned for the correct answer to a test item reflect the conditional probability that includes estimated testee psychological trait level (such as intelligence, personality etc.) and difficulty of the test item. Different statistical models can be used to represent the test item and testee characteristics. In two-parametric IRT model (2PL) the probability of giving the correct answer to an item q_i is expressed by the following logistic function:

$$P(q_i \mid \theta) = \frac{1}{1 + e^{a_i(\theta - b_i)}} \tag{2}$$

where θ is testees' estimated psychological trait level, a_i is the item discrimination and b_i is the item difficulty. Test score in IRT is expressed on the standardized scale with zero mean and standard deviation 1. The score above 1 is interpreted as high, below -1 as low while score in between that range is interpreted as average.

The main drawbacks of IRT application in practice is the requirement of large datasets necessary for parameter estimation. The measured psychological trait needs to follow the assumption of unidimensionality – meaning that measured trait is homogenous. Multiple measured characteristic can be seen from different aspects (called dimensions). The aim is to focus on only one.

2.2   Traditional Approaches to Compress the Test Content

The most intuitive idea for psychological tests curtailment referring to CTT (item reduction) was introduced in [15, 16]. It consists in taking every second or every third item from the test. The final score is a respective multiplication of the score obtained in the shortened version.

Other approaches for shortening psychological tests use statistical methods such as regression, Factor Analysis (FA) or considering item difficulty (defined as the percentage of people who gave the correct answer) [17, 18]). The FA is especially popular, usually used to construct tests and build various scales. It is possible to develop the scheme using FA to shorten the test regarding its original content, similarly to the idea proposed in this paper. The factor loadings, which are results of computations, can point at the significance of each question and find redundancies between them, even in various scales. This way the questions may be grouped into subsets with similar influence on the resulting parameter. This method can be used as the tool in our approach. Contrary to the rules-based approach, it is not easily interpreted by the human. Applying the rules generation we get the handy tool to explain, how the expert system operates and why. This justifies our approach.

2.3 Computer Adaptive Testing (CAT)

Computer adaptive tests (CAT) have customized length and item selection to each testee individually. Every CAT shares the same general structure. One of its basic elements is an *item bank* (or *item pool*), being a large set of test items. CAT parameters include algorithms for selecting starting item from the *item bank*, selecting the following items and test stopping criteria.

The most popular method for creating the CAT refers to the IRT framework [19–21]. Such algorithms for the item selection are usually based on the Fisher [22], Kullback-Leibler [23] or mutual information [24] criteria. The test stopping condition in most times uses the cut-off score and confidence interval technique [25].

The drawback of the IRT-based CAT is that the data needs to meet rigorous statistical assumptions. Large data samples are required to provide stable estimations of IRT model parameters. That's why new ideas for creating adaptive tests are required.

There have been attempts to apply artificial intelligence (AI) methods like Bayesian networks to adaptive testing [26–28]. The latter was criticized for low efficiency in test longer than 20 items [29]. The simulated annealing algorithm has also been applied to CAT design [30] as well as tree-based approach [31, 32].

3 Proposed Methodology

The rule-based approaches are well established in the computational intelligence. Since late seventies the most popular algorithms were proposed to solve multiple problems requiring the expert knowledge. Next to the artificial neural networks, they were the basis of the human-oriented expert systems, containing machine learning, decision making and knowledge explanation procedure. Unlike other approaches, the rule-based ones are popular because of the form of stored knowledge, easily readable and modifiable by the human being [33]. This is especially important in the questionnaire analysis, where there is the need to explain, which test items are significant, leading to the compressed test. In most cases the discrete data are processed by these methods, which is the case in the problem presented here. The single rule is a dyad:

$$\text{premises} \rightarrow \text{conclusions} \tag{3}$$

where the premises part contains the set of questions (from the original set Q) and answers required to fulfill all conditions to draw the corresponding conclusions. In the presented case the former are answers to test items selected from the item pool, while the latter are discrete category values, determining the intensity of the analyzed feature. The rule example may be as follows:

$$a_1(q_3) \wedge a_3(q_7) \wedge a_2(q_{11}) \rightarrow c_2 \tag{4}$$

Here $a_i(q_j)$ is the i–th answer a for the j-th question q, while c_k is the k-th decision category c. It is assumed that the set of answers for each question is discrete (usually 3 or 4 answer options are available). Therefore, the example above contains natural numbers, being the numbers of answers for subsequent questions. If the testee gave response No. 1 to the third question, No. 3 to the seventh question and No. 2 to the eleventh question, the rule will be activated (fired), producing the category c_2 as the output of the system. The decision-making module contains the number of rules as above, making the decisions after firing all rules fulfilling conditions defined in the *premises* part.

In the presented problem, the aim is the rules generation itself, not their operation to make the decision. This requires focusing on the machine learning, during which the rules are created from the learning data set L. The latter has the form (5), containing n examples (filled questionnaires), i.e. vectors of answers for subsequent questions with the one of k categories describing the testee:

$$L = \begin{bmatrix} a(q_1) & \cdots & a(q_m) & c_1 \\ a(q_1) & & a(q_m) & c_2 \\ \vdots & \ddots & \vdots & \vdots \\ a(q_1) & & a(q_m) & c_k \end{bmatrix} \tag{5}$$

Here it is assumed that each testee fills the complete questionnaire of m questions. All entries in L are discrete. It is assumed that in the set L there is the redundancy, which can be used by the learning algorithm to extract knowledge. Generation of rules leads to selecting the most important questions, which are then processed. Depending on the answers to these questions, the testees may be assigned to the particular categories. Because the same question may be present in multiple rules, the analysis of premises leads to ordering them according to their frequency of occurrence.

The proposed approach is illustrated in Fig. 1, where the subsequent stages are presented. The first step is to select a calibration sample (i.e. the training set L) form the analyzed dataset. The sample contains answers given by a group of testees to all test items. Every testee is assigned to one of the classification categories. The sample has the form of the set (5).

Next, the rules induction algorithm is used to generate the collection of rules, which will be processed to find out the importance of particular questions from L. Although multiple approaches can be used for this purpose, the AQ algorithm [34] was applied here. It is a relatively simple, but memory demanding and generates rules covering the whole data set.

Questions contained in the generated rules are counted and sorted in the descending order, according to the frequency of their occurrence. Subsequently, they can be grouped into subsets, from which the test items will be selected to the shortened questionnaire version. Each subset is generated independently for each category c, which means to distinguish the particular characteristics different sets of items are required. The redundant questions are eliminated, so that all subsets contain only one

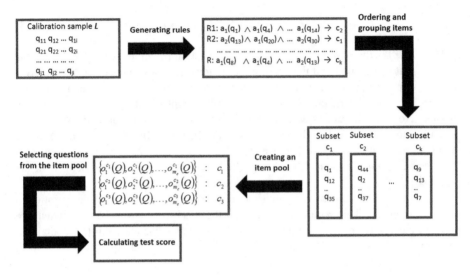

Fig. 1. General scheme of the proposed method.

copy of the particular test item. Finally, the subsets are used for every testee to generate the individual questionnaire.

The subsequent stages in Fig. 1 may be implemented using alternative approaches to the ones proposed by us (for instance, FA can also be used to group questions regarding their importance for describing the analyzed feature). This opens a possibility to improve the scheme by selecting the method most suitable for the particular task.

The next step in implementing the methodology is related with its intense verification. To do that, the measurement techniques must be implemented (checking for the validity or internal consistency of the shortened test, according to [35, 36]).

4   Application of Rule Induction in the Intelligence Test Compression

This section presents the application of the rules induction for generating the compressed test. First the AQ algorithm is briefly introduced. Next, the analysis of the rules' set structure to generate the subsets and use them to generate the tests is described.

The rules induction algorithm is based on the sequential covering scheme. The training examples (rows) from L are processed iteratively to generate rules fitting for the selected ones. In each iteration, one example (positive seed) is selected, for which the rules will be generated. The premises (multiple sets of questions allowing to distinguish this example from all other) for it are then created. The process starts with the most general premise, fitting this example, but also all other. The premise is then specialized by introducing subsequent answered questions eliminating as many examples belonging to other categories as possible. Finally, each premise should allow for distinguishing as many examples belonging to the same category as the positive seed as possible. Because for each example multiple combinations of questions are possible, only one or multiple premises (and subsequently rules) can be selected. In the presented case their number should be large, to allow for the statistical evaluation of the test items in the knowledge structure. The process of creating rules for the set L is concluded when for all examples such sets are generated. Example of three rules created for examples belonging to one of three categories based on the processing of two questions is in Fig. 2. Although it is possible to differentiate premises, preferring some rules over others, in the presented work they are all treated equally, as the greatest number of generated rules is beneficial.

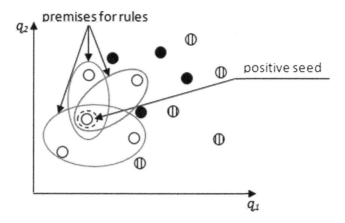

Fig. 2. Illustration fro the rules-induction method result operating on the examples belonging to three categories and distinguishable by two features (questions q_1 and q_2).

After generating the complete set of rules R, they are analyzed to determine the frequency of subsequent questions occurrence. This allows for sequencing them in the descending order, which represents their importance for determining the category of the testee. Because for each category in L different questions are important, the ordering should be performed for each category separately. Therefore R is first divided into the number of subsets, according to the number of categories, $C = \{c_1, c_2, ..., c_k\}$. Next the ordering takes place in each subset separately. Because the rules (3) contain various numbers of answered questions, the additional weighting is performed. The counter (significance factor) for the i-th question $|q_i|$ is incremented as follows:

$$|q_i| = |q_i| + \frac{1}{|q_r|} \tag{6}$$

where $|q_r|$ is the number of questions in the premise of the rule r. Finally, questions are ordered according to the values of the corresponding counters. The result is the sequence:

$$\left\{ o_1^{c_i}(Q), o_2^{c_i}(Q), \ldots, o_m^{c_i}(Q) \right\} \tag{7}$$

where $o_i(Q)$ is the *i-th* position of the question from the set Q in the ordered sequence. Because the counters can have different ranges for each category (depending on the number of rules and their contents in each subset), normalization is performed. It consists in dividing each counter by the maximum value obtained for the particular category:

$$|q_i| = \frac{|q_i|}{\max\left(|q_j| : j = 1, \ldots, m\right)} \tag{8}$$

The example of the ordered sequence for categories present in *L* is in Fig. 3.

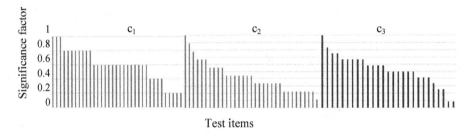

Fig. 3. Ordered sequence for items assigned to c_1, c_2 and c_3 categories

After the ordering, from each sequence m_c questions are selected to create the compressed pool of questions for the CAT, from which the questionnaire will be individually created. To have equal chances for determining each category, the number of questions in each subset should be the same, leading to the overall number of test items in the pool to $k \cdot m_c$. The size of the pool is the algorithm parameter. To avoid repeating questions in the subsets, each test item is unique for the whole pool.

The generation of the compressed test consists in randomly selecting m_s question sfrom each subset being part of the pool. To ensure the appropriate randomness in the test (avoiding repeating the particular questions too often), it is required to have $m_s \ll m_c$. For instance, if three categories are to be distinguished (c_1, c_2, c_3), the following subsets become the part of the pool:

$$\begin{aligned}
\left\{ o_1^{c_1}(Q), o_2^{c_1}(Q), \ldots, o_{m_c}^{c_1}(Q) \right\} &\quad : \quad c_1 \\
\left\{ o_1^{c_2}(Q), o_2^{c_2}(Q), \ldots, o_{m_c}^{c_2}(Q) \right\} &\quad : \quad c_2 \\
\left\{ o_1^{c_3}(Q), o_2^{c_3}(Q), \ldots, o_{m_c}^{c_3}(Q) \right\} &\quad : \quad c_3
\end{aligned} \tag{9}$$

The usage of the compressed set requires scaling the overall score obtained by the testee solving such a shorter version of the questionnaire. Results available for classical IRT and CTT approaches must be comparable with our approach, therefore the outcome from the compressed questionnaire must be multiplied to ensure that the maximum number of obtainable points is equal to the traditional methods. Therefore, score s_{ci} obtained for questions from each subset (related with the category c_i) from the pool must be multiplied by various weights w_i:

$$s = \sum_i w_i \cdot s_{c_i} \tag{10}$$

The weights w_i must be selected to ensure the proper maximal value of the overall score s. For example, the test exploiting 20 percent of the overall number of rules allowing for distinguishing between three categories (each represented by the same number of test items) would be scaled as:

$$s = 7 \cdot s_{c_1} + 5 \cdot s_{c_2} + 2 \cdot s_{c_3} \tag{11}$$

This means that half of the points from the test can be obtained after answering all questions from subset c_1, about 35 percent are given for answering questions from the subset c_2, and the rest is awarded for answering questions from the set c_3.

The particular values of weights depend on the questionnaire to be analyzed. For instance, in the intelligence test they would be related with the difficulty of the question. This way the more difficult items would be rewarded higher values in points.

The final problem is the division of the available data into the training and testing sets. The cross-validation allows for the repeated division of the data into both types of subsets and implementing the presented method for them, to verify their generalization abilities. In the presented research the simplest 50:50 partitioning was used, leading always to training and testing sets of equal sizes.

5 Dataset Description

The dataset used for the presented analysis comes from the Polish normalization study of the Culture Fair Intelligence Test (CFT-20R). The study was conducted in 2011 on the representative sample of the Polish population. The complete sample consisted of 3196 testees whose age ranged from 7 to 59 y/o. The age of the testees is wide so the test results are interpreted separately for different groups. To eliminate the effect of age differences on the results, only a subset of the normalized dataset was used. The group of 879 testees aged 15−19 y/o was selected for further computations. Female participants constituted 51 % of the subset, while male participants 49 %. Testees represented different Polish geographical regions.

CFT-20R is an intelligence test, based on the non-verbal material, used in professional psychological diagnostics. In every test item the testee needs to identify the rule combining presented graphical elements and decide which of the available answer options fits the rule and supplements the graphical sequence. The full test consists of 101 items. Every question has only one correct answer. The testee chooses the answer out of five available answer options. All participants from the normalization study took the whole test in the standardized fully controlled testing conditions [36]. Omitted test items were treated as incorrect and were assigned 0 points.

Every person taking part in the study was assigned the category based on their IQ level, which is a standard measure used in psychological diagnostics. Testees assigned to c_1 category showed IQ lower than 85. The ones assigned to the c_2 category demonstrated the IQ between 85 and 115. Testees assigned to the c_3 category presented IQ higher than 115.

6 Experimental Results

The experimental procedure was repeated 50 times for every analyzed test length option. The score calculated after answering all test items was treated as a benchmark for comparisons. Also, random item selection from the item pool was performed.

All experiments were calculated using R CRAN statistical software (version 3.3.1). Package *RoughSets* [37] was used for the rule induction procedure.

Obtained results are illustrated on Figs. 4 and 5. As expected, the modulus difference between test scores obtained using answers to all test items and test scores calculated using only subset of the item pool is systematically decreasing as the test becomes longer. This is true for both IRT and CTT approach when random item selection or the proposed rule-induction based method is implemented. In all analyzed cases the proposed method for psychological test curtailment proves to be better than the random item selection.

Scores obtained using IRT models are expressed on a standardized scale with zero mean and standard deviation 1. This is why observed differences between are relatively smaller than the differences observed for CTT score calculations. For the scores obtained using CTT methodological framework maximal score in a test is 101 points.

It is also worth noticing that the standard deviation of the observed results is significantly smaller for the scores obtained using proposed method. Shortening test by

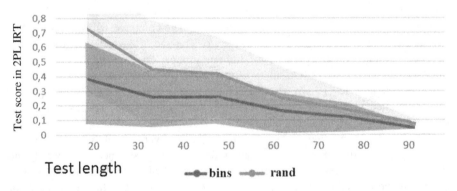

Fig. 4. Modulus difference between the test score calculated using proposed method (*bins*) or items selected at random (*rand*) and the test score calculated after answering all questions using IRT 2PL model. Continuous lines represent the mean from 50 iterations. Shaded area represents standard deviation.

one fifth (final test length equal 20 test items) differs from the test score calculated for the whole test (101 test items) by only 15 points on average (± 6 points). However, the choice of the optimal test length is not always obvious. In some applications even such difference can be considered as too big.

In the analyzed case it can be seen that the 30- items test version gives more precise total score estimation than the 20-item one with similar standard deviation values (see: Fig. 5). Further extension of the test to 40 (up to 90) items does not significantly change the average total score in 50 iterations. However, in longer versions of the test the precision of measurement (reflected by the standard deviation) is better.

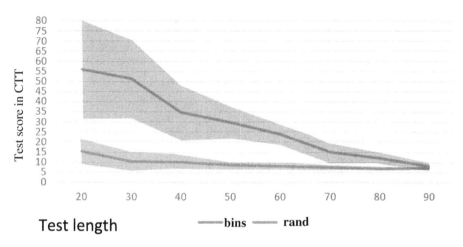

Fig. 5. Modulus difference between the test score calculated using proposed method (*bins*) or items selected at random (*rand*) and the test score calculated after answering all questions using CTT methodology. Continuous lines represent the mean from 50 iterations. Shaded area represents standard deviation.

The analysis of the demographic groups within the tested data confirms the well-known dependence between the intelligence quotient and the age of the testee. This confirms the intelligence measured by the test depends on the age of tested people. However, to confirm this trend, the additional work with larger data sets is required.

7   Conclusions and Future Work

Obtained results show that proposed method for psychological test-curtailment in preparation of computer adaptive test has advantages over the already existing approaches. The rule-induction based item selection gives significantly better results than the random item selection. This difference is particularly evident in the case of analyzing test results according to CTT. Such result is optimistic as statistical require-ments concerning data in CTT are far less strict than the ones for the IRT. On the other hand, IRT results also show the advantage of the proposed method over random item selection.

However, much work needs to be done. For example, more complex algorithm for the item selection and test stopping rule need to be defined. Also, the impact of the number of testee category on obtained results should be examined. The scheme is generic enough to be used for the analysis of tests belonging to different domains. However, it is important to verify, which parameters are universal and which must be tuned individually. Additionally, other algorithms designed for rule induction should be tested.

It is worth noticing that the presented results confirm the usefulness of the scheme in shortening the test. To correctly measure its usefulness, especially for more generic analysis (of other psychological traits), the typical quality measures must be considered (such as validity or internal consistency). Such concepts were intensively studied in the psychometrics field [38] and should be used to verify the correctness of our approach.

As the intelligence testing is a very specific research area, it would be interesting to apply proposed method to psychological questionnaires where every answer option can be interpreted separately.

References

1. Kaptein, M., Markopoulos, P., de Ruyter, B., Aarts, E.: Personalizing persuasive technologies: explicit and implicit personalization using persuasion profiles. Int. J. Hum Comput Stud. **77**, 38–51 (2015)
2. Ziemkiewicz, C., Ottley, A., Crouser, R.J., Chauncey, K., Su, S.L., Chang, R.: Understanding visualization by understanding individual users. IEEE Comput. Graph. Appl. **32**(6), 88–94 (2012)
3. Schneider, N., Schreiber, S., Wilkes, J., Grandt, M., Schlick, C.M.: Foundations of an age-differentiated adaptation of the human-computer interface. Behav. Inf. Technol. **27**(4), 319–324 (2008)
4. Mussi, S.: User profiling on the web based on deep knowledge and sequential questioning. Expert Syst. **23**(1), 21–38 (2006)
5. Purkait, S., De Kumar, S., Suar, D.: An empirical investigation of the factors that influence Internet user's ability to correctly identify a phishing website. Inf. Manage. Comput. Secur. **22**(3), 194–234 (2014)
6. Youyou, W., Kosinski, M., Stillwell, D.: Computer-based personality judgments are more accurate than those made by humans. Proc. Natl. Acad. Sci. **112**(4), 1036–1040 (2015)
7. Nunes, M.A.S.: Towards to psychological-based recommenders systems: a survey on recommender systems. Sci. Plena **6**(8) (2010)
8. Tzouveli, P., Mylonas, P., Kollias, S.: An intelligent e-learning system based on learner profiling and learning resources adaptation. Comput. Educ. **51**(1), 224–238 (2008)
9. Levy, J.C.: Adaptive Learning and the Human Condition. Routledge (2015)
10. Wong, L.H., Looi, C.K.: Swarm intelligence: new techniques for adaptive systems to provide learning support. Interact. Learn. Environ. **20**(1), 19–40 (2012)
11. Oxman, S., Wong, W.: White Paper: Adaptive Learning Systems. Integrated Education Solutions (2014)
12. Riva, G., Vatalaro, F., Davide, F., Alcañiz, M.: 2 The Psychology of Ambient Intelligence: Activity, Situation and Presence (2005)

13. Novick, M.R.: The axioms and principal results of classical test theory. J. Math. Psychol. **3**(1), 1–18 (1966)
14. Lord, F.M., Novick, M.R., Birnbaum, A.: Statistical Theories of Mental Test Scores (1968)
15. Adams, R.L., Smigielski, J., Jenkins, R.L.: Development of a Satz-Mogel short form of the WAIS—R. J. Consult. Clin. Psychol. **52**(5), 908 (1984)
16. Kaufman, A.S.: A short form of the Wechsler Preschool and Primary Scale of Intelligence. J. Consult. Clin. Psychol. **39**(3), 361 (1972)
17. Nunnally, J.C., Bernstein, I.H., Berge, J.M.T.: Psychometric theory, vol. 226. McGraw-Hill, New York (1967)
18. Coste, J., Guillemin, F., Pouchot, J., Fermanian, J.: Methodological approaches to shortening composite measurement scales. J. Clin. Epidemiol. **50**(3), 247–252 (1997)
19. van der Linden, W.J.: Constrained adaptive testing with shadow tests. In: van der Linden, W.J., Glas, C.A.W. (eds.) Computer-Adaptive Testing: Theory and Practice, pp. 27–52. Kluwer, Boston (2000)
20. Yen, Y.C., Ho, R.G., Chen, L.J., Chou, K.Y., Chen, Y.L.: Development and evaluation of a confidence-weighting computerized adaptive testing. Educ. Technol. Soc. **13**(3), 163–176 (2010)
21. Chang, S.H., Lin, P.C., Lin, Z.C.: Measures of partial knowledge and unexpected responses in multiple-choice tests. Educ. Technol. Soc. **10**(4), 95–109 (2007)
22. Van der Linden, W.J.: Linear Models for Optimal Test Design. Springer, New York (2005). doi:10.1007/0.387.29054.0
23. Eggen, T.J.H.M.: Item selection in adaptive testing with the sequential probability ratio test. Appl. Psychol. Meas. **23**, 249–261 (1999). doi:10.1177/01466219922031365
24. Weissman, A.: Mutual information item selection in adaptive classification testing. Educ. Psychol. Measur. **67**, 41–58 (2007). doi:10.1177/0013164406288164
25. Sie, H., Finkelman, M.D., Bartroff, J., Thompson, N.A.: Stochastic curtailment in adaptive mastery testing improving the efficiency of confidence interval-based stopping rules. Appl. Psychol. Meas. **39**(4), 278–292 (2015)
26. Vomlel, J.: Building adaptive tests using Bayesian networks. Kybernetika **40**(3), 333–348 (2004)
27. Tselios, N., Stoica, A., Maragoudakis, M., Avouris, N., Komis, V.: Enhancing user support in open problem solving environments through Bayesian network inference techniques. Educ. Technol. Soc. **9**(4), 150–165 (2006)
28. Morro, C.G.: Different approaches to computer adaptive testing applications. http://cs.joensuu.fi/pages/whamalai/sciwri/cristina.pdf
29. Liu, C.L.: Using mutual information for adaptive item comparison and student assessment. Educ. Technol. Soc. **8**(4), 100–119 (2005)
30. Lu, P., Cong, X., Zhou, D.: The research on web-based testing environment using simulated annealing algorithm. Sci. World J. (2014)
31. Šerbec, I. N., Žerovnik, A., Rugelj, J.: Adaptive Assessment Based on Machine Learning Technology. http://www.icl-conference.org/dl/proceedings/2008/finalpaper/Contribution 336_a.pdf
32. Yan, D., Lewis, C., von Davier, A.A.: A tree-based approach for multistage testing. In: Yan, D., von Davier, A.A., Lewis, C. (eds.) Computerized Multistage Testing: Theory and Applications, pp. 169–188. Chapman and Hall/CRC, Boca Raton (2014)
33. Bilski, P., Winiecki, W.: The rule-based method for the non-intrusive electrical appliances identification. In: 2015 IEEE 8th International Conference on Intelligent Data Acquisition and Advanced Computing Systems: Technology and Applications (IDAACS), vol. 1, pp. 220–225. IEEE, September 2015

34. Michalski, R.: The AQ15 inductive learning system: an overview and experiments. Reports of Machine Learning and Inference Laboratory (1986)
35. DeVellis, R.F.: Scale Development, 2 edn. Sage Publications, Thousand Oaks (2003)
36. Neuman, W.L.: Social Research Methods: Qualitative and Quantitative Approaches. Allyn & Bacon (2003)
37. Riza, L.S., Janusz, A., Bergmeir, C., Cornelis, C., Herrera, F., Śle, D., Benítez, J.M.: Implementing algorithms of rough set theory and fuzzy rough set theory in the R package "roughsets". Inf. Sci. **287**, 68–89 (2014)
38. Carmines, E.G., Zeller, R.A.: Reliability and Validity Assessment, vol. 17. Sage Publications (1979)

A Language-Centric Study of Twitter Connectivity

Priya Saha$^{(\boxtimes)}$ and Ronaldo Menezes

Department of Computer Sciences, Florida Institute of Technology,
Melbourne, FL, USA
psaha2010@my.fit.edu, rmenezes@fit.edu

Abstract. One factor influencing human online connectivity, which only recently has been receiving attention, is the language used by the user in his activities. This paper uses Twitter (a popular online social network) to shed light on the effect of language to the online connectivity of people. Using techniques from Network Science, our work shows that Twitter users have a stronger preference to connect to people who use a common language, but more importantly, that this preference is stronger than the trend of connecting to people with similar popularity. Furthermore, we also show that the connecting patterns between users of different languages vary considerably; we use the concept of entropy to measure the degree of variation in the connecting patterns for each language.

Keywords: Assortativity · Social networks · Languages · Entropy

1 Introduction

Written language has allowed societies to thrive because information can be passed with ease between generations; writing is probably one of the most important human achievements [25]. The native language (mother tongue) of an individual is of prime importance to group formation; in reality, language together with religion and skin color has been shown to be of great significance to the patterns of interactions within societies [9].

Social networks are a relatively new phenomena and thus far not much is known about the effect of language to user associations (friendship formation) in online social networks. Questions such as "Are there different patterns of connectivity for users of different language?" or "Is the distribution of connections between users of different languages organized in some way?" are yet to be answered. In order to move towards an understanding of language in the context of online social networks, we studied the structure of the connectivity of several users as a function of language they use on Twitter, a free online social network site that connects approximately 320 million monthly registered users[1]. Twitter defines the concept of *followers* as a passive connection in which a user

---

[1] https://about.twitter.com/company.

© Springer International Publishing AG 2016
E. Spiro and Y.-Y. Ahn (Eds.): SocInfo 2016, Part II, LNCS 10047, pp. 485–499, 2016.
DOI: 10.1007/978-3-319-47874-6_33

subscribes to receive posts (tweets) from another user. Twitter also offers a feature called *retweet* consisting of a tweet (a message) that a user can share from another user hence forwarding it to his own followers. Features of users such as gender, religion, and political views, play a role in the formation of connections; people tend to link to others who share similar views, this is conceptualized as assortativity or the so-called "birds of a feather fly together" phenomenon [16]. Assortativity can measure the extent of mixing in a network using the Pearson's correlation between the properties of adjacent nodes [19]. It is therefore not surprising if we find that people tend to connect to others who use the same language on Twitter; language is expected to be a catalyst for the formation of connection. Yet, to our knowledge, the extent of the role played by language in connection formation has not received much attention in the context of online social networks.

In addition to the idea of assortativity and in order to understand the characteristic of the connections between users of specific languages, we looked into the association patterns of users. We used the concept of entropy (Shannon entropy) to quantify the organization among the users. Shannon entropy is a well-known concept in Information Theory and is used to understand the disorder of the system [2]. Using concepts from Network Science [3,6] and Information Theory, this work delves into analyzing the social interactions of Twitter users as a function of their languages.

We organize our study as follows: after some discussion on our motivation and related work in Sect. 2, we describe our data collection approach as well as network generation in Sect. 3. Given that our data is a sample of what is available on Twitter, we demonstrate in Sect. 4, that our networks, even though small, exhibit the characteristics of the real-world networks. We follow with detailed analysis of our findings in Sect. 5. We have two main contributions in this paper: first is to understand the regularities in connectivity of Twitter users and second is to estimate the entropy of connection of users; both in the context of languages and are supported by statistical evidences. We finish this paper in Sect. 6, with a summary of our main contributions and hints about possible future works.

2   Motivation and Related Work

Some works have been published on the topic of our research. If we start from structural works, Kwak et al. argued that the follower-following network topology exhibits a non-power-law distribution, a short diameter as well as low reciprocity [15]. Myers et al. found that the Twitter follower network exhibits the characteristics of both information network as well as social network [18]. Another study by Bild et al. showed that the retweet network is more assortative in comparison to the follower network and display small-world properties [4]. Hence, the retweet network is a better representation of the real-world relationships. Kang et al. found that Twitter users demonstrate some assortativity with respect to the topics of interests [13]. Users on Twitter also tend to group together according to their emotional states [5]; users who are happy connect to others who also tend to be happy and the same is true for unhappy users.

While there are several works that shed light on the characteristics of the networks formed by the users, the analysis of Twitter users from the perspective of their languages have only recently been explored by network scientists. Hong et al. studied 62 million tweets to understand the distribution of languages on Twitter. Almost half of the tweets were in English and rest of the tweets were in other languages. The study found that there are differences among the ten most popular languages, in terms of adopting Twitter features like hashtags, URLs, mentions, replies and retweets [10]. Weerkamp observed that German tweets are more likely to contain hashtags than any other language; one in every four tweets in German have hashtags [29]. Japanese and Indonesian tweets are unlikely to have hashtags; one in every twenty-five Japanese tweets contains a hashtag. Poblete et al. studied the top ten countries in Twitter and the three most common languages in each of the top countries [23]. English was found to be the common language in all the countries. Mocanu et al. mapped the world languages through Twitter. His work was focused on understanding the linguistic homogeneity at scales ranging from country-level aggregation to city-level neighborhoods [17]. Mocanu also found that the Twitter penetration in countries is related to socio-economic factors and can give insight on the mobility patterns of humans. Nguyen et al. studied the relation between language and age of the Dutch users in Twitter. It was found that the use of capitalized words, word length, tweet length, links and hashtags on Twitter can indicate the age of the Dutch users [22]; the aforementioned user activities vary to a great extent among the young users and quite stable among the old users. Java et al. showed that language plays an important role in shaping the social networks of users. Users from Japan and Spanish speaking world generally connect to others who speak the same language [11]. Monolingual people cluster better than bilingual people on Twitter [14]. Despite the language-related works above, it seems clear that not much attention has been given to relations among the users and the regularities that emerge from patterns of the languages used by these users.

In this paper we focus on two concepts that need to be defined in more detailed: Assortativity and Entropy. These concepts form the basis of the findings we describe in Sect. 5.

2.1 Assortativity

Assortativity is a concept in Network Science that describes the tendency of individuals to associate to those who are similar to themselves [7,16,21]. In directed networks, the mixing can be characterized by an asymmetric matrix e_{ij}, which refers to the fraction of edges that connect vertex of type i (as in, language type) to vertex of type j (also language type). The sum rules that are being satisfied are,

$$\sum_j e_{ij} = a_i \qquad\qquad \sum_i e_{ij} = b_j \qquad\qquad \sum_{ij} e_{ij} = 1,$$

where, a_i and b_i are the fraction of each type (language being used to measure similarity) of end of an edge that is attached to vertices of type i. On undirected graphs, the ends of edges are of same type and hence $a_i = b_i$. The assortativity of a network can be quantified by the Pearson's correlation between nodes that are linked in the network and can be measured by the following equation:

$$r = \frac{\sum_i e_{ii} - \sum_i a_i b_i}{1 - \sum_i a_i b_i}, \tag{1}$$

The value of r ranges from $-1 < r < 1$, meaning dissasortative to assortative respectively [20].

2.2   Entropy

In Information Theory, entropy measures the level of disorganization in the system [2]. Although entropy can be quantified by many different ways, Shannon entropy is one of the most commonly used metric [27]. The Shannon entropy of a probability distribution p_m that can be measured by the following equation [12]:

$$s = -\sum_m p_m ln(p_m), \tag{2}$$

where the sum extends over all possible outcomes m. For a given pair of nodes (i, j), p_m can be expressed as the probability of m edges between i and j. Higher entropy in our case indicates a more disorganized distribution between a specific language and other, while a lower entropy indicates the distribution is more organized (more predictable).

3   Data Collection and Network Generation

Unlike other social networks, Twitter does not require approval for a user to follow another. Twitter enables the pulling of its data through an API. We used the Twitter API to collect our data. Given that extracting the entire Twitter data is a costly and time consuming procedure, we used one of the popular network sampling strategies discussed by Hanneman et al., the ego-centric (with alter connections) approach [8]. In the aforementioned network sampling, we need to identify a few focal nodes or egos (people in the social network) and collect their alters (or friends). In many cases, it is not possible to track down all the alters of the egos, so we collect some of them randomly. In Sect. 4, we describe the statistics of our sample network and show that it exhibits the real-world network characteristics.

3.1   Dataset 1: Follower Network

Indo-European languages were the first languages to spread throughout Europe and many parts of the world. The Indo-European language family has the largest number of speakers in the world (with over 2.6 billion people or 45 % of the world population) as well as the widest dispersion around the world[2]. We chose to work with English, German, Russian and Spanish as starting languages because they have a large number of speakers from the Indo-European family. We selected the top 20 famous people of each of the 4 languages (English, Spanish, Russian and German) by using the Carousel feature of Google[3], which is an extension to Google's knowledge graph [26] and displays results based on the reviews, pictures and numerous other factors. Given that, 93.6 % of Twitter users have less than 100 followers[4], we collected a random sample of up to 100 followers of every famous person (1-step-neighborhood) and up to 100 followers of every follower of a famous person (2-step-neighborhood). We chose to keep our number of followers for every user consistent in both the neighborhoods so that the results do not get biased based on the difference in the number of followers. Though the famous people chose their Twitter profile language as one of the 4 languages we started with, their followers may have different profile languages. When we collected the languages of the followers, we were able to have a representation of many languages currently available on Twitter (59 languages). For our work, 80 famous people (20 per language we studied) with up to 100 one-step followers and up to 10,000 two-step followers had their profile language collected. We generated an ego network for every famous person. We built 80 ego networks with their one-step-neighborhood and two-step-neighborhood followers. In our ego networks, every node represents a Twitter user and every edge represents a follower relationship. We merged the ego networks to form a Twitter *follower network*. Our follower network consists of 170,082 nodes and 237,588 edges. It is to be noted that we did not generate our follower network based on the reciprocal relationships, where egos and alters follow each other, unlike many previous works. If any such relation happened to be in our network, we also included it for our study. Since, in general, very few users have reciprocal relationships, we decided to keep our network as a small representation of the Twitter network. Figure 1 depicts a network of a few famous Russian people and their followers. We colored the nodes based on their languages. The size of a node represents the number of followers of the user.

3.2   Dataset 2: Retweet Network

We collected 10,663,736 tweets containing one or more of the last names (in English) of leaders of the G-20[5], for a period of 30 days. The users who posted

---

[2] http://aboutworldlanguages.com/indo-european-language-family.

[3] https://searchenginewatch.com/sew/how-to/2299454/4-google-carousel-optimization-tips.

[4] https://sysomos.com/inside-twitter.

[5] http://g20.org.

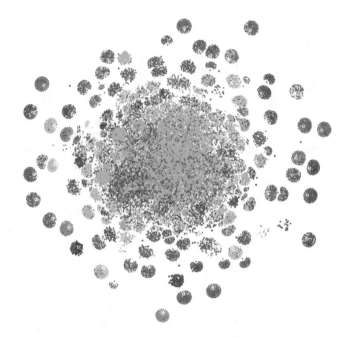

Fig. 1. A small sample Twitter follower network of a few famous Russian (green nodes) people. Colors of nodes represent languages and size of nodes represent degree. (Color figure online)

the tweets were expected to be from different nations and users of different languages. We collected the user of every retweet and the user of the original tweet. We used the information to build a network called *retweet network*. We also collected the language of the users on Twitter. Our retweet network consists of 1,922,815 nodes and 4,131,866 edges. In this retweet network, every node represents a user and every edge between the two users represents that they have shared an information.

4   Network Validation

Considering our networks are smaller compared to the real Twitter network, we made a few validations to be sure that the networks exhibited some characteristics of a real network. In Table 1, we show that our networks are similar to many other real-world networks (technological). We generated the in-degree and the out-degree distributions of the Twitter networks, and assessed the goodness-of-fit of the distributions using log-likelihood ratio [1]. We compared the degree distributions with log-normal, power-law, truncated power law and exponential distributions. The comparative study helps us to identify the best possible distribution. Tables 2 and 3 show the analysis of the degree distributions of our follower network and retweet network respectively. The Candidate Distributions column shows the distributions that were tested to find out the best fit of our data. The corresponding value of the fit is represented by a log-likelihood ratio column with a significance value (p-value). The log-likelihood value is positive if the data is more likely in the first distribution and negative if the data is more likely in the second distribution. We found that the power-law is not the best fit of the degree distributions as discussed by Kwak et al. [15]. In addition, we show that the log-normal is a better fit of the degree distributions with a significance level of $p < 0.05$. Significant research has been done on online social networks that exhibit log-normal distributions [24,28].

Table 1. The characteristics of the networks are similar to the real-world networks.

Characteristic	Follower network	Retweet network
Nodes	170,082	1,922,815
Edges	237,588	4,131,866
Type	Directed	Directed
Transitivity	0.018	0.003
Mean degree	2.864	4.299
Average path length	3.356	9.953

Table 2. The log-likelihood comparative study shows that log-normal best describes the in-degree and out-degree distributions of the follower network.

In-degree fitting		Out-degree fitting	
Candidate distributions	(likelihood, p)	Candidate distributions	(likelihood, p)
(Power-law, Log-normal)	(-358.22, $p < 0.05$)	(Power law, Log-normal)	($-2,482.09$, $p < 0.05$)
(Power-law, Truncated Power law)	(-152.59, $p < 0.05$)	(Power law, Truncated Power law)	($-2,220.68$, $p < 0.05$)
(Power-law, Exponential)	($17,046.39$, $p < 0.05$)	(Power law, Exponential)	($-2,324.18$, $p < 0.05$)
(Log-normal, Exponential)	($17,404.61$, $p < 0.05$)	(Log-normal, Exponential)	(157.90, $p < 0.05$)
(Log-normal, Truncated Power law)	(205.63, $p < 0.05$)	(Log-normal, Truncated Power law)	(261.41, $p < 0.05$)

5   Experimental Results

5.1   Mixing Patterns in Twitter: Degree and Language

In this work, we first study the extent to which the users connect to each other based on their degree (number of friends) and language. In high degree assortative

Table 3. The log-likelihood comparative study shows that log-normal best describes the in-degree and out-degree distributions of the retweet network.

In-degree fitting		Out-degree fitting	
Candidate distributions	(likelihood, p)	Candidate distributions	(likelihood, p)
(Power law, Log-normal)	$(-739.49,$ $p < 0.05)$	(Power law, Log-normal)	$(-1,866.67,$ $p < 0.05)$
(Power law, Truncated Power law)	$(-693.89, p < 0.05)$	(Power law, Truncated Power law)	$(-11.29,$ $p < 0.05)$
(Power law, Exponential)	$(388,404.99,$ $p < 0.05)$	(Power law, Exponential)	$(767,937.82,$ $p < 0.05)$
(Log-normal, Exponential)	$(389,144.48,$ $p < 0.05)$	(Log-normal, Exponential)	$(769,804.49,$ $p < 0.05)$
(Log-normal, Truncated Power law)	$(45.59,$ $p > 0.05)$	(Log-normal, Truncated Power law)	$(1,855.38,$ $p < 0.05)$

mixing networks, the users who have high degree connect to others with high degree and the users who have low degree connect to others with low degree. Although the degree assortativity of social networks is a widely studied subject, we measured it here because the expectation is that popular users are followed mostly by unpopular users leading to low assortativity or even disassortativity. The degree assortativity coefficient of our follower network is -0.18 while for the retweet network is -0.06, which indicates that both the networks are disassortative by degree. The negative assortativity confirms our expectation that the users in Twitter do not connect to other users who have similar degrees. They usually tend to connect to other users who are already popular on Twitter (have high degree). For example, users usually follow their favorite celebrities to remain up-to-date about their activities. On the other hand, celebrities do not connect to many other celebrities probably due to some competition for followers between them.

It is human nature to connect to other people whom they think are similar and there has been evidence that such propinquity to connect to alike may be beneficial or detrimental depending on the particular situation [16]. Although in today's world, multilingualism has become very popular, language is still a very important social drive to connect because a common language leads to more efficient communication. Multilingual people collect information in one language and may diffuse among the users of other languages, acting as bridges. We collected the total number of users in every language in our datasets and sorted in descending order by the number of users. Figure 2 shows the twenty most popular languages in the follower and the retweet networks.

We found that the ten most popular languages in the follower network are also the ten most popular languages in the retweet network (with the exception of Arabic and Indonesian). However, the order of popularity of the languages vary in both the networks. Arabic ranks eighth in the follower network and eleventh in the retweet network. On the other hand, Indonesian ranks twelfth in the follower network and ninth in the retweet network. In both the networks, English has the highest number of users, followed by Spanish. Other languages such as German, Russian, French, Portuguese, Italian, Arabic, Turkish, Japanese and Indonesian are also prominently used on Twitter and occupy intermediary positions in the social network. It is important to note that our rankings of languages do not reflect the ranking of the real-world or Twitter as reported by Statista[6] (perhaps because the time of our data collection is not same as the time Statista made its analysis) and hence reflect a different time period on Twitter; further analysis could be done to understand the dynamics of these language distributions (how the distribution is changing), however, this falls outside the scope of this work. Ethnologue reports that Chinese is the most spoken language in the world[7]. Twitter penetration in the countries (usage among the users) is one of the major reasons for the disparities. For example, Twitter is not very popular in Russia and blocked in China. However, the disparities do not hinder in extracting insights about the users in countries where Twitter is widely used. It is also important to keep in mind that we are trying to derive regularities in the user relations on Twitter social network, with a focus on language (Table 4).

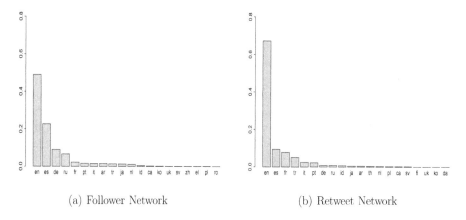

(a) Follower Network   (b) Retweet Network

Fig. 2. The top 20 languages according to the percentage of Twitter users in the two networks studied; that is, the popularity of the language is represented here.

To estimate the extent to which a user tends to connect to a friend who uses his language, we used the pairwise connection of users in relation to the languages they use on Twitter. We measured the assortativity coefficient of the follower network and the retweet network according to the language they declared in the user profile; the assumption is that the language they declare is their preferred language. The language assortativity coefficient of the follower

---

[6] http://www.statista.com/statistics/267129/most-used-languages-on-twitter/.
[7] https://www.ethnologue.com/statistics/size.

Table 4. The table below shows that the networks are disassortative by degree and assortative by language.

Assortativity	Follower network	Retweet network
By degree	−0.18	−0.06
By language	0.56	0.74

network is 0.56. The high positive coefficient in the follower network indicates that Twitter users display a strong preference for people using the same language as them. The assortativity coefficient of the retweet network is 0.74. The high assortativity coefficient of the retweet network also shows that Twitter users display strong association by languages while tweeting. Note however that the difference between one and the other is quite significant and indicates that despite the user's preference to people with the same language, the preference is stronger when we look at the information that is being transmitted (retweet). Users are a lot more likely to pass on a tweet in the language they prefer than just follow a person in that same language. In other words, a user may be willing to receive information in another language but not as likely to pass this information on (retweet the message).

Entropy of Language Associations. Although the assortativity coefficient indicate that similar language users have a higher chance to connect and that when it comes to transfer of information the chance is even more accentuated, the number of languages connected to users of a particular language may also influence the results. Hence, we studied the user connections for every language separately. The questions we address here are: *(i)* if most of the English users tend to connect to themselves, do users of other languages also have the same tendency, given that there are more English users in the network? *(ii)* to what extent can one be sure about the connecting patterns of users of the languages? To answer the aforementioned questions we investigated the characteristics of the networks with respect to the languages, a particular language connects to.

We created a vector of language exposure of every user in the network. For example, if a network is formed by an English user A and his 3 friends who use Spanish, 2 friends who use German, 3 friends who use English (let's say B, C and D) and 1 friend who uses Swahili, then A's language exposure can be represented by a vector,

$$V_{A_{English}} = [3, 2, 3, 1],$$

where A is exposed to 4 languages. We generated the language exposure vector for every user in the network. Then we combined all the users of a particular language to have the language exposure vector of the particular language in the network. Considering the example above, let's say, we generated $V_{A_{English}}$, $V_{B_{English}}$, $V_{C_{English}}$ and $V_{D_{English}}$ to have a language exposure vector $V_{English}$ (for English) of the network. We normalized the vectors to have a unit vector of every language in the network. Our follower network has users who use

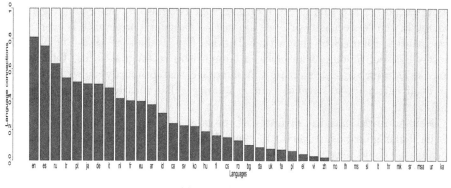

(a) Follower network

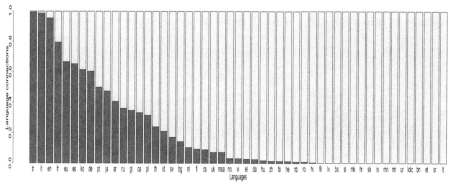

(b) Retweet network

Fig. 3. The darker part of the column represents the total number of edges that connect users who speak the same language (indicated in the x-axis). The lighter part of the column represents the total number of edges in the network that connect the users who speak other languages (different from the one in the x-axis).

59 different languages and our retweet network has users who use 51 different languages. One may ask here, while combining the users to generate the language exposure vector, the higher degree users may influence the lower degree users. Note that it is unlikely to be the case in our study, because we limited our number of followers to 100 people maximum in the follower network. At first, we measured the extent of diversity of the languages in both of our networks. We measured the extent of diversity of a language by calculating the ratio between the connections of the language to itself and the sum of its connections to other languages, as extracted from the language exposure vector. Figure 3(a) shows the diversity of languages in the follower network. We observe that some languages are more uniform (have less edges to other languages), or in other words, less diverse. Languages like English (en), Spanish (es) or Russian (ru) usually exhibit strong preference to connect within themselves; in contrast to

Georgian (ka), Urdu (ur) or Serbian (sr), which connect to other languages primarily. In the retweet network, we found that language preference is even more accentuated. Figure 3(b) shows that Turkish (tr), Italian (it) and English (en) mostly connect among themselves in comparison to languages like Lithuanian (lt), Serbian (sr) or Estonian (et). A language that primarily shows preference for itself is much less diverse than a language that mostly connects to other languages.

Next to investigate the diversity of the languages, we computed the entropy of the language exposure vector of every language. Note that this is important because we could have a language that connects to mostly others. For instance, we could have a language which evenly connects to several other languages or we could have a language which heavily connects to only one language. Although in both the cases, the languages are considered diverse, the diversity of the former language in the example is higher than the latter, or the association of the former language to other languages is much more disordered than the latter.

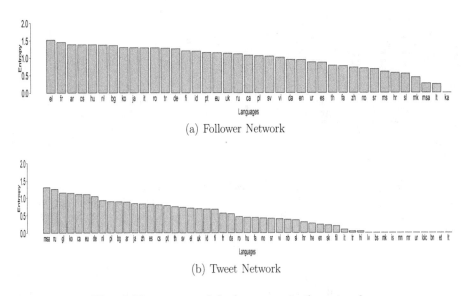

(a) Follower Network

(b) Tweet Network

Fig. 4. The entropy of the languages in the networks.

In Fig. 4(a), we demonstrate the entropy of languages in the follower network. Languages such as Greek (el) or French (fr) show high entropy, indicating that their association with other languages are very disordered. According to the French language exposure vector, French associates with itself as well as English almost evenly; Greek exhibits connections with several languages such as English, Spanish, Portuguese, German and others. On the other hand, a few languages like Lithuanian (lt) or Malay (msa) have low entropy; meaning their association patterns are less disordered. Neither Lithuanian nor Malay exhibit any preference towards themselves, but Lithuanian shows a strong association with English and

Malay shows a strong association with French. However, the retweeting and the follow patterns of the languages vary. In the retweet network, as demonstrated in Fig. 4(b), we observe that Malay (msa), Russian (ru) or Galician (gl) have very high entropy, which means their connecting patterns are disordered. Malay connects to several other languages such as Chinese, Japanese, English, Spanish and French in a random fashion. On the other hand, languages with low entropy like Hindi (hi), Turkish (tr), Italian (it) are less disordered in the network. Low entropy of a language may have two explanations. First, languages like Turkish and Italian show very high preference to retweet among themselves; hence, such languages tend to not connect to other languages arbitrarily. Second, languages like Hindi, associate strongly with English although it does not show preference to connect with itself.

There could be many different explanations for the different connecting patterns of the languages. We believe that *(i)* the language declared in the profile may not necessarily be used as the *de facto* language in the tweet (further experiment need to be performed to understand if this is the case), *(ii)* celebrities and G-20 leaders in our case are popular in different areas; thus people with different interests put extra effort to understand the information they are "broadcasting" to the world, or *(iii)* the preference for who we follow is not transferred to the preference of who we use the information (retweets).

6   Conclusion and Future Work

The aim of this work was to understand the social connectivity of users on Twitter and the effect of language to the connectivity. Twitter is a good platform to understand such user behavior because of its wide popularity, the number of people registered on the platform, and the number of languages available for these users. The task involved many theoretical and computational challenges. Since we worked with the real-world data collected from Twitter, instead of synthetic data, we believe our findings have interesting implications on the social dynamics of the human beings.

In this paper, we show that the social connectivity of users in Twitter best fits a log-normal distribution. The users usually like to connect to people who are already popular on Twitter (based on their degree) and show strong connectivity according to their languages both during following and retweeting. The association pattern of the languages on Twitter vary. Some languages show stronger association towards themselves, which may be because of the global popularity of the languages; users of such languages do not need to understand another language for information. Other language users show less association among themselves; such language users need to understand another language in order to receive information in the network. Due to the variation in the connection patterns, some languages are less disordered and some are more disordered on Twitter.

Note that in our work, we found many interesting facts about the languages of users on Twitter that we cannot confirm from a linguistic point of view as it falls outside our expertise. We hope to collaborate with linguists in the future to search for explanation for several of our findings. Also, we intend to generate our networks using different sampling methods to understand if our findings are biased due to our data collection strategy (last names of some leaders are very common terms in English, for example, Park). This also may shed light on the sparsity of our networks. Connecting patterns of users can be used to relate to the dynamics and evolution of social graphs. We also aim to compare the Twitter language dependencies with the real-world to understand if such association patterns are universal or are influenced by geographic proximity.

References

1. Alstott, J., Bullmore, E., Plenz, D.: powerlaw: a python package for analysis of heavy-tailed distributions. PloS One **9**(1), e85777 (2014)
2. Anand, K., Bianconi, G.: Entropy measures for networks: toward an information theory of complex topologies. Phys. Rev. E **80**(4), 045102 (2009)
3. Barabási, A.-L.: Network science. Philos. Trans. R. Soc. Lond. A Math. Phys. Eng. Sci. **371**(1987), 20120375 (2013)
4. Bild, D.R., Liu, Y., Dick, R.P., Mao, Z.M., Wallach, D.S.: Aggregate characterization of user behavior in twitter and analysis of the retweet graph. ACM Trans. Internet Technol. (TOIT) **15**(1), 4 (2015)
5. Bollen, J., Gonçalves, B., Ruan, G., Mao, H.: Happiness is assortative in online social networks. Artif. Life **17**(3), 237–251 (2011)
6. Börner, K., Sanyal, S., Vespignani, A.: Network science. Ann. Rev. Inf. Sci. Technol. **41**(1), 537–607 (2007)
7. Centola, D., Gonzalez-Avella, J.C., Eguiluz, V.M., San, M.: Miguel: homophily, cultural drift, and the co-evolution of cultural groups. J. Conflict Resolut. **51**(6), 905–929 (2007)
8. Hanneman, R.A., Riddle, M.: Introduction to social network methods (2005)
9. Hechter, M.: Group formation and the cultural division of labor. Am. J. Soc. **84**, 293–318 (1978)
10. Hong, L., Convertino, G., Chi, E.H.: Language matters in twitter: a large scale study. In: ICWSM (2011)
11. Java, A., Song, X., Finin, T., Tseng, B.: Why we twitter: understanding microblogging usage and communities. In: Proceedings of the 9th WebKDD and 1st SNA-KDD 2007 Workshop on Web Mining and Social Network Analysis, pp. 56–65. ACM (2007)
12. Johnson, S., Torres, J.J., Marro, J., Munoz, M.A.: Entropic origin of disassortativity in complex networks. Phys. Rev. Lett. **104**(10), 108702 (2010)
13. Kang, J.H., Lerman, K.: Using lists to measure homophily on twitter. In: AAAI Workshop on Intelligent Techniques for Web Personalization and Recommendation, Citeseer (2012)
14. Kim, S., Weber, I., Wei, L., Oh, A.: Sociolinguistic analysis of twitter in multilingual societies. In: Proceedings of the 25th ACM Conference on Hypertext and Social Media, pp. 243–248. ACM (2014)
15. Kwak, H., Lee, C., Park, H., Moon, S.: What is twitter, a social network or a news media? In: Proceedings of the 19th International Conference on World Wide Web, pp. 591–600. ACM (2010)

16. McPherson, M., Smith-Lovin, L., Cook, J.M.: Birds of a feather: homophily in social networks. Ann. Rev. Sociol. **27**, 415–444 (2001)
17. Mocanu, D., Baronchelli, A., Perra, N., Gonçalves, B., Zhang, Q., Vespignani, A.: The twitter of babel: mapping world languages through microblogging platforms. PloS One **8**(4), e61981 (2013)
18. Myers, S.A., Sharma, A., Gupta, P., Lin, J.: Information network or social network?: the structure of the twitter follow graph. In: Proceedings of the 23rd International Conference on World Wide Web, pp. 493–498. ACM (2014)
19. Newman, M.E.: Assortative mixing in networks. Phys. Rev. Lett. **89**(20), 208701 (2002)
20. Newman, M.E.: Mixing patterns in networks. Phys. Rev. E **67**(2), 026126 (2003)
21. Newman, M.E.: The structure and function of complex networks. SIAM Rev. **45**(2), 167–256 (2003)
22. Nguyen, D.-P., Gravel, R., Trieschnigg, R., Meder, T.: "how old do you think i am?" a study of language and age in twitter (2013)
23. Poblete, B., Garcia, R., Mendoza, M., Jaimes, A.: Do all birds tweet the same?: characterizing twitter around the world. In: Proceedings of the 20th ACM International Conference on Information and Knowledge Management, pp. 1025–1030. ACM (2011)
24. Radicchi, F., Fortunato, S., Castellano, C.: Universality of citation distributions: toward an objective measure of scientific impact. Proc. Natl. Acad. Sci. **105**(45), 17268–17272 (2008)
25. Robinson, A.: The Story of Writing: Alphabets, Hieroglyphs & Pictograms. Thames & Hudson (2007)
26. Roth, M., Ben-David, A., Deutscher, D., Flysher, G., Horn, I., Leichtberg, A., Leiser, N., Matias, Y., Merom, R.: Suggesting friends using the implicit social graph. In: Proceedings of the 16th ACM SIGKDD International Conference on Knowledge Discovery and Data Mining, pp. 233–242. ACM (2010)
27. Shannon, C.E.: A mathematical theory of communication. ACM SIGMOBILE Mobile Comput. Commun. Rev. **5**(1), 3–55 (2001)
28. Van Mieghem, P., Blenn, N., Doerr, C.: Lognormal distribution in the digg online social network. Eur. Phys. J. B **83**(2), 251–261 (2011)
29. Weerkamp, W., Carter, S., Tsagkias, M.: How people use twitter in different languages (2011)

Investigating Regional Prejudice in China Through the Lens of Weibo

Xi Wang[1]($\boxtimes$), Zhiya Zuo[1], Yang Zhang[2]($\boxtimes$), Kang Zhao[3], Yung-Chun Chang[4], and Chin-Shun Chou[4]

[1] Interdisciplinary Graduate Program in Informatics,
The University of Iowa, Iowa City, IA, USA
xi-wang-1@uiowa.edu
[2] Department of Political Science, The University of Iowa, Iowa City, IA, USA
yang-zhang@uiowa.edu
[3] Department of Management Science, The University of Iowa, Iowa City, IA, USA
[4] Institute of Information Science, Academia Sinica, Taipei, Taiwan

Abstract. Regional prejudice is prevalent in Chinese cities where native residents and migrants from other parts of China lack mutual trust. Weibo users actively discuss and argue about the issue of migration, which provides a good source of data to examine the communication network regarding regional prejudice. We are interested in the posts and reposts related to the topic on migrants. In a Weibo repost, one can add new content in addition to the original post. Then both original and new content as a whole can be read by others. In particular, we focus on the reposts in response to native residents' complaints about migrants. Based on the sentiment (negative or non-negative) and the direction (native resident→migrant or migrant→native resident), we classify the reposts into four categories. We find evidence of homophily in regional prejudice in the Weibo communication network: 72.7 % of the time, native residents' complaints trigger more complaints from other native residents. What interests us most are the socioeconomic factors that can reverse the sentiment or direction of the original posts. A multinomial regression model of the reposting patterns reveals that in a city with better housing security and a larger migrant population, migrant Weibo users are much more likely to argue with native residents who hold a negative view about migrants. One important implication from our findings is that a secure socioeconomic environment facilitates the communication between migrants and native residents and helps break the self-reinforcing loop of regional prejudice.

Keywords: Weibo · Regional prejudice · Social network · Sentiment analysis

1   Introduction

On New Year's Eve of 2015, 49 people were injured and 36 died in a stampede in Shanghai when more than a million visitors rushed to the observation

© Springer International Publishing AG 2016
E. Spiro and Y.-Y. Ahn (Eds.): SocInfo 2016, Part II, LNCS 10047, pp. 500–513, 2016.
DOI: 10.1007/978-3-319-47874-6_34

deck nearby Chen Yi Square on the Bund. This accident caught great attention from the public and triggered heated discussions on Weibo (a Chinese Twitter-equivalent) – the topic was mentioned 559,120 times within 10 days[1]. Although the criticisms were mostly cast toward local officials for their inadequate preventive actions, many Shanghai natives blamed visitors from other regions. They also complained that migrants brought a number of social problems to Shanghai. Shanghai, one of the most developed cities in China, is infamous for the problem of regional prejudice. Regional prejudice is rooted in distrust and conflicts between native residents and migrants, which is also a problem prevalent in Beijing and many other Chinese cities.

Using data from Sina Weibo, this study investigates regional prejudice among Chinese people. As the most popular micro-blog in China, Weibo's monthly active users reached 222 million in September, 2015, with 100 million active users on a daily average[2]. Like Twitter, Weibo allows users to publish short and instant posts to share personal stories and exchange opinions on various topics. With such valuable data of public opinion, we focus on Weibo posts that are relevant to the topic of migrants in Chinese cities. Our previous work [2] developed a machine learning algorithm to identify posts that express regional prejudice. Based on these posts, we build a repost network among users who publish posts on regional prejudice.

In this study, we investigate from Weibo data the spread of regional prejudice and explore the factors associated with opinion change. Specifically, we intend to answer the following three research questions:

- Regional prejudice can be detected by looking at three dimensions of Weibo posts: Who publishes a post, a native resident or a migrant? Does this post talk about migrants, native residents, or both? What is the sentiment of the post, negative or not? Our first research question is to find the frequent patterns of these three dimensions.
- One can include new content in a repost in addition to the reposted content. When a Weibo post is reposted, will the direction and sentiment change? In other words, will the reposter continue or reverse the original author's opinion?
- What socioeconomic factors are likely to reverse the sentiment or the direction of the original posts with regard to migrants?

The paper is organized as follows. The next section reviews related works on Weibo. Then, we introduce a repost network regarding regional prejudice. We proceed to explore the socioeconomic factors that are likely to affect the reposting patterns. Finally, we conclude by summarizing and discussing our findings.

2   Related Works

Weibo has drawn much attention from researchers who study public opinion [5,13,14,17]. Weibo posts are not independent because a post is likely to be

---

[1] http://data.Weibo.com/report.
[2] http://ir.Weibo.com/.

triggered by another post from a friend, opinion leader, or organization. Previous research has shown that opinion diffuses via Weibo users networks [4]. The mostly studied type of social media networks is "follower-followee" network [6,15,20]. Other networks that have been well studied include bipartite networks of Weibo users and posts [10] and "post-repost" network of bloggers [11]. One limit of these studies is that they only focus on basic network attributes such as in/out degrees and the number of follower/followees but ignore the rich content of Weibo posts.

Unlike Twitter, where only 35 percent of its posts are retweets, 65 percent of Weibo are reposted contents [19]. Repost in Weibo allows a user to publish a message up to 140 words. Including 140 words of the original post, 280 words in total can be read by followers of the reposter as a new post. For better understanding the spread of information on this platform, it is more reasonable to study a post-based network. In other words, we hope to capture the path of information flow by observing reposting behavior. Many studies have analyzed the reposting behavior of Weibo users [15], but failed to consider the shift of opinions in posts.

Beliefs vary with Weibo users' demographic and socioeconomic backgrounds. For example, in central and eastern China, higher socioeconomic status leads to a larger number of active Weibo users than the other areas. Wang, Paul, and Dredze [17] analyzed air-pollution-related Weibo posts from 74 cities, and found that the number of pollution-related posts is significantly correlated with the particle pollution rate reported by the local government.

Our study is the first one to explore the relationship between regional prejudice and socioeconomic status via Weibo data. Regional prejudice represents unreasonable resentment and distrust towards people from a different place. It is a widespread problem across Chinese cities. Analyzing data from Weibo provides a valuable alternative to traditional social science research methods, such as surveys and experiments, in studying prejudice and other public opinion questions. People are more likely to convey their true opinions on Weibo than in a face-to-face social settings. The concepts of prejudice, discrimination, and intolerance are hard to be directly measured with good precision, because they are subject to social desirability effects [8]. By contrast, Weibo users can post controversial or provocative posts under an anonymous identity, so they are less restrained to say what they truly believe.

3   Repost Network of Regional Prejudice

To investigate regional prejudice content and its diffusion pattern on Weibo, we construct a network to capture information flow between individuals.

3.1   Data

To the best of our knowledge, there is no publicly available corpus for regional prejudice. We compiled our own corpus. To gather Weibo posts and user information, we built a Weibo webpage crawler. Based on 13 key phrases related to migrants: "native" (本地), "permanent population" (常住人口), "census regis-ter" (户籍), "registered permanent residents" (户口), "resident permit" (居住证), "floating population" (流动人口), "settle in a new place" (落户), "migrant work-ers" (农民工), "non-native" (外地), "coming from a different town" (外来), "peo-ple from other provinces" (外省人), "transient population" (暂住人口), and "temporal residential permit" (暂住证), we retrieved related posts over four months from December 14, 2014 to April 15, 2015. These 13 key phrases can be divided into two groups. One group of phrases, including "census register," "resident permit," and "temporal residential permit," are related to internal migration policies. The other group are labels indicating residential status. It is interesting that some of the labels, such as "permanent population," "floating population," and "transient population," were created by the Chinese govern-ment. Other labels, such as "native," "non-native," and "migrant workers" are common words when people mention migrants. We kept the posts that contained at least 1 of the 13 key phrases. In total, we collected 4,641,398 Weibo posts. To build a repost network, we only kept the reposts following other original posts. As a result, we obtained 285,707 posts, including 34,187 original posts and 251,520 reposts.

3.2 Detecting Regional Prejudice from Texts

Due to the high volume of data, we have to rely on automatic text classifica-tion. We randomly selected 5,000 posts and asked five coders to annotate them via Crowdsdom[3], a Chinese annotation platform similar to Amazon Mechanical Turk. The Kappa statistic of the labeling process was 0.63, indicating a reliable result. In our previous paper [2], we proposed a new approach, Distributed Key-word Vectors, to recognize polarity and direction of Weibo posts. Performance and practicability of the sentiment classifier and direction recognition classifier were elaborated in that paper. With the classifiers, we obtained three types of labels for each post as follows:

- Owner Type (OT): Who published this post, native residents (NR) or migrants (M)?
- Direction (DR): What is the direction of this post, towards migrants (M) or native residents (NR)?
- Sentiment (SEN): What is the sentiment of this post, negative (Neg) or non-negative (Non-neg)?

For each label, there is an "unknown" category. For example, a Weibo post published by a government account talking about a new migration policy is labeled as "unknown" for both OT and SEN. Some posts cut multiple categories. For instance, a post aiming at both native residents and migrants has its DR labeled as "NR&M". We split such a post into two separate posts–one with DR as NR and the other with DR as M. Table 1 offers 4 examples of post labels. The distribution of assigned labels across different categories is shown in Table 2.

---

[3] http://crowdsdom.com/.

Table 1. Example of Weibo posts and their labels

Weibo post	Translation	OT	DR	SEN
#上海踩踏事件调查报告#网上的心态啊，其实只要做到一点灾难就不会发生，那就是，让外地人回外地，哪儿来的回哪儿去，把上海留给上海人。。一切都完美了。。。	#The Shanghai Trampling Report# What are people thinking on the Internet! This tragedy could have been avoided by just doing one thing, that is, have the migrants go back to wherever they came from and give Shanghai back to the Shanghaiese... Everything will be perfect ...	–	M	Neg
早高峰高速路匝道堵了，都是外地车害的，大楼着火了，都是外地临时工害的，外滩踩踏了，估计也有不少外地人，外地人、外地车你们什么时候不是害上海之马！？	Because of migrants, there are traffic jams. Because of migrant workers, the buildings were on fire. There must have been many migrants in the trampling on the Bund too. When will you stop bringing Shanghai down, migrants and your cars?	NR	M	Neg
外地人，本地人，来到温江一家人。	Whether you are migrants or native residents, we are families when you come to Wenjiang.	NR	–	Non-neg
同在一座城，请彼此包容也许你来自远方，也许你从一出生就在这里，也许你是深圳人，也许你是外来的新深圳	We are living in the same city. Please be tolerant of one another. Maybe you come here from far away. Maybe you were born here. Maybe you are a Shenzhen native. Maybe you are a newcomer to this city.	–	NR&M	Non-neg

Table 2. Distribution of posts across different categories

	Labels	Number of posts
OT	NR	75,481
	M	16,884
	–	193,342
DR	NR&M	13,987
	NR	6,873
	M	18,460
	–	246,387
SEN	Neg	114,477
	Non-neg	171,230

3.3   The Repost Network

To understand how information spreads on a social media platform, one way is to build a directed user network and observe who replies or makes a repost. Reposts appear in the feeds board and thus can also be seen by followers of the reposter. The goal of our study is to discover the diffusion pattern of regional prejudice on Weibo. The repost network is utilized for our analysis of regional prejudice in Chinese cities.

Different from many previous studies that focus on users as vertices, we built a "repost network." This approach can rule out the impact of users' versatility. A user might have published multiple Weibo posts with different labels, but no single post is able to represent the user's general attitude. For example, a local user in Beijing may show sympathy towards migrants in the Shanghai stampede accident, but cast criticisms towards migrants in Beijing. In addition, we only kept posts with complete labels in all three categories and removed those with one or more "unknown" labels.

All the networks examined are ego-networks. We take an original Weibo post as a focal vertex (ego), linked to one or more reposts (alters). Figure 1 is an example of ego-network. In an ego-network, 1-hop neighbors are direct reposts, and 2-hop neighbors are reposts of 1-hop neighbors. Table 3 shows the attribute distribution of original posts, 1-hop reposts, and 2-hop reposts. Along with reposting, negative sentiment becomes increasingly more prevalent: 81.54 percent of the original posts contain a negative tone, while the percentage increases to 82.84 percent among the 1-hop reposts and 89.80 percent among the 2-hop reposts). More than 80 percent of posts examined aim at migrants. Weibo users from different cities may repost posts of each other. Figure 2 illustrates the distribution of the number of cities involved in a repost ego network. It shows that more than 75 percent of the repost ego networks cross multiple cities.

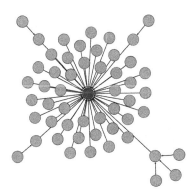

Fig. 1. Example of Ego-network (Green: Ego; Red: Alter) (Color figure online)

Table 3. Total numbers of posts in each category

	%	OT (NR)	DR (NR)	DR (M)	SEN (Neg)
Original	–	69.01 %	69.23 %	83.96 %	81.54 %
1-Hop	–	73.30 %	56.19 %	88.83 %	82.80 %
2-Hop	6 %	89.80 %	73.47 %	89.80 %	89.80 %

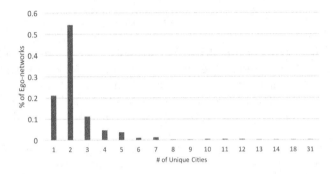

Fig. 2. Distribution of ego-networks with different number of unique cities

4   Socioeconomic Environments and Regional Prejudice

4.1   Reposting Patterns

In this section, we analyze two dimensions of reposts – direction and sentiment. Since prejudice is unreasonable mistrust between different social groups [1], we are interested in the exchange of posts between native Weibo users and the users who identify themselves as migrants, rather than the communication within a single group. We focus on the original posts that are published by native residents with a negative attitude towards migrants. Such posts are closely related to regional prejudice and most prevalent in our collected data. Furthermore, we investigate how socioeconomic environments affect people's attitudes toward migrants and shift the patterns of online discussion on the issue of migration. After being filtered, 909 pairs of posts and reposts remained in our sample.

Based on sentiment and direction, the reposts can be classified into four types (Fig. 3): (1) a native resident complaints about migrants (NR→M, Neg); (2) a native resident discusses migrants with a neutral or positive tone (NR→M, Non-neg); (3) a migrant criticizes native residents (M→NR, Neg); (4) and a migrant holds a neutral or positive view about native residents (M→NR, Non-neg). The first type of reposts are consistent with the original posts and show a clear pattern of regional prejudice. Such reposts are dominant, making up 72.7 percent in the filtered sample. Negative sentiment toward migrants is more likely to be sustained than reversed, as the reversal type of reposts only constitutes 11.1 percent. Unfortunately, the discussion of migrants on Weibo lacks voices from migrants. Among the filtered reposts, only 16.1 percent come from migrants,

Fig. 3. Regional prejudice patterns (Color figure online)

among which 1.5 percent contain a non-negative tone towards native residents while 14.6 percent show regional resentment.

The exchange of migrant-related posts on Weibo supports previous research, finding that the online community is highly polarized and segmented [12]. Compared to face-to-face settings, the Internet provides many platforms where a person can easily find a group that shares his or her beliefs and ideologies, and block or avoid those with conflicting opinions. People have a tendency to seek evidence that justifies their established opinions while neglecting challenging evidence [9]. In the case of Weibo, homophily is a major mechanism underlying the link formation of the repost network. Homophily refers to the tendency of individuals to bond with each other. On one hand, native residents' complaints about migrants often trigger more complaints from other native residents. On the other hand, migrants and the native residents who get along well with migrants tend to avoid arguments with those holding strong regional prejudice. In consequence, regional prejudice becomes a self-reinforcing process.

Distrust between natives and migrants is likely to be a result of inadequate communication. Previous research has shown that inter-group contact and communication can help overcome racism [18] and cultural conflicts [16]. Although most reposts retain native Weibo users' criticisms of migrants, we are more interested in the reposts that alter either the sentiment or the direction of an original post.

The reposting patterns reflect the dynamics of regional prejudice. We proceed to identify the factors that can explain the different patterns of reposting. The reposting pattern is thus the dependent variable in our explanatory model. Since there are a very few reposts in which migrants speak positively about native residents, these reposts are excluded, leaving 895 cases for analysis. These cases are most relevant to our research interest on regional prejudice in an online communication network.

We regard the reposting pattern as a nominal variable with three categories, NR→M (Neg), NR→M (Non-neg), and M→NR (Neg). We model the three reposting patterns using multinomial regressions. Naturally, NR→M (Neg) is treated as the baseline category, because it is consistent with the original post. Concerned with regional heterogeneity, in making statistical inferences, we rely on standard errors clustered on the pairs of the locations of the original posts and reposts.

4.2   Socioeconomic Environment

Socioeconomic contexts have substantial impacts on the relationship between different social groups. Interpersonal trust is likely to prevail in a wealthy society where people have no worry about food, housing, and other basic needs for survival [7]. Some economic resources are limited and indivisible, and the competition for such resources resembles a zero-sum game. Therefore, when one's current political power or economic well-being is threatened, he or she is likely to show hostility toward new competitors and out-group members. However, increased presence of migrants does not necessarily lead to a more intense relationship between native residents and migrants. In a region with a good socioeconomic environment, when a large number of newcomers arrive, local people may not feel insecure, but instead they may become more tolerant of diverse social groups via frequent interactions.

In explaining the reposting patterns, the key independent variables capture the differences in general socioeconomic status across Chinese cities. Socioeconomic status means the social standing of an individual or group. We retrieve six indicators of general socioeconomic status from the 2010 Census of China, the most recent census. The first indicator is %urban that measures the percentage of urban residents in a city. The level of educational attainment is measured by %high school graduate and %college graduate. %unemployed measures the percentage of unemployed adults in the population that is over 16 years old. Two separate indicators are used to measure the percentage of house owners, because rural residents often build a house whereas urban residents often buy or rent houses.

We collect Weibo authors' profiling locations as their cities and compare the indicators of socioeconomic status between the city where a repost was written and the city where the original post was sent. Specifically, we construct an independent variable by subtracting the latter city's indicators from the former city's indicators. For the cases missing the city identifier, we use the provincial indicators of socioeconomic status as a proxy for the city-level indicators.

The issue of regional prejudice is most salient in cities with a large migrant population. Therefore, we control for the percentages of migrants in the exploratory model. The 2010 Census distinguishes between two types of migrants. Within-province migrants live in a new city but in their home province. Cross-province migrants have left their home provinces to find a job. The two types of migrants have similar impacts on the local economy and labor market. However, due to distinct dialects, living styles, and even looks, cross-province migrants are more likely to receive higher prejudice from native residents. In addition, we control the distance between locations of posts and reposts using three binary variables: same province, neighboring provinces, and non-contiguous provinces.

We focus on the regional variations at the city level. In general, cross-city differences outweigh within-city differences. Moreover, urban residents often travel across districts in a city, and actual district boundaries are often not clear. Urban residents' perceptions towards migrants are thus more likely to be based on the migrant population at the city level instead of the district level.

4.3   Multinomial Regression Analysis of the Reposting Patterns

The multinomial regression model has the following functional form:

$$Pr(y_i = k) = \frac{exp(w_k^T x_i)}{exp(w_0^T x_i) + exp(w_1^T x_i) + exp(w_2^T x_i)} \tag{1}$$

where $k = 0, 1, 2$. w_0, w_1, and w_2 correspond to the coefficients for the reposting patterns NR→M (Neg), NR→M (Non-neg), and M→NR (Neg), respectively. To guarantee that w_1 and w_2 are identifiable, the constraint $w_0 = \mathbf{0}$ is added to the model.

Table 4 shows the results of the multinomial regression analysis. Among all the indicators of general socioeconomic status, only Δ%House Built has a statistically significant effect on the reposting pattern NR→M (Non-neg). The effect of Δ%House Built, however, is in an unexpected direction: as this independent variable increases, native residents are less likely to hold a neutral or positive view of migrants.

With regard to the reposting pattern M→NR (Neg), urbanization level, education level, and unemployment rate do not have a statistically significant impact. But in a city with more secure housing, migrants are more likely to criticize native residents in response to their negative posts about migrants. As Table 4 shows, both Δ%House Built and Δ%House Bought raise the probability of M→NR (Neg) in comparison to the probability of NR→M (Neg), and the effects are statistically significant at the 95 % confidence level.

We find that migrants are empowered by their numbers. With regard to M→NR (Neg), Δ%Within-Province Migrant and Δ%Cross-Province Migrant are statistically significant at the 90 % confidence level. In cities with a larger migrant population, migrant users become more engaged in the debate on Weibo with local users who complain about migrants. The magnitudes of the coefficients for the two migrant-related variables are fairly close. It appears that regardless of their origins, migrants living in the same city form a common identity as opposed to the native. We also find that resentful reposts toward native residents are most likely to be sent by Weibo users in a neighboring province.

Table 4 presents the statistical significance of the independent variables which shows whether the observed effects are systematic or largely by chance. But a statistically significant effect may be trivial, only accounting for a tiny portion of the variation in the dependent variable. Thus, it is also necessary to compare the substantive significance, which refers to whether an observed effect is large enough to be meaningful. There are multiple metrics for substantive significance, such as first difference and marginal effect. Figure 4 compares the marginal effects of each independent variable in the multinomial regression model. Margin effect is defined as $\frac{\partial Pr(y=k)}{\partial x_j}$. For a multinomial regression, a marginal effect is such that it is dependent on the combination of the values of all the independent

Table 4. Multinomial regression analysis of locations and the dynamics of regional prejudice

	NR→M	M→NR
	(Non-neg)	(Neg)
Δ%Urban	−0.06	0.05
	(0.07)	(0.09)
Δ%High School Graduate	0.09	−0.22
	(0.10)	(0.21)
Δ%College Graduate	0.01	0.04
	(0.04)	(0.06)
Δ%Unemployed	0.37	−0.53
	(0.51)	(0.60)
Δ%House Built	−0.24*	0.28**
	(0.14)	(0.13)
Δ%House Bought	−0.16	0.20**
	(0.12)	(0.10)
Δ%Within-Province Migrant	−0.17	0.24*
	(0.17)	(0.13)
Δ%Cross-Province Migrant	−0.18	0.20*
	(0.14)	(0.11)
Same Province (Baseline Category) Neighboring Province	−0.06	1.58*
	(1.63)	(0.88)
Other Province	−0.40	0.09
	(0.34)	(0.27)
Constant	−1.82***	−1.79***
	(0.08)	(0.09)
N	895	
Log-Likelihood	-641.10	

* $p < 0.05$, ** $p < 0.01$, *** $p < 0.001$

variables. Average marginal effect takes the average of the marginal effects at all the data points. Therefore, we rely on average marginal effect to measure the overall substantive effect of a variable.

In Fig. 4, the blue lines represent the confidence intervals of the average marginal effects at the 90% confidence levels; the vertical red line indicates a zero average marginal effect. If the confidence interval is broken by the red line, the corresponding average marginal effect is not statistically significant. Marginal effect is meaningful only when it is systematic across random samples. Here we focus on the four average marginal effects (Δ%House Built, Δ%House Bought, Δ%Within-Province Migrant, and Δ%Cross-Province Migrant) with regard to the reposting pattern NR→M (Neg). The average marginal effects

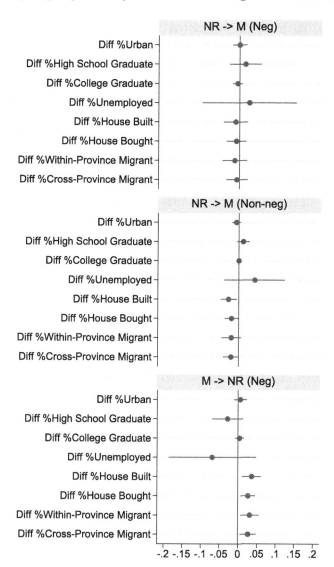

Fig. 4. Average marginal effects of socioeconomic differences on reposting patterns

are approximately the same, with 0.036 for Δ%House Built, 0.027 for Δ%House Bought, 0.031 for Δ%Within-Province Migrant, and 0.027 for Δ%Cross-Province Migrant. On average, as the gap in one of these four socioeconomic indicators increases by one percent, the probability of the reposting pattern NR→M (Neg) rises by about 3 percent, which is substantial.

5 Discussion and Conclusion

We are interested in understanding how Weibo users respond to the posts related to regional prejudice. First, we show the evidence of regional prejudice expressed in social media. In particular, we focus on the reposts of Weibo posts that contain a clear tone of regional prejudice – native residents' complaints about migrants. These reposts are classified based on the direction (native resident→migrant or migrant→native resident) and the sentiment (negative or non-negative). First, we found that homophily is the major mechanism impacting users' opinion in the repost network. This finding is only limited to our observation of the repost network, while cross-group and confrontational communications are more likely to occur through communication activities such as directly replying [3].

Second, we discovered the frequent patterns of regional prejudice on Weibo: native residents' complaints about migrants lead to more negative posts about migrants from other native residents, and only a few reposts changed the initial direction by migrants who criticized native residents. Last but not least, we found the sentiment and direction of posts about migrants are related to socioeconomic status of the author's location. Secure socioeconomic environments breed mutual trust among different social groups. Specifically, in the regions with housing security and a large migrant population, migrant Weibo users are much more likely to get engaged in the argument with native residents who hold a negative view about migrants.

In many cases, regional prejudice is a result from inadequate communication between native and migrant residents. Regional prejudice is thus likely to be weakened when both native residents and migrants get engaged in the debate over the issue of migration. Further studies are needed to understand the factors that encourage communication across different social groups. Moreover, public opinion is dynamic and big social events often ignite heated online discussion. We will continue to explore regional prejudice via Weibo and treat time as an important factor.

References

1. Allport, G.W.: The Nature of Prejudice. Basic Books, Baron (1979)
2. Chang, Y.C., Chou, C.S., Zhang, Y., Wang, X., Hsu, W.L.: Sentiment analysis of chinese microblog message using neural network-based vector representation for measuring regional prejudice. In: The 20th Pacific Asia Conference on Information Systems (PACIS 2016)
3. Conover, M., Ratkiewicz, J., Francisco, M.R., Gonçalves, B., Menczer, F., Flammini, A.: Political polarization on twitter. ICWSM **133**, 89–96 (2011)
4. Fan, R., Zhao, J., Chen, Y., Xu, K.: Anger is more influential than joy: sentiment correlation in weibo. PLoS One **9**(10), e110184 (2014). http://dx.doi.org/10.1371/journal.pone.0110184
5. Guan, W., Gao, H., Yang, M., Li, Y., Ma, H., Qian, W., Cao, Z., Yang, X.: Analyzing user behavior of the micro-blogging website SinaWeibo during hot social events. Phys. A: Statist. Mech. Appl. **395**, 340–351 (2014). http://www.sciencedirect.com/science/article/pii/S0378437113009369

6. Guo, Z., Li, Z., Tu, H.: Sina microblog: an information-driven online social network. In: 2011 International Conference on Cyberworlds (CW), pp. 160–167, October 2011

7. Inglehart, R.: Trust, well-being and democracy. In: Warren, M.E. (ed.) Democracy and Trust. Cambridge University Press, Cambridge (1999)

8. Kuklinski, J.H., Cobb, M.D., Gilens, M.: Racial attitudes and the "new south". J. Polit. **59**(02), 323–349 (1997)

9. Kunda, Z.: The case for motivated reasoning. Psychol. Bull. **108**(3), 480 (1990)

10. Lei, K., Zhang, K., Xu, K.: Understanding Sina Weibo online social network: A community approach. In: 2013 IEEE Global Communications Conference (GLOBECOM), pp. 3114–3119, December 2013

11. Li, F., Lin, N.: Social network analysis of information diffusion on Sina Weibo micro-blog system. In: 2015 6th IEEE International Conference on Software Engineering and Service Science (ICSESS), pp. 233–236, September 2015

12. Morozov, E.: The Net Delusion: The Dark Side of Internet Freedom. PublicAffairs, New York (2012)

13. Nip, J.Y.M., Fu, K.W.: Networked framing between source posts and their reposts: an analysis of public opinion on China's microblogs. Inf. Commun. Soci. **19**(8), 1127–1149 (2016). http://dx.doi.org/10.1080/1369118X.2015.1104372

14. Qu, Y., Huang, C., Zhang, P., Zhang, J.: Microblogging After a Major Disaster in China: A Case Study of the 2010 Yushu Earthquake. ACM Press, San Jose (2011). http://portal.acm.org/citation.cfm?doid=1958824.1958830

15. Ronggui Huang, X.S.: Weibo network, information diffusion and implications for collective action in China. Information **17**(1), 86–104 (2014)

16. Ting-Toomey, S., Chung, L.C.: Understanding intercultural communication. Oxford University Press, New York (2012)

17. Wang, S., Paul, M.J., Dredze, M.: Social media as a sensor of air quality and public response in China. J. Med. Int. Res. **17**(3), e22 (2015)

18. Welch, S.: Race and Place: Race Relations in an American City. Cambridge University Press, New York (2001)

19. Yu, L.L., Asur, S., Huberman, B.A.: Artificial in ation: the real story of trends and trend-setters in sina weibo. In: Privacy, Security, Risk and Trust (PASSAT), 2012 International Conference on and 2012 International Confernece on Social Computing (SocialCom), pp. 514–519, September 2012

20. Zhang, H., Zhao, Q., Liu, H., Xiao, K., He, J., Du, X., Chen, H.: Predicting retweet behavior in weibo social network. In: Wang, X.S., Cruz, I., Delis, A., Huang, G. (eds.) WISE 2012. LNCS, vol. 7651, pp. 737–743. Springer, Heidelberg (2012). doi:10.1007/978-3-642-35063-4_60

Author Index

Aberer, Karl   II-225
Abernethy, Jacob   I-438
Abro, Altaf Hussain   II-385
Adaji, Ifeoma   II-3
Agarwal, Arvind   II-194
Agrafiotis, Ioannis   I-185
Akkiraju, Rama   II-337
Ali, Mohammed Eunus   II-454
Almaatouq, Abdullah   I-407
Almeida, Virgilio   I-419
Alrawi, Danya   I-3
Anthony, Barbara M.   I-3
Araújo, Camila Souza   I-419
Arendt, Dustin   I-312
Azab, Mahmoud   I-438

Balasuriya, Lakshika   I-527
Bashir, Masooda   II-347
Bilski, Piotr   II-471
Bohr, Jeremiah   II-347
Bosse, Tibor   II-439
Butner, Ryan   I-481

Capra, Licia   II-179
Cesare, Nina   I-155
Cha, Meeyoung   II-297
Chang, Yung-Chun   II-500
Chavoshi, Nikan   I-58, II-14
Chen, Bo-Chiuan   I-330
Chetviorkin, Ilia   I-312
Chou, Chin-Shun   II-500
Chung, Christine   I-3
Chung, Wen-Ting   I-168
Corley, Court   I-481
Corley, Courtney   I-510
Cudré-Mauroux, Philippe   II-225
Cuong, To Tu   II-142

Davis, Clayton   II-269
de Gier, Michelle   II-259
Debnath, Madhuri   I-347
Del Vigna, Fabio   I-494
Diaz, Fernando   I-41
Ding, Hao   I-361

Dittus, Martin   II-179
Dong, Cailing   II-194, II-400, II-421
Donovan, Bryan   II-337
Doran, Derek   I-527

Eisenstein, Jacob   I-41
Elmasri, Ramez   I-347
Endo, Masaki   I-389

Faloutsos, Michalis   I-75
Farzan, Rosta   I-24
Ferrara, Emilio   I-330, II-22
Flammini, Alessandro   II-22
Frias-Martinez, Vanessa   I-240
Fukuda, Ichiro   I-125
Fushimi, Takayasu   II-40

Galstyan, Aram   II-22
Giasemidis, Georgios   I-185
Goel, Rahul   I-41
Goel, Sonal   I-206
Golbeck, Jennifer   I-454, I-468
Goyal, Naman   I-41
Greetham, D.V.   I-185
Gudehus, Christian   II-210
Gupta, Divam   I-206

Hajibagheri, Alireza   II-55
Halappanavar, Mahantesh   II-133
Halford, Susan   II-116
Hamooni, Hossein   I-58, II-14
Han, Kyungsik   I-510
Han, Sifei   II-307
Hang, Huy   I-75
Hasan, Mahmud   I-224
Hayashi, Ryota   II-323
Hirota, Masaharu   I-389
Hodas, Nathan O.   I-481
Höllerer, Tobias   II-279
Hong, Lingzi   I-240
Hürriyetoğlu, Ali   II-210

Ishikawa, Hiroshi   I-389

Jin, Hongxia   II-400

Kando, Noriko I-138, II-40
Kang, Byungkyu II-279
Karamshuk, Dmytro I-257
Kavuluru, Ramakanth II-307
Kazama, Kazuhiro II-40
Kim, Erin Hea-Jin I-376
Kim, Heechul II-297
Kim, SuYeon I-376
Kim, Wonjoon II-297
Klein, Michel C.A. II-385
Knijnenburg, Bart P. II-400
Kooti, Farshad II-71
Kowalik, Grzegorz II-87
Krafft, Peter I-290
Kumaraguru, Ponnurangam I-206

Lakkaraju, Kiran II-55
Lau, Wing Cheong II-97
Lee, Hedwig I-155
Lerman, Kristina II-71
Li, Guanchen II-97
Li, Guangyu I-361
Li, Tai-Ching I-75
Lim, Ee-Peng I-92
Lin, Yu-Ru I-168
Liotsiou, Dimitra II-116
Liu, Yong I-361
Liu, Zhe II-337
Lo, David II-368
Lokot, Tetyana I-257

Mahmud, Jalal II-337, II-454
Maity, Mrinmoy I-330
Mavlyutov, Ruslan II-225
McCormick, Tyler I-155
Medeiros, Lenin II-439
Meira Jr., Wagner I-419
Menczer, Filippo II-269
Menezes, Ronaldo II-485
Mihalcea, Rada I-438
Mirowski, Tom I-273
Mitomi, Keisuke I-389
Moreau, Luc II-116
Morishima, Atsuyuki II-323
Moro, Esteban II-71
Mueen, Abdullah I-58, I-75, II-14
Mukta, Md. Saddam Hossain II-454
Müller-Birn, Claudia II-142
Murdopo, Arinto I-92

Nahm, Alison I-290
Nielek, Radoslaw II-87
Nurse, Jason R.C. I-185

O'Donovan, John II-279
Obradovic, Zoran I-273
Oentaryo, Richard J. I-92
Oostdijk, Nelleke II-210
Orgun, Mehmet A. I-224

Paparrizos, John I-41
Pentland, Alex I-290, I-407
Petrocchi, Marinella I-494
Pilgrim, Alan I-185
Prasetyo, Philips K. I-92
Prieto-Castrillo, Francisco I-407
Prokofyev, Roman II-225
Proskurnia, Julia II-225
Pryymak, Oleksandr I-257

Quach, Tu-Thach I-110
Quattrone, Giovanni II-179

Rafalak, Maria II-471
Resnik, Philip I-240
Reyes Daza, Brayan S. II-244
Roberts, Margaret E. II-269
Roychoudhury, Shoumik I-273

Sachdeva, Niharika I-206
Saha, Priya II-485
Saito, Kazumi II-40
Salcedo Parra, Octavio J. II-244
Sanders, Eric II-259
Sastry, Nishanth I-257
Sathanur, Arun V. II-133
Satoh, Tetsuji I-138, II-40
Savage, Saiph I-24
Saviaga, Claudia Flores I-24
Schoudt, Jerald II-337
Schwitter, Rolf I-224
Sharma, Abhishek II-368
Sheth, Amit I-527
Shimizu, Nobuyuki II-323
Shoji, Yoshiyuki I-389
Singleton, Colin I-185
Soni, Sandeep I-41
Spiro, Emma S. I-155
Strickland, Beth II-347

Suárez-Serrato, Pablo   II-269
Subramanyam, A.V.   I-206
Sukthankar, Gita   II-55
Sulistya, Agus   II-368

Takano, Masanori   I-125
Tesconi, Maurizio   I-494
Tintarev, Nava   II-279
Tommasi, Alessandro   I-494
Torres, Julián Giraldo   II-244
Treur, Jan   II-157
Tripathi, Praveen Kumar   I-347

van den Bosch, Antal   II-210, II-259
Van Durme, Benjamin   I-312
Varol, Onur   II-22
Vassileva, Julita   II-3
Volkova, Svitlana   I-312, I-510

Wada, Kazuya   I-125
Wallach, Hanna   I-41
Wang, Wen-Qiang   II-22

Wang, Xi   II-500
Wang, Yi   II-337
Wei, Kai   I-168
Wen, Xidao   I-168
Wendt, Jeremy D.   I-110
Wierzbicki, Adam   II-471
Wijeratne, Sanjaya   I-527
Willis, Chris   I-185

Xu, Anbang   II-337

Yamamoto, Shuhei   I-138
Yang, Weiwei   I-240
Yang, Xinxin   I-330
Yokoyama, Shohei   I-389
Yu, Chenguang   I-361

Zavattari, Cesare   I-494
Zhang, Yang   II-500
Zhao, Kang   II-500
Zhou, Bin   II-421
Zhou, Fang   I-273
Zuo, Zhiya   II-500